STRATEGIES FOR ORGANIC DRUG SYNTHESIS AND DESIGN

STRATEGIES FOR ORGANIC DRUG SYNTHESIS AND DESIGN

DANIEL LEDNICER
Rockville, MD

A Wiley-Interscience Publication
John Wiley & Sons, Inc.
New York / Chichester / Weinheim / Brisbane / Singapore / Toronto

This book is printed on acid-free paper. ∞

Library of Congress Cataloging-in-Publication Data

Lednicer, Daniel, 1929–
Strategies for organic drug synthesis and design/ Daniel Lednicer.
p. cm.
"A Wiley-Interscience Publication."
Includes index.
ISBN 0-471-19657-6 (cloth : alk. paper)
1. Pharmaceutical chemistry. 2. Drugs--Synthesis. 3. Drugs--Design. 4. Organic compounds--Synthesis. I. Title.
[DNLM: 1. Drug Design. 2. Chemistry, Pharmaceutical--methods. 3. Drugs--chemical synthesis. 4. Chemistry, Organic--methods. QV 744 L473s 1998]
RS403.L386 1998
615'.19--dc21
DNLM/DLC
for Library of Congress 97-11813
CIP

Printed in the United States of America.

10 9 8 7 6 5 4 3

This book is dedicated to those whose names appear in the bibliographies for each of the chapters that follow as well as chemists whose compounds never quite made the clinic. They experienced the excitement of the chase for new drugs; the chemistry they elaborated provided me with the material for this volume.

CONTENTS

PREFACE

"One of the most interesting aspects of organic chemistry is that dealing with the building up of complex substances from simpler ones. The synthesis of organic compounds whether for scientific or industrial purposes, has been very important in the development of the science and is still of greatest importance today."

Those words, set down 70 years ago as the opening for a chapter on organic synthesis in Conant's pioneering textbook *Organic Chemistry*,* still very aptly describe the important role held by that aspect of the discipline. The use of organic transformations for the preparation of new compounds with more or less complex structures has had a profound influence on both organic chemistry and more importantly on modern civilization. One need only bring to mind medicinal agents at one extreme and, at the other, the monomers used for the plethora of polymers that have provided the basis for a whole new materials science. The practice of organic synthesis covers an extremely broad range, from the highly practical, economically driven, preparation of a tonnage chemical to a multistep very elegant enantiospecific synthesis of a complex natural product. This very diversity may account for the relative paucity of books devoted specifically to the subject.

The manipulations used for the preparation of therapeutic agents seems to offer a middle ground between the extremes in complexity noted above. The published syntheses for these agents are typically relatively short, seldom exceeding 10 or so steps. The target compounds for these syntheses do, however, cover a very wide range of structural types, encompassing both carbocyclic and heterocyclic compounds.

*Conant, James B. *Organic Chemistry*, Macmillan, New York, 1928, p. 117.

Moreover, the chemistry includes a very broad selection of organic reactions. The published syntheses most often describe the route that was used in the initial discovery of some new compound. Some quite exotic and versatile reagents are used since reaction conditions are not circumscribed by their applicability to plant processes. The syntheses of therapeutic agents thus offer a good didactic tool.

This book represents a selection of syntheses from the five volume series *The Organic Chemistry of Drug Synthesis*, which were chosen to illustrate the strategy and the organic transformations that were used to prepare the various structural classes investigated as drugs.

One of the main motivations that led to the writing of the original book entitled *The Organic Chemistry of Drug Synthesis* was curiosity as to how various classes of drugs were in fact prepared. The enormous number of compounds reported in the literature as potential drugs led to an early decision to restrict the book to those agents that had been granted nonproprietary names. This filtering mechanism was based on the assumption that, in the judgment of the sponsor, the compound in question showed sufficient activity to merit eventual clinical evaluation. Within a few years of publication of *The Organic Chemistry of Drug Synthesis*, a follow-up volume was issued to bring the coverage up to date and to make up gaps in the coverage in the original book. Between them, the two books included the large majority of compounds that had been granted generic names up to that time. The subsequent three volumes of what became a series appeared at roughly semidecennial intervals in order to cover the syntheses of compounds granted generic names during those intervals. All but the last volume of the series I hasten to add were written in collaboration with Les Mitscher.

Anecdotal evidence over the years indicated that at least Volume 1 of the series was in use as an adjunct text in organic chemistry courses. That book provided the core for several graduate organic courses that I taught in the mid-1980s, with then available Volumes 2 and 3 used as occasional references. As the series aged, new and more relevant chemistry appeared in the more recent volumes. Sheer bulk, to say nothing of expense, mitigate against the use of the full set of five volumes in any course, even as supplementary references. The present book, as noted above, is intended to present the chemical highlights from the series.

The focus of this book differs from that of the series in that it is aimed more specifically at the organic chemistry used for the preparation of the drugs in question. Drugs have been selected mainly for the illustrative value of the chemistry used for their synthesis; hence too the inclusion of the rather extensive "Reaction Index." The structures in chemical schemes have been drawn with special attention to clarifying the individual reactions; rearrangement starting materials and products, for example, are shown in similar views. The very brief discussions of medicinal chemistry are intended to provide the reader with a feel for the activity and occasionally the mechanism of action of various drugs. Salient principles of drug action are presented in capsule form at appropriate points; by the same token the claimed therapeutic effect of each agent is noted along with the discussion of its preparation. The pharmacological presentations are thus abbreviated over those that occur in the series. Interested readers should consult any of a wide selection of medicinal chemistry or pharmacology texts such as *Burger's Medicinal Chemistry* for fuller and more authoritative discussions.

DANIEL LEDNICER
Rockville, MD
January, 1997

CHAPTER 1

PROSTAGLANDINS, PROTEASE INHIBITORS, AND RETINOIDS

1. PROSTAGLANDINS

It is highly likely that those not themselves involved in scientific research perceive the development of new knowledge within a given area of science as a linear process. The popular image is that the understanding of the specific details of any complex system depends on prior knowledge of the system as a whole. In turn, this knowledge is believed to derive from the systematic stepwise study of the particular system in question. The piecemeal, almost haphazard, manner in which the details of the existence and later the detailed exposition of the arachidonic acid cascade was put together is much more akin to the assembly of a very complex jigsaw puzzle. This particular puzzle had the added complication of including many pieces that did not in fact fit the picture as finally revealed; the pieces that would, in the end, fit were found at very different times.

The puzzle had its inception with the independent observation in the early 1930s by Kurzok and Lieb[1] and later by von Euler[2] that seminal fluid contained a substance that caused contraction of isolated guinea pig muscle strips. von Euler named this putative compound prostaglandin in the belief that it originated in the prostate gland; the ubiquity of those substances was only uncovered several decades later. The discovery remained an isolated oddity until the mid-1960s at which time methods for chromatographic separation of complex mixtures of polar compounds and spectroscopic methods for structure determination were finally up to the characterization of humoral substances that occur at very low levels. The isolation and structural assignment of the first two natural prostaglandins, PGE_1 and PGF_2, was accomplished by Bergstrom and co-workers[3] at the Karolinska Institute. [The letter that follows PG probably initially referred to the order in which the compounds were isolated—E refers to 9-keto-11-hydroxy compounds, F refers to 9,11-diols; the subscripts refer to the number of double bonds; the carbon atoms of the hypothetical fully saturated otherwise unsubstituted carbon skeleton, prostanoic acid, are numbered sequentially starting with the carboxylic acid as 1; the carbon atoms bearing the side chains are numbered as well (8,12)].

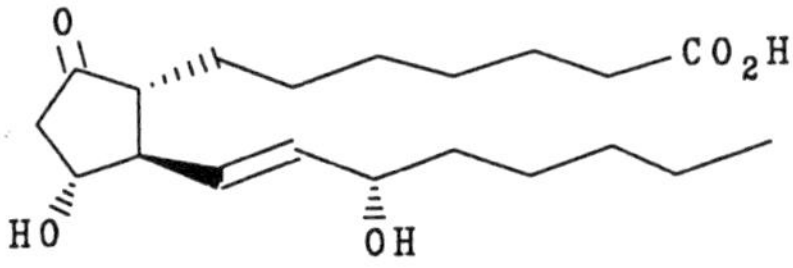

O CO$_2$H HO OH

PGE$_1$

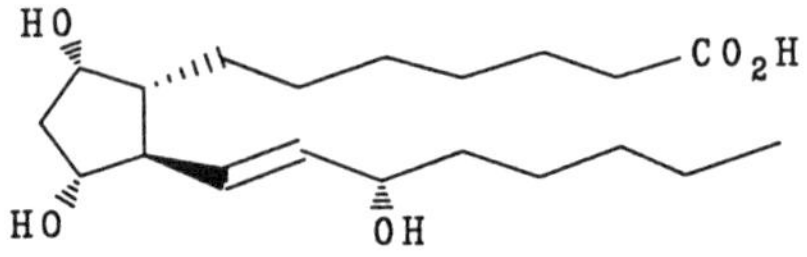

HO CO$_2$H HO OH

PGF$_{1\alpha}$

The identification of these two prostaglandins, in combination with their very high potency in isolated muscle preparations, suggested that these compounds might be the first of a large class of new hormonal agents. Extensive research in the laboratories of the pharmaceutical industry had successfully developed a large group of new steroid based drugs from earlier similar leads in that class of hormones; this finding encouraged the belief that the prostaglandins provided a lead that would culminate in a broad new class of drugs. Like the steroids, exploration of the pharmacology of the prostaglandins was initially constrained by the scarcity of supplies. The low levels at which the compounds were present as well as their limited stability forced the pace in developing synthetic methods for those compounds; the anticipated need for analogues served as a further incentive for elaborating routes for their synthesis.

Further work on the isolation of related compounds from mammalian sources, which spanned several decades, led to the identification of a large group of related substances. Investigations on their biosynthesis made it evident that all eventually arise from oxidation of the endogenous substance, arachidonic acid. The individual products induce a variety of very potent biological response with inflammation predominating. Arachidonic acid, once freed from lipids by the enzyme phospholipase A_2, can enter one of two branches of the arachidonic acid cascade[4] (Scheme 1). The first pathway to be identified starts with the addition of two molecules of oxygen by reaction with cyclooxygenase (sometimes called prostaglandin synthetase) to give PGG_2. One oxygen adds across the 9,11-positions to give a cyclic peroxide while the other adds to the 14-position. This reaction is reminiscent to that of singlet oxygen to give a hydroperoxide at the 14-position with the resulting shift of the olefin to the 12-position with concomitant isomerization to the trans configuration. The initial hydroperoxide is readily reduced to an alcohol to give the key intermediate PGH_2. Reductive ring opening of the oxide leads to the PGF series; while an internal rearrangement leads to the very potent inflammatory thromboxanes. Later, it was found that aspirin and indeed virtually all non-steroid antiinflammatory drugs (NSAIDs) owe their efficacy to inhibition of cylcooxygenase.

On the other hand, reaction of arachidonic acid with the enzyme lypoxygenase leads to attack at the 5-position and rearrangement of the double bonds to the 7,9-*trans*-11-

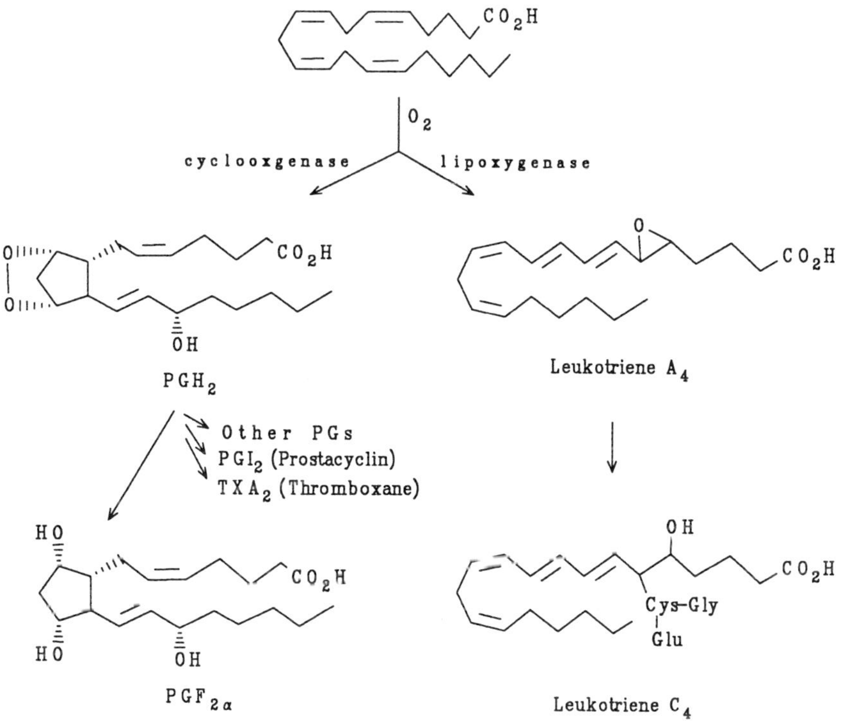

Scheme 1 Arachidonic acid cascade

cis array typical of leukotrienes; the initial product closes to an epoxide thus yielding leukotriene A_4. The reactive oxirane in this compound in turn reacts with endogenous glutathione to give leukotriene C_4. This compound and some of its metabolites, it turned out, constitute the previously well-known "slow reacting substance of anaphylaxis" (srs-A), involved in allergic reactions and asthma.

Much of the early work on this class of compounds focused on developing routes for producing the agents in quantities sufficient for biological investigations. There was some attention paid to elaborating flexible routes, as it was expected that there might be some demand for analogues not found in nature. This work was hindered by a relative dearth of methods for elaborating highly substituted five-membered rings that allowed control of stereochemistry. The unexpected finding of a compound with the prostanoic acid skeleton in a soft coral, the sea whip *plexura homomalla*,[5] offered an interim source of product; the group at Upjohn in fact developed a scheme for converting this compound to the prostaglandin that they were investigating in detail.[6] The subsequent development of a practical total syntheses in combination with ecological considerations led to the eventual replacement of that source.

The methodology developed by E. J. Corey and his associates at Harvard provides the most widely used starting material for prostaglandin syntheses. This key intermediate, dubbed the "Corey lactone," depends on rigid bicyclic precursors for control of stereochemistry at each of the four functionalized positions of the cyclopentane ring. Alkylation of the anion from cyclopentadiene with chloromethylmethyl ether under conditions designed to avoid isomerization to the thermodynamically more stable isomer gives the diene **1**. In one approach, this compound is then allowed to react with α-chloroacrylonitrile to give the Diels–Alder adduct **2** as a mixture of isomers. Treatment with aqueous base affords the bicyclic ketone **3**, possibly by way of the cyanohydrin derived from displacement of halogen by hydroxide. The Bayer–Villiger oxidation of the carbonyl group with peracid gives the lactone **4**; the net outcome is to establish the cis relationship of the hydroxyl that will occupy the 11-position in the product and the side chain that will be at the 9-position. Simple saponification then gives hydroxyacid **5**; the presence of the carboxyl group provides the means by which this compound can be resolved by conventional salt formation with chiral bases. Reaction of the last intermediate with base in the presence of iodine results in formation of iodolactone; the reaction may be rationalized by positing the formation of a cyclic

CH_2OCH_3 1 → (CN, Cl) → CH_3OCH_2 CN Cl 2 → OH^- → CH_3OCH_2 O 3 → RCO_3H → CH_3OCH_2 O O 4 → OH^- → CO_2H HO CH_2OCH_3 5 → 1. OH^-, I_2 2. Ac_2O → O O R AcO CH_2OCH_3

6, X = I
7. X = H

→ BBr_3 → O O AcO CH_2OH 8 → CrO_3:PyH → O O AcO CH_2OH 9

iodonium salt on the open face of the molecule; attack by the carboxylate anion will give the lactone with the observed stereochemistry. Acetylation of the hydroxyl gives **6**; halogen is then removed by reduction with tributyltin hydride (**7**). The methyl ether on the substituent at the future 11-position is then removed by treatment with boron tribromide to give **8**. Oxidation of the primary hydroxyl by means of the chromium trioxide/pyridine complex (Collins reagent) gives the Corey lactone **9** as its acetate.[7]

A slightly more direct route to the Corey lactone, which was developed quite recently, depends on a radical photoaddition–rearrangement reaction as the key step. The scheme starts with the Diels–Alder addition of α-acetoxyacrylonitrile to furan to give the bridged furan **10** as a mixture of isomers. Hydrolysis by means of aqueous hydroxide gives the ketone **11**; the reaction may proceed through the intermediate cyanohydrin. This cyanohydrin is in fact produced directly by treatment of the mixture of isomers with sodium methoxide in a scheme for producing the ketone in chiral form. The crude intermediate is treated with brucine. Acid hydrolysis of the solid "complex" that separates affords the quite pure dextrorotary ketone **11**[8]; this complex may consist of a ternary imminium salt formed by sequential reaction with the cyanohydrin function. Irradiation of the ketone in the presence of phenylselenylmalonate leads to the rearranged product **14** in quite good yield. The structure can be rationalized by postulating homolytic cleavage of the C–Se bond in the malonate to give the intermediate **12** as the first step; the resulting malonate radical would then add to the olefin; acyl migration would then give the rearranged carbon skeleton of **13**. Addition of the phenylselenyl radical to **13** will then give the observed product. Reduction of the carbonyl group by means of sodium borohydride gives the product from approach of hydride from the more open exo face (**15**). Decarboxylation serves to remove the superfluous carboxyl group to afford **16**; treatment with *tert*-butyldimethylsilyl chloride (*t*-BDMSCl) in the presence of imidazole gives the protected intermediate **17**, which contains all the elements of the Corey lactone with the future aldehyde, however, they are in the wrong α configuration. Saponification of the ester followed by acid hydrolysis in fact gives the all-cis version of the lactone.[9] The desired trans isomer **18** can be obtained by oxidizing the selenium with hydrogen peroxide in the presence of sodium carbonate.[10]

The biological investigations made possible once supplies of prostaglandins became available, revealed the manifold activities of this class of agents. The very potent effect of $PGF_{2\alpha}$ on reproductive function was particularly notable. Ovulation in most mammalian species is marked by the formation on the ovary of a corpus luteum that produces high levels of progesterone if a fertile ovum has been implanted in the uterus; administration of even low doses of $PGF_{2\alpha}$ was found to have a luteolytic effect with loss of the implant due to withdrawal of progestin. This prostaglandin was in fact one of the first compounds in this class to reach the clinic under the *United States Adopted Names* (USAN) name **dinoprost**. Development of drugs for use in domestic animals tends to be faster and much less expensive than those that are to be used in humans. This is particularly true if the animals are not to be used as food since this dispenses with the need to study tissue residues. Consequently, it is of interest that one of the early prostaglandins to reach the market is **fluprostenol (25)**. This compound differs from $PGF_{2\alpha}$ in that the terminal carbon atoms in the lower side chain are replaced by

10 $\xrightarrow{OH^-}$ 11 $\xrightarrow[h\nu]{C_6H_5SeCH(CO_2Me)}$ [12] → [13] → 14

14 $\xrightarrow{NaBH_4}$ 15 $\xrightarrow[NaI]{Me_2SO}$ 16 $\xrightarrow{tBDMSCl}$ 17 $\xrightarrow[2.\ H_2O_2]{1.\ OH^-}$ 18

10: CN, OAc

12: MeO_2C, CO_2Me

13: MeO_2C, CO_2Me, SeC_6H_5

14: CO_2Me, CO_2Me, SeC_6H_5

15: HO, CO_2Me, CO_2Me, SeC_6H_5

16: HO, CO_2Me, SeC_6H_5

17: $tBuMe_2SiO$, CO_2Me, SeC_6H_5

18: $tBuMe_2SiO$, CH=O

the trifluromethyl phenoxy group; this modification markedly enhances potency as well as stability. This drug is marketed under the name Equimate® for controlling fertility in race mares, a species in which costs are probably of little consequence.

Reaction of the anion from the phosphonate **19** with ethyl *m*-triflurophenoxymethylacetate (**20**) results in acylation of the phosphonate by displacement of ethoxide and formation of **21**. Condensation of the ylide from this intermediate with the biphenyl ester at the 11-position of the Corey lactone **22** leads to the enone **23**, with the usual formation of a trans olefin expected for this reaction. The very bulky biphenyl ester comes into play in the next step; the reduction of the side chain ketone by means of zinc borohydride proceeds to give largely the 15α alcohol as a result of the presence of this bulky group. The ester is then removed by saponification and the two hydroxyl groups are protected as their tetrahydropyranyl (THP) ethers (**24**).

The next step in the sequence involves conversion of the lactone to a lactol; the carbon chain is thus prepared for attachment of the remaining side chain while revealing a potential hydroxyl group at the 9-position. This transform is effected by treating **24** with diisobutylaluminum hydride at –78°C; overreduction to a diol occurs at higher temperatures. Wittig reactions can be made to yield cis olefins when carried out under carefully defined, "salt free" conditions.[11] Condensation of the lactol **25** with the ylide from 5-triphenylphosphoniumpentanoic acid under these conditions gives the desired olefin. Treatment with mild aqueous acid serves to remove the protecting groups. Thus **fluprostenol** (**26**) is formed.[12]

Prostaglandins have been called hormones of injury since their release is often associated with tissue insult. Consequently, most of these agents exhibit activities characteristic of tissue damage; many cause vasoconstriction and a consequent increase in blood pressure as well as the platelet aggregation that precedes blood clot formation. Thromboxane A_2 is in fact one of the most potent known platelet aggregating substances. Prostacyclin (PGI_2) one of the last cyclooxygenase products to be discovered, constitutes an exception; the compound causes vasodilation and inhibits platelet aggregation. This agent may be viewed formally as the cyclic enol ether of a prostaglandin that bears a carbonyl group at the 6-position of the upper side chain. This very labile functionality contributes to the short half-life of PGI_2. The fact that this value is measured in single digit minutes precludes the use of this agent as a vasodilator or as an inhibitor of platelet aggregation.

The analogue in which carbon replaced oxygen in the enol ring should of course avoid the stability problem. The synthesis of this compound initially follows a scheme similar to that pioneered by the Corey group. Thus, acylation of the ester **27** with the anion from trimethyl phosphonate yields the activated phosphonate **28**. Reaction of the ylide from this intermediate with the lactone **22** leads to Compound **29**, which incorporates the lower side chain of natural prostaglandins. This compound is then taken to lactone **30** by sequential reduction with the use of zinc borohydride, removal of the biphenyl ester by saponification, and protection of the hydroxyl groups as tetrahydropyranyl ethers.

The first step in building the carbocyclic ring consists in effect of a second acylation on trimethyl phosphonate; thus, addition of the anion from this reagent to the lactone carbonyl in **30** leads to the product as its cyclic hemiketal (**31**); it should be noted that

$(CH_3O)_2\overset{O}{P}CH_3$ + C_2H_5O ... CF_3

19 20

$(CH_3O)_2\overset{O}{P}$ ... CF_3

21

$C_6H_5C_6H_4CO$ ... CH=O

22

$C_6H_5C_6H_4CO$ ... CF_3

23

1. $ZnBH_4$
2. OH^-
3. DHP

THPO ... OTHP ... CF_3

24

THPO ... OH ... OTHP ... CF_3

25

1. $(C_6H_5)_3P{=}$ ... CO_2^-
2. H_3O^+

HO ... CO_2H ... HO ... OH ... CF_3

26

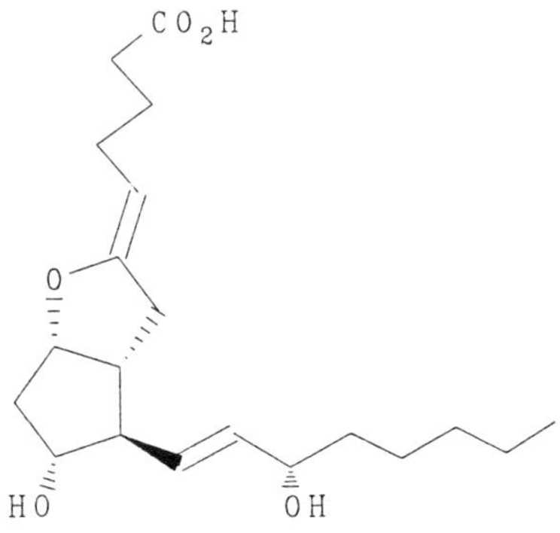

PGI_2

this last step now incorporates an activated phosphonate group. Oxidation of this compound with Jones' reagent gives the diketone **32**. The ylide prepared from this compound by means of potassium carbonate in aprotic media adds internally to the ring carbonyl group to give fused cylopentenone (**33**). Conjugate addition of a methyl group to the enone by means of the cuprate reagent from methyl lithium occurs predominantly on the open β face of the molecule to afford **34**. The counterpart of the upper side chain is added to the molecule by condensation with the ylide from triphenylphosphoniumpentanoic acid bromide. The product is obtained as a mixture of (*E*) and (*Z*) isomers about the new olefin due to the absence of directing groups. Removal of the tetrahydropyran protecting groups with mild aqueous acid completes the synthesis of **ciprostene** (**36**).[13] This compound has the same platelet aggregation inhibitory activity as PGI_2, although with much reduced potency.

Much of the early detailed biology of $PGF_{2\alpha}$ was worked out by a group of pharmacologists at the Upjohn Company using a compound prepared by the chemists in those laboratories. This specific compound under the USAN name **dinoprost** was also the first to be approved for clinical induction of labor. It has received some recent publicity because of its use as an adjunct in RU-486 (**mifepristone**, see Chapter 4) induced abortions. Although initial supplies of $PGF_{2\alpha}$ were obtained by partial synthesis from soft coral starting materials, this source was supplanted by total synthesis product. The reported synthesis, like those noted above, relies on a rigid fused bicyclic starting material for determination of the relative configuration of the substituents on the cyclopentane ring.

The sequence starts by epoxidation of bicycloheptadiene **37** with peracid, a reaction that had been found earlier to proceed to aldehyde **39** rather than stopping at the epoxide. This rearrangement, which will control stereochemistry at the 11-, 12-, and 15-positions in one fell swoop, is related conceptually to the i-steroid rearrangement discovered at least a decade earlier; This reaction relies in effect on the mobile equilibrium between a cyclopropylcarbinyl carbocation and its homoallyl partner:

$(CH_3O)_2\overset{O}{P}CH_3$ + EtO_2C → $(CH_3O)_2\overset{O}{P}$

19 27 28

THP =

$C_6H_5C_6H_4CO$ CH-O

22

$C_6H_5C_6H_4CO$

29

1. $ZnBH_4$
2. OH^-
3. DHP

THPO OTHP

30

HO
$PO(OMe)_2$
O
THPO
OTHP
31

CrO_3
(Jones')

O
$PO(OMe)_2$
O
THPO
OTHP
32

K_2CO_3
18-Crown-6

O
THPO
OTHP
33

Me_2CuLi

O
H_3C
THPO
OTHP
34

$(C_6H_5)_3\overset{+-}{P}CH(CH_2)_4CO_2^-$

CO_2H
H_3C
RO
OR

35; R = THP
36; R = H

RCO_3H

H
O
+

37
38
39

HO
HO

$1. CCl_2{=}C{=}O$
2. Zn

Ephedrine
H_3O+

H_5C_6
NCH_3
O
O
O

42
41
40

The rearrangement can be visualized as starting with protonation of the initially formed epoxide to **38**. This compound could then first ring open to an alcohol; the observed product (**39**) would be obtained by Wagner–Meerwin rearrangement of the resulting carbocation. The same product would be formed by the concerted reaction shown in the scheme below. The aldehyde is then protected as its acetal (**40**) with 2,2-dimethylpropylene glycol. The two carbon atoms that will form the upper side chain are then incorporated by electrocyclic addition of dichloroketene; the chlorine atoms are removed by reduction with zinc to give **41**. Leaving the all important resolution step to late in the synthesis of chiral compounds involves the penalty of carrying useless inactive enantiomer through a large number of transformations; efficient syntheses either incorporate the separation early or, better yet, start with chiral compounds. An unusual method is used to effect the resolution in the case at hand. Thus, condensation of fused cyclobutanone (**41**) with l-ephedrine affords a pair of diastereomeric oxazolidines (**42**) the higher melting of which providentially corresponds to the desired isomer. Separation followed by hydrolysis over silica give **41** with the prostaglandin stereochemistry.

The cyclobutanone is then lactonized by means of Bayer–Villiger oxidation; treatment with dilute acid serves to remove the acetal group to afford lactone-aldehyde **43**. The next step consists in incorporating the remaining carbon atoms required for the lower side chain; Wittig condensation of the aldehyde with the ylide from triphenylphosphoniumhexyl bromide under salt free conditions affords the cis olefin

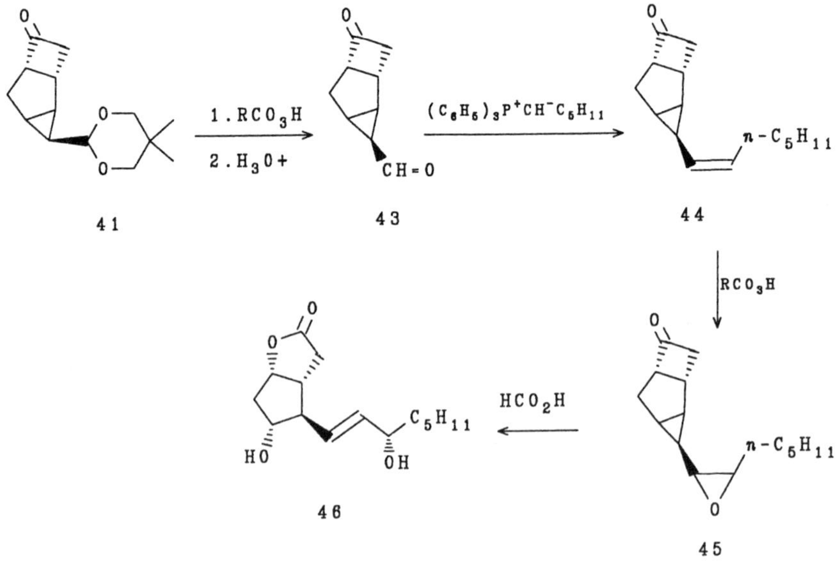

1. RCO_3H 2. H_3O+ ; $(C_6H_5)_3P^+CH^-C_5H_{11}$; RCO_3H ; HCO_2H

41 43 44 45 46

44; this compound is converted to epoxide **45** by means of peracid. Solvolysis of **45** in formic acid gives compound **46** accompanied by significant amounts of glycols (**49**); the mixture is recycled to give **46** in modest yield.

These rearrangements, which are in effect the reverse of those used to form the cyclopropyl ring in **39**, can be visualized as starting with the protonated epoxide **47**; this compound can then go on to rearrange via the homoallyl ion **48**; the observed stereoselective formation of the 11-hydroxyl argues for a concerted reaction. Solvolysis of the diol byproduct **49** may also go through carbocation **48** or through a more concerted transition state. The product **46** is finally taken on to $PGF_{2\alpha}$ by a sequence very similar to the one used to first add the lower side chain to **30**, and after suitable protection of the hydroxyl groups, elaboration of the upper side chain in **26**.[14,15]

It has been known for some time that a mucus layer secreted by gastric cells protects the lining of the stomach from noxious agents including its own digestive agents. Studies on the pharmacology of the prostaglandins revealed that these compounds had a cytoprotective effect on the gastric mucosa by maintaining the mucus layer. The recognition that aspirin and the pharmacologically related NSAIDs owed their action to inhibition of cyclooxygenase offered an explanation for their well-recognized injurious effect on gastric mucosa; inhibition of that enzyme leads to a decrease in prostaglandin synthesis and a consequent increased vulnerability to irritants including normal stomach acid. This prostaglandin deficit is difficult to remedy due to the manifold activity of most agents, their very short biological half-life, and poor oral bioavailability. The finding that biological activity is retained when the side chain hydroxyl group is moved from the prime site of metabolism, 15, to the 16 position

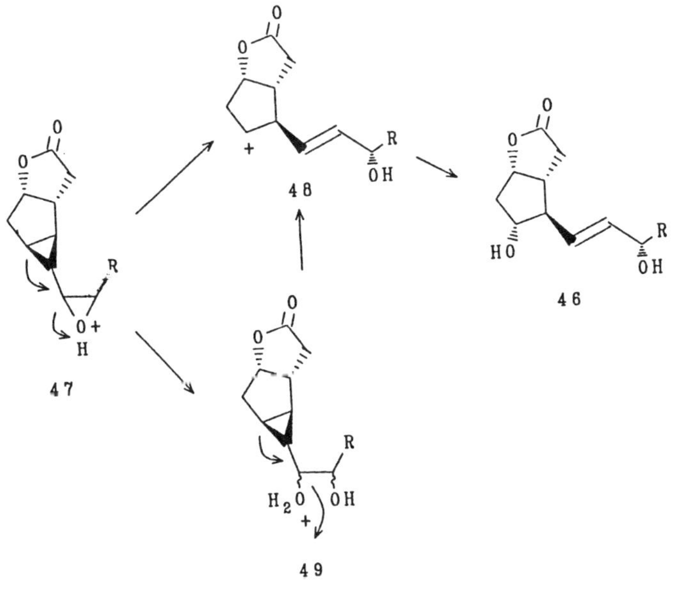

47 48 46 49

resulted eventually in the development of **misoprostol** (**62**), a drug approved for prevention of NSAID induced ulcers.

The synthesis of this compound represents a notable departure from those discussed above. The presence of the carbonyl group at the 9-position of the cyclopentane ring, which classes this compound as a PGE, serves to slightly reduce the stereochemical complexity of the synthesis; more importantly, this ketone introduces the possibility of attaching the lower side chain by means of a 1,4-addition reaction; the trans relationship of the two side chains should be favored by thermodynamic considerations. The very unusual functionality of the required Michael acceptor, that of a cyclopent-2-en-4-ol-1-one, leads to a rather lengthy albeit straightforward synthesis for that intermediate. The scheme starts by activation of monomethylazeleiate (**50**) as its imidazole amide by means of thionyl bisimidazole (Im_2SO). Condensation of this product with the bis anion from the reaction of the lithium salt of monomethyl malonate gives acetoacetate **51**; the first formed tricarbonyl compound decarboxylates on workup. The two terminal methyl ester groups are then saponified to the corresponding acids; the group that is β to the carbonyl group decarboxylates to a methyl

HO_2C ... CO_2CH_3 (50) → [1. $(Im)_2SO$; 2. $LiO_2CCH_2CO_2CH_3$; 3. H_3O^+] → CH_3O_2C ... CO_2CH_3 (51)

51 → [1. OH^-; 2. H_3O^-] → ... CO_2H (52)

52 → [CH_3O–CO–CO–OCH_3; t-BuOK] → [CH_3O_2C, CO_2CH_3, CO_2H (53)]

53 → CH_3O_2C ... CO_2H (54)

54 → [H_3O^+] → OH ... CO_2H (55)

55 → [1. H_2; Pd/C; 2. $(CH_3)_2C(OCH_3)_2$] → CH_3O, HO ... CO_2CH_3 (56)

56 → [LAH] → HO ... CO_2CH_3 (57)

ketone on acidification to afford **52**. Acylation of **52** with dimethyl oxalate leads to addition of an oxalyl group to each carbon flanking the ketone to give an intermediate such as **53.** (Compounds **53** and **54** are depicted as their unlikely all ketone tautomers in the interest of clarity.) Intermediate **53** cyclizes to the triketocyclopentane **54** under reaction conditions. Treatment with acid leads to scission of the superfluous pendant oxalyl group. The product (**55**) probably exists as a mixture of the two possible enolates. Hydrogenation in the presence of palladium on charcoal interestingly leads to reduction of the single carbonyl group not involved in that tautomerism to give the future prostaglandin 11 hydroxyl. Reaction of the product with acetone dimethyl acetal in the presence of acid initially leads to the formation of enol ethers; these compounds can be forced to **56** because of its lower solubility in ether. Reduction of **56** with lithium aluminum hydride (LAH) or Vitride at –60°C leads on workup to the enone **57.**

Preparation of the reagent required for addition of the lower side chain involves a series of metal interchanges carried out as a one-pot reaction. The sequence starts by stereospecific stannylation of the acetylene **58** by means of tributyltin hydride.

CH_3 $(CH_3)_3SiO$ 58 $\xrightarrow{HSnBu_3}$ CH_3 $SnBu_3$ $(CH_3)_3SiO$ 59

1. BuLi
2. CuC≡CPr

CH_3 CuC≡CPr $(CH_3)_3SiO$ 60

O CO_2CH_3 61

O CO_2CH_3 CH_3 HO OH 62

Reaction of **59** with butyllithium gives the corresponding vinyl lithio reagent, where the tin is replaced with retention of configuration. The lithium is then replaced by organocopper by reaction with copper pentyne to give the cuprate reagent **60**. Addition of **60** to the cyclopentenone as its silyl ether **61** gives **misoprostol** (**62**) as a mixture of enantiomers.[16,17]

As noted above, NSAIDs inhibit the inflammatory and, to some extent, the platelet aggregating activities of prostaglandins by inhibiting the enzyme cylooxygenase, which catalyzes their formation. One of the few nitrogen-containing prostaglandin analogues, **vapiprost** (**71**), is reported to be an inhibitor of thromboxane A_2 induced platelet aggregation. This congener is a potentially more specific inhibitor of platelet aggregation, the prelude to thrombus formation, than the NSAIDs in that it blocks thromboxane A_2 at the receptor site. Treatment of the chiral adduct **63** of ketene and cyclopentadiene with bromodimethylhydantoin in acetic acid results in the formation of bromoacetate (**64**), which results from the formal addition of hydrobromous acid. The stereochemistry of the product probably results from formation of the initial bromonium ion on the more open face of the molecule. Treatment with piperidine leads to rearrangement to a 2,2,1-bicycloheptane with incorporation of nitrogen on the new one-carbon bridge. The structure of the product can be rationalized by postulating an intermediate, or transition species, such as **65** along the reaction pathway; saponification of the initially formed product gives the ketoalcohol **66**. This alcohol is acylated to **67** by means of *p*-phenylbenzoyl halide, a group used in other prostaglandin syntheses for directing the stereochemistry of reductions. Bayer–Villiger oxidation with peracid gives a bridged version (**68**) of a Corey lactone; reduction with diisobutylaluminum hydride (DIBAL-H) in the cold leads to the hydroxyaldehyde **69**, here isolated in open form. The aldehyde is then homologated first with methoxymethyl phosphorane to give **70**. A second Wittig condensation with the ylide from triphenylphosphonium butyrate completes construction of the side chain that differs from the one in natural prostaglandins in so far as the olefin is moved one atom closer to the acid. The next two steps consist of inversion of the stereochemistry of the "11"-hydroxyl group to the unnatural β configuration. Thus, Swern oxidation of the initial product followed by reduction with DIBAL-H gives vapiprost (**71**).[18] The stereochemistry of the reduction is probably guided by the very bulky *p*-phenylbenzoyl group at the "9"-position.

2. PROTEASE INHIBITORS

The central role of polypeptides as regulators of life is of course very generally recognized. An important class of these regulators consist of enzymes, virtually all of which are made up of chains of amino acids. It is an interesting fact that these compounds, whose assembly is mediated by transcription of RNA, are not synthesized directly in their final form. Instead, they quite often first appear as part of a much larger peptide; a specialized class of enzymes, dubbed proteases, cut the chain at specific locations so as to excise the enzyme in its active form. Renin was one of the first of the proteases to be investigated in some detail. This polypeptide specifically cleaves the large peptide angiotensinogen to excise therefrom the decapeptide angiotensin I. Angiotensin I yields the potent vasoconstrictor octapeptide angiotensin II on reaction with yet another protease, angiotensin converting enzyme (ACE). A series of nonpeptide compounds that blunted the action of this enzyme, known as ACE inhibitors, have proven useful in treating hypertension by decreasing levels of vasoconstricting angiotensin II by lowering levels of ACE. Considerable effort has been devoted to the search for drugs that block this cascade upstream at the level of renin in the search for antihypertensive agents that would avoid some of the shortcomings of the generally well-tolerated ACE inhibitors.

Proteases, like many enzymes, act by stabilizing a relatively high-energy transition state; in this case, the initial adduct of a hydroxyl group, or its functional equivalent, to the carbonyl carbon; this addition causes the geometry of that center to change from trigonal to tetrahedral. Known inhibitors consist of molecules that mimic an essential stretch of the protease recognition site and most importantly provide a sequence that duplicates the transition center sterically without, however, including a cleaveable bond. The fermentation product, pepstatin, a peptide-like inhibitor of pepsin, provided a clue for the synthesis of protease inhibitors; the central portion of this molecule provides a 1,3-hydroxyamide sequence that is thought to act as a transition state analogue from a peptide bond; note that a methylene replaces one amide nitrogen. The active moiety, statine, has been prepared by a total synthesis involving aldol-like condensation of isoleucylaldehyde with the lithio carbanion from acetate.[19]

Preparation of the renin inhibitor **terlakiren** (**77**), starts with the reaction of the *S*-methyl ether of cysteine, which is protected as its *tert*-butylcarbonyl (Boc) amide (**73**), with the statine analogue **72** in which cyclohexyl replaces the isobutyl group; the coupling reaction is catalyzed by dicyclohexylcarbodiimide (DCC). The amino group in the product **74** is then deprotected by treatment with trifluoroacetic acid; this reagent leads to elimination of isobutylene from the *t*-Boc group followed by decarboxylation of the now unstable free carbamic acid. The coupling sequence is now repeated using the phenylalanine derivative **76** to yield the desired product **77**.[20]

The functional simplicity of viruses combined with the fact that they require a living host for their replication has made them an unusually difficult therapeutic target. Human immunodeficiency virus (HIV), in common with most viruses consists of a packet of genetic information encoded in this case in RNA and an outer protein coat. One of the final steps in viral replication involves synthesis of the coat peptide. Production of this coat peptide involves scission of the initially produced much larger protein by means of an aspartyl protease; virus lacking the correct coat is not functional. The research that led to the HIV protease inhibitors detailed below represents a new era in drug development. The availability of a full structure of the protease, obtained by X-ray diffraction, made possible the use of computer based modeling programs for designing inhibitors that best fit the target enzyme. This finding largely accounts for the fact that the inhibitors, which to some extent must mimic a polypeptide, include at most only one of the naturally occurring amino acids.

The statine like moiety in this case is a transition state mimic for cleavage of phenylalanylprolyl and tyrosylprolyl sequences. Construction starts with protection of the amino group of phenylalanine as its phthaloyl derivative (Phth) by reaction with

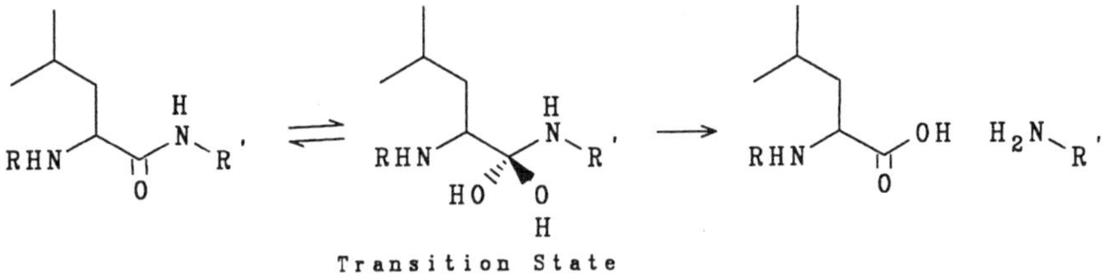

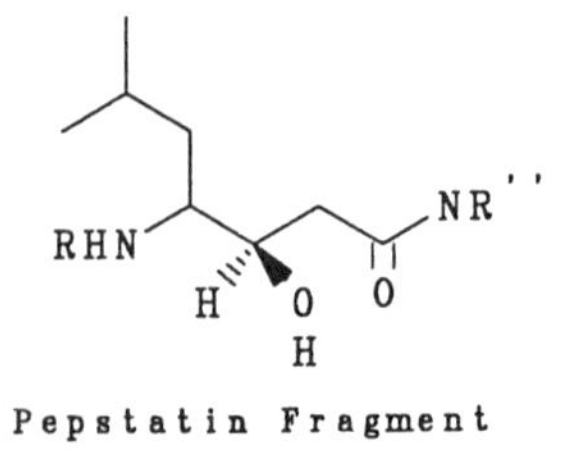

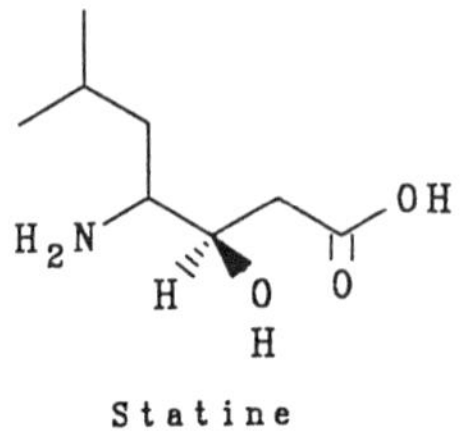

RHN H N R' O ⇌ RHN H N R' HO O H → RHN OH O H_2N R'

Transition State

RHN NR'' H O H O

Pepstatin Fragment

H_2N OH H O H O

Statine

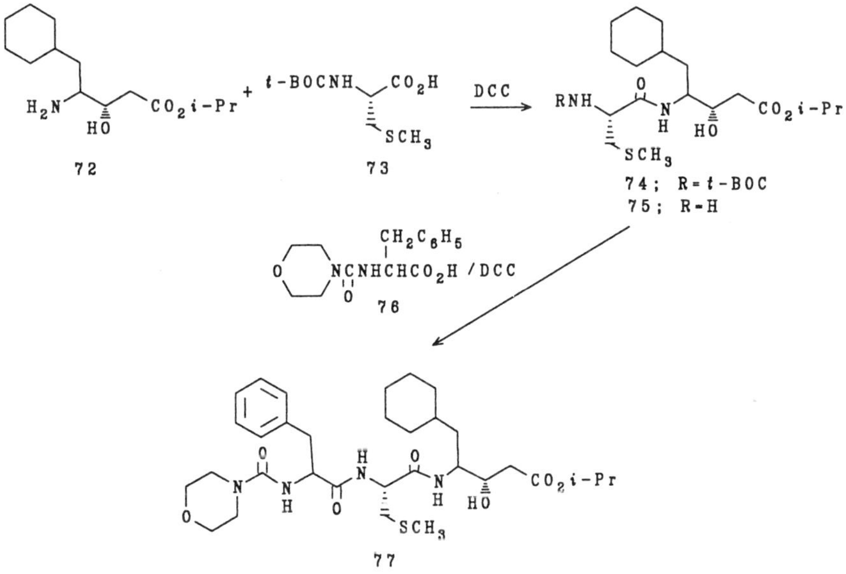

phthalic anhydride; this compound is then converted to an acid chloride. The chain is then extended by one carbon using a Friedel–Crafts-like reaction. The required reagent, **79**, is prepared by reaction of the enolate obtained from the bis-silyl ether **78** of glyoxylic acid and lithiohexamethyldisilazane (LiHMDS) with trimethylsilyl chloride (TMSCl).[21] Uncatalyzed reaction of the acid chloride **80** with **78** gives the chain-extended product **82** directly on acidification; the first formed β-carbonyl compound **81** apparently decarboxylates spontaneously. The terminal alcohol is then protected as a tetrahydropyranyl ether by addition to dihydropyran; reduction of the ketone with sodium borohydride occurs enantioselectively due to the presence of the adjacent chiral center. Reaction with methanesulfonyl chloride then gives the intermediate mesylate **83**, which is not isolated. The pyranyl ether is then removed by acid catalyzed exchange with ethanol to give **84**. On treatment with potassium *tert*-butoxide, the alkoxide formed from the terminal hydroxyl internally displaces the adjacent mesylate to form epoxide **85** in which the configuration of the former alcohol carbon is inverted.[22]

The other major fragment consists of a decahydroisoquinoline that may be viewed as a rigid analogue of an amino acid. Methanolysis of the adduct **86** from butadiene and maleic anhydride in basic methanol gives the half-ester **87**; the obligate cis stereochemistry of these adducts determines the configuration of the perhydroisoquinoline ring fusion. This compound is resolved as its salt with *l*-ephedrine. The desired enantiomer is then converted to the acid chloride **88**; hydrogenation of **88** under Rosenmund conditions, palladium in charcoal in the presence of quinoline, leads to the aldehyde **89**. The next step essentially involves addition of methyl glycinate to the aldehyde group. Conversion of this compound to its benzal derivative **90** serves to

TMSO O OTMS 78 LiHMDS TMSCl TMSO OTMS OTMS 79

C_6H_5 PhthN COCl 80 1. 79 2. H_3O^+ [C_6H_5 PhthN CO_2H OH O 81] $-CO_2$ C_6H_5 PhthN OH O 82

1. DHP 2. $NaBH_4$ 3. CH_3SO_2Cl

C_6H_5 PhthN O 85 t-BuOK C_6H_5 PhthN OMs OH 84 p-TsA, EtOH [C_6H_5 PhthN OMs OTHP] 83

p-TSA-p-toluenesulfonic acid

remove the more acidic amino protons and at the same time activates the protons on the methylene group. Condensation of the lithium salt from this compound with aldehyde **89** may be envisaged as first forming an adduct such as **91**. The acidic workup serves to dehydrate the β-hydroxyester, hydrolyze the Schiff base, and to cyclize the ester with the newly revealed amine, although not necessarily in that order. The first product isolated is in fact the lactam **92**. Reaction with diborane in the presence of propylamine serves to reduce both the lactam and the olefin conjugated with the ester to afford **93**. Displacement of the ester methoxyl by means of dibutylaluminum-*tert*-butylamide gives the decahydroquinoline **94**.[23]

The last stage in this convergent synthesis consists in assembly of the individual units. Ring opening of epoxide **85** by the secondary amino group on perhydroisoquinoline **94** gives the alcohol **95**. The phthaloyl protecting group is then removed by traditional treatment with hydrazine or alternatively with methylamine; the latter being more suitable to large scale work. The free amino group in **96** is then condensed with the carbobenzyloxy (Cbz) derivative **97** of the monoamide from aspartic acid to give amide **98**. Hydrogenation over palladium on charcoal reductively removes the benzyl group from the Cbz derivative; the unstable carbamic amide that remains decarboxylates to afford the amine **99**. Condensation with quinoline-2-carboxylic acid catalyzed by DCC forms the last amide bond.[22] Thus, the HIV protease inhibitor **saquinovir** (**100**) is obtained.

Saquinovir has shown activity in treatment of acquired immunodeficiency syndrome (AIDS) in several clinical trials and has been made available for large scale use in patients. The other HIV protease inhibitor that shows promise, **indinavir**, differs markedly in its structural components and is notable because it does not include a

86 → ($MeOH$, K_2CO_3) → 87 (CO_2H, CO_2CH_3) → (1. *l*-ephedrine, 2. SO_2Cl) → 88 ($COCl$, CO_2CH_3) → (H_2, Pd) → 89 ($CH=O$, CO_2CH_3)

89 → ($C_6H_5CH=NCH_2CO_2CH_3$, 90) → [91] (OH, CO_2CH_3, N=, CO_2CH_3, C_6H_5)

91 → (H_3O^+) → 92 (CO_2CH_3, NH, O) → (B_2H_6, *n*$PrNH_2$) → 93 (CO_2CH_3, NH) → (Bu_2AlNHt-Bu) → 94 (O, CN*t*-Bu, H, NH)

PhthN 85 + HN CONH*t*Bu 94 → PhthN OH N CONH*t*Bu 95

$MeNH_2$ or H_2NNH_2

H_2N OH N CONH*t*Bu 96

CbzNH CO_2H $CONH_2$ 97, DCC

RHN O N H $CONH_2$ OH N CONH*t*Bu

98; R=Cbz

99; R=H

($Cbz = C_6H_5CH_2OCO$)

N O N H O N H $CONH_2$ OH N CONH*t*Bu

100

101 102 103 104 105

1. LiHMDS
2. OTs

Ts- p-toluenesulfonyl

single natural α-amino acid.[24] Construction of this compound starts with reaction of resolved 1-amino-2-indanol with acetone to afford the cyclic carbinolamine derivative **102**, which will act as a protecting group for both functions. Acylation of this acetonide with hydrocinnamyl chloride (**101**) gives the amide **103**. One of the key transformations in the sequence involves alkylation of the carbanion that is obtained on treatment of **103** with lithiohexamethyldisilazane with the toluenesulfonate derivative **104** of chiral glycidol. The enantioselective course of the alkylation reaction leading to **105** can be attributed to the proximity of the two chiral centers on the indan.

Catalytic reduction of the *tert*-butylamide (**106**) of pyrazine carboxylic acid gives the corresponding piperazine; this compound is then resolved as its camphorsulfonate salt. The amine at the 4-position is then selectively protected as its *t*-Boc derivative **108**. The less steric bulk about this amino group as well as possible hydrogen bonding of the amine at the 1-position with the adjacent carbonyl group contribute to the selectivity of this acylation step.

Condensation of intermediate **108** with the large fragment **105** leads to attack of the free amino group of the piperazine on the epoxide with consequent ring opening and formation of the alcohol **109**; this reaction proceeds with the expected retention of configuration of the chiral center bearing the hydroxyl group. The *tert*-butoxycarbonyl protecting group is then removed by exposure of the intermediate to acid; the

1. H_2
2. CSA

$(t\text{-BOC})_2O$

CONHtBu CONHt-Bu CONHt-Bu

106 107 108

CSA- camphorsulfonic acid

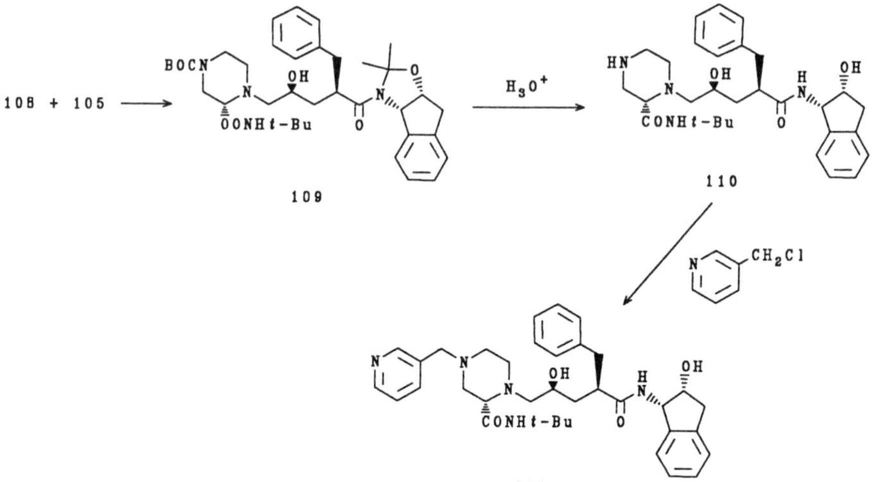

108 + 105 → 109 → (H_3O^+) → 110 → (3-chloromethylpyridine) → 111

acetonide hydrolyzes under reaction conditions to give **110** as the overall reaction product. Alkylation of the newly revealed piperazine nitrogen with 3-chloromethylpyridine affords the protease inhibitor **indinavir** (**111**).[25]

3. RETINOIDS

Vitamin A consists of a mixture of two polyene diterpenes, retinol and its biologically active oxidation product, retinal, shown as its trans isomer, **112**. Retinal forms a crucial link in vision; light induced isomerization of the double bond adjacent to the carbonyl, conjugated with vision proteins as a Schiff base from cis to trans, plays a key role in transduction of light to visual perception. The metabolite from oxidation of retinal, retinoic acid (**113**), has potent biological activity in its own right. The compound is a ligand for receptors involved in epithelial differentiation. All-*trans* retinoic acid, under the USAN name **tretinoin (113)**, has, as a result of that activity, found clinical use in the treatment of skin diseases such as acne; its off-label use as a skin rejuvenation agent was the focus of some recent attention in the press. Considerable research has been prompted by data suggesting that the agent may have an effect on cancer progression.

The majority of published syntheses of **tretinoin** (**113**) start with the readily available β-ionone **114**, a compound that already incorporates the highly substituted cyclohexene ring as well as four of the side chain atoms. Condensation with the

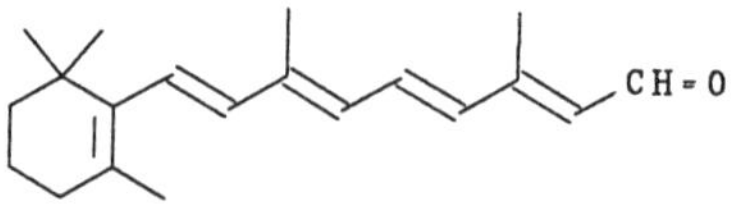

112 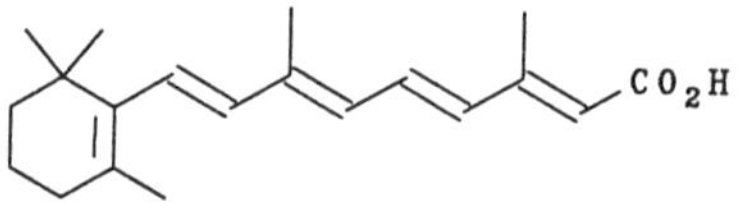 113

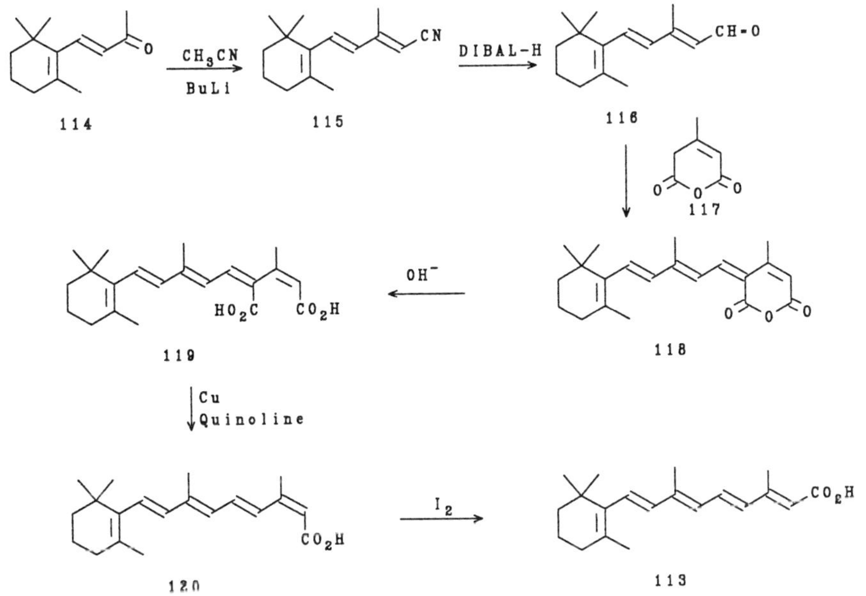

carbanion from acetonitrile followed by dehydration of the initially formed carbinol gives intermediate **115**. Reduction of the cyano group by DIBAL-H leads to the corresponding imine; this compound hydrolyzes to aldehyde **116** during the acid workup.[26] Base catalyzed aldol condensation of this aldehyde with β-methylglutaconic anhydride (**117**) involves condensation with the activated methylene group of the anhydride and leads to product **118**, in which the remainder of the side chain has been added. The anhydride is then hydrolyzed to the vinylogous β-dicarboxylic acid **119**; the superfluous carboxyl group is removed by heating the compound in quinoline in the presence of copper to give **120**. The terminal double bond is then isomerized by any of several methods to give tretinoin (**113**).[27]

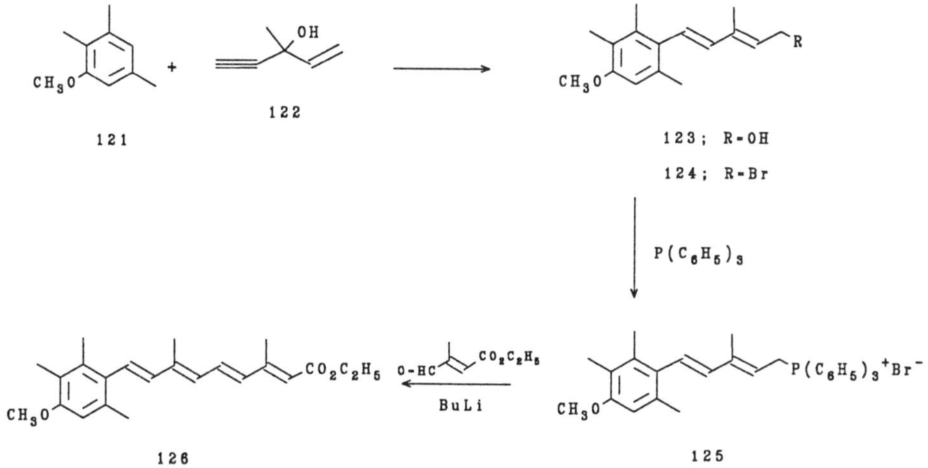

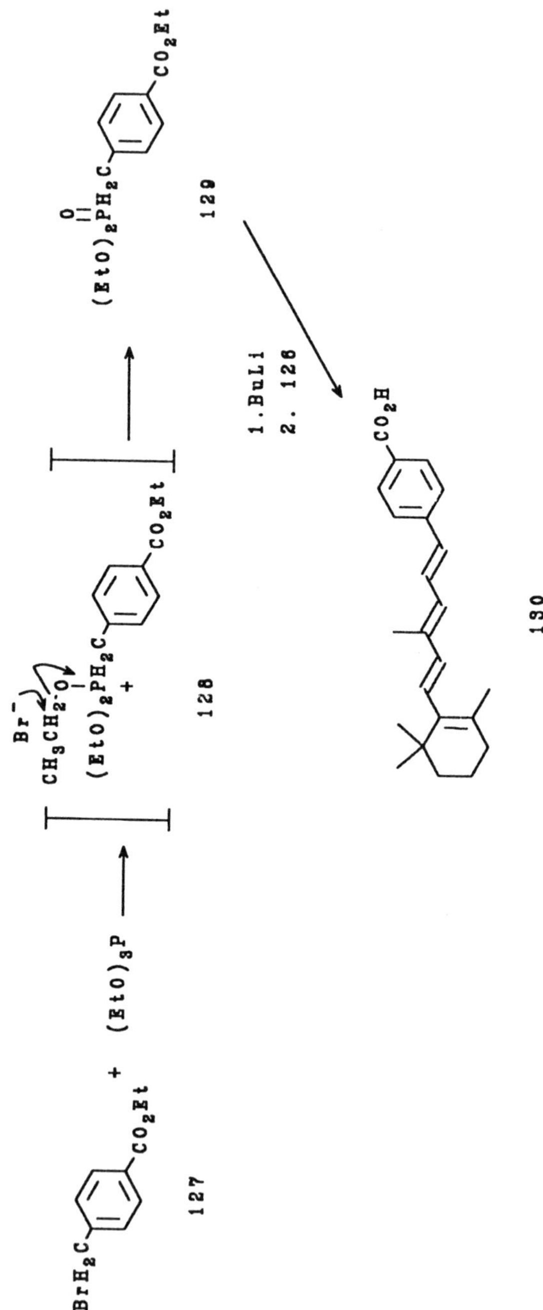
BrH_2C
CO_2Et
127
+
$(EtO)_3P$
Br^-
CH_3CH_2-O
$(EtO)_2PH_2C$
+
CO_2Et
128
O
$(EtO)_2PH_2C$
CO_2Et
129
1. BuLi
2. 128
CO_2H
130

Replacement of the cylohexene ring by an aromatic moiety is interestingly quite consistent with retinoid-type activity. Construction of this analogue starts by reaction of 2,3,5-trimethyl anisole (**121**) with the acetylene **122**, which is obtainable from acetylene and methylvinyl ketone with a strong Lewis acid under Friedel–Crafts conditions. The ambident carbocation from **122** reacts with the aromatic ring at the acetylene; the alkylated intermediate then adds back a hydroxyl at the terminal position to give the terpene-like intermediate **123**; the resulting allylic alcohol is then converted to its bromide **124**. Displacement of halogen by means of triphenylphosphine leads to the phosphonium salt **125**. Wittig condensation of the ylide from this last intermediate with the half-aldehyde derivative of α-methylfumaric ester adds the remainder of the retinoid side chain to give etretinate (**126**).[28]

Activity is largely retained in a compound in which one of the olefinic bonds in the side chain of **tretinoin (131)** is replaced by an aromatic ring. The key reaction in the construction of this compound consists in Wittig condensation of the ylide from the phosphonate **129** with the chain-extended aldehyde **116** used in the synthesis of tretinoin (**131**) itself. The Arbuzov rearrangement provides ready access to the phosphonate. Reaction of ethyl *p*-bromomethylbenzoate (**127**) with triethyl phosphite probably proceeds by initial displacement of halogen by phosphorus to give a transient intermediate charged species such as **128**. Displacement of one of the ethyl groups on phosphorus by bromide ion from the first displacement followed by bond reorganization leads to the phosphonate **129** and ethyl bromide as a byproduct. Saponification of the ylide condensation product yields **pelretin (130)**.[29]

The breadth of the structural tolerance in this class of compounds is suggested by the fact that all but one of the five double bonds can be replaced by aromatic rings. It is of note that the terminal carboxyl group can be replaced by an alkylsulfonyl function, a moiety that is quite polar but one whose p*K* is orders of magnitude removed from that of a carboxylic acid. The construction of this compound also demonstrates a reversal of the strategy use to prepare **pelretin** in that the ylide is prepared from the nucleus. The synthesis starts with the acylation of tetramethyltetralin (**131**) with acetyl chloride in acetic anhydride to give **132**. Successive reduction with sodium borohydride to the alcohol (**133**), conversion to a bromide (**134**), and displacement of halogen

Ac_2O, AcCl

131 132

130, R=OH
134; R=Br
135; $R{-}P^+(C_6H_5)_3$

1. BuLi
2. $p\text{-}O{=}HCC_6H_5SO_2CH_3$

SO_2CH_3

136

with triphenylphosphine gives the phosphonium salt **135**. Condensation of the ylide obtained from this compound with *p*-methylsulfonylbenzaldehyde leads directly to, **136**, **sumarotene**.[30]

REFERENCES

1. Kurzok, R.; Lieb, C.C. *Proc. Soc. Exp. Biol. Med.* **1930**, *28*, 268.
2. von Euler, U.S. *J. Physiol. (London)* **1937**, *4*, 213.
3. Abrahamson, S.; Bergstrom, S.; Samuelsson, B. *Proc. Chem. Soc.* **1962**, 352.
4. For a detailed account see Lewis, A.L., Ed.; In *CRC Handbook of Eicosanoids: Prostaglandins and Related Lipids*; CRC Press: Boca Raton, FL, 1987.
5. Weinheimer, A.J.; Spraggins, R.L. *Tetrahedron Lett.* **1969**, 5185.
6. Bundy, G.L.; Lincoln, F.H.; Nelson, N.A.; Pike, J.E.; Schneider, W.P. *Ann. N. Y. Acad. Sci.* **1971**, *180*, 76.
7. (a) Corey, E.J.; Schaaf, T.K.; Huber, W.; Koelliker, V.; Weinshenker, N.M. *J. Am. Chem. Soc.* **1970**, *92*, 397. (b) For a review see Axen, U.; Pike, J.E.; and Schneider, W.P. In *The Total Synthesis of Natural Products*, ApSimon, J.W., Ed.; Total Synthesis of Prostaglandins, Wiley, New York, 1973, Vol. 1, p. 81.
8. Black, K.A.; Vogel, P. *Helv. Chim. Acta* **1984**, *67*, 1612.
9. Renard, P.; Vionet, J.P. *J. Org. Chem.* **1993**, *58*, 5893.
10. Renard, P.; Vionet, J.P. *Helv. Chim. Acta* **1994**, *77*, 1781.
11. Corey, E.J.; Hamanaka, E. *J. Am. Chem. Soc.* **1967**, *89*, 2758.
12. Binder, D.; Bowler, J.; Brown, E.D.; Crossley, N.S.; Hutton, J.; Senior, M.; Slater, L.; Wilkinson, P.; Wright, N.C.A. *Prostaglandins* **1974**, *6*, 87.
13. Aristoff, P.A.; Johnson, P.D.; Harrison, A.W. *J. Org. Chem.* **1983**, *48*, 5341.
14. Kelly, R.C.; Van Rheenen, V. *Tetrahedron Lett.* **1973**, 1709.
15. Kelly, R.C.; Van Rheenen, V.; Schletter, I.; Pillai, M.D. *J. Am. Chem. Soc.* **1973**, *95*, 2746.
16. Collins, P.W.; Dajani, E.Z.; Driskill, D.R.; Bruhn, M.S.; Jung, J.J.; Pappo, R. *J. Med. Chem.* **1977**, *20*, 1152.
17. Collins, P.W. In *Chronicles of Drug Discovery*, Lednicer, D., Ed.; ACS Books: Washington, DC, 1993, Vol. 3, p. 101.
18. Lumley, H.; Finch, H.; Collington, E.W.C.; Humphrey, P.P.A. *Drugs Future* **1990**, *15*, 1087.
19. Rich, D.H.; Sun, T.E.; Boparai, A.S. *J. Org. Chem.* **1978**, *43*, 3624.
20. Hoover, D.J.; Webster, R.T.; Rosati, R.L. Eur. Patent Application 1988, EP 311012; *Chem. Abstr.* 1989, *110*, 154889.
21. Wissner, A. *J. Org. Chem.* **1979**, *44*, 4617.
22. Parkes, K.E.B.; Bushnell, D.J.; Crackett, P.H.; Dunsdon, S.J.; Freeman, A.C.; Gunn, H.P.; Hopkins, R.A.; Lambert, R.W.; Martin, J.A.; Merrett, J.H.; Redshaws, S.; Spurden, W.C.; Thomas, G.J. *J. Org. Chem.* **1994**, *59*, 2656.
23. Houpis, I.N.; Molina, A.; Reamer, R.A.; Lynch, J.E.; Volante, R.P.; Reiden, P.J. *Tetrahedron Lett.* **1993**, *34*, 2593.

24. Dorsey, B.D.; Levin, R.B.; McDaniel, S.L.; Vacca, J.P.; Guare, J.P.; Drake, P.L.; Zugay, J.A.; Emini, E.A.; Schleif, W.A.; Quintero, J.C.; Lin, J.H.; Holloway, M.K.; Fitzgerald, P.M.D.; Axel, M.G.; Ostovic, D.; Anderson, P.S.; Huff, J.R. *J. Med. Chem.* **1994**, *37*, 3443.
25. Askin, M.D.; Eng, K.K.; Rossen, K.; Purick, R.M.; Wells, K.M.; Volante, R.P.; Reider, P.J. *Tetrahedron Lett.* **1994**, *35*, 673.
26. Cainelli, G.; Cardillo, G.; Contento, M.; Grasseli, P.; Ronchi, A.U. *Gazz. Chim. Ital.* **1973**, *103*, 117.
27. Lewin, A.H.; Whaley, M.G.; Parker, S.R.; Carroll, F.I.; Moreland, C.G. *J. Org. Chem.* **1982**, 1799.
28. Soukup, M.; Broger, E.; Widmer, E. *Helv. Chim. Acta* **1989**, *72*, 370.
29. Dawson, M.L.; Hobbs, P.D.; Dedzinski, K.; Chan, R.L.S.; Grber, J.; Chan, W.; Smith, S.; Thies, R.W.; Schift, L.J. *J. Med. Chem.* **1984**, *27*, 1516.
30. Klaus, M.; Bollag, W.; Huber, P.; Kueng, W. *Eur. J. Med. Chem.* **1983**, *18*, 425.

CHAPTER 2

DRUGS BASED ON A SUBSTITUTED BENZENE RING

Benzene rings and other aromatic systems abound among compounds used as therapeutic agents. The 50 or so drugs described below represent a very small sample of the hundreds of agents that are centered on substituted benzene rings. These rings play a manifold role in drugs ranging from just providing simple steric bulk to forming an integral part of the pharmacophore. Most, but not all, of the drugs discussed in this chapter fall into the latter category and have been chosen for discussion here for their illustrative value.

1. ARYLETHANOLAMINES

One of the most reliable sources of leads for new drugs consists of the endogenous compounds that act as messengers for various vital functions. **Adrenaline** (**1**) and **noradrenaline** (**2**), two closely related arylethanolamine that play a key role in homeostasis, were isolated and characterized structurally in the mid-1930s. It was recognized at the time that these two agents are intimately associated with the sympathetic branch of the involuntary or autonomic nervous system. These compounds, which transmit nerve signals across synapses in this system, play a key role in regulating blood pressure, heart rate, constriction or dilation of bronchioles, and a host of other involuntary functions. Consequently, this is also known as the adrenergic system.

Epinephrine (**1**) is often one of the first drugs used in treating trauma because of its cardiostimulant and bronchodilating actions. Simple replacement of the methyl group on nitrogen by isopropyl gives an agent, **isoproterenol** (**5**) with longer lasting action. Each of these drugs is available in racemic form by a relatively short, straightforward synthesis. Friedel–Crafts acylation of catechol with chloroacetyl chloride leads to the chloroketone **3**. Displacement of halogen with isopropylamine gives aminoketone **4**; hydrogenation over platinum reduces the carbonyl group to give racemic isoproterenol (**5**). The same sequence using methylamine leads to epinephrine (**1**); resolution of **1** as its tartrate salt gives **l-epinephrine**, which is identical to the natural product.[1]

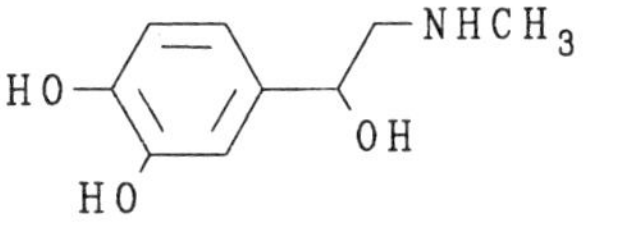

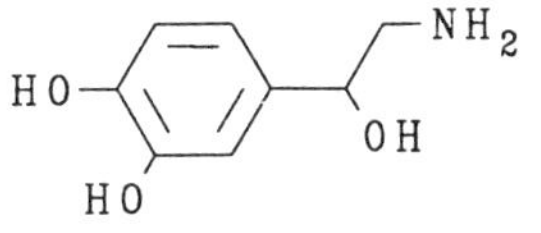

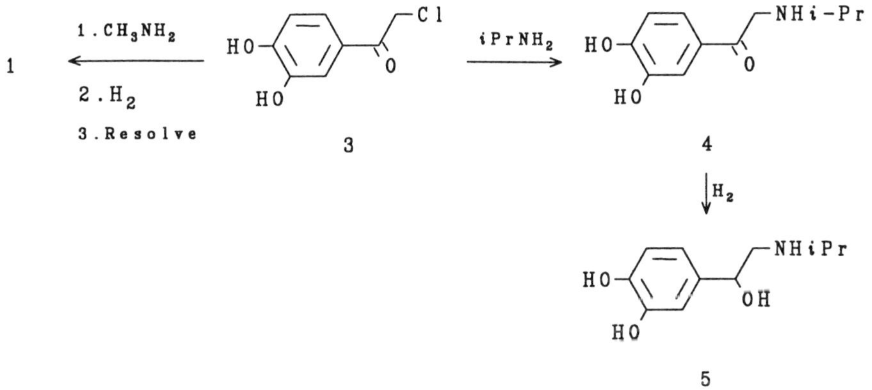

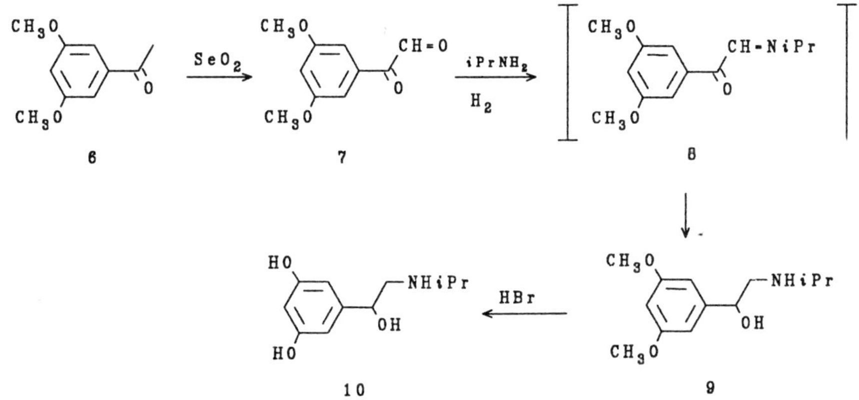

The isomer of **5**, **metaproterenol** (**10**), in which both phenolic hydroxyl groups occupy the meta position, retains bronchodilating activity. The synthesis begins with treatment of substituted acetophenone (**6**) with selenium dioxide; the methyl group is thus oxidized to the corresponding aldehyde to give the glyoxal **7**. Reductive amination with isopropylamine can be envisaged to proceed through the imine **8**. Hydrogenation of the imine and carbonyl group then yields the aminoalcohol **9**. The phenolic methyl ethers are then cleaved by means of hydrogen bromide to give metaproterenol (**10**).[2]

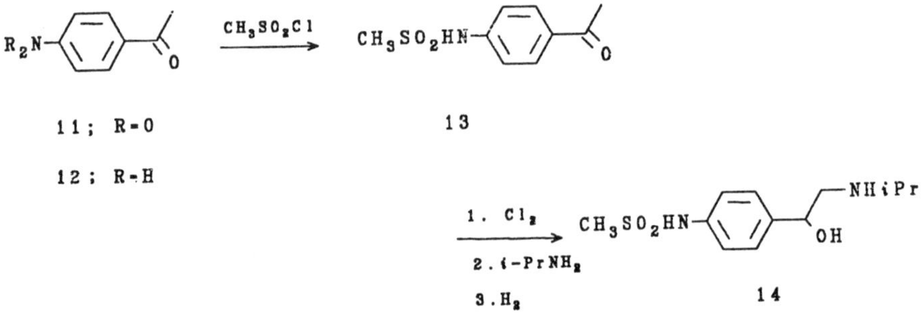

The adrenergic nervous system is itself divided into two broad categories, which are denoted as the α and β branches. Drugs such as **metaproteronol** and **deterenol**, a congener of **isoproterenol** lacking the meta hydroxyl group, act largely as β-adrenergic agonists. The fact that the proton in a sulfonylanilide should have a p*K* in the same range as a phenol encouraged the preparation of the **deterenol** congener **sotalol** (**14**). Note that although this compound interacts with β-adrenergic receptors, it does so as an antagonist; this compound was in fact one of the first β-blockers. One of several routes to this compound begins with the reduction of the readily available *p*-nitroacetophenone (**11**) to the corresponding aniline **12**, by a method specific to nitro groups such as iron and hydrochloric acid. Reaction with methanesulfonyl chloride give the sulfonanilide **13**. This intermediate is then carried on to sotalol (**14**), by the same series of reactions used to prepare **isoproterenol** (**5**).

The history of drug discovery aptly illustrates the important role played in this process by serendipity. A short digression illustrates that this route is still a viable approach to finding new structural types for drugs. Clinical investigations on **sotalol** (**14**) revealed that the agent had pronounced activity as an antiarrhythmic agent, an action that could be, and was, logically attributed to the compound's β-blocking action. The observation that both enantiomers seemed to have equal potency, however, cast some doubt on this explanation for the antiarrhythmic activity. More recent work, perhaps spurred by this discrepancy, has in fact shown that sulfonanilides that lack the phenethanolamine side chain show quite good antiarrhythmic activity in their own right. The first of these agents, **ibutilide** (**17**), incorporates a vestige of the ethanolamine side chain, in the form of a 1,4-aminoalcohol. Preparation starts with the Friedel–Crafts acylation of methanesulfonylanilide with succinic anhydride to give the ketoacid **15**. Reaction of the corresponding acid chloride with ethylheptylamine gives the amide **16**. Reaction with lithium aluminum hydride in the cold serves to reduce both the amide and ketone to afford ibutilide (**17**).[3] Further work shows that activity is retained when the hydroxyl group is replaced by polar groups such as an amide or even a nonenolizable sulfonamide. Ester interchange of the mesylate from ethyl *p*-aminobezoate (**18**) with *N*,*N*-diethylethylenediamine gives the antiarrhythmic agent **19 sematilide**.[4] In a similar vein, reaction of sulfonyl chloride (**20**) (from reaction of methanesulfonylanilide and chlorosulfonic acid) with *N*,*N*′-di-*iso*-propylethylenediamine give **risotilide** (**21**).[5]

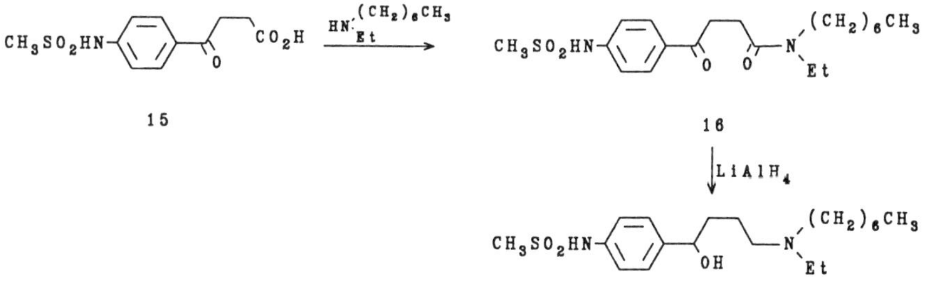

The β-adrenergic system is itself divided into several different branches; receptors for these subsystems show different ligand structural preferences. The cardiovascular system is responsive largely to β_1 adrenergic agents; activation leads to an increase in blood pressure and heart rate. Bronchioles constitute an important target for β_2 adrenergic agonists; activation leads to relaxation and resolution of bronchospasm. Use of the classical β agonist **isoproterenol** for treatment of asthma is limited by the side effect due to poor selectivity for β_2 receptors. Compounds that exhibit β_2 adrenergic agonist activity have proven very useful in the treatment of asthma. The compounds discussed below represent a very small selection from the dozens of antiasthma compounds that have been investigated in the clinic. It is noteworthy that while replacement of the para hydroxyl group of a phenylethanolamine by a sulfonanilido, as found in the compound **sotalol**, leads to an antagonist, the corresponding change at the meta position leads to an adrenergic agonist that shows selectivity for β_2 receptors. The synthesis of the agent **soterenol** (**27**) starts with the nitration of *p*-benzyloxyacetophenone (**22**); reduction of the intermediate nitro compound **23** with hydrazine in the presence of Raney nickel gives the corresponding aniline **24**. This compound is then converted to the sulfonamide (**25**) by reaction with methanesulfonyl chloride. Bromination of the methyl group of the ketone followed by displacement with isopropylamine leads to the intermediate **26**. Reduction of the ketone to an alcohol followed by hydrogenolysis of the benzyl protecting group affords soterenol (**27**).[6]

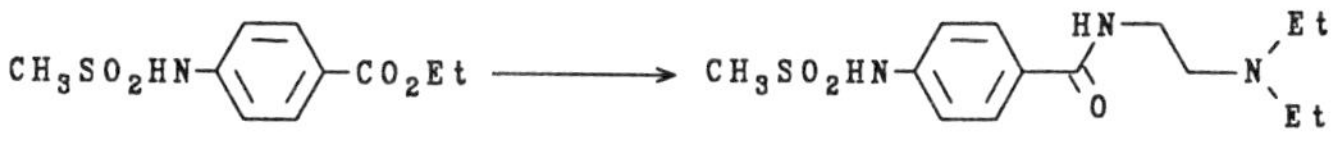

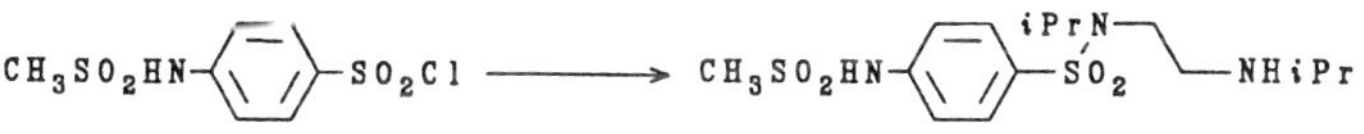

A simple aliphatic alcohol at the meta position is actually sufficient for conferring β_2 agonist activity to a phenolethanolamine as demonstrated by the very widely used drug **albuterol** (**31**), formerly known as **salbutamol**. The acetylation product **28** from methyl salicylate provides the starting material. The usual amination sequence using *tert*-butylbenzylamine gives the corresponding aminoketone **29**. Reduction by means of lithium aluminum hydride converts the ester to a carbinol and the ketone to the requisite alcohol in a single step to give **30**. The benzyl protecting group is then removed by catalytic reduction to afford albuterol (**31**).[7] Compound **32**, the analogue in which the amino group is primary, provides the starting material for a significantly more lipophilic β agonist. Construction of the side chain for this compound involves monoalkylation of 1,6-dibromohexane with 2-phenylethanol to give bromide **33**. Alkylation of **32** with that halide gives **salmeterol** (**34**) in a single step.[8]

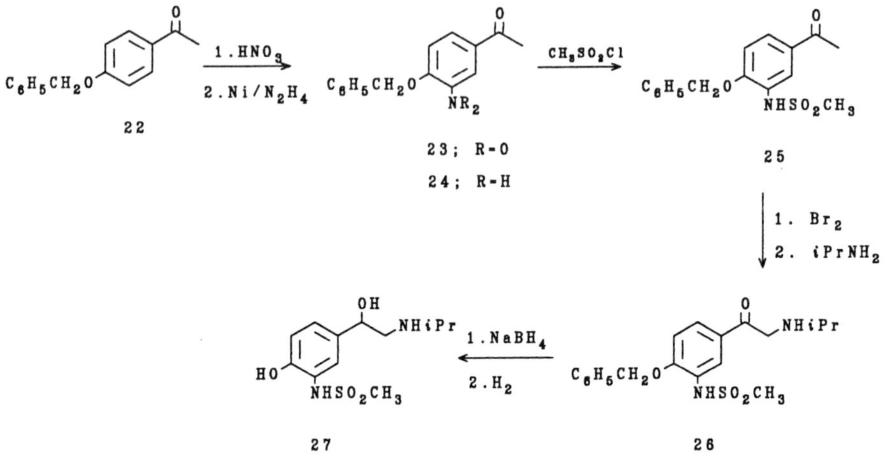

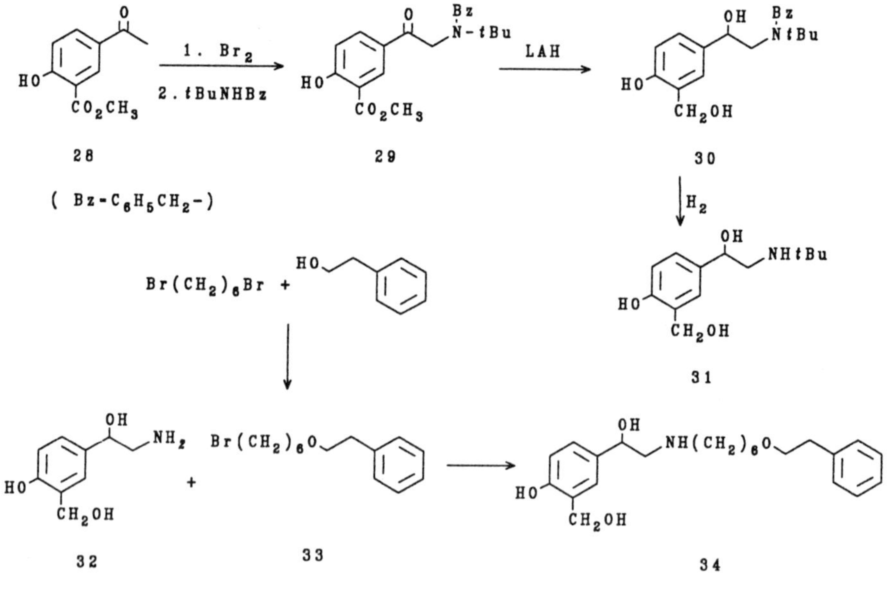

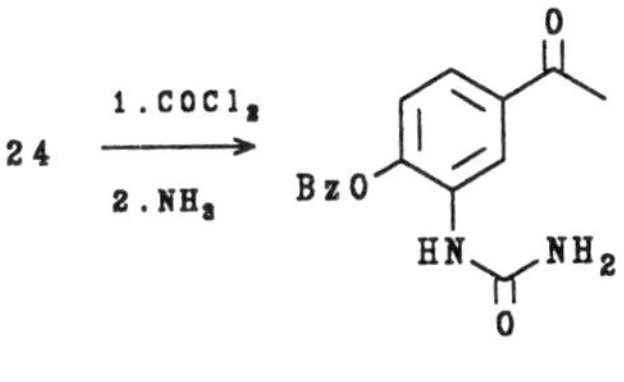

Selective β_2 agonist is retained when the phenol at the meta position is replaced by urea. Sequential reaction of the **soterenol** intermediate **24** with phosgene and ammonia leads to urea derivative **35**. The by now familiar bromination–amination sequence gives the aminoketone **36**. The ketone is then reduced to an alcohol with a sodium borohydride and the benzyl protecting group is removed by hydrogenolysis to give **carbuterol** (**37**).[9]

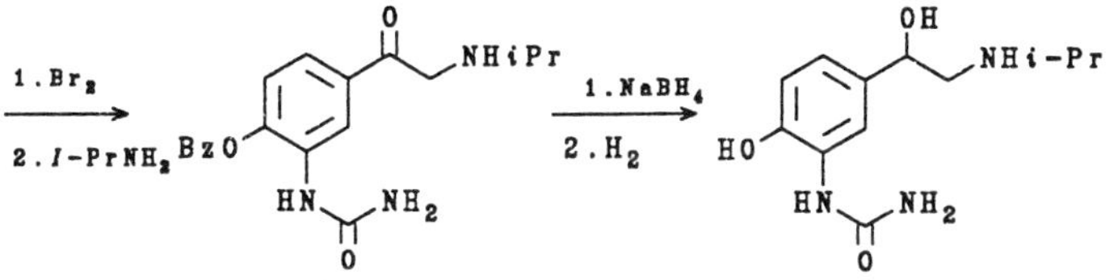

The activity of β-blockers as antihypertensive agents is discussed in greater detail in Section 2; however, in this discussion it is relevant to note that some of the shortcomings of these drugs can be overcome to some extent by incorporating a degree of α-blocking activity into the compound. The prototype combined α/β-blocker **labetolol** (**43**), incorporates an amide group on the phenylethanolamine moiety. Friedel–Crafts acetylation of salacilamide (**38**) gives substituted acetophenone **39**; this compound is then converted to the bromoketone **40**. Use of **40** to alkylate 4-phenyl-butyl-2-amine (**41**) gives the aminoketone **42**. The ketone is then reduced to an alcohol by catalytic hydrogenation.[10] The resulting compound, **labetolol** (**43**), consists of a mixture of two diastereomers because of the presence of two chiral centers.

Chloramphenicol (**49**), which can at least formally be classified as a phenylethanolamine derivative, was one of the first orally active antibacterial agents. The one time heavy usage of this agent declined with the recognition of its propensity to cause blood discrasias and the availability of safer alternatives; this compound is, however, still in wide use as a topical antibacterial agent. The relatively simple structure of this product from *Streptomyces venezuela* fermentation, initially known as **chloromycetin**, led early on to its production by total synthesis. The relatively short and straightforward route presented in the first synthesis does, however, suffer from lack of steric control.[10] The first step in the synthesis consists of aldol condensation of benzaldehyde with 2-nitroethanol to give a mixture of all four enantiomers of nitropropanediol **44**; the total mixture is reduced catalytically to the corresponding mixture of aminodiols. The threo isomer is then separated by crystallization and resolved as a diasteromeric salt to give the D(−) isomer (**45**). Acylation with dichloroacetyl chloride initially gives the triacetate; saponification gives the desired product **46**. The free hydroxyl groups are then converted to the acetates by means of acetic anhydride and the resulting product (**47**) is nitrated with the traditional nitric–sulfuric acid mixture. Saponification of **48** removes the acetate protecting groups and affords chloramphenicol (**49**).

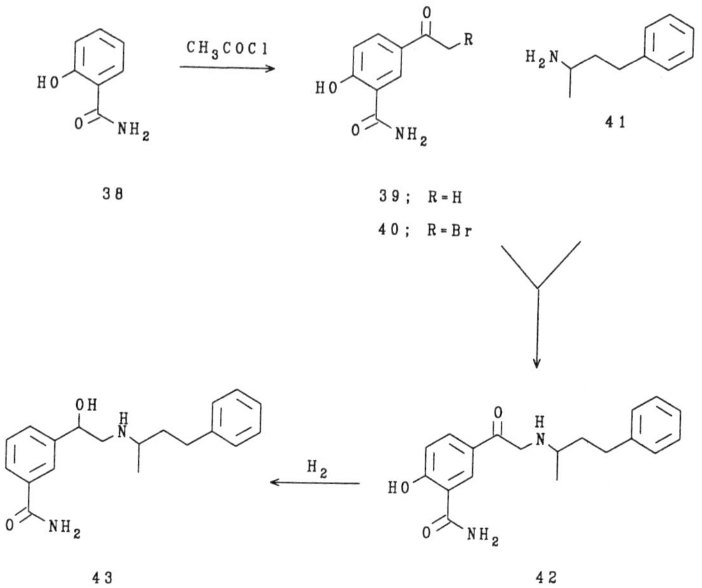

The recent upsurge in interest in enantioselective synthesis combined with the availability of methods and reagents for achieving such transformations have led to a reexamination of the syntheses for many drugs that are formulated as pure enantiomers. One novel approach to a **chloramphenicol** intermediate begins with oxidation of E-cinnamyl alcohol with the Sharpless reagent [*tert*-butyl hydroperoxide, titanium isopropoxide, L(+)diisopropyl tartrate] to give the enantiomerically pure epoxide **50**. Ring opening of the epoxide with benzoic acid in the presence of titanium isopropoxide gives the diol benzoate **51** with inverted configuration at the central side chain alcohol;

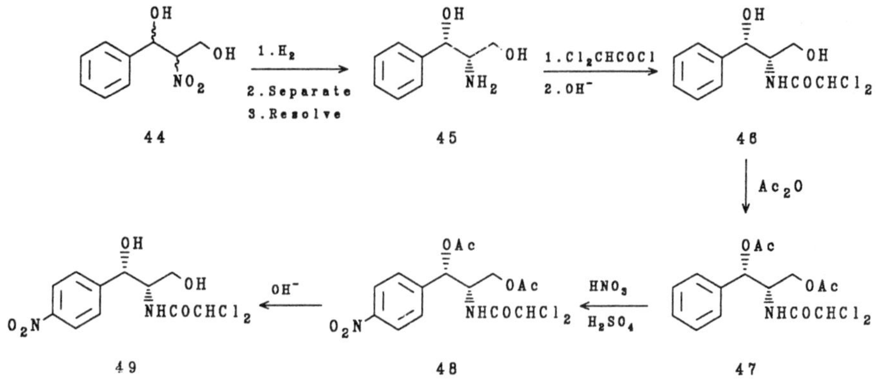

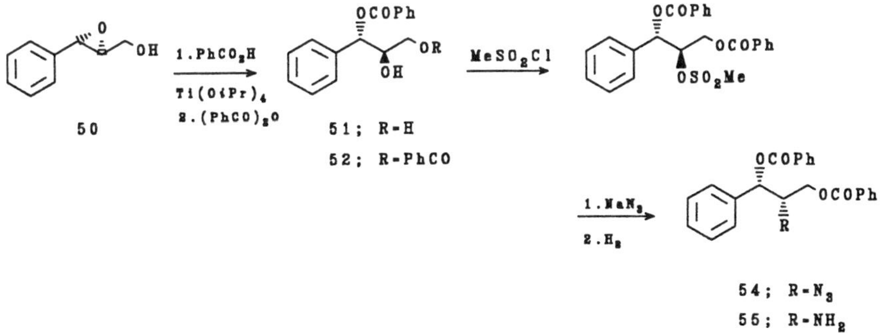

carefully controlled benzoylation leads to **52** in which the second alcohol remains free. This compound is then converted to a methanesulfonate (**53**) and that group is displaced by azide ion to give **54**; this last reaction proceeds with inversion of configuration to give the desired stereochemistry. Catalytic reduction of the azido group to a primary amine gives an enantiomerically pure intermediate (**55**), which is equivalent to **45** in the original scheme.[11]

2. ARYLOXYPROPANOLAMINES

A. β-Blockers

The discovery that β-sympathetic blocking agents (e.g., sotalol) seemed to have useful clinical activity in treating the symptoms of cardiovascular disease such as angina and arrhythmias, engendered considerable interest in this class of agents. The finding that β-blocking activity was retained when an oxymethylene group was interposed between the aryl group and the ethanolamine side chain made access to this class of compounds much easier. One of the first drugs of this new structural class, **propranolol** (**56**), found extensive clinical use. This use led to the unexpected finding that the drug caused a decrease in blood pressure among those patients whose disease was complicated by hypertension. The usefulness of the drug in treating heart disease was of course attributed to a decrease in stimulation of cardiac β-receptors by endogenous epinephrine. This activity would, however, be expected to increase blood pressure by blocking the largely relaxant effect of this neurotransmitter on the vasculature. The seemingly paradoxical action of β-blockers is now attributed to a decrease in force of cardiac contraction caused by these drugs. This serendipitous finding opened an enormous market for β-blockers as antihypertensive drugs with a consequent increase in research on this class of drugs. The following discussion covers only a fraction of the enormous number of aryloxypropanolamine that have been reported, or for that matter the very large number of drugs on the market. The early β-blockers such as propranolol showed some tendency to exacerbate bronchoconstriction in patients who also had asthma; an effect attributed to blockade of the relaxant effect of epinephrine on bronchioles. The finding that the vasculature is populated by β_1 receptors while those in the lungs consist mainly of β_2 receptors has led to emphasis on so-called β_2-selective drugs.

The key, and usually final, sequence in the synthesis of β-blockers consists in adding the propanolamine side chain. The customary approach, shown in generalized form in the scheme below, consists of initial alkylation of the appropriate phenoxide with epichlorohydrin (ECH in the scheme below). As one of the two possible reaction pathways, the phenoxide can react initially at the oxirane; the resulting alkoxide will then displace the adjacent chlorine to form a new epoxide ring. Alternatively, the phenoxide may displace halogen directly in an S_N2 reaction to give the same glycidic ether. It is noteworthy that the central chiral carbon retains its configuration in both schemes, an important consideration for the synthesis of enantiomerically defined drugs. Ring opening of the epoxide ring in the glycidic ether with an appropriate amine, usually isopropyl amine or *tert*-butylamine, leads to the aryloxypropanolamine compound; these compounds always consist of a secondary amine, which is generally recognized to be a rigid structural requirement for β-adrenergic blocking activity.

The synthesis of a typical β-blocker starts with the monoalkylation of catechol to give the ether **57**. Application of the standard side chain building sequence leads to the nonselective β-blocker **oxprenolol** (**58**)[12] (the -olol ending is approved USAN nomenclature for β-blockers). **Atenolol** (**61**) is one of the most widely used β_1 selective agents. The requisite phenol **60** can be obtained by ester interchange of methyl 4-hydroxyphenylacetate (**59**) with ammonia. Elaboration of intermediate **60** via the customary scheme then affords **atenolol** (**61**).[13]

Injectable β-blockers have found an important use in the treatment of cardiac infarcts as a means of reducing the demands on the injured heart muscle; this strategy

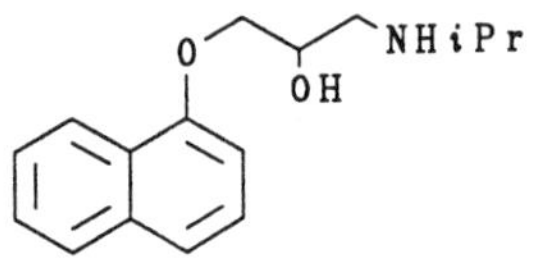

NHiPr
OH
O

56

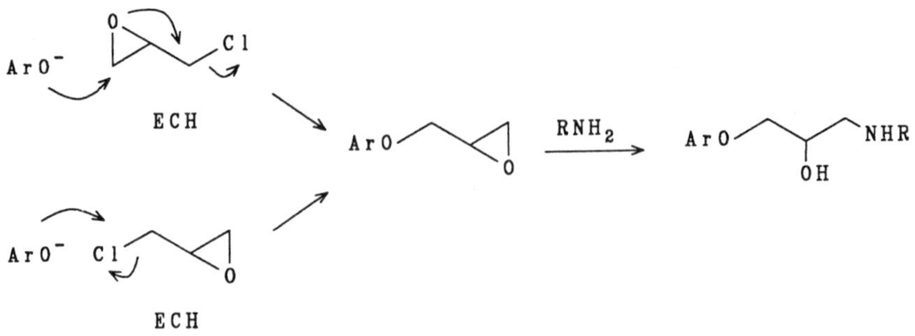

ArO⁻
O
Cl
ECH

ArO⁻ Cl
O
ECH

ArO
O
RNH_2
ArO
NHR
OH

Aryloxypropanolamine Construction

57 → 58 (1. ECH, K_2CO_3; 2. iPrNH_2)

59; R = OCH_3
60; R = NH_2
→ 61 (1. ECH, K_2CO_3; 2. iPrNH_2)

62 → 63 (1. ECH, K_2CO_3; 2. iPrNH_2)

carries with it, however, the hazard that excessive blood levels of drug cannot be quickly withdrawn in those cases where heart failure sets in. An injectable β-blocker with a very short half-life in the circulation was designed to address this problem; the terminal ester group in this compound, **esmolol (63)**, is very quickly hydrolyzed to the carboxylic acid by serum esterases; this last compound lacks β-blocking activity. This drug is prepared by subjecting 4-hydroxyhydrocinnamic acid to the usual side chain forming sequence.[14]

Interposition of the oxymethylene moiety is not itself a sufficient condition for changing an ethanolamine from agonist to antagonist. The analogue of epinephrine lacking the meta hydroxyl group is known to be a reasonable potent adrenergic agonist; the local vasoconstricting activity of **synephrine** accounts for its use in nasal decongestants. Interposition of the oxymethylene (and replacement of *N*-methyl by isopropyl) leads to **prenalterol** (**71**), a β-sympathetic agonist that shows selectivity for β_2 receptors. The enantioselective synthesis of this compounds incorporates the required chiral carbon in the first step of the synthesis by using a carbohydrate derived intermediate. Note that the central carbon on the epoxide is the sole chiral carbon retained in the final product; a more modern synthesis of this compound would probably depend on glycidic ester formation with commercially available chiral epichlorohydrin. In any case, opening of the epoxide **64**, which is obtained in several steps from D-glucofuranose, with the monobenzyl ether **65** from hydroquinone leads to the intermediate **66**. Scission of one of the 1,2-glycol linkages with periodate gives the hydroxyaldehyde **67**, a compound now relatively inert to the reagent. Reduction

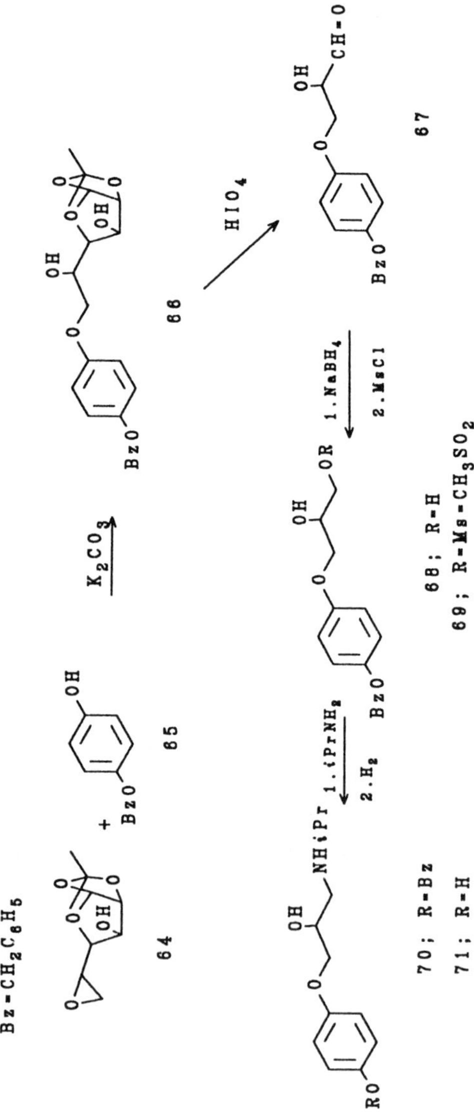
$Bz = CH_2C_6H_5$
64
65
K_2CO_3
66
HIO_4
67
1. $NaBH_4$
2. MsCl
68; R = H
69; R = Ms = CH_3SO_2
1. $iPrNH_2$
2. H_2
70; R = Bz
71; R = H

with sodium borohydride gives the diol **68**; the more reactive primary alcohol reacts with methanesulfonyl chloride to afford the hydroxy mesylate **69**. Displacement of this leaving group by isopropylamine completes construction of the aminoalcohol (**70**). Hydrogenation over palladium on charcoal removes the benzyl protecting group to afford finally, **prenalterol** (**71**).[15]

B. Nontricyclic Antidepressants

The development of the tricyclic antidepressant drugs in the late 1950s followed hard on the heels of the discovery of the structurally closely related antipsychotic agents. A discussion of the chemistry of these drug classes will be found in Chapter 3. As expected, very widespread use of antidepressant drugs uncovered a series of undesired effects. The most troubling of these side effects involved occasional findings of cardiotoxicity; the fact that this occurred with varied compounds suggested that this might be due to the presence in these compounds of a tricyclic nucleus. The development of very active open-chain antidepressant compounds made available drugs devoid of this limitation. The effects, true and/or imagined, of the first of these compounds to be marketed (**fluoxetine, 75**) have been widely publicized under its commercial soubriquet, Prozac®. The first published synthesis of this compound starts with the Mannich base **72** from the reaction of acetophenone, formaldehyde, and dimethylamine. The ketone is then reduced to an alcohol (**73**). This alcohol (**73**) is converted to the chloride **74** by any of several methods such as the reaction with hydrogen chloride in chloroform. Displacement of halogen with the phenoxide from treatment of *p*-trifluoromethylphenol leads to the corresponding *O*-alkyl ether. One of the methyl groups on nitrogen is then removed by treatment of the intermediate with cyanogen bromide followed by hydrolysis (von Braun reaction) or with the recently

$N(CH_3)_2$ 1. B_2H_6 2. HCl $N(CH_3)_2$

72 73; R-OH 74; R-Cl

1. CF_3 OH 2. CNBr

1. OH OCH_3 2. CNBr

1. OH CH_3 2. $ClCO_2Et$ 3. (+)mandelic acid

CF_3 O $NHCH_3$ CH_3O O $NHCH_3$ H_3C O $NHCH_3$

75 76 77

developed modification that uses ethyl chloroformate. The same sequence using the monomethyl ether of catechol leads to **nisoxetine** (**76**).[16] The antidepressant action of this class of compounds is known to be due to their interaction with receptors in the brain that inhibit reuptake of neurotransmitters (serotonin or norepinephrine) from the synaptic cleft. One of the two enantiomers is a good deal more potent, as would be expected from agents that bind to inherently chiral receptors. The current trend to formulate drugs that consist solely of active isomers is reflected in the fact that the analogue **tomexetine** (**77**) consists of the pure levorotatory isomer. In this case, the product from the standard sequence starting with *o*-cresol is resolved by salt formation with D-(+)mandelic acid.[17]

N-Demethylation is a well-recognized drug metabolism pathway that more often than not leads to inactivation. Consequently, it is of interest that this hypothetical **fluoxetine** metabolite shows the same activity as the parent. The synthesis of this agent reverses the ether formation step. Thus, displacement of fluorine from 4-fluorotrifluoromethylbenzene in an aromatic nucleophilic replacement reaction with the alkoxide from **78** (Phth = phthaloyl) affords the ether **79**. Removal of the phthaloyl protecting group by reaction with hydrazine gives the antidepressant drug **seproxetine** (**80**).[18]

3. ARYLSULFONIC ACID DERIVATIVES

A. Antibacterial Sulfonamides

The quantum leap in human life expectancy that has been observed since the beginning of the twentieth century is most commonly attributed by epidemiologists to the decreased mortality and morbidity from infectious disease. Improvements in sanitation and the availability of antibacterial drugs are the greatest factors leading to this decrease. The first of the synthetic antibacterial agents was in fact discovered due to a set of adventitious events. Intrigued by the observation that certain organic dyes showed strong affinity for specific bacteria, Domagk and his collaborator Klarer in Germany set up a synthesis and screening program to test the antibacterial action of such dyes in the early 1930s. The fact that all compounds were tested in vivo in mice,

Prontosil 1,2,4-TAB Sulfanilamide

rather than in vitro as was then, and is now again, far more customary, proved crucial for the discovery. The dramatic curative action of a red dye dubbed **prontosil** in infected mice attracted immediate attention. The dye became available for clinical use when the activity was found to hold up in humans as well. Puzzled by the observation that prontosil failed to show activity in any of the then current in vitro antibacterial assays, Bovet and his colleagues in France considered the possibility that the activity was in fact due to a metabolite. Work based on this premise demonstrated that one of the metabolites from cleavage of the N–N azo link, **sulfanilamide**, accounted for all the activity of prontosil both in vivo and in vitro; the other metabolite, 1,2,4-triaminobenzene (1,2,4-TAB), was devoid of activity by either route. Sulfanilamide quickly replaced the dye as the drug of choice and gained widespread use just in time to save innumerable lives of wound victims in World War II.

Elucidation of the mechanism of action of the sulfonamides served to clarify both their activity and marked selectivity for bacteria. In contrast to mammals that obtain the folates involved in nucleotide synthesis in their diet, bacteria must synthesize these compounds de novo. *para*-Aminobenzoic acid (PABA) is an important structural moiety of folates. It has been rigorously demonstrated that sulfonamides act as competitive inhibitors for the bacterial enzyme that incorporates PABA, dihydropteroate synthetase; the enzyme presumably recognizes the acidic sulfonamide proton as a carboxylate hydrogen. The rather strict structural requirements in this class directly reflect the mode of action: The presence of a primary aniline group and at least one sulfonamide proton are mandatory for activity; additional substituents on the ring decrease activity by interfering with recognition.

The synthesis of **sulfanilamide**, **82**, is a straightforward exercise in aromatic chemistry. (It is of interest to note, that the preparation of this drug, starting from benzene, was at one time a standard assignment in beginning undergraduate organic chemistry lab courses; this exercise probably set more than one medicinal chemist on his or her career path.) The key reaction involves chlorosulfonation of acetanilide to yield the sulfonyl chloride (**81**). Reaction with ammonia followed by acid catalyzed hydrolysis of the acetamide amide gives sulfanilamide itself. This same general reaction with other amines or heterocyclic amines leads to a host of other drugs that have virtually the same antibacterial spectrum but may differ in their pharmacokinetic properties. The 1991 edition of the *USP Dictionary of Names* lists well over 50 different sulfonamides that have been assigned US Adopted Names.

$H_2N-C_6H_4-SO_2NH_2$ ← (1. NH_3; 2. H_3O^+) $AcHN-C_6H_4-SO_2Cl$ → (1. RNH_2; 2. H_3O^+) $H_2N-C_6H_4-SO_2NHR$

82 81

81 + H_2N– ... → 82 → 83 → HONO → [84] → 85 ← 86; R=H / 87; R=CH_3 — 1. HONO 2. MeSalicylate 3. H_3O^+ → 88

A sulfonamide that seemingly departs from the rule requiring a free anilino group, **sulfasalazine** (**85**), has proved useful for the treatment of ulcerative colitis, a poorly understood and often fatal disease of the colon. This compound undergoes the same metabolic cleavage by bacteria in the gut as does prontosil, that is, cleavage of the azo linkage. In this case, however, there is good evidence that the active moiety is in fact the 4-aminosalicylic acid (**86**) metabolic product. Sulfasalazine thus seems to be a prodrug for delivering this compound directly to the disease site. The starting material for this agent, **sulfapyridine** (**83**), is prepared by reaction of 2-aminopyridine with sulfonyl chloride (**81**). The aniline function is then converted to a diazonium salt by reaction with nitrous acid. Coupling of the salt with salicylic acid proceeds at the 4-position to give sulfasalazine (**85**).[19] **Olsalazine** (**88**), which was designed after the mode of action of the parent agent, represents a more direct approach for delivering the active moiety to the lower intestine, with both halves of the molecule providing 4-aminosalicylic acid on reductive cleavage of the azo linkage. The compound is prepared by coupling of the diazonium salt from methyl 4-aminosalicylate (**87**) with methyl salicylate followed by hydrolysis of the esters.[20]

B. Diuretic Agents

Widespread use of the sulfonamide antibacterial agents uncovered a series of minor side effects. An increase in urine output when the drugs were administered at high doses was among these effects. This adventitious observation was the spur for work in modifying the molecule so as to optimize what had been a side effect, since the only diuretic drugs then available were several organomercurials whose use was limited by their toxicity. One of the first successes lay in the finding that compounds in which a second sulfonamide group was added at the meta position showed reasonable diuretic activity; these compounds are devoid of antibacterial activity since they now do not show the slightest resemblance to *p*-aminobenzoic acid.

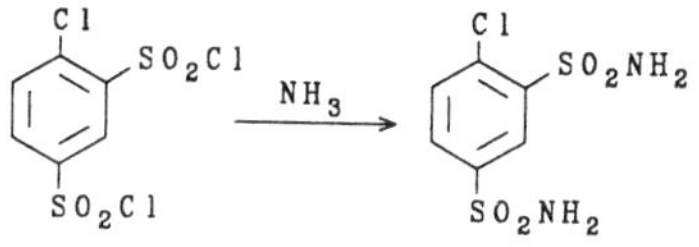

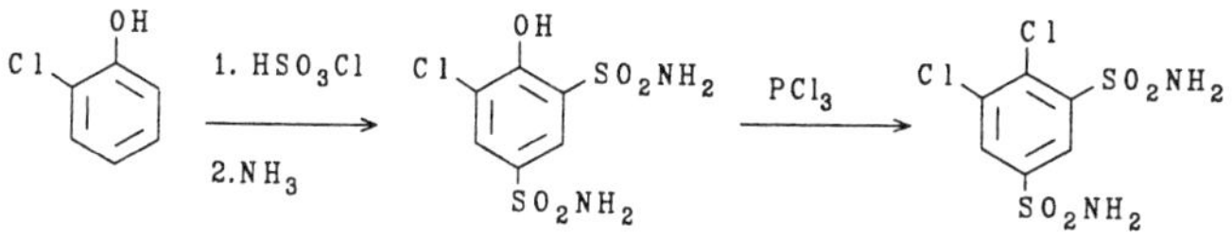

Treatment of chlorobenzene with chlorosulfonic acid under forcing conditions leads to the meta disubstituted sulfonyl chloride (**89**); ammonolysis of **89** leads to the diuretic agent **chlorphenamide** (**90**).[21] In a similar vein, *o*-chlorophenol (**91**) yields bis-sulfonamide (**92**) on sequential reaction with chlorosulfonic acid and ammonia. Hydroxyl groups in heterocyclic compounds behave very much like enol forms of carbonyl groups; they can thus be replaced by chlorine. The presence of the two strongly electron-withdrawing sulfonamides meta to the hydroxyl group in **92** seems to make that group assume this enol character as well. Reaction of **92** with phosphorus trichloride thus leads to formation of the dichloro compound; therefore **dichlorphenamide** (**93**) is obtained.[22] Note that the simple bis-sulfonamide diuretics have been largely displaced by heterocyclic thiazides (Chapter 11) and the so-called high-ceiling agents.

Though the terms "potency" and "activity" are often used interchangeably, albeit erroneously, in the literature they in fact denote different aspects of a given compounds' biological action. The dose of a given agent required to produce a stated effect, for example, 25% inhibition of an enzyme, is correctly termed as its potency; the maximal effect, in the same case the highest percent inhibition achievable with the same agent, is its activity. The simple disulfonamide as well as the thiazide diuretics are often termed "low-ceiling" compounds because increasing doses will not lead to increased diuresis above a threshold level. The "high-ceiling" compounds cause dose related increases in diuresis beyond those achievable with their counterparts. The two high-ceiling diuretics, **furosemide** (**96**), and **azosemide** (**100**), include a heterocyclic ring connected through an aminomethyl link; one of the sulfonamides in each is replaced by a carbon based acid moiety. The synthesis of furosemide begins with the chlorosulfonation–ammonolysis reaction sequence starting with 2,4-dichlorobenzoic acid (**94**). For reasons that are not immediately evident, the chlorine para to the sulfonamide group is preferentially activated over that at the ortho position toward nucleophilic aromatic displacement. Reaction of **95** with furfurylamine (2-methylaminofuran) thus leads to **furosemide** (**96**).[23] In an analogous scheme, chlorosulfonation of substituted benzonitrile (**97**) followed by ammonolysis of the product gives the sulfonamide **98**. The regiochemistry of the displacement reaction can be attributed in this case to the better leaving group properties of fluoride over chloride. Reaction with 2-methylaminothiophene thus gives **99** as the product. There is ample precedent to indicate that tetrazoles are bioisosteric with carboxylic acids, with the two groups showing quite comparable pK_a values. Treatment with sodium azide and hydrochloric acid leads to 1,3-addition of the elements of hydrazoic acid to the nitrile and formation of a tetrazole ring. This reaction yields the high-ceiling diuretic agent, **azosemide** (**100**).[24]

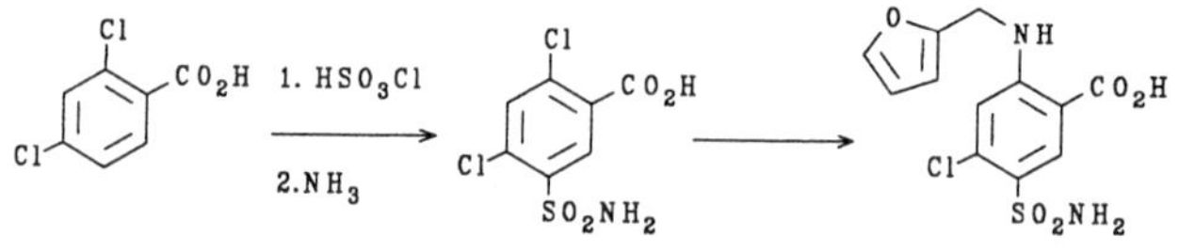

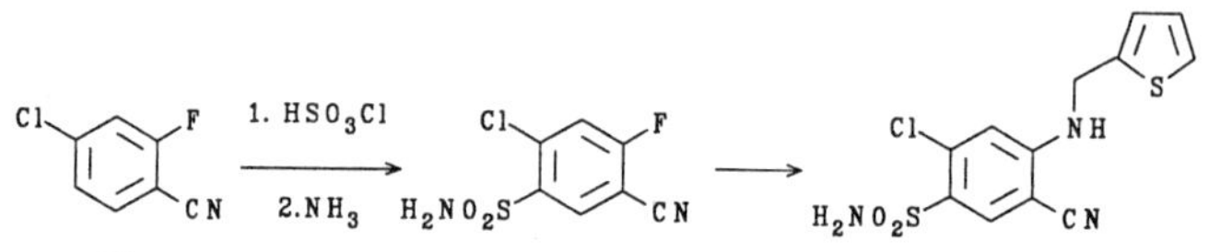

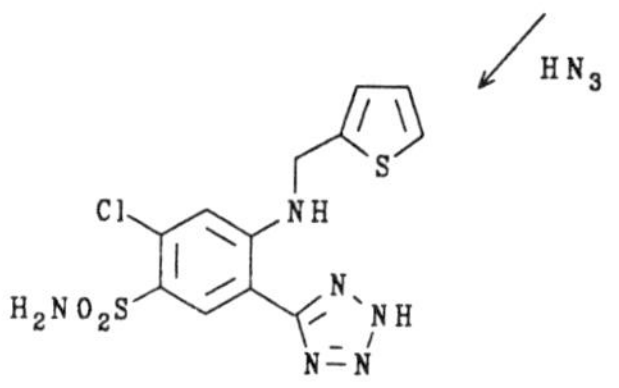

Incorporation of a perchloroethylene side chain interestingly leads to a disulfonamide which is used as a veterinary antiparasitic agent. As expected, the reaction of *m*-nitro(perchloroethyl)benzene (**101**), with iron powder in acid reduces the nitro group to an aniline; the reagent in addition removes two chorines from the side chain to give the corresponding olefin **102**. By subjecting **102** to the standard chlorosulfonation, ammonolysis sequence then gives **clorsulon** (**103**).[25]

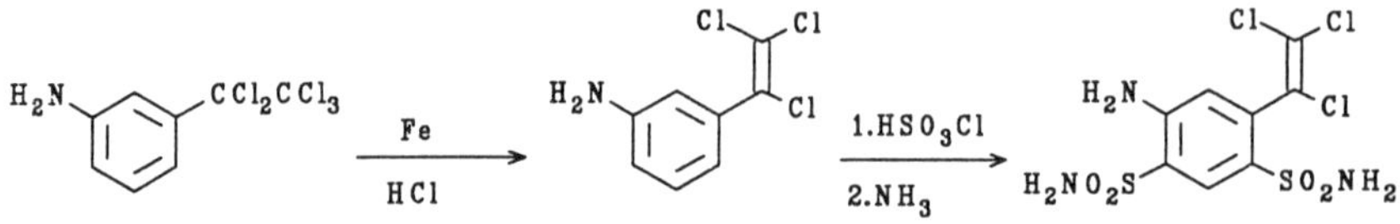

C. Oral Hypoglycemic Sulfonylureas

The peptide hormone insulin is intimately involved in glucose turnover. Disruptions in insulin levels or insulin receptors are manifested as diabetes. So-called juvenile onset diabetes results from failure to secrete adequate levels of this hormone; this form of the disease, also dubbed Insulin Dependent Diabetes, can only be treated by administration of insulin itself. The far more common form of the disease, which typically strikes in middle age, may be due to a number of causes that result in either insufficient levels of insulin or decreased response of cellular insulin receptors. This disease, also known as Non-Insulin Dependent Diabetes (NIDD) can be treated by strict diets, by administration of insulin, or most conveniently with a series of drugs that lower the elevated glucose levels due to insulin deficiency. These drugs, most of which are arylsulfonylureas, also trace their parentage to the sulfonamide antibacterials and a tendency of these drugs to lower blood sugar at very high doses.

A number of different routes are available for the preparations of **tolbutamide** (**106**), the first oral hypoglycemic agent to be used clinically. The shortest route involves simple addition of *p*-toluenesulfonamide (**104**) to butyl isocyanate (**105**).[26] An alternate route is required for the preparation of a drug that includes a tertiary urea nitrogen. The same starting material (**104**) is converted to its carbamate (**107**) with ethyl chloroformate in the presence of base. Heating this intermediate with *N*-aminohexamethyleneimine leads to displacement of the ethoxy group and formation of **tolazamide** (**108**).[27]

The very low potency of first generation sulfonylureas required daily intake of doses measured in grams. Incorporation of complex side chains on the sulfonyl-bearing benzene ring led to orders of magnitude increases in potency. (This compound was memorialized by the Upjohn trade name Micronase® for **glyburide**.) Reaction of the acetamide from 2-phenethylamine (**109**) with chlorosulfonic acid results in formation of the para sulfonyl chloride; ammonolysis of this intermediate followed by base catalyzed removal of the acetamide gives the free phenethylamine (**110**). This compound is then acylated with the acid chloride from salicylate (**111**) to give the amide **112**. Condensation of **112** with cyclohexyl isocyanate gives the sulfonylurea glyburide (**113**).[28]

There is evidence from further investigation of the structure–activity relationship (SAR) in this series that the sulfonylurea function does not need to be attached to an aromatic ring. The synthesis of this compound starts with nitrogen interchange between substituted piperidine (**114**) and sulfamide; the phthaloyl protecting group is then removed by reaction with hydrazine to afford the primary amine **115**. Acylation

H_3C–C6H4–SO_2NH_2 (104) + O=N=C–C_4H_9 (105) ⟶ H_3C–C6H4–$SO_2NHCONHC_4H_9$ (106)

104 ↓ $ClCO_2C_2H_5$

H_3C–C6H4–$SO_2NHCO_2C_2H_5$ (107) —(H_2NN)⟶ H_3C–C6H4–$SO_2NHCONH$–N (108)

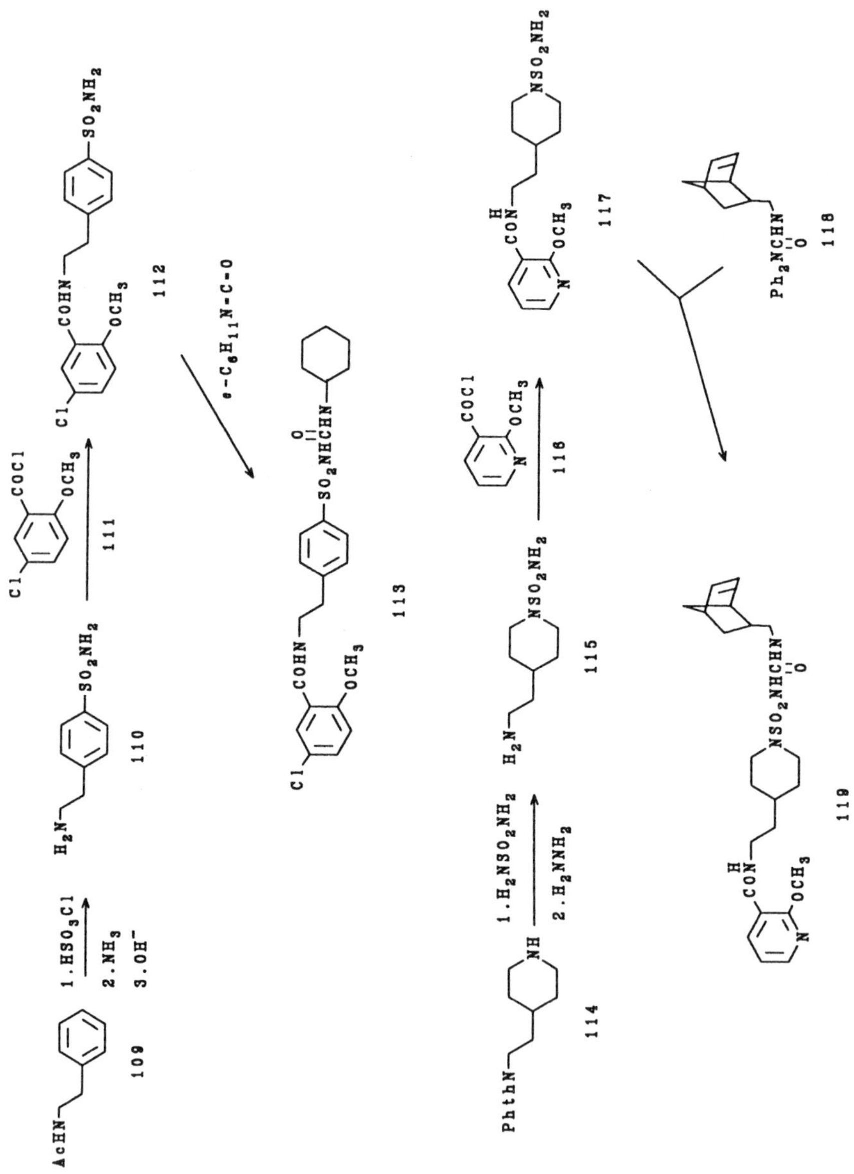

with the 2-methoxynicotinyl chloride (**116**) gives the corresponding amide (**117**). Nitrogen interchange between the sulfonamide group in **117** and the urea function in the bicyclic reagent **118** results in displacement of diphenylamine and formation of a sulfonylurea function. **Gliamilide** (**119**) is thus obtained.[29]

Oral hypoglycemic activity is interestingly retained when the urea function is cyclized in the guise of a pyrimidine ring. Acylation of substituted 2-aminopyrimidine (**121**) with the product **120**, from reaction of methyl phenylacetate with chlorosulfonic acid, gives the sulfonamide **122**. The terminal ester is then hydrolyzed and the resulting acid is converted to an acid chloride with thionyl chloride. Reaction of this last intermediate with the substituted aniline **123** leads to the hypoglycemic agent **glicetanile** (**124**).[30]

D. Thromboxane Receptor Antagonists

As noted in the discussion of the arachidonic acid cascade, thromboxane A_2 (TXA_2), one of the products from this cascade, is one of the most potent known platelet aggregating and inflammatory substances. Though cyclooxygenase inhibitors such as the NSAIDs do decrease levels of TXA_2, these agents often exhibit deleterious effects on the GI tract. A pair of structurally closely related compounds that contain sulfonamide moieties have recently been described. These compounds compete with TXA_2 for its binding sites and thus offer the possibility of more specific drug action. Preparation of **daltroban** (**126**) involves formation of the sulfonamide by reaction of the primary amine in **125** with benzenesulfonyl chloride. Synthesis of the possibly more accessible analogue starts by alkylation of the phenol group in **127** with ethyl bromoacetate in the presence of base. Hydrolysis of the alkylation product with strong acid removes both the amide and ester groups to yield the free amino acid **128**. This compound is converted to its benzenesulfonamide, as shown above, to afford **sultroban** (**129**).[31]

4. ARYLACETIC AND ARYLPROPIONIC ACIDS

The antiinflammatory and antipyretic salicylates and pyrrazoles were among the first synthetic organic drugs to find extensive clinical use. Their widespread use led to a growing awareness of the side effects associated with these drugs, particularly when used in the high doses required for alleviating the symptoms of osteoarthritis. The finding that arylacetic and arylpropionic acids showed the same activity led to intensive research in the area and the introduction of dozens of new NSAIDs. All these compounds share the same mechanism of action: inhibition of arachidonic acid cyclooxygenase.

A. Arylacetic Acid "Fenacs"

The prototype for this class of compounds is **ibufenac** (**132**), which was developed by a research group at Boots in the United Kingdom. This drug was to be quickly superseded by its α-methylated congener **ibuprofen**, which was made in the same

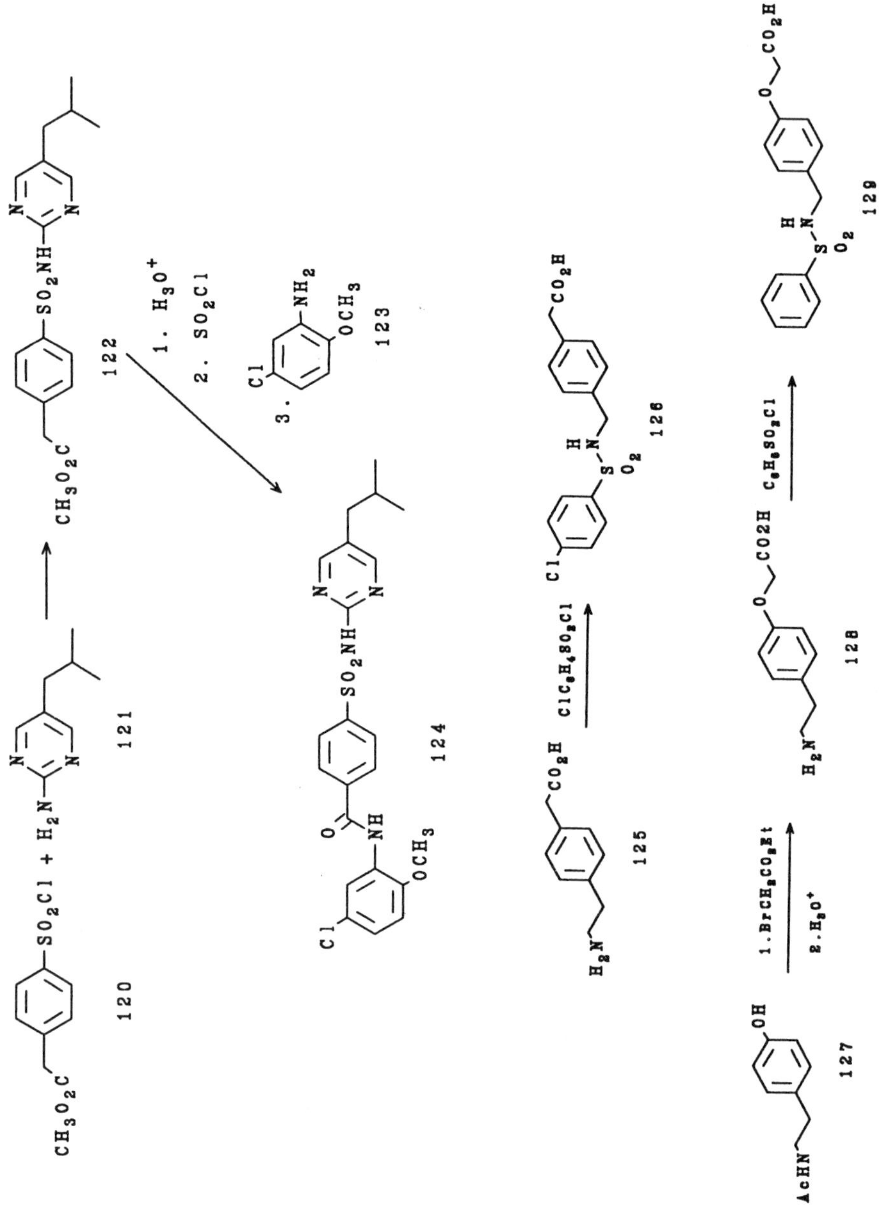

120
121
122
1. H_3O^+
2. SO_2Cl
3.
123
124
125
$ClC_6H_4SO_2Cl$
126
127
1. $BrCH_2CO_2Et$
2. H_3O^+
128
$C_6H_5SO_2Cl$
129

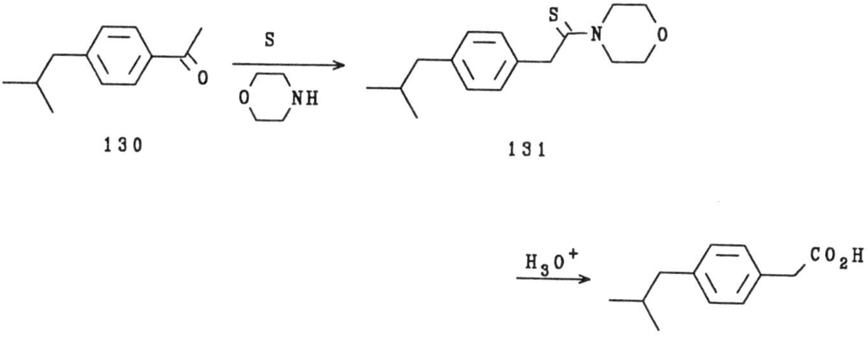

130 131 132

laboratory.[32] The chemistry of ibuprofen is discussed in Chapter 2 Section 4.B. The mechanistically very complex Wilgerodt reaction constitutes the key to the preparation of this compound. Thus, reaction of the acetylation product **130** from isobutyl benzene and acetyl chloride with sulfur and morpholine leads to transposition of the oxidized function to the terminal carbon and formation of thiomorpholide (**131**). Hydrolysis of the thioamide in acid results in concomitant replacement of sulfur by oxygen to give a carboxylic acid. Thus ibufenbac (**132**) is obtained.

Both potency and duration of action are markedly increased by addition of the appropriate substituents to the benzene ring. The routes for preparing such compounds differ markedly from those used to prepare ibufenac (**142**). The synthesis of one of the more recently approved NSAIDs, **diclofenac** (**137**), starts by careful acylation of diphenylamine (**133**) with one equivalent of oxalyl chloride. Friedel–Crafts ring closure of the obtained acid chloride **134** leads to isatin (**135**). The two carbonyl groups in **145** differ in that one is an amide while the one next to the benzene ring has more ketone-like reactivity. Treatment of the isatin with hydrazine and potassium hydroxide under Wolff–Kischner conditions effect reduction of the ketone function and forma-

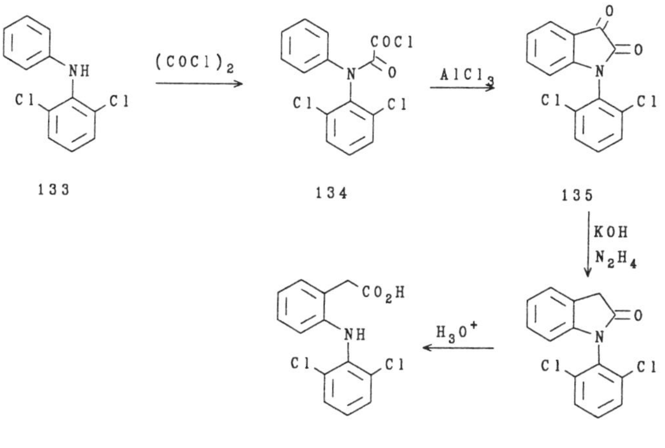

133 134 135 136 137

tion of the lactam **136**. Hydrolysis of the amide function then affords diclofenac (**137**).[33]

Good activity is interestingly retained when a carbonyl group is added to the ring and the aniline becomes primary. The synthesis of two of these agents resemble those above by involving an intermediate indolone ring. The starting *N*-amino-2-indolone (**138**) can be obtained by reduction of the *N*-nitrosation product of indolone itself. Reaction of this intermediate, which is in essence an *N*-acylhydrazine with phenylacetone (**139**), gives the corresponding hydrazone **140**. The hydrazone is then treated with ethanolic hydrogen chloride; Fischer indole formation (see Chapter 10, pages 289ff) would in theory initially lead to the fused bicyclic indole **141**. The observed product **142** can be rationalized by assuming that the labile bicyclic *N*-acyl indole opens to an ester in the presence of ethanolic hydrogen chloride. Ozonization of the double bond in the heterocyclic ring then leads to *N*-acylated benzophenone (**143**), the acetyl group arising from the ring atom bearing a methyl group. Acid hydrolysis of **153** removes both the amide and the ester group to afford **amfenac** (**144**).[34]

Construction of the closely related NSAID **bromefenac** (**150**) depends on the Gassman indolone synthesis[35] for incorporation of the acetic acid chain. This reaction involves an anion initiated electrocyclic rearrangement related conceptually to the little known Hauser ortho substitution rearrangement. The simplest example of the latter depends on formation of a carbanion by abstraction of one of the acidic protons from a benzyltrimethyl quaternary salt to give **I** (abstraction of a more acidic benzyl proton gives a stable anion that simply reverts to starting material). The resulting anion then adds to the aromatic ring to start the electrocyclic reaction depicted below (**I** → **II**); the net effect after bond reorganization (**III**) is migration of a dimethylaminomethyl group to the ortho position.

The key sulfonium reagent for the Gassman synthesis (MCSS) is obtained by chlorination of ethyl 2-methylthioacetate. Displacement of chlorine from this reagent by nitrogen in aminobenzophenone (**145**) gives the corresponding sulfonium salt **146**. The reaction proceeds with surprisingly mild base; the anion from treatment of **146** with triethylamine, which add to the benzene ring, can start an electrocyclic reaction analogous to the one shown below. The initial product **147** is not observed. Instead,

O
=O
+ C_6H_5
N
NH_2
138
139
=O
N
N
C_6H_5
140
HCl
EtOH
=O
N
C_6H_5
141
CO_2H
NH_2
C_6H_5
O
144
H_3O^+
CO_2Et
NHAc
C_6H_5
O
143
O_3
CO_2Et
NH
C_6H_5
142

$CH_3SCH_2CO_2Et + Cl_2 \longrightarrow CH_3\overset{Cl}{S}CH_2CO_2Et\ Cl^- + MCSS$

triethylamine is a sufficiently stronger base than aniline to ionize the latter. The product from this process (**158**) would then displace ethoxide from the adjacent ester group to give the observed indolone **149**. The thiomethyl group is then removed with Raney nickel; hydrolytic cleavage of the amide completes the synthesis of **bromfenac** (**150**).[36]

B. Arylpropionic Acid "Profens"

Further work on the arylacetic acids by investigators at Boots revealed that incorporation of a methyl group on the acid chain improved the potency of ibufenac (**132**). The resulting NSAID ibuprofen (**154**) found very widespread use since it showed improved potency and tolerance over aspirin, which had for decades been the mainstay drug for the same indications. Direct methylation of the carbanion from **151** would at first sight seem an attractive route to this drug. In practice, however, this reaction leads to a difficultly separable mixture of desired product, bis-alkylated byproduct and starting material. Conversion of the acetate side chain to a malonate blocks overalkylation and at the same time facilitates deprotonation. Thus, reaction of the ethyl ester **151** from ibufenac with diethyl carbonate in the presence of sodium ethoxide leads to the malonate carbanion **152**. This compound is not isolated but quenched in situ with methyl iodide to give the alkylation product **153**. The malonate is then hydrolyzed in acid; the free malonic acid decarboxylates under reaction conditions to afford ibuprofen (**164**).[37]

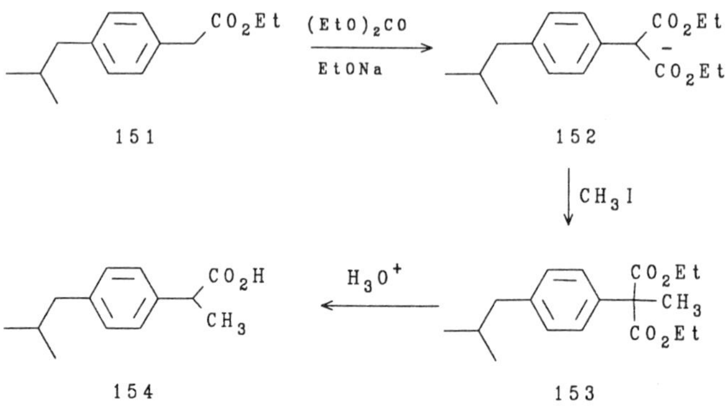

The huge demand for ibuprofen, combined with the relatively high dosage, which can amount to grams per day, led to intense efforts to develop other synthetic approaches. One alternate scheme for preparing this compound involves a modification of the Darzens glycidic ester synthesis. Addition of the carbanion from treatment of 2-chloroacetonitrile with sodium ethoxide initially leads to the adduct **155**. As in the Darzens reaction, the alkoxide displaces chloride on the adjacent carbon to form an epoxide to give the observed product **156**. Treatment of **156** with a Lewis acid such as lithium perchlorate leads to opening of the oxirane ring with concomitant hydrogen migration. Thus, the α-cyanoketone **157**, a structure that may be viewed as a variant on an acid chloride is obtained. Treatment with aqueous base converts this intermediate to ibuprofen (**154**).[38]

An interesting variant on the Wilgerodt reaction offers a simple three-step procedure that avoids the wastage involved in the previous schemes, which require incorporation of an extra carbon atom that must later be eliminated. The scheme starts with the acylation of isobutylbenzene with propionyl chloride to give propiophenone **158**.

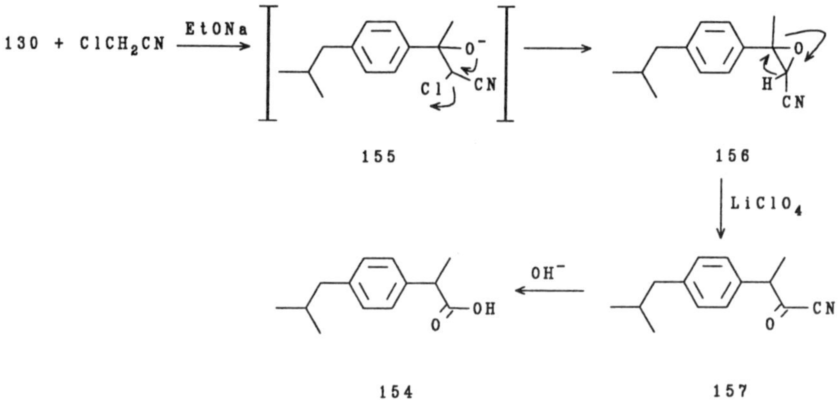

$Tl(NO_3)_3$
$CH(OMe)_3$
MeOH

158 159

1. Wilgerodt
2. Hydrolyze
3. EtONa/HCO_2Et

160 161

1. CH_3I; 2. H_3O^+

162

Reaction of **158** with thallium III nitrate and methyl orthoformate in methanol leads to the methyl ester **159** of **ibuprofen** in high yield.[39] This scheme would be the method of choice for preparing the drug but for two unfortunate facts: the extreme toxicity of thallium and the very high sensitivity of analytical methods for detection of metals. It proved, in practice, to be virtually impossible to produce samples that showed zero residues of thallium.

Yet a further increase in potency is observed when the *p*-isobutyl group is replaced by a benzene ring. One published synthesis for this compound is quite analogous to the malonate route to the parent drug. Acetyl biphenyl **160** is thus converted to the corresponding arylacetic acid by reaction with sulfur and morpholine followed by hydrolysis of the first obtained thiomorpholide. This compound is then esterified and converted to the malonate anion **161** with sodium ethoxide and ethyl formate. The anion is quenched with methyl iodide; hydrolysis of the esters followed by decarboxylation yields the NSAID **flubiprofen** (**162**).[40]

It is of interest that the isobutyl function may also be replaced by a heterocyclic ring. The route **pirprofen** (**168**) starts with direct methylation of unesterified 4-nitrophenylacetic acid (**163**). The presumable selectivity in this case may reside in the structure of the dianion, whose most stable form is presumably depicted in **164**. Catalytic reduction of the product **165** gives the corresponding aniline (**176**); this compound is then converted to its acetanilide derivative with acetic anhydride. Treatment with chlorine followed by hydrolysis gives the chloroaniline **167**. Double alkylation of intermediate **167** with 1,4-dichlorobut-2-ene (depicted as the cis isomer for aesthetic reasons) forms the dihydropyrrole ring. Thus the NSAID **pirprofen** (**168**) is obtained.[41]

The rather different scheme used to prepare the propionic acid side chain in **cicloprofen** (**173**) leads to the inclusion of this tricyclic compound in a chapter dealing with monocyclic compounds. The synthesis begins with the Friedel–Crafts acylation

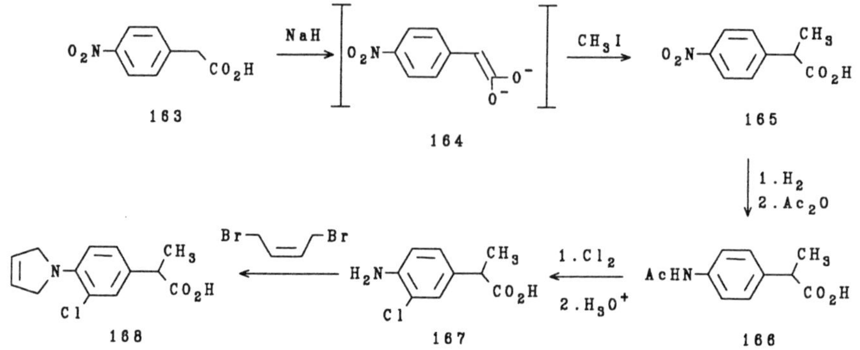

163 164 165 166 167 168

of the hydrocarbon fluorene (**169**) with the half-ester of oxalyl chloride to give the α-ketoester **170** as product. The required side chain methyl group is then added by reaction of **180** with methylmagnesium bromide to give **171**; this apparently proceeds selectively at the ketone function to give the tertiary carbinol **172**. The benzylic tertiary alcohol readily dehydrates on treatment with acid to a methylene group. Catalytic reduction of **182** followed by hydrolysis of the ester leads to cicloprofen (**173**).[42]

The fact that the profen NSAIDs owe their activity to inhibition of the enzyme cyclooxygenase would lead to the expectation that activity would reside in mainly a single enantiomer since one of these isomers should more closely fit an interacting site in the chiral enzyme than the other. The enzyme is of course asymmetric because of the chiral amino acids of which it is composed. Extensive work has demonstrated that activity of NSAIDs is due in virtually all cases to the (*S*) enantiomers. It has also been shown, as an interesting sidelight, that the less active (*R*) isomers of many NSAIDs are converted in vivo to their (*S*) enantiomers. The current emphasis on developing drugs that consist of the more active enantiomers has seen a counterpart in the development of technology for producing single isomers. This reaction is of course more economical than resolution of a racemate, since the latter process implies discarding at least 50% of the product. A recently disclosed scheme for producing pure

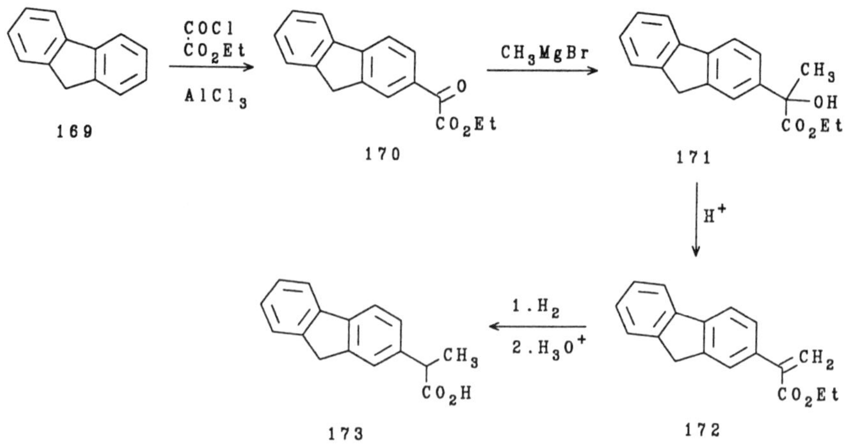

169 170 171 172 173

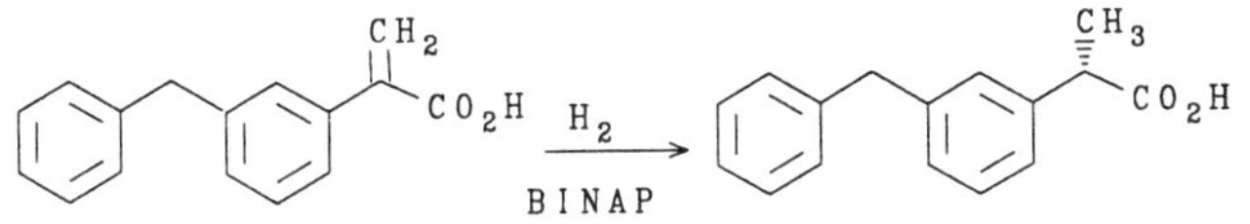

(*S*) ketoprofen provides a good example of stereospecific reduction.[43] Only the last of the several schemes for introducing the chiral center located at the side chain methyl group offers a method for controlling the enantiomeric identity for this position. Several asymmetric Wilkinson-type catalysts have been developed over the past few years that are chiral by reason of binaphthyl asymmetry. Reduction of the α-methylene substituted *m*-benzylphenylacetic acid (**174**) leads to a single isomer of the propionic acid **175**. Oxidation of the benzyl group to a ketone gives **(*S*)-(+)-ketoprofen (176)**. (Attempted reduction of the corresponding methylene ketoprofen is said to give far inferior selectivity.)

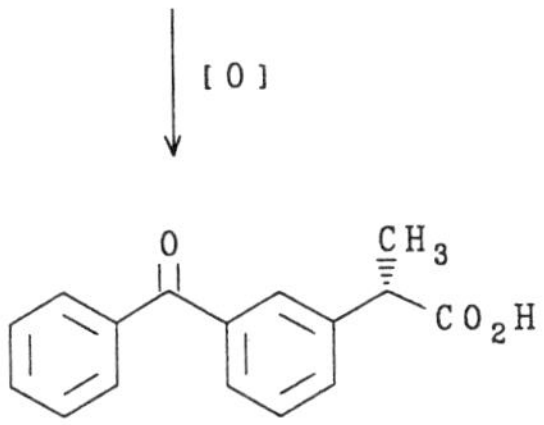

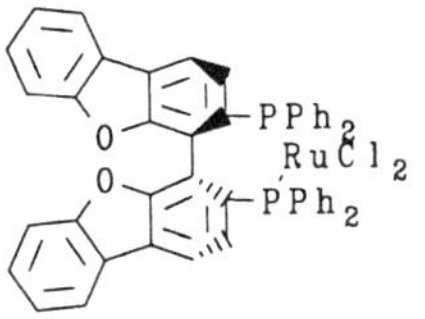

C. Arylacetamide Antiarrhythmic Compounds

A small series of drugs that are used clinically as antiarrhythmic agents are based on alkylated derivatives of phenylacetamide. The first of these, **disopyramide** (**179**), bears an interesting structural resemblance to the opioid analgesic **methadone** (**180**), which represents the ultimate structural simplification of morphine. Both compounds incorporate the minimal structural requirements for opioid activity posited by the Becket–Casey rule: A tertiary nitrogen atom at the equivalent of two carbon atoms' removed from an aromatic ring attached to a quaternary center (see Chapter 7 for a more detailed discussion of opiate analgesics). The syntheses for these antiarrhythmic drugs rely heavily of carbanion alkylation chemistry. The synthesis of the first of these drugs starts with nucleophilic aromatic displacement of bromine in 2-bromopyridine by the carbanion from phenylacetonitrile to give the intermediate **177**. Alkylation of the carbanion from this product with *N*,*N*-diisopropyl-2-chloroethylamine gives the highly substituted nitrile **178**. Hydrolysis of **186** with sulfuric acid stops at the amide stage to give disopyramide (**179**).[44]

Replacement of the pyridine ring by a more strongly basic ethylpiperidine moiety leads to the antiarrhythmic drug **disobutamide** (**183**). The synthesis of **183** also involves successive carbanion alkylation reactions. Thus, reaction of the anion from *o*-chlorophenylacetonitrile (**181**) with *N*-(2-chloroethyl)piperidine gives the intermediate **182**; alkylation of the anion from this compound leads to **183**. Hydrolysis with sulfuric acid completes the preparation of disobutamide (**184**).[45]

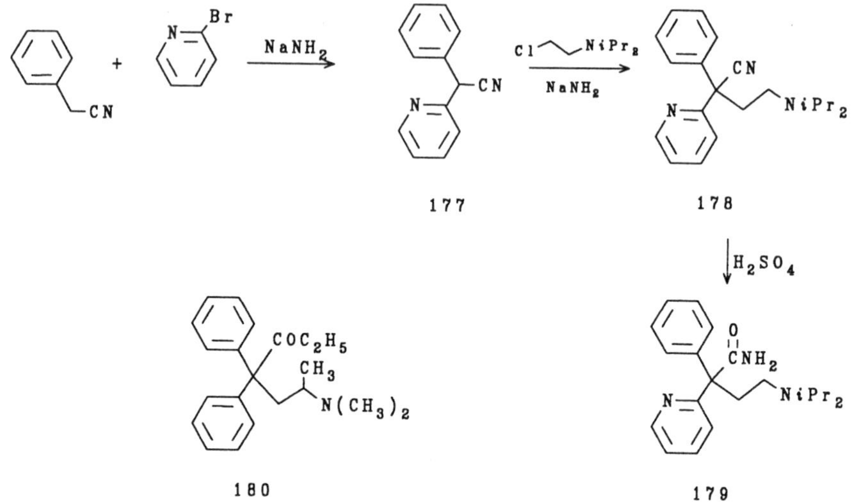

One of the more common pathways for metabolic transformation of tertiary amines involves *N*-dealkylation to a secondary amine; the fact that these products often show the same biological activity as the parent drug in many cases confounds the issue of the identity of the chemical species responsible for drug action. Because the demethylation product of disobutamide shows antiarrhythmic activity in its own right prompted the synthesis of the acetyl derivative of this dealkylation product; this agent may be considered a latent form of the active metabolite. This compound is prepared by first repeating the penultimate step in the disobutamide (**184**) synthesis using (*N*-benzyl-*N*-isopropyl)-2-chloroethylamine instead of the diisopropyl intermediate. The product from this reaction (**185**) is then hydrolyzed to the amide with sulfuric acid. Sequential

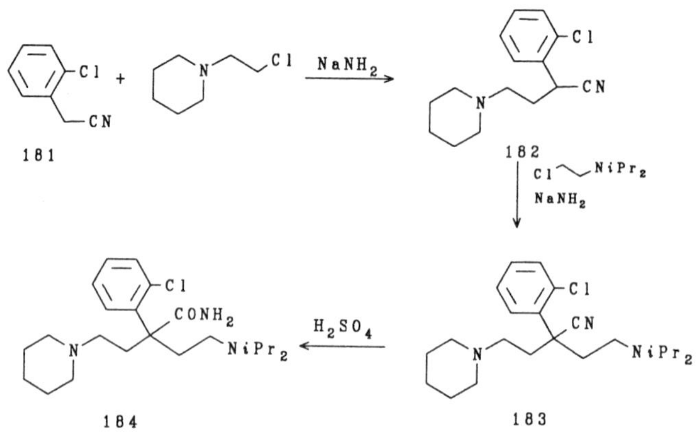

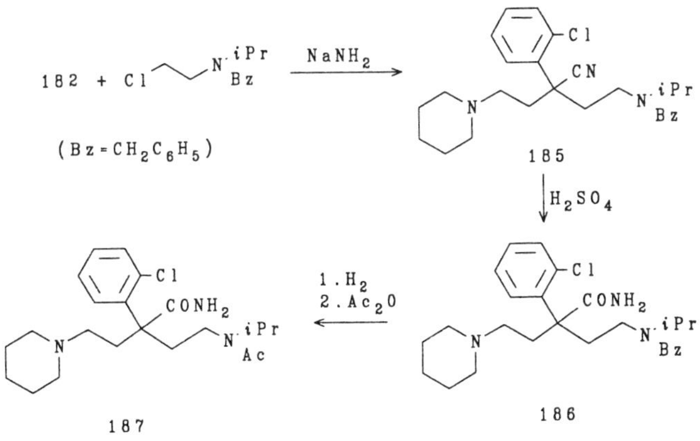

182 + Cl–CH$_2$CH$_2$–N(iPr)Bz —NaNH$_2$→ 185
(Bz = $CH_2C_6H_5$)
185 —H_2SO_4→ 186 —1. H_2, 2. Ac_2O→ 187

hydrogenolysis of the benzyl group in the resulting amide **186**, followed by acetylation of the secondary amine with acetic anhydride gives **bidisomide** (**187**).[46]

D. Miscellaneous Compounds

Carbanion alkylation reactions also play an important role in the synthesis of a pair of calcium channel blocking agents used in the treatment of angina which is caused by insufficient blood supply to cardiac muscle. The activity on the coronary vasculature of the first of these compounds, **verapamil** (**192**), was recognized over 30 years ago; elucidation of the detailed mechanism of action, however, awaited recognition of the role of cellular calcium channels on vascular tone. A first step in the convergent

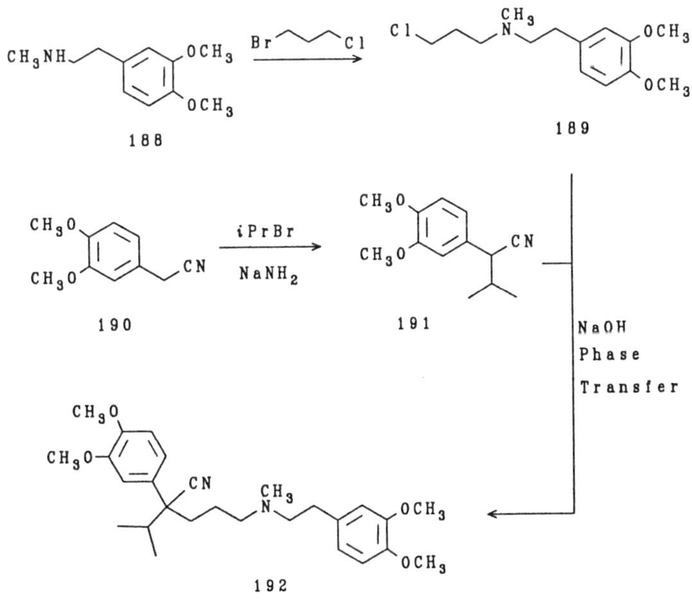

188 —Br(CH$_2$)$_3$Cl→ 189
190 —iPrBr, NaNH$_2$→ 191
189 + 191 —NaOH, Phase Transfer→ 192

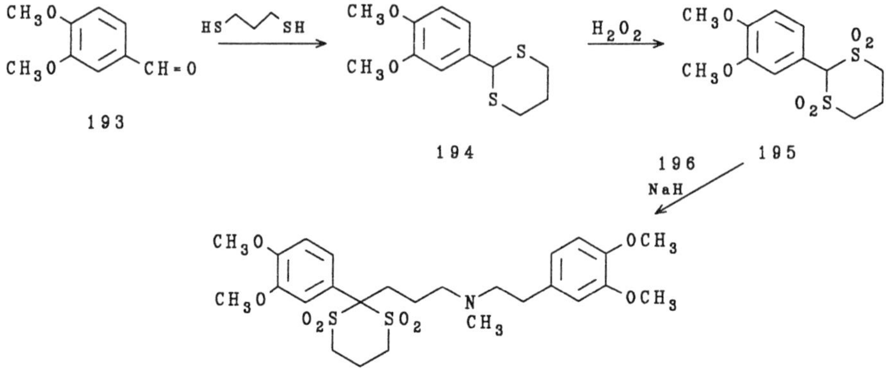

synthesis consists in alkylation of homoveratrylamine (**188**) with 1-bromo-3-chloropropane to give **189**; alkylation of the carbanion from dimethoxyacetonitrile with isopropyl bromide gives **191**. A recent modification of the last alkylation step involves phase-transfer catalysis like conditions; thus, reaction of **189** and **191** with solid powdered sodium hydroxide and potassium carbonate and sodium iodide in toluene in the presence of tetrabutylammonium bromide affords verapamil (**192**).[47]

A recent modification on verapamil (**192**) uses a spiro-dithiane moiety to supply the quaternary center. Reaction of veratraldehyde (**193**) with propane 1,3-dithiol leads to dithiane **194**; reaction with hydrogen peroxide oxidizes the ring sulfur atoms to the corresponding sulfones (**201**). Alkylation of the carbanion from the now quite acidic acetonitrile with the verapamil intermediate **189** gives the antianginal agent **tiapamil** (**196**).[48]

REFERENCES

1. Loewe, H. *Arzneim.-Forsch.* **1954**, *4*, 586.
2. Anonymous Belg. Patent **1961**, 611502; *Chem. Abstr.* **1962**, *57*, 13678.
3. Mais, D.E.; Mohamadi, F.; Dube, G.P.; Kutz, W.L.; Brune, K.A.; Utterback, B.G.; Spees, M.M.; Jubakowski, J.A. *Eur. J. Med. Chem.* **1991**, *26*, 821.
4. Hester, J.B.; Gibson, J.K.; Cimini, G.B.; Emmert, D.E.; Locker, P.K.; Perricone, S.C.; Skaletzky, L.S.; Sykes, J.K.; West, B.E. *J. Med. Chem.* **1991**, *34*, 308.
5. Lumma, W.C.; Wohl, R.A.; Davey, D.D.; Argentieri, T.M.; DeVita, R.J.; Gomez, R.P.; Jain, V.K.; Marisca, A.J.; Morgan, T.K.; Reiser, H.J.; Sullivan, M.E.; Wiggins, J.; Wong, S.S. *J. Med. Chem.* **1987**, *30*, 755.
6. Larsen, A.A.; Gould, W.A.; Roth, H.R.; Comer, W.T.; Uloth, R.H.; Dungan, K.W.; Lish, P.M. *J. Med. Chem.* **1967**, *10*, 462.
7. Collin, D.T.; Hartely, D.; Jack, D.; Lunts, L.H.C.; Press, J.C.; Ritchie, A.C.; Toon, P. *J. Med. Chem.* **1970**, *13*, 674.

8. Skidmore, I.F.; Lunts, L.H.C.; Finch, H.; Naylor, A. German Offen. **1984**, 3414752; *Chem. Abstr.* **1986**, *102*, 95383.
9. Lunts, L.H.C.; Collin, D.T. German Offen. **1970**, 2032642; *Chem. Abstr.* **1971**, *75*, 5520.
10. Controulis, J.; Rebstock, M.C.; Crooks, H.M. *J. Am. Chem. Soc.* **1949**, *71*, 2463.
11. Boliang, L.; Zhang, Y.; Dai, L. *Chem. Ind.* **1993**, 249.
12. Yale, H.L.; Pribyl, E.J.; Braker, W.; Bergeim, F.H.; Lott, W.A. *J. Am. Chem. Soc.* **1950**, *72*, 3710.
13. Barrett, A.M.; Hull, R.; LeCount, D.J.; Squire, C.J.; Carter, J. German Offen. **1970**, 2007751; *Chem. Abstr.* **1970**, *73*, 120318.
14. Erhardt, P.W. In *Chronicles of Drug Discovery*, Lednicer, D. Ed.; 1993, ACS Books, Washington DC. 1993; Vol. 3, p. 191.
15. Jaeggi, K.A.; Schroeter, H.; Ostermayer, F. German Offen. 2503968; *Chem. Abstr.* **1976**, *84*, 5322.
16. Molloy, B.B.; Schmiegel, K.K. German Offen. 2500110; *Chem. Abstr.* **1975**, *83*, 192809.
17. Foster, B.J.; Lavagnino, E.R. Eur. Patent Appl. **1982**, 52492; *Chem. Abstr.* **1982**, *97*, 215718.
18. Fuller, R.W.; Robertson, D.W.; Wong, D.T. Eur. Patent Appl. **1990**, 369685; *Chem. Abstr.* **1990**, *113*, 190895.
19. Korkuczanski, A. *Prezemsyl Chem.* **1958**, *37*, 162.
20. Agback, K.H.; Nygren, A.S. Eur. Patent Appl. **1981**, 36636; *Chem. Abstr.* **1982**, *96*, 122401.
21. Novella, F.C.; Sprague, J.M. *J. Am. Chem. Soc.* **1957**, *79*, 2028.
22. Schultz, E.M.; U. S. Patent **1958**, 2835702; *Chem. Abstr.* **1958**, *52*, 17184.
23. Sturm, K.; Siedel, W.; Weyer, R. German Offen. **1962**, 1122541; *Chem. Abstr.* **1962**, *56*, 14032.
24. Popelak, A.; Lerch, A.; Stach, K.; Roesch, E.; Hardebeck, K. **1970**, German Offen. 815922; *Chem. Abstr.* **1970**, *73*, 45519.
25. Mrozik, H.; Bochis, R.J.; Eskola, P.; Matzuk, A.; Wakamunski, F.S.; Olen, L.E.; Schwartzkopf, Jr., A.; Grodski, A.; Linn, B.O.; Lusi, A.; Wu, M.T.; Schunk, C.H.; Peterson, L.H.; Mikowski, J.D.; Hoff, D.R.; Kulsa, P.; Ostlind, D.A.; Campbell, W.C.; Riek, R.F.; Harmon, R.E. *J. Med. Chem.* **1977**, *20*, 1225.
26. Ruschig, H.; Avmuller, W.; Korger, G.; Wagner, H.; Scholtz, J., Bander, A. U. S. Patent **1961**, 2976317; *Chem. Abstr.* **1962**, *56*, 12771.
27. Wright, J.B.; Willette, R.E.; *J. Med. Chem.* **1962**, *5*, 927.
28. Hsi, R.S.P. *J. Labeled Comp.* **1973**, *9*, 91.
29. Sarges, R. *J. Med. Chem.* **1976**, *19*, 695.
30. Rufer, C. *J. Med. Chem.* **1974**, *17*, 708.
31. Mais, D.E.; Mohamadi, F.; Dube, G.P.; Kutz, W.L.; Brune, K.A.; Utterback, G.B.; Spees, M.M.; Jakubowski, J.A. *Eur. J. Med. Chem.* **1991**, *26*, 821.
32. For a review of research leading to these drugs see Nicholson, J.S. in *Chronicles of Drug Discovery,* Bindra, J.S.; Lednicer, D., Eds.; Wiley, New York; 1982, Vol. 1, p. 149.
33. Sallman, A.; Pfister, R. German Offen. **1969**, 1815802; *Chem. Abstr.* **1970**, *72*, 12385.
34. Welstead, Jr., W.J.; Moran, H.W.; Stauffer, H.F.; Turnbull, L.B.; Sancillio, L.F.; *J. Med. Chem.* **1979**, *22*, 1074.

35. Gassman, P.G.; Van Bergen, T.J.; Gruetzmacher, G. *J. Am. Chem. Soc.* **1973**, *95*, 6508.
36. Walsh, D.A.; Moran, H.W.; Shamblee, D.A.; Uwaydah, I.M.; Sancillio, L.F.; Dannenburg, W.N. *J. Med. Chem.* **1984**, *27*, 1379.
37. Nicholson, J.S. In *Chronicles of Drug Discovery,* Bindra, J.S.; Lednicer, D., Eds., Wiley, New York; 1982, Vol. 1, p. 164.
38. White, D.R.; Wu, D.K. *J. Chem. Soc. Chem. Commun.* **1974**, 988.
39. Taylor, E.C.; Robey, R.L.; Liu, K.T.; Favre, B.; Bozimo, H.T.; Conley, R.A.; Chiang, C.S.; McKillop, A.; Ford, M.E. *J. Am. Chem. Soc.* **1976**, *98*, 3037.
40. Adams, S.S.; Bernard.; Nicholson, J.S.; Blancafort, A.B. U. S. Patent **1973**, 3755427; *Chem. Abstr.* **1973**, *79*, 104952.
41. Carney, R.W.J.; DeStevens, G. German Offen. **1970**, 2012327. *Chem. Abstr.* **1971**, *74*, 141553.
42. Fieser, L.F.; Schirmer, J.P.; Archer, S.; Lorenz, R.R.; Pfafenbach, P. *J. Med. Chem.* **1967**, *10*, 513.
43. Laue, C. quoted in *Chem. Eng. News* **1995**, *Oct., 9*, 72.
44. Cusic, J.W.; Sause, H.W. U. S. Patent **1965**, 3225054. *Chem. Abstr.* **1966**, *64*, 6625.
45. Youan, P.K.; Novotney, R.L.; Woo, C.M.; Prodan, K.A.; Hershenson, F.M. *J. Med. Chem.* **1980**, *23*, 1102.
46. Desai, B.N.; Chorvat, R.J.; Rorig, K.J. Eur. Patent Appl. **1986**, 170901, *Chem. Abstr.* **1986**, *105*, 97320.
47. Zwierak, A.; Gajda, T.; Koziara, A.; Zawadski, S.; Osawa-Pacewicka, K.; Olejniczak, B.; Wasilewska, W.; Kotlicki, S.; Cichon, J. Pol. Patent **1992**, 156975; *Chem. Abstr.* **1993**, *119*, 138862.
48. Ramuz, H. *Arzneim.-Forsch.* **1978**, *28*, 2048.

CHAPTER 3

INDENES, NAPHTHALENES, AND OTHER POLYCYCLIC AROMATIC COMPOUNDS

Chapter 2 illustrated cases in which benzene rings served as structural elements in drugs where its function ranged from forming part of a pharmacophore to simply providing a framework to support needed functionality. Much the same is true for carbocyclic fused-ring systems. These ring systems may serve, for example, as surrogate substituted benzene rings. The extra fused ring in **propranolol** (**1**) provides bulk at the position ortho to the oxypropanolamine side chain for this β-blocker. In much the same vein, the indene ring system in the hypoglycemic agent **glyhexamide** (**2**) replaces the benzene ring found in many other sulfonylureas of this class. This chapter focuses, however, on drug classes characterized by polycyclic rings systems and those compounds that illustrate interesting synthetic methods.

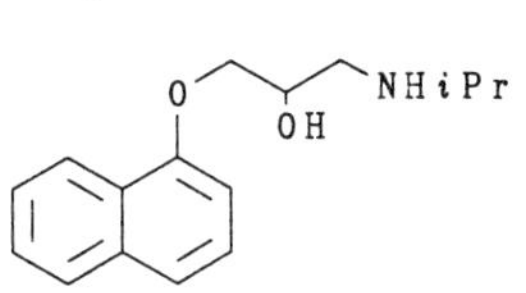

O NHiPr OH SO_2NHCNH O

1 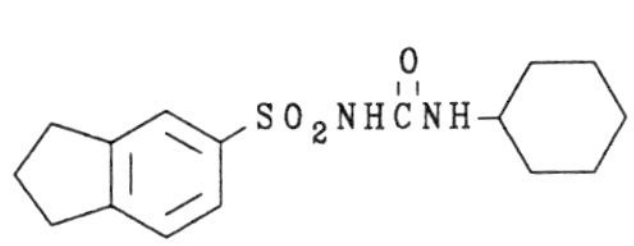 2

1. INDENES

It is now widely recognized that the majority of drugs owe their action to binding with receptors on target organ cells. The shape of such a ligand will be expected to have a direct effect on binding efficiency, since receptors themselves exist in relatively defined configurations. A molecule that closely complements the shape of a receptor should bind more efficiently than one that needs to undergo conformational changes prior to binding. That provides the motivation for preparing so-called rigid or constrained analogues, molecules that are locked in those close-fit conformations.

The acryloketone side chain of the non-thiazide "high-ceiling" diuretic agent **ethacrynic acid** (**3**) can rotate quite freely; the ortho chloro substituent would in fact be expected to favor an orthogonal arrangement to avoid nonbonding interaction. **Indacrinone** (**9**) is an analogue of this compound in which the motion of this side chain is constrained, and interestingly it shows slightly better activity than the parent; a direct comparison is complicated by the fact that the two agents show different electrolyte and urate excretion patterns. The synthesis starts with the Friedel–Crafts acylation of 2,3-dichloroanisole (**4**) with phenylacetyl chloride to give ketone **5.** The side chain exomethylene group is introduced by the Mannich reaction with the preformed carbinolamine from formaldehyde to give **6**. Reaction of **6** with concentrated sulfuric acid leads to internal alkylation and formation of the indanone **7**. This ketone is then alkylated by means of a strong base and methyl iodide; demethylation of the methyl ether with strong acid affords the corresponding phenol (**8**). The phenoxide ion from treatment of this product with base is then alkylated with ethyl bromoacetate. Saponification of the ester leads to the free acid indacrinone (**9**).[1]

Another illustration of the importance of steric considerations for biological activity comes from the finding that the indene **sulindac** (**16**) shows qualitatively the same NSAID activity as the older compound **indomethacin** (**10**). The indanone ring in intermediate **12** is formed by internal Friedel–Crafts acylation of acid **11** by means of polyphosphoric acid (PPA). Reformatskii reaction on this ketone with methyl bromoacetate and zinc leads to addition to the carbonyl group and formation of a tertiary

4 → ($C_6H_5CH_2COCl$, $AlCl_3$) → **5** → ($CH_2=O$, Me_2NH) → **6** → (H_2SO_4) → **7** → (1. CH_3I/tBuOK; 2. HBr) → **8** → (1. $BrCH_2CO_2Et$; 2. OH^-) → **9**

3

F, CO_2H — **11** $\xrightarrow{\text{PPA}}$ F, O — **12** $\xrightarrow[\text{Zn}]{BrCH_2CO_2Me}$ F, HO, CO_2Me — **13** $\xrightarrow{\text{TSA}}$ F, CO_2Me — **14**

$\xrightarrow[\text{NaOMe}]{p\text{-}MeSC_6H_4CH\text{=}O}$ CH_3S, F, CO_2H — **15** $\xrightarrow{NaIO_4}$ CH_3S(O), F, CO_2H — **16**

Cl, O, N, CH_3O, CO_2H — **10**

carbinol (**13**); this compound readily dehydrates to the corresponding indene **14** on treatment with *p*-toluenesulfonic acid (p-TSA). The protons on the remaining benzylic position are reasonably acidic due to the potential stabilization of an anion by the adjacent olefin and benzene ring. Thus, reaction of **14** with *p*-thiomethylbenzaldehyde in the presence of sodium methoxide leads to formation of a condensation product; hydrolysis of the ester then gives **15** as a mixture of isomeric olefins where the (*Z*) isomer predominates. Isolation of this isomer, followed by oxidation of sulfur by reaction with sodium metaperiodate, completes the synthesis of **sulindac** (**16**).[2] There is some evidence to indicate that the active agent is in fact the sulfide **15**, which is formed by an unusual metabolic reduction of the sulfone **16**.

2. NAPHTHALENES

A. Antifungal Compounds

Two small series of antifungal drugs with markedly different functionality include benzocycloalkane, or hydronaphthalene, rings. The absence in the literature of simple benzenoid analogues may indicate that the bicyclic moiety plays a role in the mechanism by which those drugs act. Members of the older series all share a thiocarbamate function based on *m*-toluidine. The starting material for the synthesis of the first of these compounds, thiocarbamoyl chloride (**16**), is obtained in straightforward fashion by the reaction of 2-naphthol with thiophosgene. Treatment of this acid chloride with *N*-methyl-*m*-toluidine gives the antifungal agent **tolnaftate** (**17**);[3] wide use of this compound demonstrated the safety and efficacy that led to its approval for use in nonprescription products. The exact analogue prepared from 5-indanol,[4] **tolindate** (**18**), is reported to be somewhat less potent.

The fused ring in a more recent member of this series, **tolciclate** (**24**), combines structural elements from both the naphthalene and indan series. The key step in the preparation of this agent involves Diels–Alder condensation of a substituted benzyne with cyclopentadiene. Reaction of iodobromo anisole (**19**) with magnesium can be envisaged to proceed initially to a Grignard reagentlike intermediate such as **20**; this could then undergo internal elimination of magnesium bromoiodide to lead to the

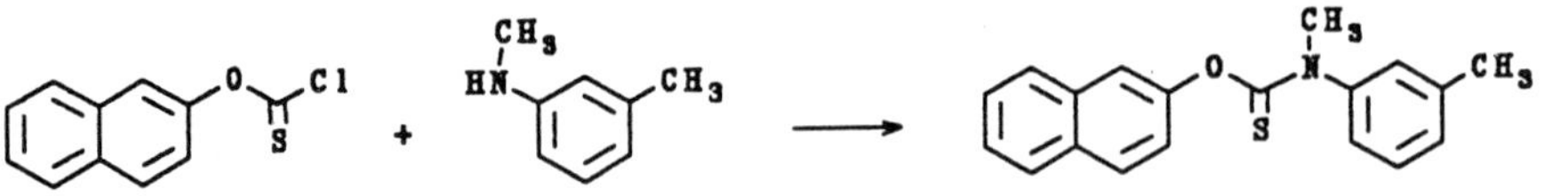

O Cl S + CH_3 HN CH_3 → O N CH_3 S CH_3

16a 17

O N CH_3 S CH_3

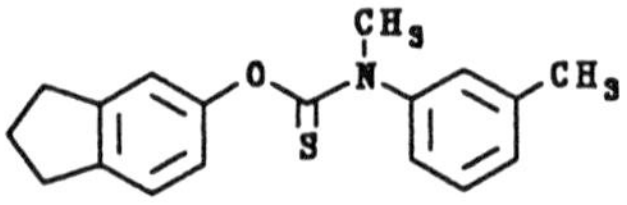

18

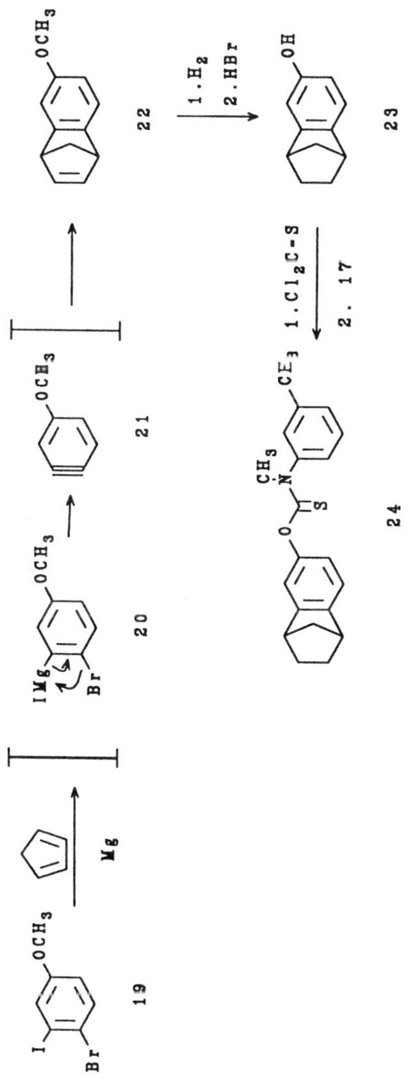

19
20
21
22
23
24
Mg
1. H2
2. HBr
1. Cl2C=S
2. 17

highly reactive benzyne **21**; 1,4-addition across cyclopentadiene will lead to the observed product, benzonorbornadiene **22**. This isolated olefinic bond in the intermediate is reduced by catalytic hydrogenation; the phenol ether is then cleaved by treatment with hydrobromic acid to give free phenol (**23**). The standard sequence, which consists of reaction with thiophosgene followed by *N*-methyl-*m*-toludine, completes the synthesis of **tolciclate** (**24**).[5]

Naphthalenes lacking any functional group beyond an α-naphthylmethylallylamine group form another small series of antifungal agents. The prototype for this series, **naftifine** (**26**), is prepared by simple alkylation of *N*-methyl-α-methylnaphthylamine (**25**) with cinnamyl bromide.[6] The trans geometry of the double bond in the starting material is retained in the product. Replacement of the phenyl group in this compound by *tert*-butylacetylene leads to a more potent compound. The preparation begins with alkylation of the common intermediate **25** with propargyl bromide. This compound (**27**) is then coupled with 1-bromo-2-*tert*-butylacetylene by means of a copper salt catalyzed reaction in the presence of a mild base such as a tertiary amine. Construction of the allylamine function takes advantage of a reaction characteristic of propargyl amines. Thus, reaction of the intermediate **28** with lithium diisopropylaluminum hydride proceeds to give selective reduction of the double bond closest to the amine group; the observed selectivity can be rationalized by assuming the intermediacy of a

CH_3 NH

$BrCH_2$ C_6H_5

25

CH_3 N C_6H_5

26

$BrCH_2$≡H

CH_3 N ≡H

27

Br≡tBu

CuBr, base

CH_3 N ≡ ≡ tBu

28

$LiAl(iPr)_2H$

tBu

CH_3 N

29

complex of the hydride reagent with basic nitrogen. Thus, the antifungal agent **terbinafine** (**29**)[7] is obtained.

B. Another β-Blocker

The oxypropanolamine is an almost invariant substituent in the multitude of β-adrenergic blocking agents. Modification on the aryl portion of these drugs to which this pharmacophore is attached can serve to modify pharmacodynamic properties of individual drugs. Partial agonist action is a common property exhibited by receptor antagonists in many pharmacological classes; appropriate substitution can decrease the intrinsic β-adrenergic agonist action shown by many β-blockers. Thus, replacement of one of the rings in the naphthalene moiety in **propranolol** (**1**) by a partly reduced ring bearing a diol has been shown to reduce β-adrenergic agonist activity.

The synthesis of this agent, **nadolol** (**36**), begins with the Birch reduction of 1-naphthol (**30**). These dissolving metal reductions usually show selectivity for the ring bearing electron-rich substituents; a ring carrying a methoxy group will usually be reduced preferentially over unsubstituted rings. In the case at hand, the phenol is protected against reduction because it bears a negative charge under very basic reaction conditions; thus treatment of **30** in liquid ammonia with lithium followed by the addition of ethanol as a proton source gives the dihydronaphthol **31**. The phenolic group is then acetylated as protection from the oxidative reagents used in the next step. The required cis diol function is then introduced by means of the Woodward modification of the Prevost oxidation with iodine and silver acetate. The reaction can be rationalized by assuming formation of the cyclic iodonium intermediate **32** as the first step. Attack by acetate anion will lead to a trans iodoacetate such as **33** as well as its regioisomer; each compound will lead to the same final product. Displacement of the reactive iodo group then leads to the triacetate **34**. This compound is not isolated but converted in situ to the triol **35** with strong base. The oxypropanolamine side chain is then introduced in the usual way by sequential reaction with epichlorohydrin (ECH) and *tert*-butylamine; ECH reacts selectively at the phenol because it is much more acidic than the aliphatic hydroxyl groups. Thus, the β-blocker nadolol (**36**)[8] is obtained.

OH, 30 — Li/NH_3, EtOH → OH, 31 — 1. Ac_2O, 2. AgOAc, I_2, 3. OH^- → [+I, OAc, 32 → AcO, I, OAc, 33] → [AcO, AcO, OAc, 34] → HO, HO, OH, 35 — 1. ECH, 2. $tBuNH_2$ → HO, HO, O, OH, NHtBu, 36

C. Aminotetralin Antidepressants

The shortcomings of the tricyclic antidepressant drugs such as **amitriptyline** and its active metabolite **nortriptyline** (**37**) were noted in the discussion of the "oxetin" compounds in Chapter 2. Since the side effects of tricyclic antidepressant drugs are not traceable to the mechanism of action, topological analogues, that is, compounds that have similar functionality supported on a different carbon skeleton, offer the possibility for designing drugs that have the same activity but may be devoid of the side effects attributed to this structural class.

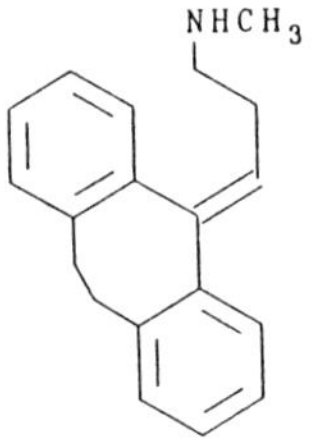

$NHCH_3$

37

The structure of **nortriptyline** (**37**) depicted in an admittedly biased projection suggested 4-aryl-1-aminotetralins as possible topological analogues by the formal opening of the seven-membered ring and cyclization of the side chain. The preparation of the first of these agents starts with 4,4-diphenylbutyric acid (**38**). Cyclization of the acid chloride with aluminum chloride gives the tetralone **39**. This compound is then converted to its *N*-methylimine (**40**) derivative by using methylamine and titanium tetrachloride. This last intermediate is then reduced with sodium borohydride to give a mixture of cis and trans aminotetralins (**41**). The trans isomer **tametraline** (**42**) is separated by fractional crystallization of the hydrochloride salt.[9] Detailed pharmacological investigation showed that this compound owes its antidepressant action to the inhibition of reuptake of dopamine and norepinephrine from the synaptic cleft.

A somewhat more elaborate scheme is used for the analogue bearing different substitution on the pendant phenyl ring. The Stobbe condensation of substituted benzophenone (**43**), which is obtainable from Friedel–Crafts acylation of benzene with 3,4-dichlorobenzoyl chloride, gives the chain extended half-ester **44** with ethyl succinate and potassium *tert*-butoxide. Treatment of this product with hydrobromic acid leads to hydrolysis of the remaining ester group with concomitant decarboxylation to give the acrylic acid **45** as a mixture of isomers. The total product is then hydrogenated to the 4,4-diarylbutyric acid **46**. Friedel–Crafts cyclization proceeds by attack on the more electron rich of the two benzene rings to afford the tetralone. This last intermediate is then converted to its *N*-methylimine **47** with methylamine in the presence of titanium tetrachloride. Hydride reduction, as shown above, leads to a mixture of cis and trans aminotetralins. The cis isomer **sertraline** (**48**) is then separated by column chromatography.[10] This compound, in contrast to the trans tetralin **tametraline** acts as an antidepressant mainly by its serotonin reuptake inhibiting activity.

CO_2H

1. SO_2Cl
2. $AlCl_3$

O

CH_3NH_2
$TiCl_4$

NCH_3

38 39 40

$NaBH_4$

$NHCH_3$

Separate

$NHCH_3$

42 41

O

$(CH_2CO_2Et)_2$
tBuOK

Cl Cl

CO_2H
CO_2Et

Cl Cl

HBr

CO_2H

Cl Cl

43 44 45

H_2

$NHCH_3$

Cl Cl

1. $NaBH_4$
2. Separate

NCH_3

Cl Cl

1. SO_2Cl
2. $AlCl_3$
3. CH_3NH_2
$TiCl_4$

CO_2H

Cl Cl

48 47 46

3. TRICYCLIC COMPOUNDS

A. Dibenzocycloheptane and Dibenzocycloheptene Antidepressants

The discovery of the antipsychotic activity of **chlorpromazine** (**49**) opened the modern era of central nervous system (CNS) pharmacology. In many laboratories, an intense effort ensued to investigate the SAR of related compounds. This effort included the preparation of what may be regarded, possibly post facto, as a carbocyclic bioisostere of a phenothiazine; the sulfur bridge of the prototype being replaced by an ethylene chain of approximately the same size and the nitrogen by a trigonal sp^2 carbon. The resulting compound, **amitriptyline** (**50**), unexpectedly showed antidepressant rather than antipsychotic activity. It is now known that the antipsychotic effect of the phenothiazines is due mainly to the fact that they are competitive antagonists for dopamine at its receptor sites; the tricyclic antidepressants on the other hand inhibit the reuptake of norepinephrine and serotonin from the synaptic cleft, in effect prolonging the action of those two neurotransmitters.

The protons on the methylene group in phtalide are surprisingly acidic; this compound thus readily undergoes aldol condensation with benzaldehyde to give the benzal derivative **51**, which is in effect an internal enol ester. Reaction with phosphorus and hydriodic acid serves to reduce the carbonyl group; thus, the stilbene carboxylic acid **52** is obtained. Catalytic hydrogenation of the double bond leads to **53**; this compound can then be cyclized to dibenzocycloheptanone (**53**) under standard Friedel–Crafts conditions, for example, treatment with PPA. Cyclization of the corresponding unsaturated acid **52** leads to the ketone **55**; compound **55** can be reduced to the corresponding hydrocarbon **56** by Wolff–Kishner reduction with hydrazine and potassium hydroxide.

One of the more direct syntheses of the prototype compounds in this series, amitriptyline (**50**), depends on the cyclopropylcarbinyl–homoallyl rearrangement derived from steroid chemistry and discussed in greater detail in Chapter 1. The key intermediate **57** is obtained by condensation of the Grignard reagent from cyclopropyl chloride with dibenzocycloheptanone **54**. Reaction of **57** with hydrogen bromide leads to rearrangement of the cyclopropyl carbinol to the homoallyl bromide **58** via an intermediate homoallyl carbocation. Displacement of halogen with dimethylamine affords amitriptyline (**50**); the corresponding reaction with methylamine affords **nortriptyline** (**37**).[11]

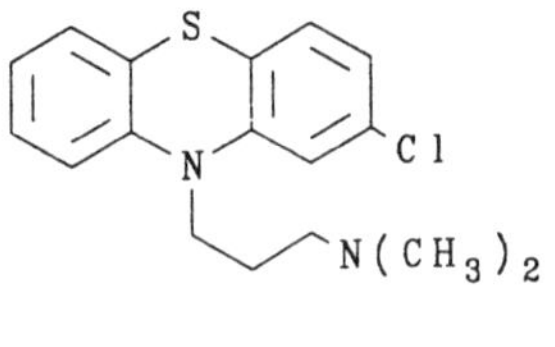

49

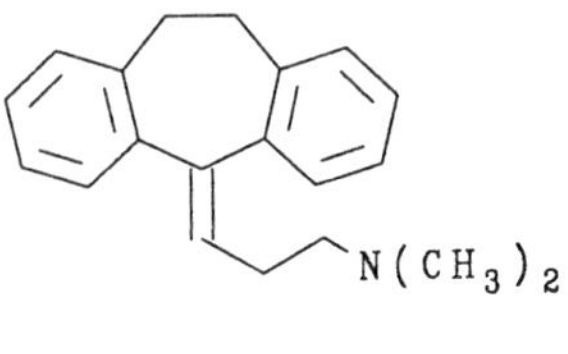

50

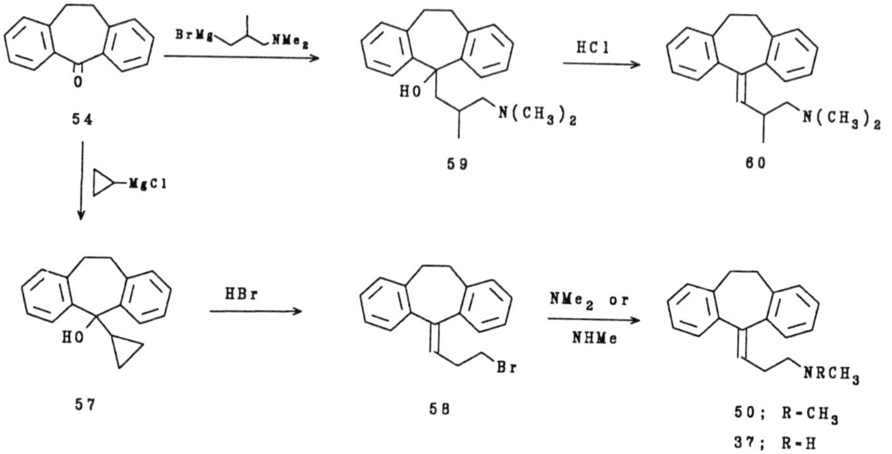

The more conventional approach used for the synthesis of the analogue incorporating a branched side chain involves additional steps, which are not shown, that are required for preparation of the starting material for this chain. Once the required *N,N*-dimethyl-3-chloro-2-methylpropylamine is at hand, it is converted to its corresponding Grignard reagent; condensation with dibenzocycloheptanone affords the amino alcohol **59**. Dehydration by treatment with strong acid affords **butriptyline** (**60**).[12]

An analogue in which the side chain is cyclized to form a piperidine ring, **cyproheptadine** (**62**), interestingly predominantly shows the antihistaminic activity that led to the preparation of the phenothiazines in the first place. This analogue is prepared in straightforward fashion by condensation of the Grignard reagent from chloropiperidine **61** with **55**; dehydration of the intermediate alcohol gives **cyproheptadine** (**62**).[13]

Extension of the side chain by interposition of the ether oxygen restores antidepressant activity even when nitrogen is included in a piperidine ring. The starting alcohol **63** is obtained by hydride reduction of dibenzocycloheptanone **54**. Reaction of the product with strong acid in the presence of *N*-methyl-4-piperidinol in all likelihood proceeds first to form the dibenzyl carbocation by loss of water from protonated **63**; reaction with oxygen on piperidinol will then give the observed product **hepzidine** (**64**).[14]

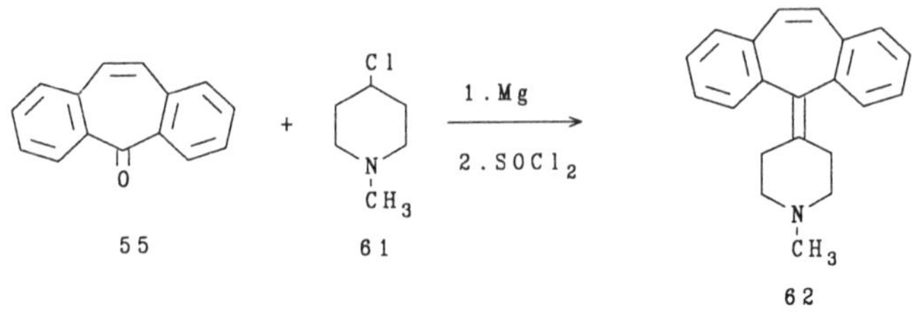

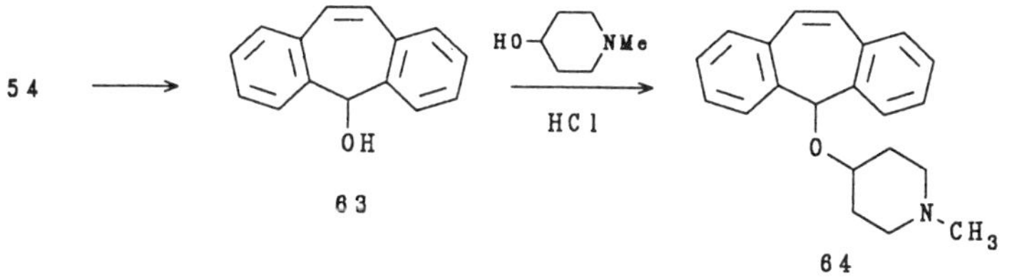

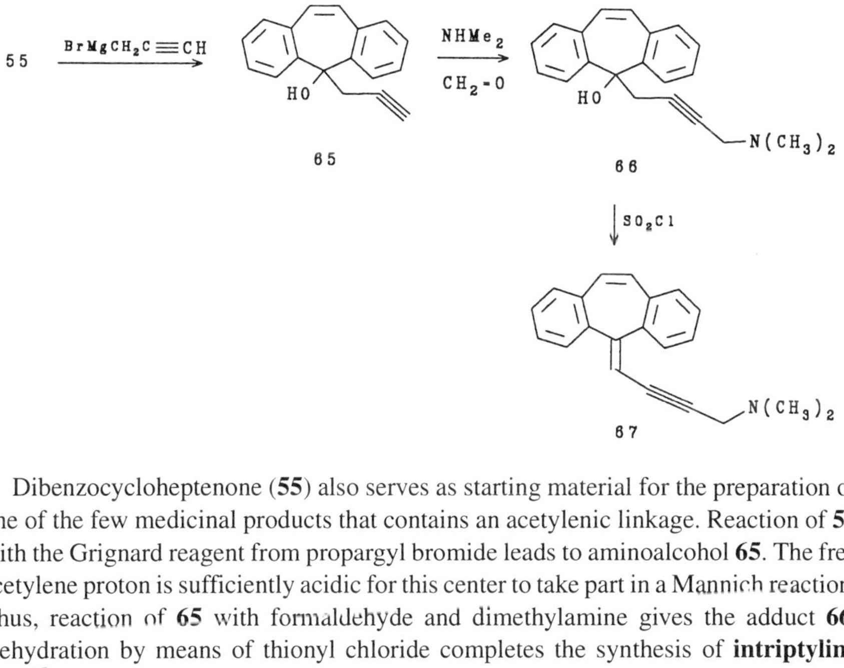

Dibenzocycloheptenone (**55**) also serves as starting material for the preparation of one of the few medicinal products that contains an acetylenic linkage. Reaction of **55** with the Grignard reagent from propargyl bromide leads to aminoalcohol **65**. The free acetylene proton is sufficiently acidic for this center to take part in a Mannich reaction. Thus, reaction of **65** with formaldehyde and dimethylamine gives the adduct **66**. Dehydration by means of thionyl chloride completes the synthesis of **intriptyline** (**67**).[15]

Antidepressant activity is retained when the cycloheptene double bond is replaced by a fused cyclopropane. The starting material for this compound is obtained by reaction of the unsaturated ketone **55** with dichlorocarbene, which is formed from the reaction of ethyl trichlororacetate acid with a strong base. In a sequence patterned after the one used for the prototype drug, compound **68** is next condensed with the Grignard reagent from cyclopropyl chloride to give carbinol **69**. The halogen is then removed by dissolving metal reduction with lithium and *tert*-butanol to give **70**. This intermediate is then subjected to the cyclopropylcarbinyl rearrangement to give a homoallyl halide; this step also does away with any potential problem posed by geometrical isomerism of **69** and **70**. Displacement of halogen with dimethylamine affords the antidepressant agent **octriptyline** (**71**).[16]

Antidepressant activity is retained when the trigonal one carbon bridge is replaced by tetrahedral carbon. Thus, reaction of hydrocarbon **56** with a metal amide in liquid ammonia leads to the corresponding carbanion (**72**). Treatment of **72** with the ethyl carbamate from *N*-methyl-3-chloropropylamine (**73**) leads to the alkylation product. The carbamate protecting group is then removed by sequential saponification with base followed by acidification. This reaction yields the antidepressant agent **protriptyline** (**74**).[17]

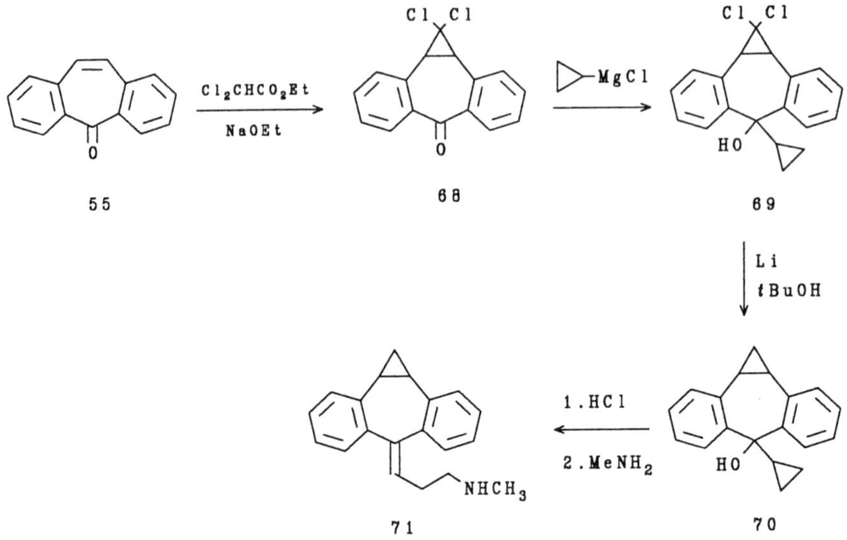

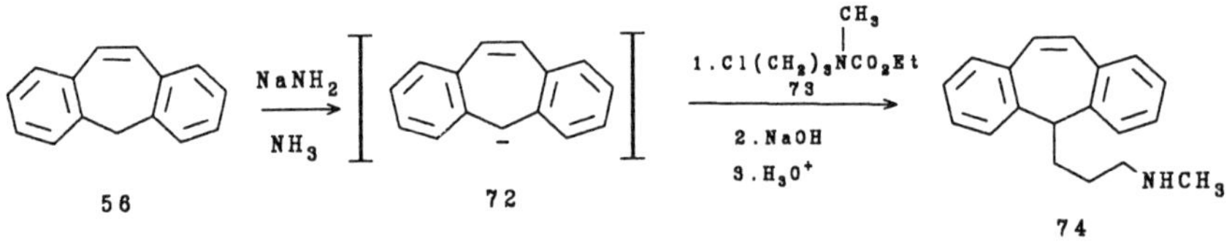

B. Antidepressants Based on Dihydroanthracenes

Antidepressant activity is retained when the central ring in the classical tricyclic compounds is contracted to a dihydroanthracene even though the resulting compounds at first sight more closely resemble the antipsychotic phenothiazines. Most of this small group of drugs owe their antidepressant activity to inhibition of reuptake of norepinephrine.

The synthesis of the simplest member of this series, **melitracen** (**77**), takes advantage of the relatively acidic protons of the methylene group in anthrone (**75**) due to activation from the carbonyl group transmitted via the aromatic rings. Thus treatment of the carbanion from anthrone with methyl iodide proceeds to the dimethyl derivative **76**. The Grignard reagent from *N*,*N*-dimethyl-3-propylamine is then added

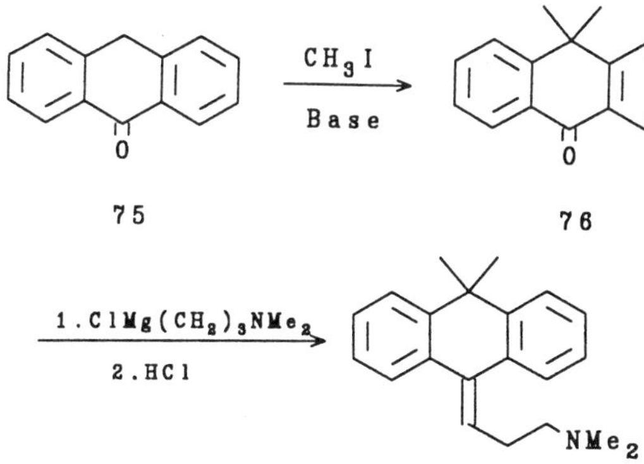

75 → (CH_3I, Base) → 76

→ (1. $ClMg(CH_2)_3NMe_2$, 2. HCl) → 77

to the carbonyl group of this intermediate. Dehydration of the resulting carbinol gives **melitracen** (**77**).[18]

A more complex scheme is required for the preparation of a derivative that bears a trifluoromethyl substituent on one of the benzene rings. The scheme starts with the condensation of the nitrile group in **78** with phenylmagnesium bromide to give the corresponding imine; treatment with aqueous acid leads to the substituted benzophenone **79**. The future methyl on one of the bridges is introduced in a sequence involving the addition of a trimethylsulfonium ylide.

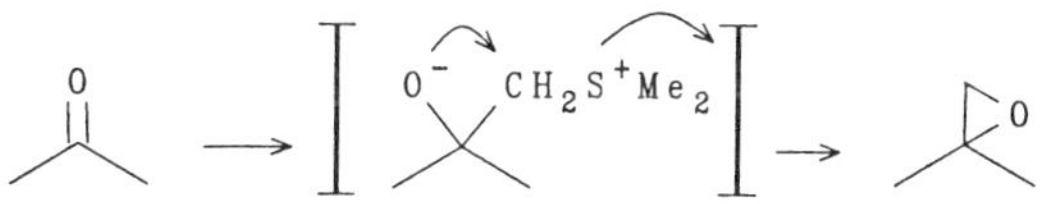

O^- $CH_2S^+Me_2$

The first step in the reaction involves straightforward reaction with the carbonyl group to give a betaine such as the one depicted above. Internal displacement of dimethyl sulfide by the alkoxide anion leads to an epoxide and, in the case of the benzophenone, intermediate **80**. Reaction with red phosphorus and hydriodic acid leads to a fully reduced methyl derivative. This reduction may be rationalized by assuming initial opening to an iodohydrin followed by subsequent reductive elimination of the benzhydrol hydroxyl and aliphatic iodine. The resulting intermediate is then reacted with cuprous cyanide; this reaction leads to displacement of bromine by cyanide and formation of nitrile **81**. This compound is then condensed with the Grignard reagent from 3-bromo-1-methoxypropane to give the imine **82**. Treatment with hydrobromic acid leads directly to an anthracene (aromatization presumably results from bond reorganization of the first formed intermediate that contains an exo double). In addition, the reaction conditions are sufficiently strenuous to cleave the terminal methyl ether and to convert the resulting alcohol to a bromide. Construction of the side chain is completed by replacement of halogen by dimethylamine to give **83**. Catalytic reduction proceeds to give the 9,10-*cis*-dihydroanthracene and thus **fluotracen** (**84**).[19]

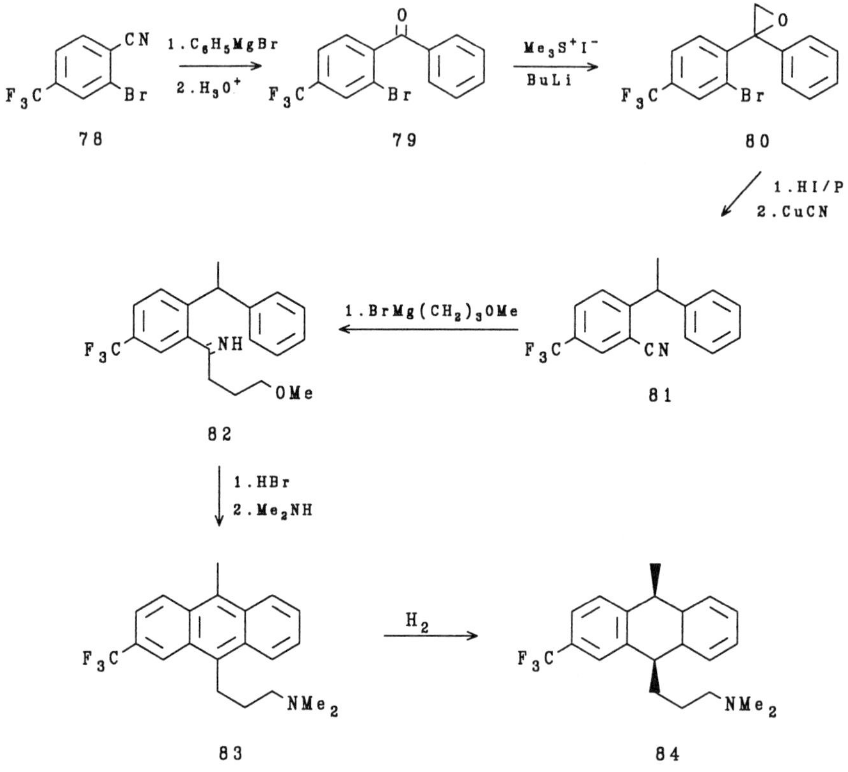

The first step in the preparation of the antidepressant **maprotiline (87)** takes advantage of the acidity of anthrone protons for incorporation of the side chain. Thus treatment of **75** with ethyl acrylate and a relatively mild base leads to the Michael adduct; saponification of the ester group gives the corresponding acid **85**. The ketone group is then reduced by means of zinc and ammonium hydroxide; dehydration under acidic conditions leads to formation of a fully aromatic anthracene **86**. Diels–Alder addition of ethylene under high pressure leads to addition across the 9,10-positions and formation of the central 2,2,2-bicyclooctyl moiety (**87**). The final steps involve construction of the typical antidepressant side chain. The acid in **87** is thus converted to an acid chloride and this function is reacted with methylamine to form the amide **88**. Reduction to a secondary amine with lithium aluminum hydride (LAH) completes the synthesis of **maprotiline (89)**.[20]

The aminoalcohol side chain in the antidepressant **oxaprotiline (94)** interestingly resembles that of a β-blocker, although no such activity has been reported for the compound. The first few steps in the synthesis leading to the key intermediate **90** parallel those for **maprotiline (85 → 87)** starting with an acetic rather than a propionic acid. The carboxyl group is then reduced to an aldehyde (**91**) by sequential conversion to its acid chloride and then hydrogenation over poisoned palladium catalyst. Reaction of the aldehyde with hydrogen cyanide leads to the corresponding cyanohydrin **92**, which now incorporates the required third side chain carbon. Reduction with LAH yields the aminoalcohol **93**.[21] The primary amino group can then be monomethylated by any of several methods such as conversion to its carbamate followed by a second hydride reduction.

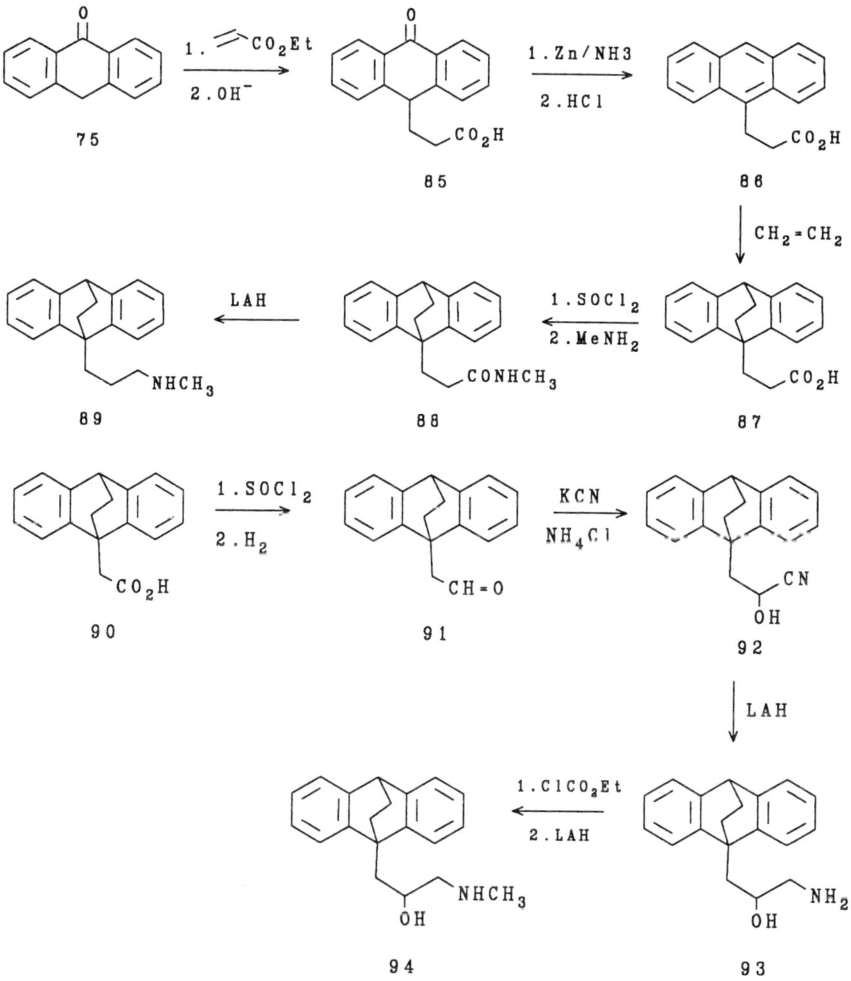

C. Anthraquinones: The "Antrone" Chemotherapy Agents

Anthraquinones have a venerable history as dyes, which dates back to the late nineteenth century. Their use as drugs is, however, much more recent, going back at most two decades. The initial discovery of the activity of this class was due to the lack of any good leads for compounds that were useful for the treatment of cancer. This circumstance led the National Cancer Institute (NCI) of the U.S. National Institutes of Health (NIH) to undertake a massive screening program for compounds that might show such activity. The NCI and its representatives scoured all possible sources worldwide for compounds to be submitted to the screens. In order to acquire the broadest possible selection of structural types, potential sources included all chemical and pharmaceutical industries as well as academia. A dark blue anthraquinone originally developed as a ballpoint ink pigment perhaps unexpectedly showed activity in the mouse leukemia screen, then used by NCI. Although this compound, **ametantrone** (**97**), never reached the clinic, its more hydrophilic congener, **mitoxantrone** (**99**), was eventually approved for the treatment of leukemias.

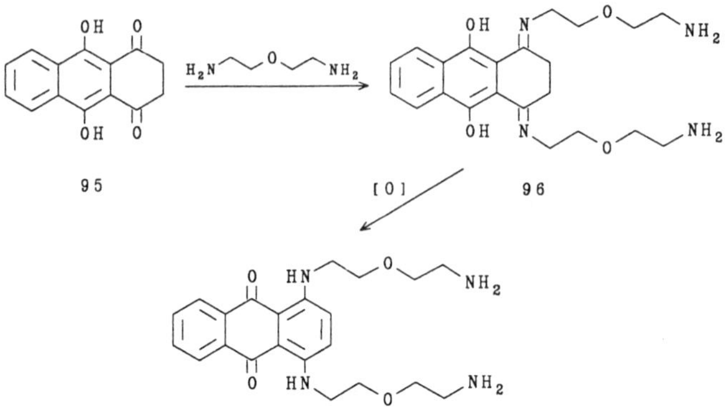

The synthesis of the first of these agents starts with leucoquinizarin (**95**), the reduction product of the dihydroxyanthraquinone, quinizarin. Condensation with bis(2-aminoethyl)ether leads to the corresponding bis-imine **96**. Air oxidation may be visualized as starting at the hydroquinone-like central ring; bond reorganization will lead to ametrantrone (**97**).[22]

In much the same vein, condensation of the tetrahydroxy intermediate **98** with *N*-(2-hydroxyethyl)ethylenediamine gives an intermediate bis-imine, and air oxidation of the intermediate then affords mitoxantrone (**99**).[23]

Good activity is retained in these compounds when one of the quinone oxygen atoms is included in a pyrrazole ring. The preparation of one of these agents, **piroxantrone** (**103**), starts by protection of the hydroxyl groups in 1,4-dichloro-5,8-dihydroxyanthraquinone as their benzyl ethers to give **100**. Condensation of **100** with the hydrazine equivalent of the reagent used to prepare **99** leads to the pyrrazole intermediate **101**; the order in which the steps involved in forming the heterocyclic ring, imine formation and nucleophilic displacement of aromatic chlorine, occur is not immediately apparent. The remaining aromatic halogen is then displaced with 1,3-pro-

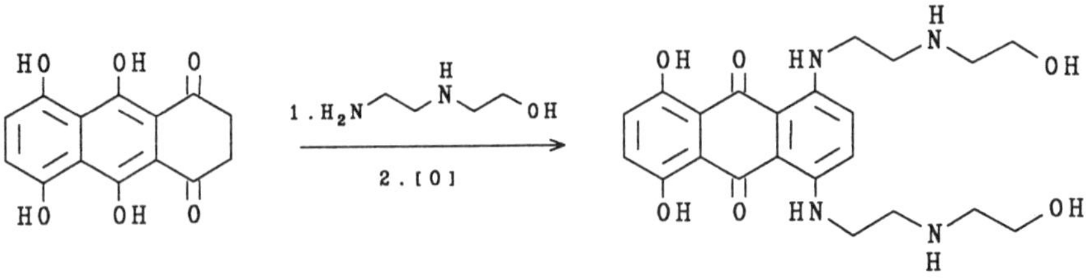

BzO O Cl
BzO O Cl
100
H_2NHN–CH₂CH₂–NH–CH₂CH₂–OH
101
$Bz = CH_2C_6H_5$
H_2N–(CH₂)₃–NH_2
102
H_2
103

pylene diamine to give **102**. Removal of the benzyl ethers by hydrogenolysis completes the synthesis of piroxantrone (**103**).[24,25]

The symmetrical substitution pattern in the foregoing anthraquinone derivatives greatly simplified the chemistry. The presence of a single hydroxyl substituent on the terminal ring of the pyrrazolo derivative **losoxantrone** (**106**) adds to the complication of its preparation. The phenol in the key intermediate in this case is protected as its somewhat more bulky 2,4,6-trimethylbenzyl intermediate. The additional methyl groups may help steer the condensation away from the adjacent carbonyl group and

$Me_3C_6H_2CH_2O$ O Cl
104
1. H_2NHN–CH₂CH₂–NH–CH₂CH₂–OH
2. tBOCCl
3. TFA
105
H_2N–CH₂CH₂–NH–CH₂CH₂–OH
106

107 $\xrightarrow{H_3O^+}$ 108 $\xrightarrow{HIO_4}$ 109

1. [O]
2. H_2NNH–(2-imidazolinyl)

110

in addition add lipophilicity, which aids in the solubility of intermediates in common organic reaction solvents. Condensation with the same hydrazine as above leads to a mixture of **105** and its isomer as a 4:1 mixture. The crude mixture is then converted to its *tert*-butoxycarbonyl derivative in order to facilitate separation. Treatment of the purified derivatized intermediate with trifluoracetic acid (TFA) removes the *t*-Boc and benzyl groups in a single step to give **105**. Displacement of the remaining chlorine by means of *N*-(2-hydroxyethyl)ethylenediamine then gives losoxantrone (**106**).[24,25]

An earlier congener, **bisantrene** (**110**), shares the linear tricyclic ring system and a wealth of nitrogen atoms with the "antrones." The unusual method used for adding a functional group that anchors to the side chains at the bridgehead positions hinges on the diene-like character of the central ring in anthracene. Thus Diels–Alder condensation of anthracene with vinylene carbonate leads to the key intermediate **107**. Hydrolysis with mild acid removes the carbonate group and affords the diol **108**. Cleavage by means of periodic acid leads to the 1,9-dialdehyde **109**. The carbonyl groups are then converted to the corresponding hydrazones by reaction with 2-hydrazinoimidazoline. Thus, the chemotherapy agent bisantrene (**110**)[26] is obtained.

REFERENCES

1. DeSolm, S.J.; Woltersdorf, D.W.; Cragoe, E.J.; Waton, L.S.; Fanelli, G.M. *J. Med. Chem.* **1978**, *21*, 437.
2. Shen, T.Y.; Winter, C.A. *Adv. Drug Res.* **1977**, *12*, 90.
3. Noguchi, T.; Hashimoto, Y.; Miyazaki, K.; Kaji, A. *J. Pharm. Soc. Jpn.* **1968**, *88*, 335.

4. Elpern, B.; Youlus, J.B. U. S. Patent **1970**, 3509200; *Chem. Abstr.* **1970**, *73*, 14546.
5. Melloni, P.; Rafaela, M.; Vechietti, V.; Logemann, W.; Castellino, S.; Monti, G.; De Carneri, I. *Eur. J. Med. Chem.* **1974**, *9*, 26.
6. Daniel, B. German Offen. **1977**, 2716943; *Chem. Abstr.* **1978**, *88*, 62215.
7. Stuetz, A.; Petranyi, G. *J. Med. Chem.* **1984**, *27*, 1539.
8. Condon, M.E.; Cimarusti, C.M.; Fox, R.; Narayanan, V.L.; Reid, J.; Sundeen, J.E.; Hauck, F.P. *J. Med. Chem.* **1978**, *21*, 913.
9. Reinhard, S. *J. Org. Chem.* **1975**, *40*, 1216.
10. Welch, W.M.; Kraska, A.R.; Sarges, R.; Koe, B.K. *J. Med. Chem.* **1986**, *27*, 1508.
11. Hoffsommer, R.D.; Taub, D.; Wendler, N.L. *J. Org. Chem.* **1962**, *27*, 4134.
12. Wintrop, S.O.; Davis, M.A.; Myers, G.S.; Gavin, J.G.; Thomas, R.; Barber, R. *J. Org. Chem.* **1962**, *27*, 230.
13. Engelhardt, E.L. *J. Med. Chem.* **1965**, *8*, 829.
14. Davis, M.A.; Winthrop, S.O.; Stewart, J.; Sunhara, F.A.; Herr, F. *J. Med. Chem.* **1963**, *6*, 251.
15. Van der Stelt, C.; Harms, A.F.; Nauta, W.T. *J. Med. Chem.* **1961**, *4*, 335.
16. Cavalla, J.F.; White, A.C. Br. Patent **1975**, 1406481; *Chem. Abstr.* **1976**, *84*, 4744.
17. Engelhardt, E.L. *J. Med. Chem.* **1968**, *11*, 325.
18. Andrews, E.R.; Fleming, R.W.; Grisar, J.; Kihm, C.; Wenstrup, D.L.; Mayer, G.D. *J. Med. Chem.* **1974**, *17*, 882.
19. Fowler, P.J.; Zirkle, C.L.; Macko, E.; Setler, P.E. *Arzneim.-Forsch.* **1977**, *27*, 1589.
20. Wilhelm, M.; Schmidt, P. *Helv. Chim. Acta* **1969**, *52*, 1385.
21. Wilhelm, M.; Bernasconi, R.; Storni, A.; Beck, D.; Schenker, K. German Offen. **1972**, 2207097; *Chem. Abstr.* **1973**, *78*, 4026.
22. Zeecheng, R.K.Y.; Cheng, C.C. *J. Med. Chem.* **1978**, *21*, 291.
23. Murdock, K.C.; Child, R.G.; Fabio, P.F.; Angier, R.B.; Wallace, R.E.; Durr, F.E.; Citavella, R.V. *J. Med. Chem.* **1979**, *22*, 1024.
24. Showalter, H.D.H.; Johnson, J.L.; Hoftiezer, J.M.; Turner, W.R.; Werbel, L.M.; Leopold, W.R.; Shillis, J.L.; Jackson, R.C.; Elslager, E.E. *J. Med. Chem.* **1987**, *30*, 121.
25. Beylin, V.G.; Colbry, N.L.; Goel, L.P.; Haky, J.E.; Johnson, D.R.; Johnson, J.L.; Kanter, G.D.; Leeds, R.L.; Leja, B.; Lewis, E.P.; Ritner, C.D.; Showalter H.D.H.; Serces, A.O.; Turner, W.P.; Uhlendorf, S.E. *J. Heterocycl. Chem.* **1989**, *26*, 85.
26. Murdock, K.C.; Child, R.G.; Lin, Y.; Warren, J.D.; Fabio, P.F.; Lee, V.J.; Izzo, P.T.; Lang, S.A.; Angier, R.B.; Citavella, R.V.; Wallace, R.E.; Durr, F.E. *J. Med. Chem.* **1982**, *25*, 505.

CHAPTER 4

STEROIDS
PART I: ESTRANES, GONANES, AND ANDROSTANES

1. INTRODUCTION

The fused tetracyclic steroid nucleus provides the carbon framework for at least four large groups of mammalian hormones; these hormones comprise the estrogens, androgens, progestins, and corticosteroids. In addition, the parent molecule, cholesterol, plays an important structural role in many membranes. This nucleus is interestingly ubiquitous in nature and occurs in both plant and animal species. The fact that it is a product of one of the isoprene pathways may in part explain this widespread distribution. The pathway for formation of cholesterol has been shown to closely follow the initial steps leading to the formation of the pentacyclic C_{30} triterpenes; the key common intermediate, squalene, is formed by head-to-head coupling of two activated C_{15} farnesol derivatives (arrow in scheme below). Cyclization of this polyene, which has been shown to proceed via an epoxide at the 2,3 terminal olefin, leads to an intermediate that contains the *cyclo*pentanoperhydrophenanthrene nucleus. Two successive methyl group migrations then give the intermediate lanosterol, which has been isolated. A series of catabolic enzymes next remove the remaining three superfluous methyl groups. The product, cholesterol, serves as the metabolic starting material for endogenous synthesis of all other steroids.

The bulk of the work on structure determination of steroids, long predated the availability of any of the modern instrumental methods. The structures of the compounds were worked out entirely by classical noninstrumental methods which relied largely on degradation reactions to known simpler compounds, Rast molecular weight

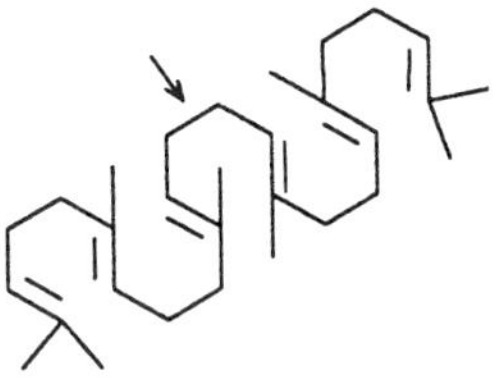

Squalene 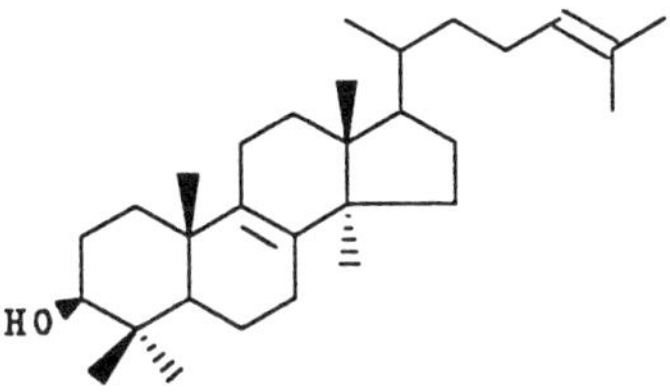 Lanosterol 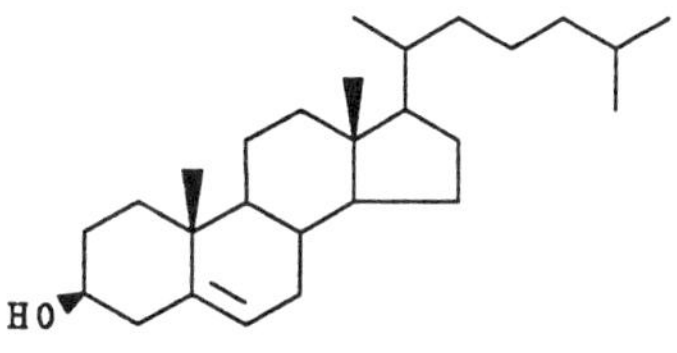 Cholesterol

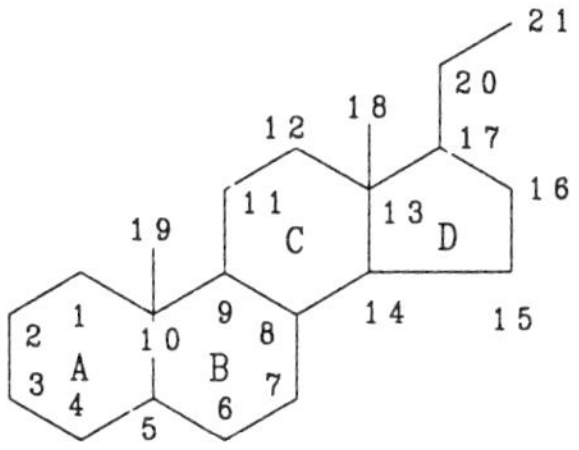

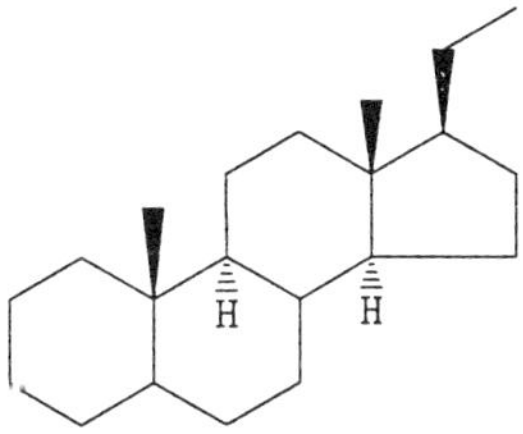

determinations and elemental analysis. Elegant chemical reasoning played an enormous role in this very difficult work. The somewhat idiosyncratic numbering system used in steroid chemistry is traceable to the fact that it was devised before the full structure of the nucleus was known.

All naturally occurring steroids, and the overwhelming majority of those used as drugs, consist of single enantiomers. When written as above, the diagrams depict the absolute configuration. Substituents below the plane of these rather flat molecules are named α while those above that face are named β, a convention that later spread to many other structural classes.

2. STEROID STARTING MATERIALS

The initial pharmacologic and clinical investigations on steroid hormones relied on supplies of those compounds that were isolated from mammalian sources. The difficult isolations involved in obtaining those supplies combined with the drive to prepare chemically modified entities led to the search for more abundant and easily accessible steroid starting materials among plant sterols. The structural and stereochemical complexity of these compounds largely rule out total syntheses. The elegant syntheses, which have been published for all steroid hormones, are not competitive with semi-syntheses from natural products. Selected estranes and gonanes, which are described below, however, are more readily prepared by total synthesis since they include modifications not accessible from plant sterol derived intermediates.

Note that one commercially important class of steroid drugs is still obtained by isolation from mammalian sources. These drugs comprise the conjugated estrogens, which consist of a mixture of 3-sulfate esters of estrone, estradiol, and other estranes indicated as postmenopausal estrogen supplements.

A. From Diosgenin

The glycoside dioscin from the Mexican wild yam root *Dioscorea* constituted the first plant source for steroid drugs. Hydrolysis of this saponin leads to scission of the trisaccharide at the 3-position and formation of the aglycone, diosgenin (**1**). Treatment of this acetal with hot acetic anhydride in the presence of *p*-toluenesulfonic acid (TSA) leads initially to protonation of one of the acetal oxygen atoms; this oxygen is then eliminated to form an enol ether. The free hydroxyl groups are acetylated under the reaction conditions to give the diacetate **2** as the product. Oxidation by means of chromium trioxide leads to preferential attack at the electron-rich enol ether double bond and formation of **3**. This transformation in effect converts the extended side chain at C-17 in diosgenin to the acetyl group required for many steroid drugs. Heating this intermediate in acetic anhydride leads to elimination of the ester grouping β to the ketone; thus 16-dehydropregnenolone acetate (**4**) is obtained.[1] As subsequently shown, the presence of the olefin at 17 allows ready entry to C-19 androstanes and provides the necessary function for synthesis of potent 16- and 16,17-substituted corticosteroids.

B. From Soybean Sterols

The chemistry for providing steroid raw materials from the far more readily available soybean sterols was developed somewhat later. The so-called unsaponifiable fraction of soybean oil consists largely of stigmasterol (**5**) and sitosterol (**10**). The original process took advantage of the olefin in the side chain of the former for degrading this chain to an acetyl group. The first step involved extraction of stigmasterol from the mixture by an ingenious leaching process. The 3-hydroxy-5-ene functionality is then

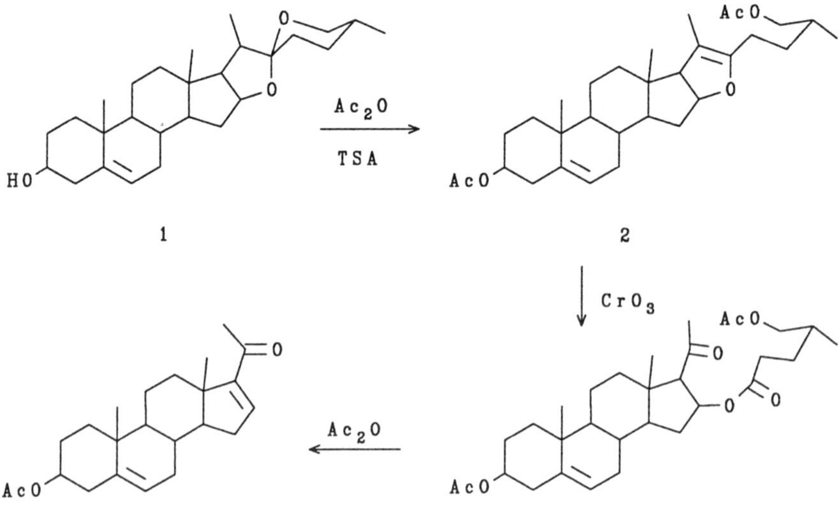

5 $\xrightarrow[\text{Cyclohexanone}]{Al(iOPr)_3}$ 6 $\xrightarrow{O_3}$ 7 $\xrightarrow{\text{HN}}$ 8 $\xrightarrow[h\nu]{O_2}$ 9

10 $\xrightarrow{\textit{Pseudomonas}}$ 11

converted to a conjugated enone in order to reduce its reactivity in subsequent steps. Thus, Oppenauer oxidation (cyclohexanone, and aluminum isopropoxide) of (**5**) leads to the enone **6**. Ozonization of **6** followed by oxidative workup of the ozonide leads to the aldehyde **7** in which the bulk of the side chain has been removed. The extraneous carbon atom is then removed by way of its enamine **8**, which is obtained by reaction of **7** with piperidine; photooxygenation of **7** leads directly to progesterone **9**.[2,3]

The byproduct sitosterol was for many years quite useless due to the lack of a chemical point of attack on the side chain that would permit its removal. Extensive efforts on the part of many laboratories led to the discovery of a *pseudomonas* microbe that efficiently effected this transformation; thus, fermentation of **10** gives a mixture of 17-keto products including dehydroepiandrosterone (**11**).[4]

3. ESTRANES

A. Synthesis of Estranes

The estranes constitute the simplest steroids in terms of structure since the aromatic A ring removes several chiral centers. The endogenous compounds, estrone **17** and the corresponding alcohol, estradiol **18**, play a central role in female reproductive endocrinology. Medical uses of pure estrogens or their derivatives are largely restricted to replacement therapy in those cases where endogenous levels are deficient; the use of conjugated estrogens to relieve postmenopausal symptoms has already been alluded to. Estranes are, however, used extensively as components of oral contraceptives and as starting materials for the so-called 19-nor steroids (gonanes).

The synthesis of estranes perforce invokes the need to perform an aromatization step that involves the unlikely expulsion of a methyl group due to the lack of abundant sources of natural products that have an aromatic A rings. The starting dehydroepiandrosterone (**11**) may be obtained from fermentation as noted above or by degradation of 16-dehydropregnenolone.[5] Oxidation of this starting material under Oppenauer conditions gives androstene-3,17-dione (**12**). Catalytic reduction of the enone proceeds from the unhindered side to give the 5α dihydro derivative (**13**); this compound is then allowed to react with bromine in acetic acid. It has been shown that production of the observed 2α,4α-dibromo derivative in fact involves a complex series of intermediate reactions. Dehydrohalogenation of the product with 2,4,6-collidine leads to the 1,4-dienone **14**. This compound can be aromatized directly to estrone (**17**), albeit in very poor yield, by pyrolysis in mineral oil at 600°C.[6] A more recent method essentially involves elimination of the quaternary methyl group as methyllithium. The first step in this sequence consists of protection of the ketone at 17 as its propylene acetal **15**. This compound (**15**) is then heated with lithium metal in tetrahydrofuran (THF); the diphenylmethane present in the reaction mixture presumably quenches the methyllithium to prevent its adding to the starting dienone. Thus, the estrone derivative **16**[7] is obtained.

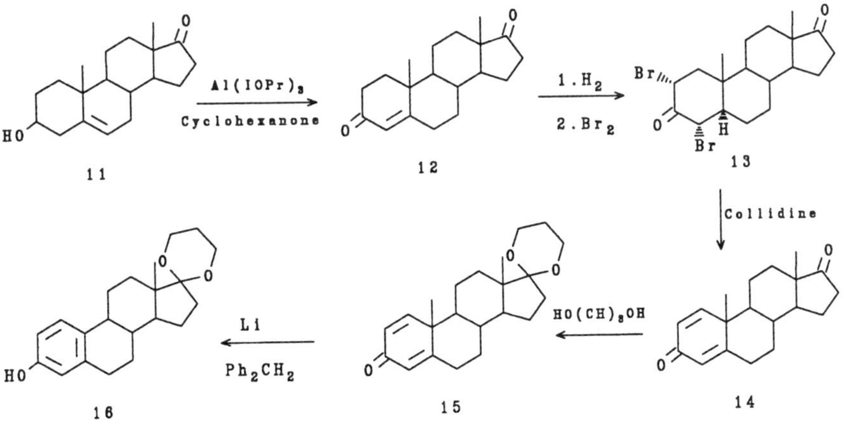

Hydrolysis of the acetal group in **16** then gives estrone (**17**). The 17 ketone can be reduced by any of several methods, such as reaction with borohydride, to give the 17β-alcohol **estradiol** (**18**).

The so called Torgov–Smith synthesis[7,8] provides a commercially practical route for the synthesis of estranes, particularly when combined with resolution of early intermediates. This approach is in fact mandatory for the synthesis of **norgestrel** (**59**), a gonane that bears an ethyl rather than methyl group at the C-13 position. The synthesis begins by addition of the Grignard reagent from vinyl chloride to the tetralone **19**. The key step involves condensation of the allylic alcohol product (**20**) with the cyclopentadione **21** in the presence of a catalytic amount of base. It is likely that this step, which gives the appearance of a simple displacement, in fact depends on the initial formation of an ambident allyl carbocation from **20**, which is catalyzed by the very acidic dione; addition of the enolate from **21** will give the observed product **22**. Intermediate **22** is then treated with TSA; cyclization leads to the tetracyclic intermediate **23**. The structure of steroids is notable for the fact that the CD ring fusion involves a thermodynamically disfavored cis hydrindane. The next step in the synthesis fortuitously establishes that stereochemistry; catalytic hydrogenation selectively reduces the 14,15 olefin; the addition of hydrogen from the more open α side of the molecule establishes the CD trans ring fusion. The ketone at the 17-position is then reduced to an alcohol with sodium borohydride. The remaining styrenoid olefin at the 9,10-positions in compound **25** is then removed by Birch reduction; it is likely that this reaction stereochemistry is determined by thermodynamics. Thus the 3-methyl ether of estradiol (**26**) is obtained.

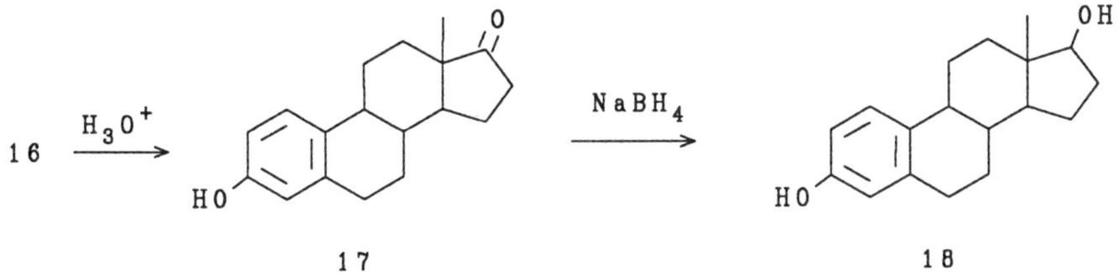

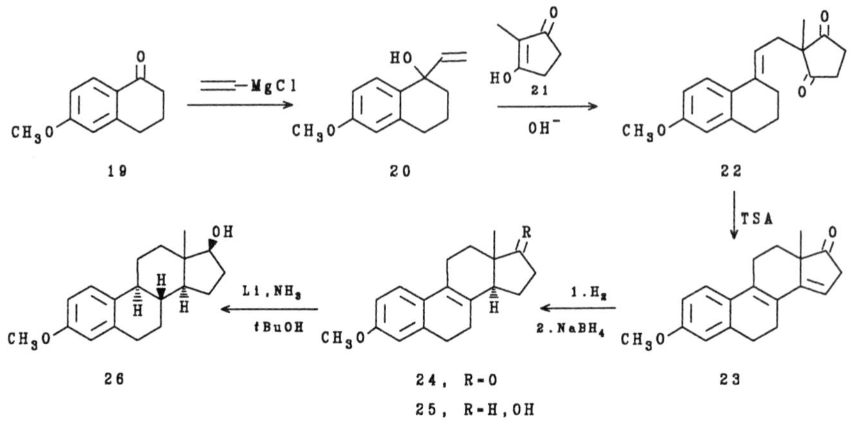

B. Drugs Based on Estranes

The finding that estrogens improved the endocrine status of postmenopausal women offered a much larger market for these compounds beyond simple replacement therapy. This indication has, however, been surrounded by controversy for several decades due to the conflicting risk/benefit ratios revealed by rival studies. The market is largely dominated by the so-called conjugated estrogens, the 3-sulfate esters from natural sources since estrone and estradiol are not suitable for oral administration due to poor absorption and ready metabolism. That ready metabolism also obtains when the latter are administered parenterally. This shortcoming is partly overcome by more lipophilic derivatives of estradiol that may themselves be devoid of estrogenic activity. These compounds form a depot of drug when injected subcutaneously in an oily medium. As the esters slowly leach into the circulation, endogenous esterases convert them to estradiol, thereby providing relatively long-lasting levels of the biologically active compound. These fatty esters are prepared by straightforward esterification of estradiol with the acid chlorides of the desired acids. Relatively strenuous conditions are required because 17β alcohol is sterically quite hindered by the adjacent 18 methyl group; the 3,17-diacyl derivatives (**27**,**28**) are consequently the initial products from this reaction. Exposure to mild aqueous base leads to saponification of the labile phenolic ester at the 3-position. The sequence with benzoyl chloride leads to **estradiol benzoate (29)**; cyclopentylpropionic acid gives **estradiol cypionate (30)**.

Absorption of orally administered, relatively lipophilic compounds, such as estrone or estradiol, occurs mainly in the intestine. The bacteria that colonize the gut are, however, particularly adept at converting those compounds to very water soluble derivatives that defy absorption by attack at the 17-position. Alkylation of this position avoids this catabolic pathway and consequently enhances bioavailability on oral administration. Reaction 17-keto steroids with nucleophiles illustrates the high degree of stereospecifity that maintains in many steroid reactions; approach of this carbonyl group from the β face is virtually forbidden by the presence of the adjacent 18 methyl; the reaction products consequently consist of almost pure isomers from attack at the α face. Reaction of estradiol with lithium acetylide thus gives **ethynylestradiol (32)**;[9]

18 $\xrightarrow{RCOCl}$ 27; $R = C_6H_5$ / 28; $R = CH_2CH_2cC_5H_9$ $\xrightarrow{OH^-}$ 29; $R = C_6H_5$ / 30; $R = CH_2CH_2cC_5H_9$

the corresponding alkylation of estradiol 3-methyl ether (**31**) leads to **mestranol** (**33**).[10] Both compounds are potent orally active estrogens; the latter, as noted later, finds widespread use as a component of oral contraceptives. In a similar vein, reaction of estradiol with the lithium reagent from 2-iodofuran gives the adduct **34** as a single isomer. Acetylation with acetic anhydride leads to the orally active estrogen **estrofurate 35**.[11]

Only very subtle differences exist between cancer cells and their normal predecessors. The small divergences in genetic material and regulatory factors have so far proven beyond the reach of therapy. The bulk of cancer chemotherapy thus relies largely on kinetics; that is, on the fact that cells within neoplasms multiply far more rapidly than do normal cells. The majority of chemotherapeutic agents disrupt DNA in multiplying cells and thus affect them in preference to resting tissue. The so-called nitrogen mustard alkylating agents, which include a dichloroethylamino group, have

17; $R = H$ / 31; $R = CH_3$ $\xrightarrow{LiC\equiv CH}$ 32; $R = H$ / 33; $R = CH_3$

34 $\xrightarrow{Ac_2O}$ 35

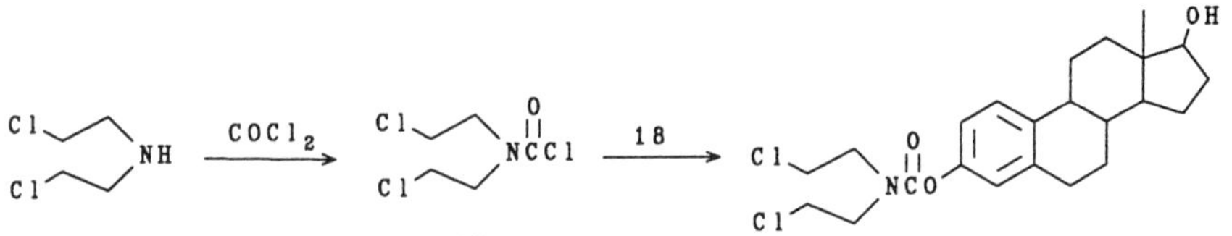

in fact been shown to react directly with DNA. Various attempts have been made over the years to achieve some organ specificity by attaching mustards to compounds that interact with tissue specific receptors. The chemotherapeutic agent **estramustine (38)** is intended to steer an alkylating moiety to estrogen sensitive tissue by invoking the affinity of estradiol for estrogen receptors. The synthesis starts with the reaction of mustard **36** with phosgene to give carbamoyl chloride **37**. Reaction of **37** with estradiol in the presence of base leads to the product **38**.[12] The preferential formation of a phenoxide over an alkoxide combined with the more hindered nature of the 17 alcohol contributes to the selectivity.

4. GONANES: THE 19-NOR STEROIDS

Steroids based on the estrane nucleus in which the formerly aromatic A ring occurs in reduced form are named as derivatives of the hypothetical hydrocarbon, gonane. This very large class of compounds, which are more familiarly known as the 19-nor steroids, has no counterpart in nature. It is thus noteworthy that this class includes some very potent androgens, progestins, and most recently progesterone and corticosteroid antagonists. The original investigations in this structural class were spurred by the increasing availability of estrone as the starting material and the discovery by A. J. Birch of the eponymous method for reducing aromatic rings by means of dissolving metals in liquid ammonia.[13]

A. Progestational Compounds

It was recognized by the 1940s that the high levels of circulating progesterone (**39**) that existed once pregnancy was established would inhibit subsequent ovulation. This theory was confirmed in rabbits where ovulation could be inhibited by injecting progesterone. The very poor absorption of orally administered progesterone frustrated attempts to use this as a method for inhibiting ovulation. The search for an oral contraceptive thus focused on finding orally active progestins.

The observation that the ethynyl derivative of testosterone, **ethisterone** (**40**), exhibited progestational activity demonstrated that the acetyl side chain at the 17-position was not an absolute requirement; it should be noted that the side chain in **40** has the opposite configuration from that of the acetyl in progesterone. Oral progestational activity had also been reported for 19-norprogesterone, which was obtained in minuscule yield by a lengthy degradation of a plant sterol. In one of those not infrequent

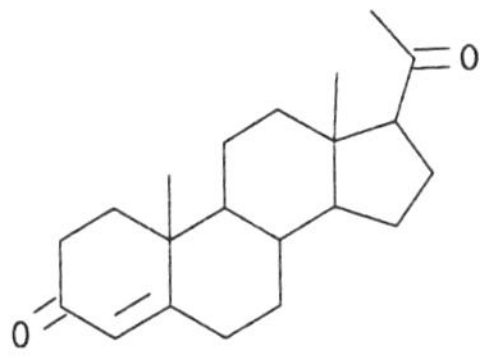

39

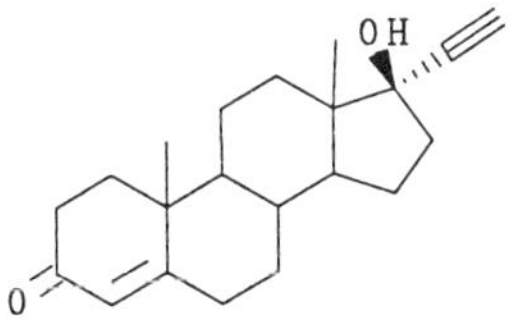

40

coincidences, chemists working quite independently in the laboratories of Searle and Syntex took these cues as incentive for the preparation of 17-ethynyl-19-norsteroids. The compounds **norethynodrel** (**45**) and **norethindrone** (**46**) both eventually found widespread use as the progestational component of oral contraceptives.

Treatment of the methyl ether (**41**) of estradiol with lithium in liquid ammonia under the conditions of the Birch reduction leads initially to a radical dianion at the 1,4-positions, since these are the sites of lowest electron density. Neutralization of the charges by addition of ethanol gives the dihydrobenzene derivative **42** in which the future 3 keto group is present as its enol methyl ether. The hydroxyl group at the 17-position is then carefully oxidized using slightly basic Oppenauer conditions to give **43**. Condensation with lithium acetylide occurs with addition of the carbanion to the more open α face of the molecule to give the carbinol **44**. Hydrolysis of the enol ether under mild conditions (e.g., oxalic acid) gives the unconjugated ketone **norethynodrel** (**45**).[14] Hydrolysis of the ethynylation product **44** with mineral acid leads the double bond to shift to the conjugated 4,5-position and formation of **norethindrone** (**46**).[15]

Clinical trials of these orally active progestins showed that they were effective as contraceptives with a success rate that exceeded 99%. These compounds were then marketed as obtained from the reaction sequence after appropriate purification. As analytical methodology improved, it became apparent that a small amount of an impurity was present in all active samples. An examination of the reaction scheme allowed ready identification of this byproduct. Any unreduced estradiol methyl ether (**41**) will go to estrone methyl ether on oxidation; this compound will then afford the potent orally active estrogen mestranol (**33**) on ethynylation. Subsequent investigation of mestranol free norethynodrel suggested that estrogen was required for full efficacy. Virtually all commercial oral contraceptives consequently consist today of a progestational component admixed with a small controlled amount of mestranol or **ethynylestradiol**.

Oral progestational activity is retained when the ketone at the 3-position is reduced to an alcohol. Treatment of norethindrone (**46**) with the bulky reducing agent lithium aluminum-tri-*tert*-butoxy hydride leads to attack from the more open α face and formation of the 3β-hydroxy derivative **47**. Acylation under forcing conditions affords the 3,17-diacetate derivative **ethynodiol diacetate** (**48**).[16]

The compound in which the 3 keto group is reduced to a hydrocarbon interestingly still acts as an orally active progestin. The preparation of this compound starts with hydrolysis of dihydrobenzene (**42**) to afford 19-nortestosterone (**49**). Reaction with

CH_3O OH **41** $\xrightarrow[\text{EtOH}]{\text{Li, } NH_3}$ CH_3O OH **42** $\xrightarrow[\text{cyclohexanone}]{Al(iPrO)_3}$ CH_3O O **43**

43 $\xrightarrow{LiC{\equiv}CH}$ CH_3O OH **44**

44 $\xrightarrow{H_3O^+}$ O OH **45**

45 $\xrightarrow{H_3O^+}$ O OH **46**

46 $\xrightarrow{Al(tBuO)_3H}$ HO OH **47** $\xrightarrow[\text{TSA}]{Ac_2O}$ AcO OAc **48**

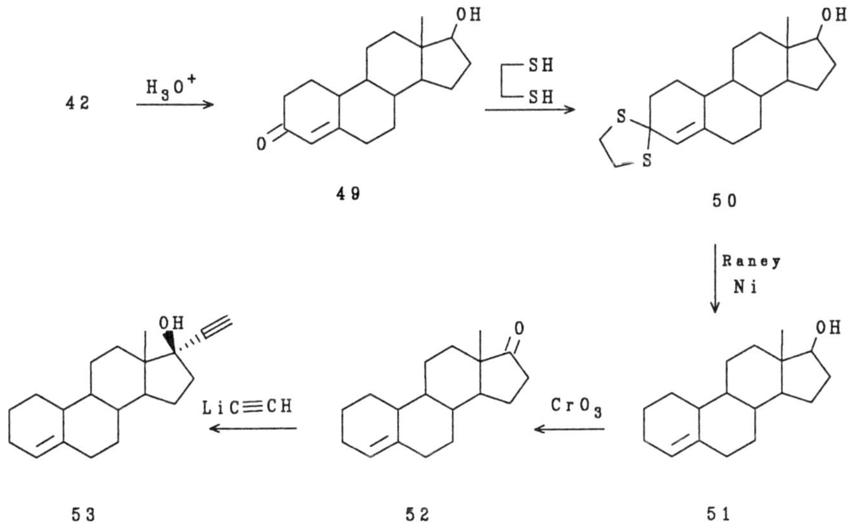

42 $\xrightarrow{H_3O^+}$ 49 $\xrightarrow{HS-CH_2CH_2-SH}$ 50 $\xrightarrow{\text{Raney Ni}}$ 51 $\xrightarrow{CrO_3}$ 52 $\xrightarrow{LiC\equiv CH}$ 53

ethane-1,2 thiol in the presence of catalytic acid leads to the cyclic thioacetal **50**. Treatment of **50** with Raney nickel in the presence of alcohol leads to the reduced desulfurized derivative **51**. The alcohol at 17 is then oxidized by any of several methods such as chromic acid in acetone (Jones' reagent), and the resulting ketone (**52**) is treated with lithium acetylide. Thus, the progestin **lynestrol (53)**[17] is obtained.

In much the same vein as above, hydrolysis of the enol ether **42** with weak acid gives the unconjugated 5,9 olefin **54**. This compound is then successively converted to a thioacetal, desulfurized, oxidized, and ethynylated. Thus, the orally active progestin **tigestol** (**55**)[17] is obtained.

The oral progestin **norgestrel** (**59**) occupies an unusual place among steroid drugs in that it incorporates an angular ethyl group at the 18-position, a modification not accessible from the usual steroid starting materials; it is also one of the rare racemic steroid drugs. This compound is prepared by a variation on the Torgov–Smith scheme. Thus, reaction of vinyl tetralol **19** with 2-ethylcyclopenta-1,3-dione (**56**) instead of the methylated dione gives the homologue **57** of the more usual intermediate. This compound is then subjected to the usual cyclization and reduction sequence to give homologue **58** of estradiol 3-methyl ether. Intermediate **58** is then taken through the same sequence of steps used to prepare norethindrone (Birch reduction, oxidation, ethynylation, and hydrolysis) to afford norgestrel (**59**).[18]

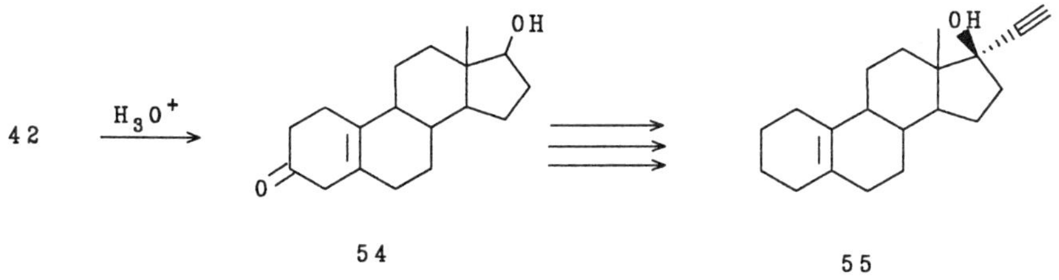

42 $\xrightarrow{H_3O^+}$ 54 $\Rrightarrow$ 55

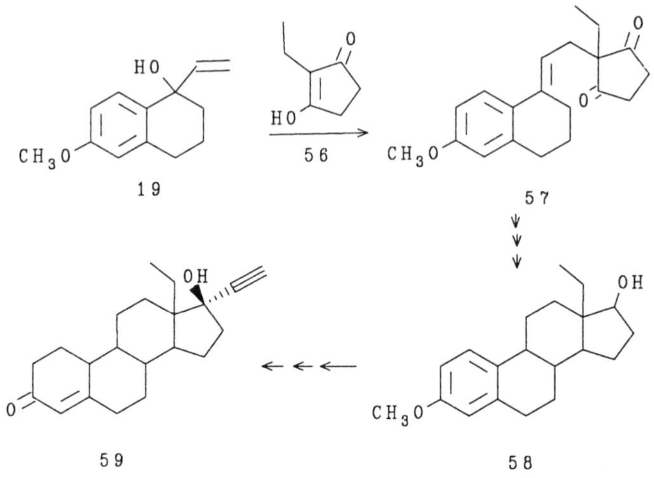

The almost immediate commercial success of the gonane based oral contraceptives spurred an impressive amount of work in this series, which was aimed at developing new market entries. Describing each of these contraceptives in detail is beyond the scope of this volume; instead, some examples chosen simply because their synthesis involves some interesting chemistry are presented below.

The key intermediate (**60**) in the preparation of **gestodene** (**64**) is obtained by Birch reduction of the methyl ether of 18-methylestradiol from total synthesis, followed by hydrolysis to the conjugated ketone, and subsequent oxidation to a diketone. The distinguishing feature of this compound, unsaturation at the 15,16, positions is introduced by biotransformation. Thus, fermentation of diketone **60** with *Penicillium raistricki* introduces a hydroxyl group at the 15α-position. Reaction of **61** with neopentyl glycol proceeds to form an acetal of the ketone at the sterically more accessible 3-position; this product consists of a mixture of unrearranged 4,5 olefin and the product in which the olefin has shifted to the 5,9-position. The alcohol at the 15-position is then converted to its mesylate (**62**). Treatment with a base such as triethylamine results in elimination to give the conjugated ketone **63**. Addition of lithium acetylide followed by removal of the acetal by hydrolysis affords gestodene (**64**).[19]

Ethynerone (**68**) differs from the prototype progestins in that it incorporates a chloroacetylene moiety as well as an extended conjugated diene. Reaction of *cis*-1,2-dichloroethylene with lithium metal involves elimination of the elements of hydrogen chloride and formation of chloroacetylene as the first step; that product again reacts with lithium to form the anion of chloroacetylene. Addition of that anion to the 17 ketone in the norethynodrel intermediate **43** gives the corresponding carbinol **65**. Hydrolysis of the acetal under mild acidic conditions affords the unconjugated ketone **66**. This intermediate is then treated with bromine and pyridine. The first step almost certainly involves addition of bromine across the 5,9 double bond to give a trans dibromide (diaxial opening of the initial α bromonium should give the 5α,10β isomer).

58
1. Birch
2. H_3O^+
3. CrO_3
60
Pen. R.
61
1.
2 MsCl
62
OSO_2Me
Et_3N
63
1. LiC≡CH
2. H_3O^+
64

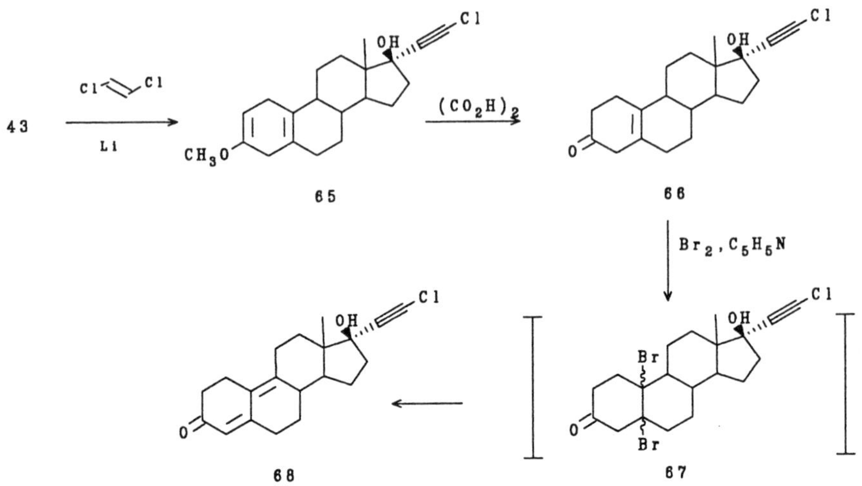

This compound loses hydrogen bromide in the presence of pyridine to give a diene and thus ethynerone (**68**).[20] The stability of this compound is surprising because it is at the oxidation state of a phenol.

The 5,9,10,11-gonane **78** was first prepared as an intermediate to a 4,9,11-triene progestin, **norgestrienone**, which was never commercialized in the United States. The synthesis is, however, of interest since **78** serves as starting material for two progesterone antagonists discussed later in this chapter. The total synthesis used to prepare this compound differs from the Torgov–Smith route in that three of the rings are constructed sequentially; the presence of an intermediate carboxylic acid in the scheme permits resolution at a relatively early stage. The synthesis starts with Robinson-type annulation of cyclopentadione with the methylvinyl ketone derivative **69** to give the hydrindene **70**. Catalytic reduction of the double bond establishes the all important trans stereochemistry for the CD ring fusion. The fragment that will form ring B is attached to an enolizeable position that will be altered later in the synthesis.[21] The cyclopentane carbonyl group in product **71** is reduced to an alcohol, while temporarily protecting the more accessible cyclohexanone as its acetal. Acylation of the newly formed hydroxyl affords benzoate **72**; this compound is then converted to cyclic enol lactone **73** with acetic anhydride. Condensation of the enol lactone with the Grignard reagent from protected 1-bromo-4-pentanone (**74**) affords a product (**75**) from a reaction that formally consists of replacement of ring oxygen by carbon.

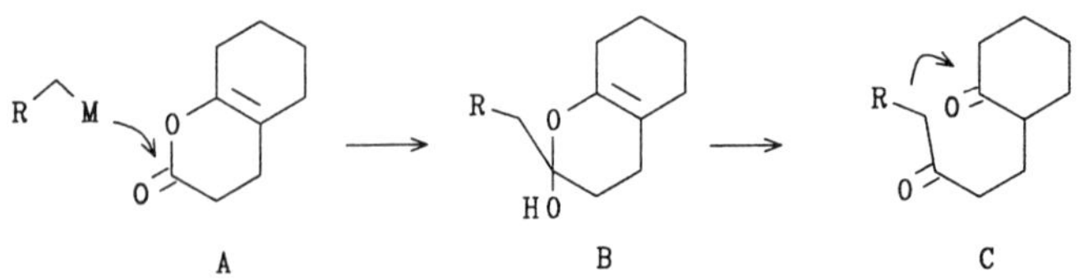

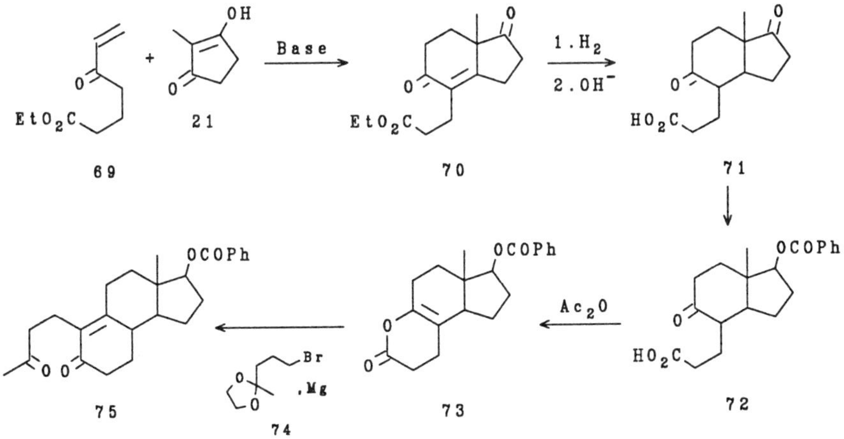

The sequence can be rationalized by assuming that only a single organometallic adds to **A** to give the lactol **B** as a magnesium salt. The usual acid workup of the product will then proceed at least transiently through lactol **B** to give diketone **C**. Acid cyclization will then afford a cyclohexanone; in this case, further hydrolysis of the side chain protecting group leads to the observed product **75**.

Reaction of the diketone (**75**) with piperidine leads to formation of an enamine **76** at the side chain carbonyl group. The nucleophilic terminal methine then adds to the ring carbonyl to effect cyclization; the double bond presumably shifts in the course of the reaction; enamine **77** is thus obtained as an isolable intermediate. Acid catalyzed hydrolysis of the enamine affords the desired steroid intermediate as a single enantiomer (**78**).[22]

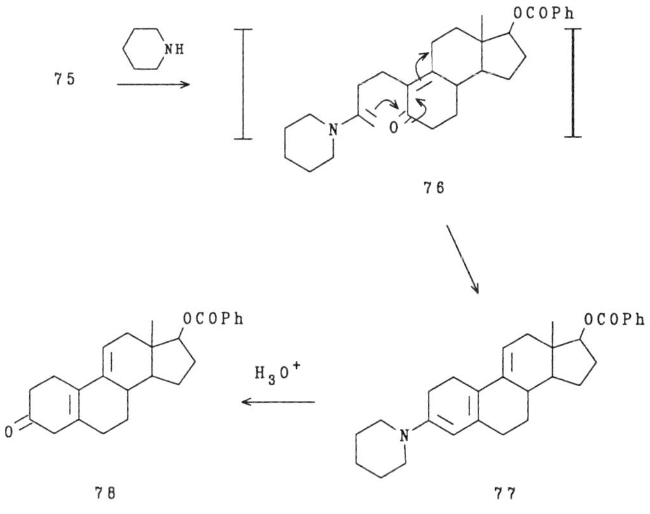

We have already seen that **ethisterone** where the side chain at the 17-position consists of an ethynyl group in the α-position combined with a β hydroxyl has the same biological activity as progesterone, which has a β-acetyl side chain. It is interesting that this analogy carries over to the gonane series as does the improved potency noted in the natural series when an α hydroxyl is added to progesterone. The synthesis of the gonane analogue starts with the diosgenin degradation product **4**. Saponification of the acetyl group at the 3-position followed by Oppenauer oxidation of the resulting alcohol affords 16-dehydropregnenolone (**79**). This compound is then converted to the 1,4-dienone (**80**) by the usual sequence involving bromination and dehydrohalogenation. This compound is then converted to its A-ring aromatic counterpart by one of several methods, such as treatment with lithium metal in the presence of diphenylmethane as a buffer. Methylation of the newly formed phenol by means of methyl iodide and base affords the methyl ether **81**. The 15,16 double bond of the enone is then converted to its epoxide (**82**) by means of basic hydrogen peroxide; the less hindered milieu of the α side determines the stereochemistry of epoxide formation. The first step in the reaction with hydrogen bromide involves protonation of the epoxide oxygen. Examination of this intermediate suggests that attack by bromide at the 16-position is highly favored for steric reasons; the fact that the product (**83**) consists of the 16β-bromo-17α-hydroxy derivative bears out this supposition.

The superfluous bromine is then removed by reduction with zinc in acetic acid. The 20 ketone is next protected against the strongly reducing conditions in the subsequent step by conversion to the ethylene glycol acetal (**85**). Birch reduction with lithium in liquid ammonia in the presence of ethanol proceeds as usual to the dihydrobenzene **86**. Treatment of **86** with mineral acid serves to hydrolyze both the enol ether at the 3-position and the acetal at the 20-position; the double bond shifts into conjugation as well. Thus, **gestonerone** (**87**) is obtained.[23] This compound is used in the clinic as its 17-caproate ester.

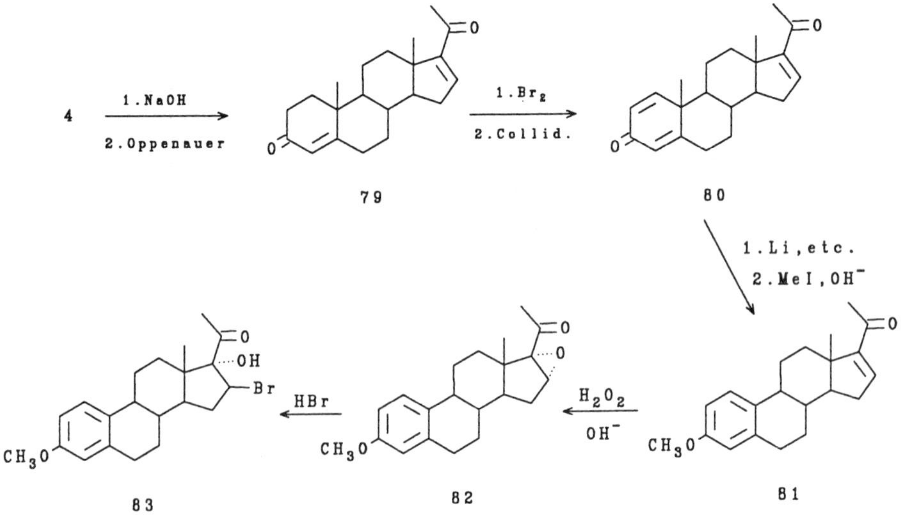

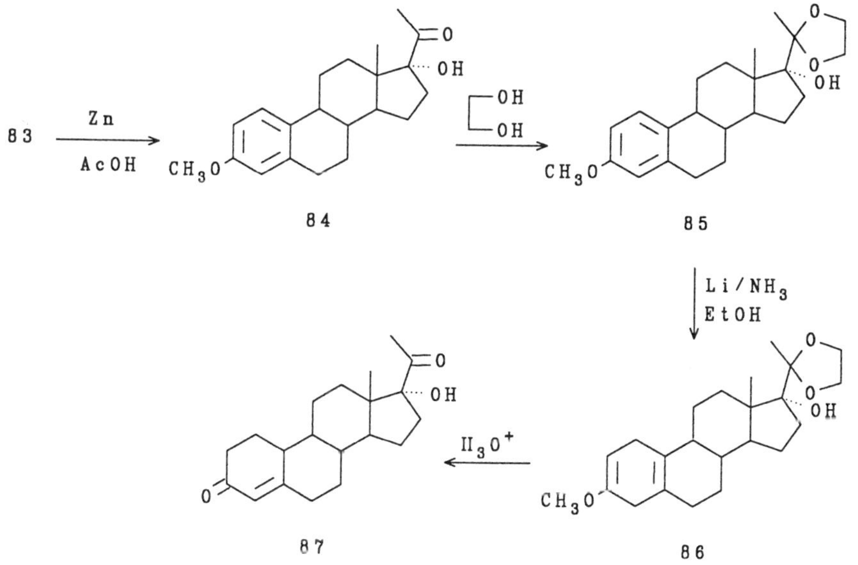

B. Progesterone Antagonists

Important drug classes have often come from the ranks of compounds that show antagonist activity against endogenous hormones or messenger substances; the β-blockers, which provide a prime example, have been discussed in Chapter 2. Until quite recently, the only examples of compounds that blocked steroid action were the estrogen antagonists; those compounds whose structures only superficially mimic steroids, are discussed in Chapter 6. The discovery of the first progesterone antagonists, which also shows some antiglucocorticoid activity, are derived from intermediates for totally synthetic gonane progestins.[24] The first clinically successful progesterone antagonist, **mifepristone** (**94**), has attracted public attention under the manufacturer's code name RU-486. Considerable controversy has been generated by the use of the drug as an abortifacient. Whereas the progestin agonists prevent ovulation, the antagonists deprive the lining of the uterus of the progestational stimulation required for successful implantation and maintenance of a fertilized ovum that in effect results in early abortion. This reaction is aided, in practice, by subsequent administration of a prostaglandin; the latter typically consists of misoprostol (see Chapter 1).

The synthesis of the first of these antagonists, mifepristone (**94**), starts by conversion of intermediate **78** to the corresponding 3,17 diketone by sequential saponification of the benzoate at 17 and oxidation of the resulting alcohol. Reaction of this compound with ethylene glycol proceeds preferentially at the 3-position to give the 3-ethylene acetal; treating this with trimethylsilyl cyanide (TMSCN) leads to formation of the silylated cyanohydrin at the 17-position to afford **88**. Oxidation of **88** with the adduct from hexafluoroacetone and hydrogen peroxide occurs selectively at the 5,9 double bond affording the α-epoxide **89**. Attack from the somewhat less hindered face of **88** accounts for the stereochemistry; the regiochemistry may be controlled by the proximity of acetal oxygen. Reaction with the Grignard reagent from 4-bromodimethylaniline results in what can be viewed as a vinylogous oxirane opening to give **91**. The organometallic thus approaches the 11 terminal of the olefin from the same side as would be required for trans diaxial opening of the oxirane; or, more probably, a magnesium salt complex of the oxirane. Bond reorganization will then lead to the observed product **91**.

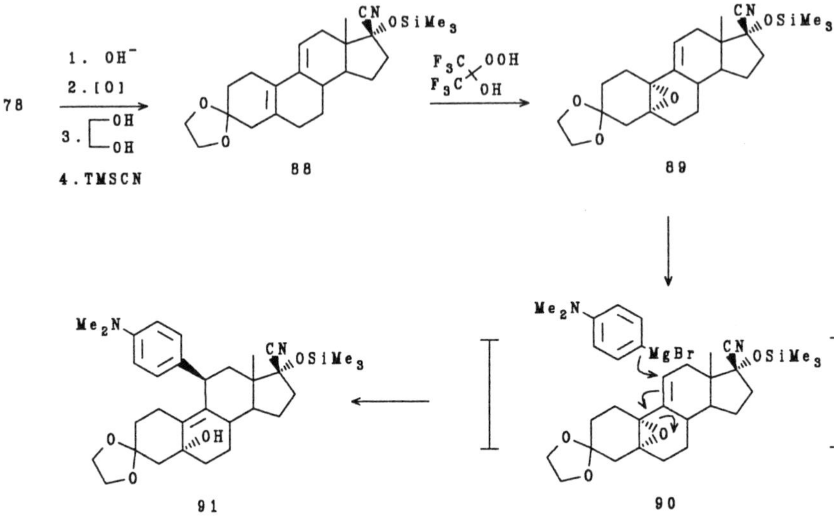

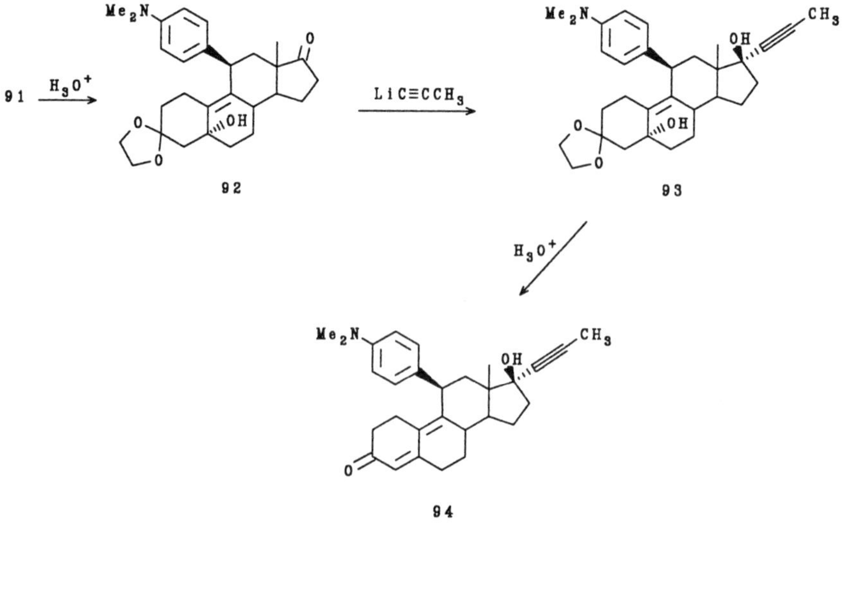

The protecting group at the 17-position is then removed by hydrolysis under weakly acidic conditions to give the corresponding 17 ketone **92**. This compound is then reacted with propargyl lithium to afford the diol **93**. Hydrolysis of **93** in the presence of strong acid first removes the acetal group at the 3-position; the resulting β-hydroxy-alcohol dehydrates to give the 4,9 conjugated dienone. Thus, mifepristone (**94**)[25] is obtained.

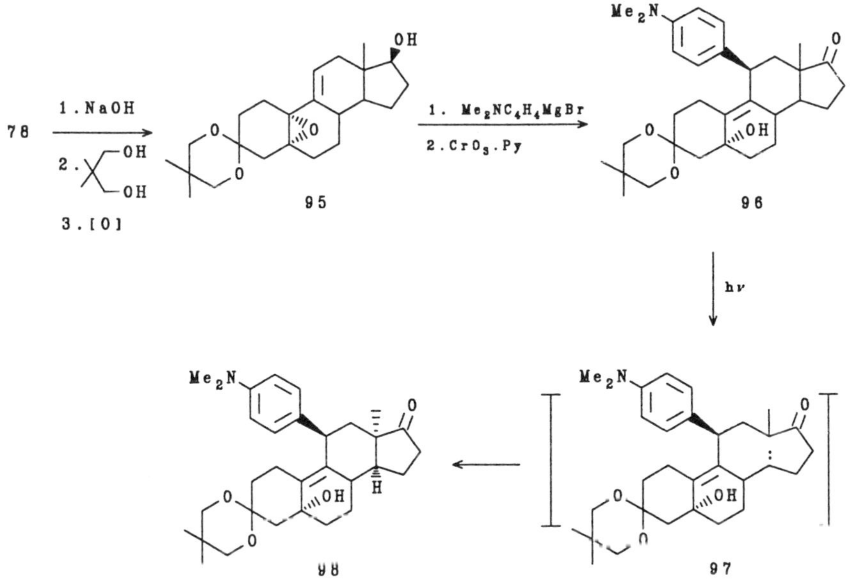

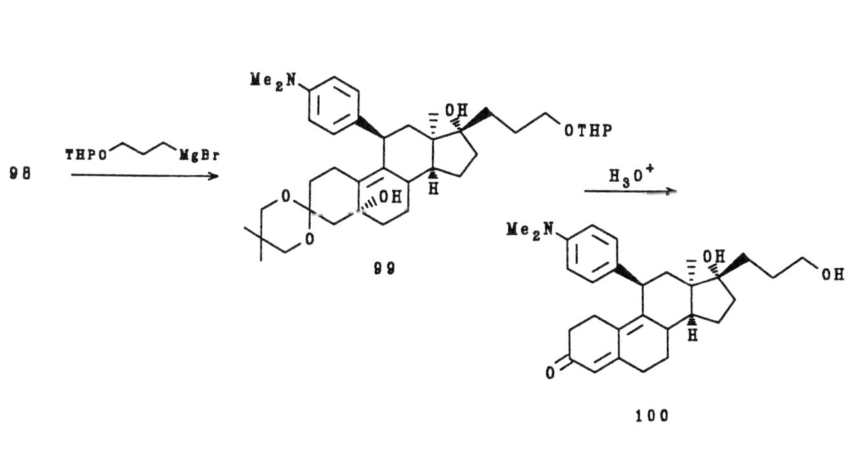

The synthesis of the progestin antagonist **onapristone** (**100**) also begins with the saponification product from **78**. In this case, the intermediate is first converted to a 3,3-dimethylpropylene acetal by reaction with neopentyl glycol; oxidation as shown above with the hydroxy hydroperoxy acetal of perfluoroacetone gives the 5α,9α-oxirane **95**. Condensation with the Grignard reagent from 4-bromo-*N*,*N*-dimethylaniline proceeds as shown above to give the 1,4-addition product **96**. The route diverges markedly from the one described above in the next step. Thus, photolysis of **96** gives a product of inversion of the 13-position; the transformation can be rationalized by assuming photolytic scission of the 13,14 bond to give an intermediate such as **97**. Ring closure to a thermodynamically favored cis hydrindanone will lead to the observed product **98**.

The altered stereochemistry about the 17 ketone interestingly reverses the course of additions to that carbonyl group. Thus, condensation of **98** with the Grignard reagent from the tetrahydropyranyl ether of 3-bromopropan-1-ol gives the product (**99**) of addition of the organometallic reagent from the β face. Hydrolysis of **99** with strong acid results in loss of the protecting groups as well as dehydration of the hydroxyl group at the 4-position; thus, onapristone (**100**)[26] is formed.

C. Androgenic Compounds

The clinical success of the progestational gonanes engendered a corresponding effort toward finding potent orally active androgens in the 19-nor steroid series. The prototype in this series is testosterone. This endogenous hormone, or more properly its active metabolite 5α-dihydrotestosterone, plays a pivotal role in supporting male reproductive function as well as integrity of sexual organs. The immediate indication for these androgens is of course replacement therapy. More important, from the commercial viewpoint, is the fact that many androgens also serve as anabolic agents; that is, they modify utilization of nutrients so as to conserve nitrogen and thus build muscle mass. Misuse of androgens, commonly referred to as "steroids" by athletes, is the subject of many news stories around the time of Olympic games.

The gonane that most directly corresponds to testosterone shows full activity as an androgen. This compound, **nandrolone**, is prepared in a straightforward manner from estradiol-3-methyl ether, which is itself obtained from the estrone derivative **31** by reduction of the 17 ketone. Birch reduction of this intermediate followed by acid hydrolysis of the first produced enol ether gives nandrolone (**101**). This compound is seldom used as such because of its ready inactivation in vivo, but it is used instead as a starting material for injectable derivatives. Esterification of the 17 hydroxyl group with any of several lipophilic acids gives oil soluble derivatives. These derivatives are administered by subcutaneous injection so as to form depots; the ester that slowly diffuses into the blood stream from the depot is saponified to **nandrolone** by serum esterases. This results in long-lasting blood levels of this drug. A typical example of one of these drugs is **nandrolone decanoate** (**102**), which is obtained by esterification with decanoyl chloride.[27]

One of the important mechanisms by which orally administered steroids are inactivated, as noted earlier, involves formation of water soluble derivatives at the 17-position, a process that is greatly reduced in 17α-alkyl-17β-hydroxy derivatives. Extensive use of the resulting orally active compounds has since revealed that 17

31 → (1. $NaBH_4$; 2. Li/NH_3; 2. H_3O^+) → 101 (OH) → ($ClCO(CH_2)_8CH_3$) → 102 ($OCO(CH_2)_8CH_3$)

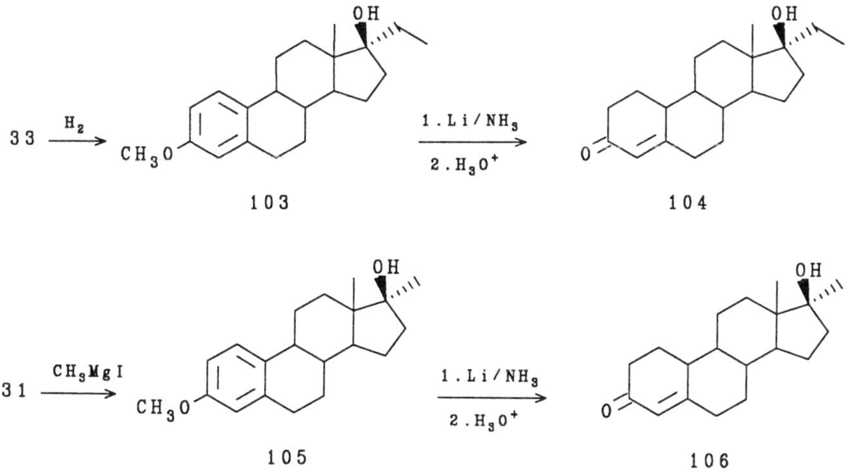

alkylation also leads to increased liver toxicity. Preparation of the first of these compounds, **norethandrolone** (**104**), begins with catalytic reduction of the side chain unsaturation in **mestranol** (**33**) to the corresponding ethyl derivative, **103**. Birch reduction, followed by acid hydrolysis, gives the orally active androgen.[10]

In a similar vein, reaction of estrone methyl ether (**31**) with methylmagnesium iodide gives the 17α-methyl derivative **105**. Birch reduction followed by acid hydrolysis leads to **normethandrolone** (**106**).[15]

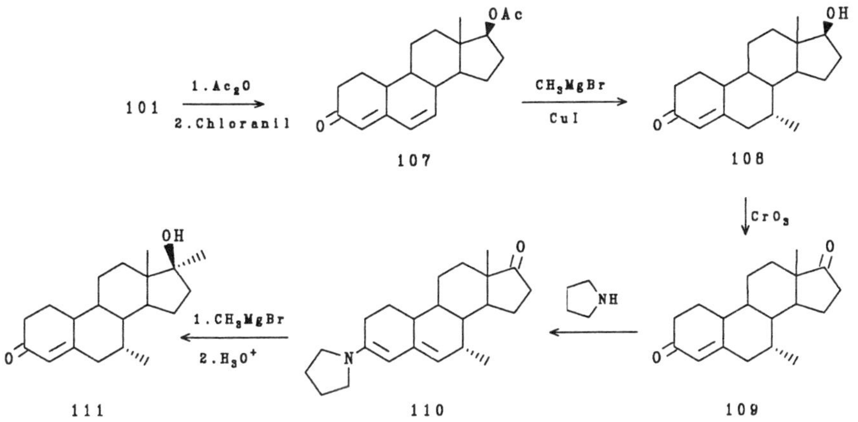

Addition of a methyl group at the 7α-position results in a major increase in potency in 19-nor androgens. The key dienone **107** is obtained by dehydrogenation of the 17-acetyl derivative of **nandrolone** with chloranil (1,2,4,5-tetrachlorobenzoquinone). Condensation of **107** with methylmagnesium bromide in the presence of cuprous iodide results in 1,6-addition of the organometallic to give the 7α-methyl derivative **108**, possibly via the intermediacy of a cuprate reagent. The stereochemistry can be rationalized by assuming approach from the more open face of the molecule. The alcohol at the 17-position is then oxidized to the corresponding ketone **109**, typically with Jones' reagent. In order to direct an additional methyl group to the carbonyl group at the 17-position, the ketone at the 3-position must first be protected. Thus, reaction with pyrrolidine results in preferential formation of an enamine at the 3-position (**110**) due to the greater steric hindrance about the 17 ketone. Reaction of **110** with methyl Grignard reagent followed by removal of the enamine by acid hydrolysis affords the potent, orally active androgen **mibolerone** (**111**).[28]

5. ANDROSTANES

Steroids that possess angular methyl groups at both the 10- and 13-positions but that are devoid of acetyl side chains at the 17-position are usually classed as derivatives of androstane, a hypothetical hydrocarbon. The endogenous male sex hormones, such as testosterone (**39**), can all be classified as androstanes; as noted below, the biological activity of compounds in this structural class is not restricted to this narrow definition.

A. Starting Materials

The androstane nucleus is sufficiently complex so as to effectively rule out total synthesis as a source for starting materials. The great preponderance of steroid natural products include carbon side chains at the 17-position; preparation of androstane starting compounds thus relies on schemes for degradation of these side chains. One route for preparation of dehydroepiandrosterone, DHEA (**115**), the key intermediate, starts from 16-dehydropregnenolone acetate (**4**), which is itself derived from *diosgenin*. (Androsterone differs from **115** in that the 3 hydroxyl has the α configuration, hence *epi* and the compound includes the 5,6 olefin, hence *dehydro*.) Reaction of compound **4** with hydroxylamine hydrochloride in the presence of base leads to the oxime **112**. Treatment of **112** with strong acid results in a Beckmann rearrangement

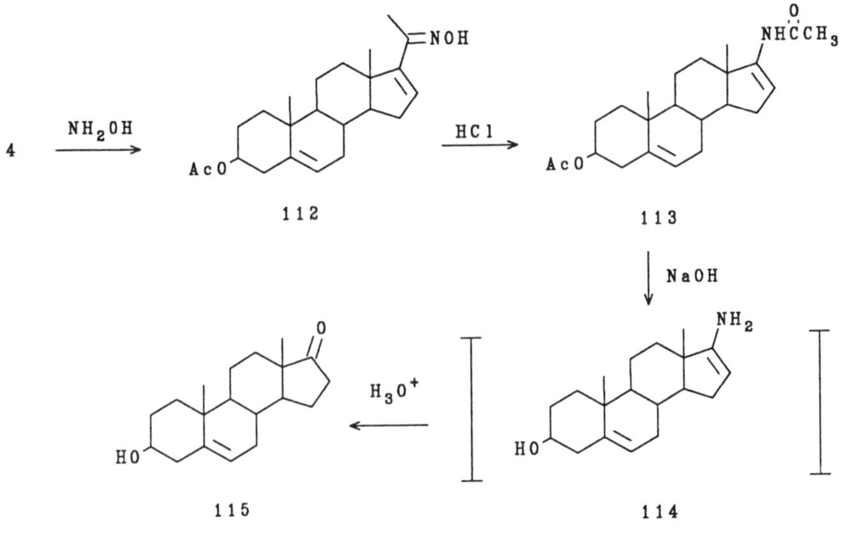

4 —NH_2OH→ 112 (AcO, =NOH) —HCl→ 113 (AcO, $NHCCH_3$ with O) —NaOH→ 114 (HO, NH_2) —H_3O^+→ 115 (HO, O)

of the isonitroso group to afford the enamine acetate **113**; saponification with base leads to formation of an intermediate primary enamine as well as loss of the acetate at the 3-position. Subsequent treatment of **114**, which may exist as the 17 imine, with mild acid leads to hydrolysis of this group to a 17 ketone, to afford dehydroepiandrosterone (**115**).[29] Note that endogenous testosterone is in fact derived from this compound by metabolic transformation; for this and other reasons, **115** has gained some popularity as a health food supplement under the label DHEA.

B. Androgenic Compounds

One of the earliest derivatives in the androgen series was designed to circumvent inactivation of orally administered compounds by metabolic transformation about the 17-position by addition of a methyl group for steric hindrance. Thus, reaction of **115** with methylmagnesium bromide gives **116** from addition of the α face. Reaction **116** with aluminum isopropoxide in the presence of cyclohexanone (Oppenauer oxidation) serves to oxidize the alcohol at the 3-position; the double bond shifts into conjugation under reaction conditions. Thus, the orally active androgen **17-methyltestosterone** (**117**)[30] is obtained.

The presence of an additional methyl group at a variety of positions increases the oral potency in the androgenic 17-methyltestosterone series as also seen to occur in the gonane series noted previously (see mibolerone, **111**). In much the same vein, dehydrogenation of the conjugated enone **117** with chloranil gives the dienone **118**.

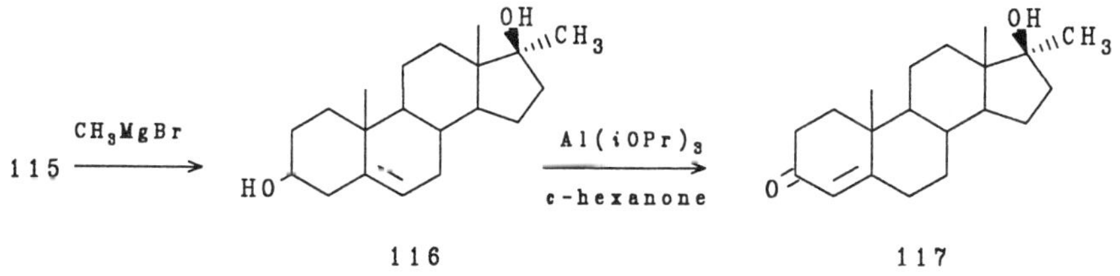

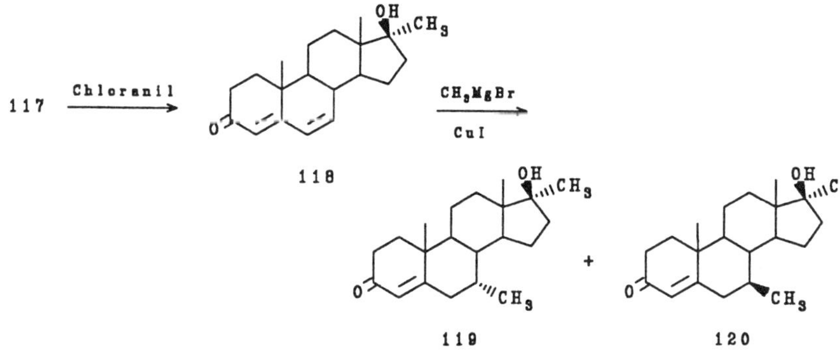

115 —CH_3MgBr→ 116 —$Al(iOPr)_3$, c-hexanone→ 117

117 —Chloranil→ 118 —CH_3MgBr, CuI→ 119 + 120

115 $\xrightarrow{H_2}$ 121 $\xrightarrow{CH_3MgBr}$ 122 $\xrightarrow{CrO_3}$ 123 $\xrightarrow[NaOMe]{HCO_2Me}$ 124

Addition of methylmagnesium bromide in the presence of cuprous iodide gives predominantly the 7α methyl derivative **bolasterone** (**119**), which shows potent androgenic and anabolic activity.[31] The 7β isomer is a minor product; this compound **calusterone** (**120**), which can be isolated from the mixture, has been used in the treatment of breast cancer.

The 2 formyl derivative from androstanone **123** provides the key intermediate for entry to androgens bearing an additional fused heterocyclic ring as well as derivatives methylated at the 1- or 2-positions. Preparation of this intermediate begins with the reduction of dehydroepiandrosterone (**115**) to give the 5α dihydro derivative **121**. This compound is then condensed with methylmagnesium bromide to give **122**; oxidation of the hydroxyl at the 3-position by one of a number of reagents such as Jones' reagent gives the dihydrotestosterone derivative **123**. Although 3 ketone can of course form enolates at either the 2- or 4-positions, reactions involving them at the 2-position are favored. Sodium methoxide mediated condensation with methyl formate exclusively affords product **124** from formylation at the 2 position.[32]

Catalytic hydrogenation of the formyl derivative **124** proceeds as usual from the more open α face of the molecule to afford the axial 2β-methyl derivative **125**. Treatment with strong base leads to epimerization to the more stable equatorial 2α-methyl isomer. Thus, the androgen **dromostanolone** (**126**)[32] is obtained. The compound is used clinically as its 17-propionate ester. Direct reaction of the formyl ketone **124** with hydrazine leads to formation of a fused pyrrazole ring;[33] this product, **stanozolol** (**127**), also shows androgenic activity.

Careful investigation of the bromination reaction of 3 keto steroids, which leads ultimately to 2,4-dibromo derivatives (see, e.g., **12** → **13**), starts with the formation of a 2α-bromo derivative, further demonstrating the preferred enolization toward the 2-position. Thus treatment of 5α-dihydrotestosterone (5α-DHT) with one equivalent of bromine under carefully controlled conditions affords the 2α-bromide **128**. The bromoketone is then dehydrobrominated by means of lithium carbonate in *N,N*-dimethylformamide (DMF) to give the unusual enone **129**. Conjugate addition of a methyl Grignard reagent in the presence of cuprous iodide goes directly to the 1β-methyl derivative **mesterolone** (**130**);[34] the reaction in this case goes directly to the energetically favored equatorial isomer. One of the standard methods for preparing fused cyclopropanes involved dipolar addition of diazomethane to an olefin followed by pyrolysis of the resulting pyrrazole. The sequence follows a different pathway for enones, with the pyrrazole losing nitrogen, leaving behind a new methyl group on the double bond. Thus, dipolar addition of diazomethane to the enone **129** leads to the pyrrazole **131**; this intermediate loses nitrogen on heating to give the 1-methyl enone **methenolone (132)**.[35]

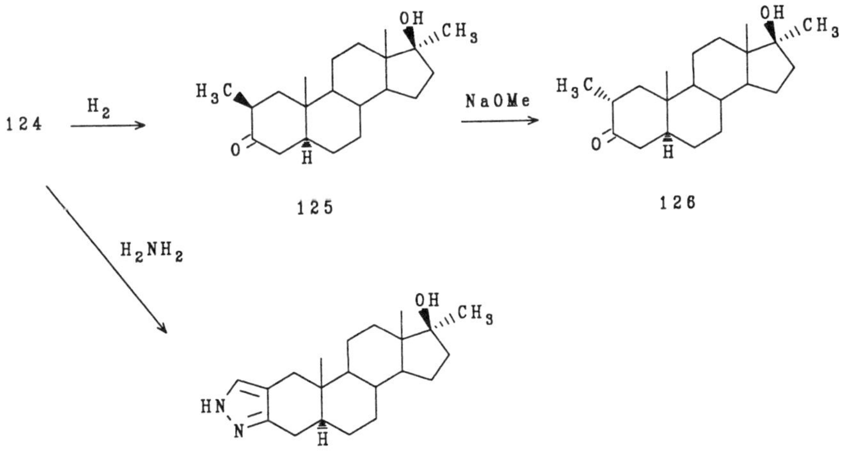

The 17-methyl analogue of unsaturated ketone **129** provides entry to a compound in which the A ring is converted to a heterocycle. The fact that this compound shows full androgenic and anabolic activity demonstrates the bioisosterism of methylene and oxygen in this series. Oxidation of the enone with lead tetraacetate may proceed through initial formation of a 1,2,3-tricarbonyl compound, which undergoes ring scission and eventual loss of carbon 2 to give the acid–aldehyde **133**. Compound **133** is in equilibrium with the lactol form **134**. Reduction of **134** with sodium borohydride leads to the lactone **oxandrolone** (**135**).[36]

The route used to prepare the unsaturated 2-methyl androgen, **stenbolone**, provides yet another illustration of the propensity for formation of enolates at the 2-position. Thus, reaction of dihydrotestosterone acetate (**136**) with formaldehyde and dimethylamine gives the Mannich product **137**. Hydrogenolysis of **137** gives the 2-methyl derivative **138**; the relatively elevated temperature used for this last reaction suggests that the reaction may proceed via the methylene product from elimination of dimethylamine. Bromination of the ketone leads to the 2-bromo derivative **139**. Dehydrobromination by means of lithium carbonate in DMF affords the androgen **stenbolone acetate** (**140**).[37]

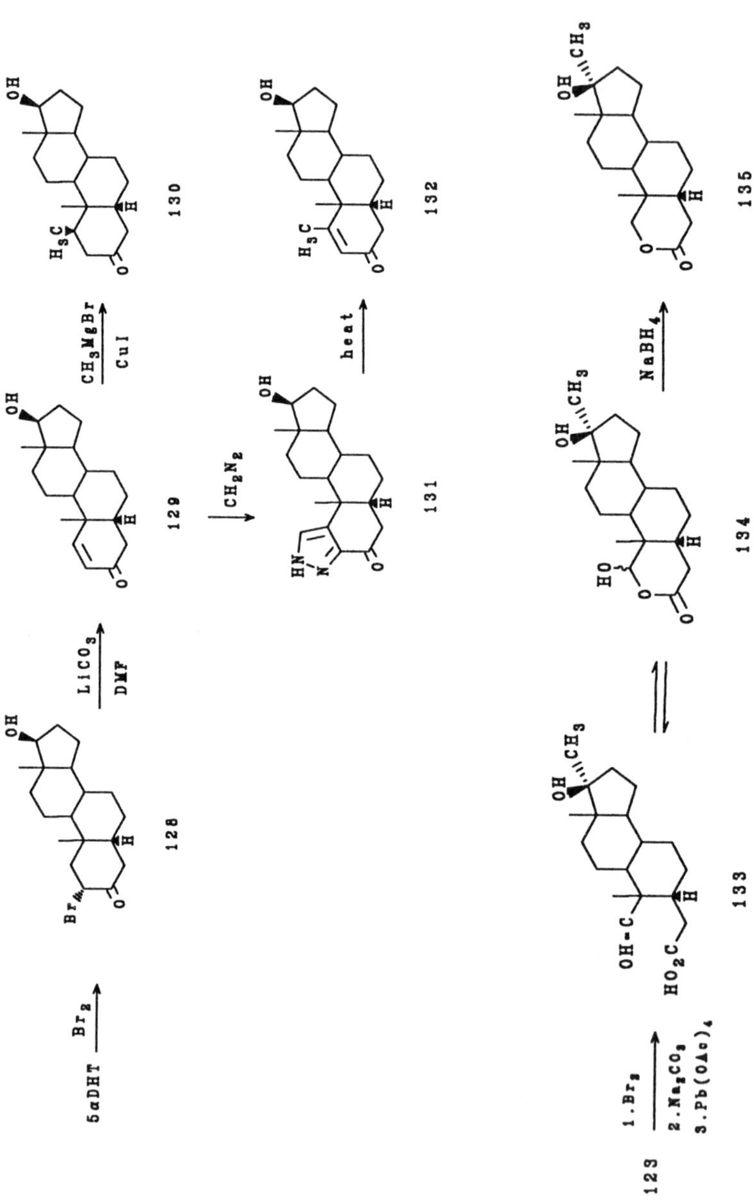
5αDHT
Br2
128
LiCO3
DMF
129
CH3MgBr
CuI
130
CH2N2
131
heat
132
123
1.Br2
2.Na2CO3
3.Pb(OAc)4
133
134
NaBH4
135

The crucial importance of the presence of oxygen at the 11-position for adrenocorticoid activity led to the development of methods for introducing this function by microbiological methods. Thus, fermentation of progesterone with soil organisms serves to oxidize the 11-position. Some of these same organisms were also found to hydroxylate simpler substrates. Thus, fermentation of andostene-3,17-dione (**141**), which was obtained by Oppenaucr oxidation of dehydroepiandrosterone, affords the corresponding 11α-hydroxyl derivative **142**. Oxidation by means of Jones' reagent leads to the 3,11,17-triketone **143**.

The difference in reactivity of carbonyl groups at the 3- and 17-positions, which are attributable to the greater steric bulk near the 17 ketone, have played a major role in the design of several of the syntheses discussed so far. Ketones at the 11-position are far less reactive than those at 17, due largely to the shielding effect of the two angular methyl groups, each of which bears a 1,3-diaxial relation to the carbonyl. Much of the work involved in steroid synthesis consists in taking advantage of these differences by juggling protecting groups so as to direct reaction to the desired sites. Thus, reaction of triketone **143**, also known as adrenosterone, with pyrrolidine, occurs selectively at the 3-position to give the enamine **144**. Condensation of **144** with methylmagnesium bromide results in formation of the 17-methyl carbinol **145**; the 11 ketone is totally inert under these reaction conditions. The potentiating effect of fluorine, particularly at the 9α-position, on biological activity was first observed in the corticoids. Inclusion of this group, starting from 11 oxygenated compounds, involves a fairly standard set of transforms; the key starting intermediate usually possesses an 11β hydroxyl group. Conversion of **145** to the alcohol requires that the 3 ketone first be protected. The enamine at 3, lost during workup, is restored by reaction with pyrrolidine; exposure to lithium aluminum hydride leads to one of the few reactions the ketone at 11 undergoes, that is, the reduction to the β alcohol yielding **146**, after removal of the enamine. The hydroxyl is then acylated with *p*-toluenesulfonyl chloride; exposure of this compound to base leads to elimination and the formation of the 9,11 olefin **147**.

Note that the hydroxyl group in the first obtained fermentation product is equatorial and would eliminate only with great difficulty, since it lacks a *transoid* proton at the adjacent position. Reaction of **147** with *N*-bromosuccinimide (NBS) in aqueous base leads to addition of the elements of hypobromous acid. The stereochemistry of the reaction is in all probability determined by initial formation of a bromonium derivative on the more open α face; diaxial opening by hydroxide leads to the bromohydrin **148**. Reaction of **148** with base results in internal displacement of bromine by alkoxide to give the β epoxide **149**. Compound **149** is then allowed to react with hydrogen fluoride in THF. Diaxial opening of the intermediate oxiranium ion by fluoride anion completes the construction of the 9α-fluoro-11β hydroxy moiety. Thus, the potent androgenic agent **fluoxymestrone** (**150**)[38] is obtained.

It has been known for many years that androgen stimulation is responsible for benign prostatic hypertrophy, a condition suffered by close to one-half of all men as they age. There is also considerable evidence to indicate that androgens exacerbated prostatic cancer. The search for an effective antiandrogen was thus the focus of considerable effort over the years. The first antiandrogen to show clinical activity,

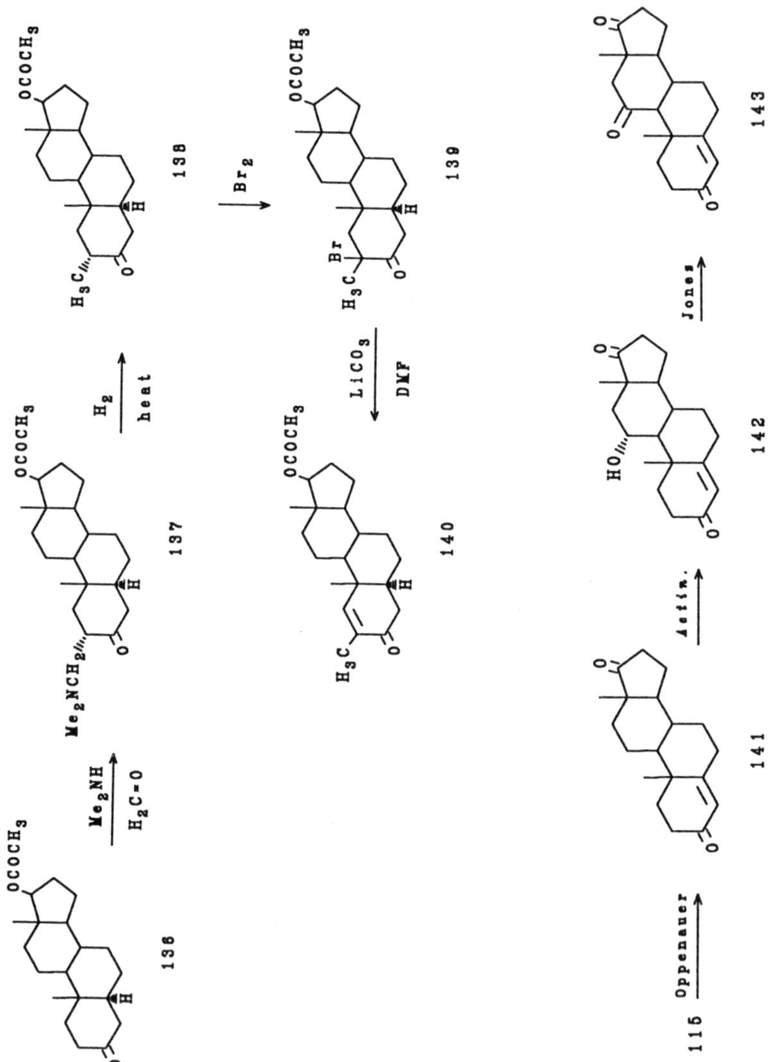

136 $\xrightarrow[H_2C=O]{Me_2NH}$ 137 $\xrightarrow[heat]{H_2}$ 138

138 $\xrightarrow{Br_2}$ 139 $\xrightarrow[DMF]{LiCO_3}$ 140

115 $\xrightarrow{Oppenauer}$ 141 $\xrightarrow{Actin.}$ 142 $\xrightarrow{Jones}$ 143

143 → (pyrrolidine) → 144 → 1. MeMgBr 2. H_3O^+ → 145 → 1. pyrrolidine 2. LAH 3. H_3O^+ → 146 → 1. TSA 2. Et_3N → 147

finasteride (**157**), involves some significant structural changes from the agonists as is also the case in the progestin antagonist mifepristone (**94**). The first step in the synthesis of **157** consists of oxidation of the acetyl side chain in progesterone (**39**) to a carboxylic acid (**151**), for example, by some modification of the haloform reaction. Oxidation of the enone function in ring A with a combination of potassium permanganate and sodium periodate leads to keto acid **152** in which C-4 has been excised.

Bromonium

Oxiranium

147 → HOBr (NBS/OH^-) → 148 → NaOH → 149 → HF THF → 150

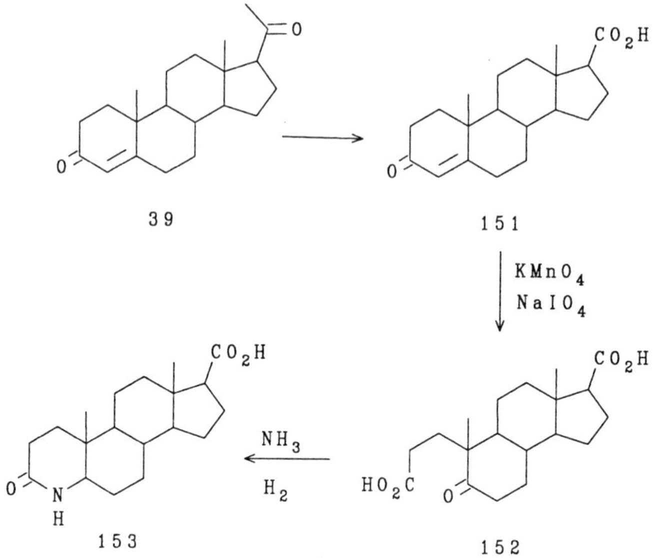

The reaction can be rationalized by assuming that permanganate first hydroxylates the double bond, cleavage of the resulting glycol by periodate will then open the ring. Further reaction then takes the 3 and 5 carbon atoms to their fully oxidized states. Reductive amination of the keto acid with ammonia results in formation of the lactam **153**. Reduction of the hypothetical initially formed 5 imino group from the more open α face will give the β amino group and consequent maintenance of steroid stereochemistry.

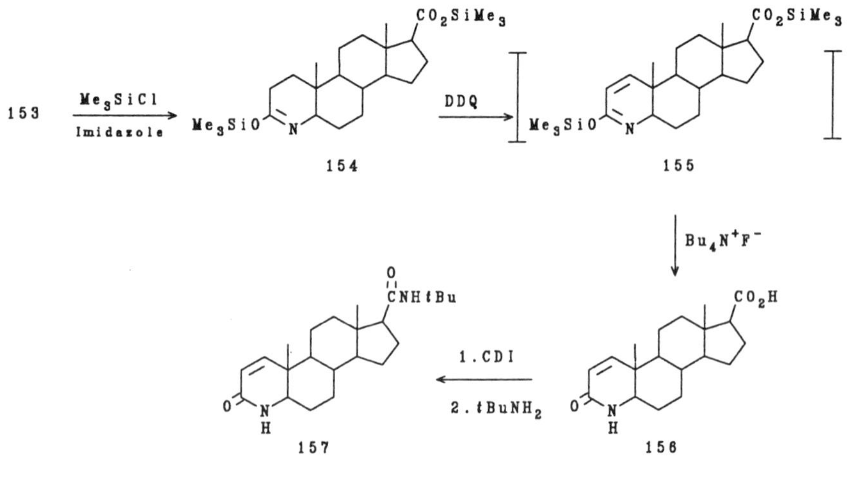

The lactam group is then converted to its iminoether by treatment with trimethylsilyl chloride in the presence of imidazole; the carboxylic acid is transformed to its silyl ester **154** under the reaction conditions. 2,3-Dicyano-5,6-dichlorobenzoquinone (DDQ) is a much more powerful oxidizing agent than its tetrachloro analogue; the use of the latter for introducing an additional double bond at the 6,7-position in 3-keto-4-enes has been described previously. The use of DDQ for introducing an olefin in the 1-position of 3-keto-4-enes will often be seen in the discussion of corticoids. Treatment of lactam enol ether **154** with DDQ introduces unsaturation at the 1-position in this compound as well (**155**). The silyl protecting groups are then removed with fluoride ion to give the lactam acid **156**. The reduced reactivity of the carboxylic acid at the 17-position due to steric hindrance necessitates that it first be converted to an activated intermediate; reaction with carbonyl dimidazole (CDI) gives the corresponding imidazolide. Condensation of this intermediate with *tert*-butylamine affords the amide and thus finasteride (**157**).[39,40]

C. Aldosterone Antagonists

The major proportion of the steroids isolated from the adrenal cortex consists of the so-called corticoids. These compounds, which include cortisone and hydrocortisone, are discussed in detail in Chapter 5. Although these compounds have profound effects on mineralite transport and balance, further research indicated that most of that activity of adrenal extracts was in fact due to the aldosterone (**158**). This unusual steroid, which is oxygenated at both the 11-position and on the angular methyl group at the 13-position, was among the last to be characterized structurally. It has since been established that this compound plays a key role in the renin–angiotensin system, which regulates body fluid volume, sodium and potassium retention, or excretion and thus ultimately blood pressure. The adventitiously discovered aldosterone antagonists are quite effective diuretic agents, by opposing agonist mediated fluid and sodium retention; they are of particular interest because they do not cause the lowering of serum potassium levels noted with the sulfonamides or thiazides. The aldosterone antagonists exert antihypertensive activity by a mechanism thought to be analogous to that of the latter drugs; they were used quite extensively for treatment of high blood pressure before the emergence of more effective and better tolerated drugs.

The diuretic activity of this structural class, it is said, was first noted with the simplest member of the series, **162**, in the course of a wide range screening program. Preparation of **162** begins with the condensation of dehydroepiandrosterone with lithium acetylide to give the intermediate **159**, which displays the usual preference for addition to the α face. Reaction of **159** with excess methylmagnesium iodide leads to

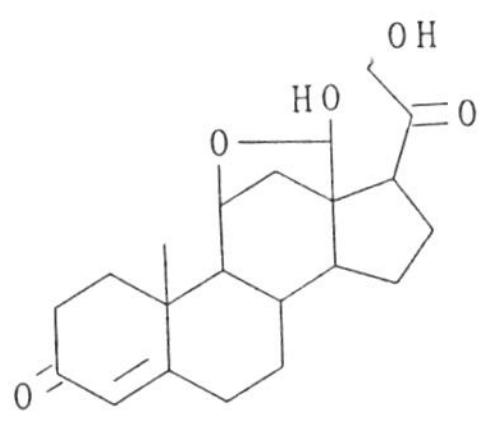

OH
HO
O
O
O

158

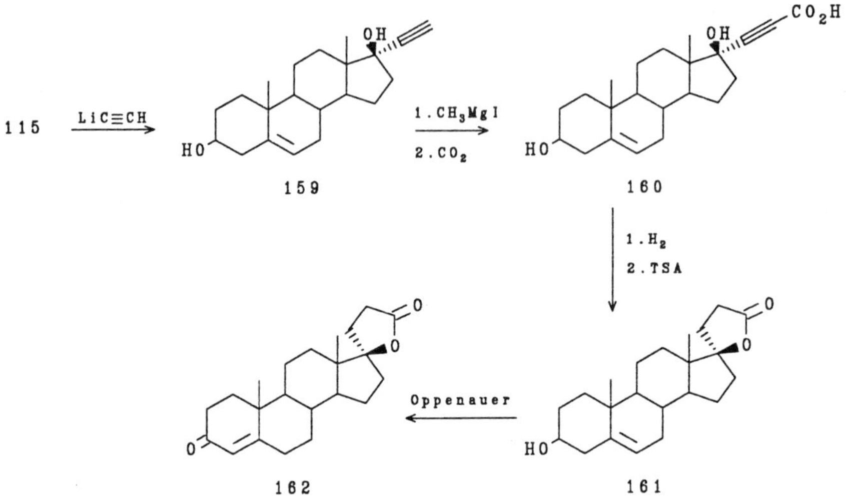

115 $\xrightarrow{LiC\equiv CH}$ 159 $\xrightarrow{1. CH_3MgI; 2. CO_2}$ 160 $\xrightarrow{1. H_2; 2. TSA}$ 161 $\xrightarrow{Oppenauer}$ 162

formation of magnesium salts at both the alcohols and, most importantly, at the acetylene terminus. Reaction of the salt with carbon dioxide in situ leads to carbonation at this position; the carboxylic acid **160** is obtained on workup. The catalytic hydrogenation over palladium then selectively reduces the triple bond; treatment with acid forms the spirocyclic lactone to give **161**. Oppenauer oxidation of the hydroxyl group at the 3-position gives the 4-ene-3-one and completes the synthesis of **162**.

The search for compounds that had improved oral activity initially led to the 7α-thioacetyl derivative **164**.[41] Dehydrogenation of the enone function in **162** by using the now familiar quinone chloranil, leads to the diene **163**. This compound undergoes 1,6-conjugate addition on treatment with the sodium salt from thioacetic acid to give

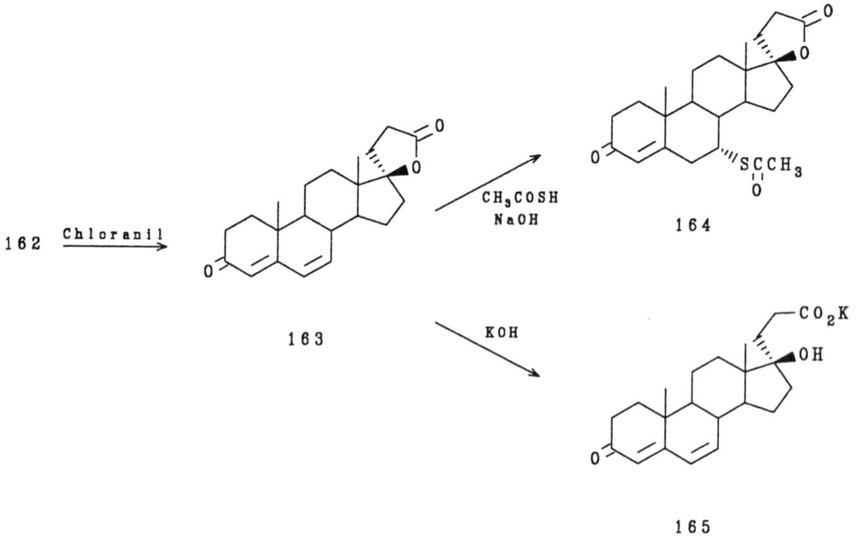

162 $\xrightarrow{Chloranil}$ 163 $\xrightarrow{CH_3COSH, NaOH}$ 164

163 $\xrightarrow{KOH}$ 165

163 $\xrightarrow[\text{NaH}]{Me_3S^+I^-}$ 166 ⟶ 167

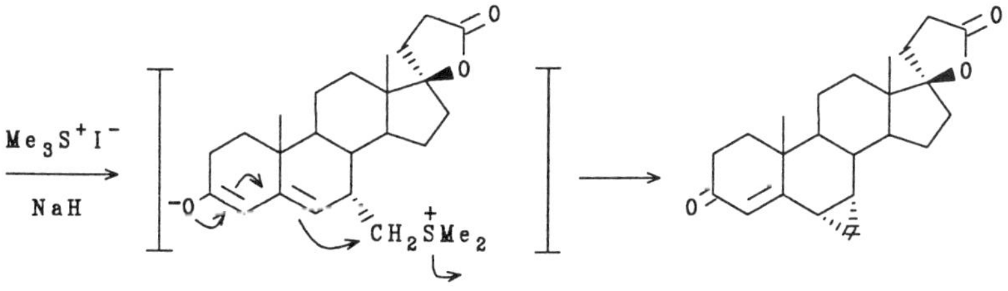

the 7α derivative **164**; this compound, under the name **spironolactone** (**164**) was the first clinical aldosterone antagonist. Studies on metabolic disposition of this drug revealed that the lactone not unexpectedly opened in vivo; the fact that the thioacetyl was eliminated was not anticipated. The observation that the metabolite showed full clinical activity led to its use as a drug. This compound, **canrenoate potassium** (**165**) is prepared by simple ring opening of the lactone in **163** with potassium hydroxide.

A reagent more commonly used for converting carbonyl groups to homologous epoxides interestingly converts the terminal olefin of a dienone to cyclopropane. Thus reaction of dienone canrenone (**163**), with the ylide from trimethylsulfonium iodide can be envisioned to afford the betaine **166** as an intermediate. Return of the negative charge results in displacement of neutral dimethyl sulfide and the consequent formation of a cyclopropyl ring. The product **prorenone** (**167**)[42] is used as its lactone ring-opened potassium salt **prorenoate potassium**.

The spironolactone analogue in which the thioacetyl group is replaced by a carboxylate ester retains the activity of the parent indicating that metabolism to a

163 $\xrightarrow{\text{KCN}}$ 168 ⟶ 169 $\xrightarrow{H_3O^+}$ 170

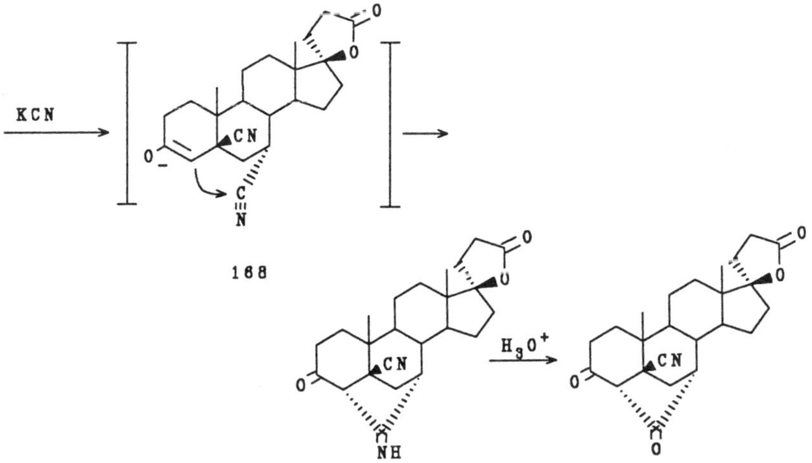

170 $\xrightarrow{NaOMe}$ [171 → 172] ↓ 173

canrenone-like product is not an absolute requirement. The actual chemical manipulations involved in the transformation, addition of cyanide, hydrolysis, and finally treatment with sodium methoxide belie the complexity of the actual chemistry. The product isolated from the first step in fact consists of the bridged bicyclic imine **169**. The initial 1,6-addition product apparently first converts to a neutral 4-ene-3-one; this compound obviously undergoes addition of a second mole of cyanide to give **168**. Addition to the β face may be due to the hindrance from the new cyano group; this also results in the A/B ring junction assuming the *cis*-decalin conformation, in effect forcing the 4-position into close proximity to the 7α nitrile. This compound closes in the basic cyanide medium to give the observed product **169**. Hydrolysis of this product with mild acid leads to the ketone **170**.

The final step involves reaction of **170** with sodium methoxide. This transform can be envisaged by first assuming addition of methoxide to the bridge ketone to give an intermediate such as **171**. Steric strain, albeit small, would drive scission of the bridge ring to give the enolate **172**. This compound is in fact equivalent to the initial intermediate from conjugate addition to the 4-ene-3-one; simple reversal gives the neutral product **173** and thus **mexrenone**.[43] Mexrenone is also used as the potassium salt of the ring-opened form, in this case, **mexrenoate potassium**.

REFERENCES

1. Marker, R.E.; Wagner, R.B.; Ulshafer, P.R.; Wittbecker, E.L.; Goldsmith, D.P.J.; Ruof, C.H. *J. Am. Chem. Soc.* **1947**, *69*, 2167.
2. Heyl, F.W.; Herr, M.E. *J. Am. Chem. Soc.* **1950**, *72*, 2617.

3. Slomp, G.; Johnson, J.L. *J. Am. Chem. Soc.* **1953**, *75*, 915.
4. Lenz, G.R. *Kirk–Othmer Encyclopedia of Chemical Technology* 3rd Ed.; Wiley-Interscience: London, 1983; Vol. 21, p. 645.
5. Hershberg, E.B.; Rubin, M.; Schwenk, E. *J. Org. Chem.* **1950**, *15*, 292.
6. Dryden, H.L.; Webber, G.M.; Weiczovek, J. *J. Am. Chem. Soc.* **1964**, *86*, 742.
7. Ananchenko, S.N.; Torgov, I.V.; *Dokl. Akad. Nauk. SSSR* **1959**, *127*, 533.
8. Hughes, G.A.; Smith, H. *Chem. Ind. (London)* **1960**, 1022.
9. Inhofen, H.H.; Logeman, W.; Hohlweb, W.; Serini, A. *Chem. Ber.* **1938**, *71*, 1024.
10. Colton, F.B.; Nysted, L.N.; Riegel, B.; Raymond, A.L. *J. Am. Chem. Soc.* **1957**, *79*, 1123.
11. Revesz, C.; Banik, U.K.; Lefebvre, Y. *J. Reprod. Fertil.* **1970**, *22*, 27.
12. Hogberg, K.B.; Fex, H.J.; Konyves, I.; Kneip, O.J. German Offen. **1967**, 1249862; *Chem. Abstr.* **1968**, *68*, 3118.
13. Birch, A.J. *Quart. Rev.* **1950**, *4*, 69.
14. Colton, F.B. U. S. Patent **1952**, 2655518; *Chem. Abstr.* **1954**, *48*, 11503.
15. Djerassi, C.; Miramontes, G.; Rosenkrantz, G.; Sondheimer, F. *J. Am. Chem. Soc.* **1954**, *76*, 4092.
16. Klimstra, P.D.; Colton, F.B. *Steroids* **1967**, *10*, 411.
17. DeWinter, M.S.; Siegman, C.M.; Szpilfogel, C.A. *Chem. Ind. (London)* **1959**, 905.
18. Douglas, G.H.; Graves, J.M.H.; Hartley, D.; Hughes, G.A.; McLaughlin, B.J.; Sidall, J.B.; Smith, H. *J. Chem. Soc.* **1963**, 5072.
19. Hofmeister, H.; Wiechert, R.; Annen, K.; Laurent, H.; Steinbeck, H. German Offen. **1977**, 2546062; *Chem. Abstr.* **1977**, *86*, 168265.
20. Fried, J.; Bry, T.S. U. S. Patent **1968**, 337426; *Chem. Abstr.* **1968**, *69*, 77608.
21. Nominee, G.; Amiardo, G.; Torelli, V. *Bull. Soc. Chim. Fr.* **1968**, 3664.
22. Velluz, L.; Nominee, G.; Bucourt, R.; Mahieu, J. *C. R.* **1964**, *257*, 5669.
23. Popper, A.; Prezewowsky, K.; Wiechert, R.; Gibian, H.; Raspe, G. *Arnezeim.-Forsch.* **1969**, *19*, 352.
24. For an account of this research see Teutsch, J.G.; Deraedt, R.; Philbert, D. *Chronicles of Drug Discovery*, Lednicer, D., Ed., ACS Books: Washington, DC, 1993, Vol. 3, p. 1.
25. Teutsch, G.; Costerousse, G.; Philbert, D.; Deraedt, R. U. S. Patent **1984**, 4386085; *Chem. Abstr.* **1984**, *101*, 130975.
26. Neef, G.; Beier, S.; Elger, W.; Henderson, D.; Wiechert, R. *Steroids* **1984**, *44*, 349.
27. DeWytt, E.D.; Overbeek, O.; Overbeek, G.A. U. S. Patent **1961**, 2998423; *Chem. Abstr.* **1961**, *55*, 9478.
28. Babcock, J.C.; Campbell, J.A. Belg. Pat. **1962**, 610385.
29. Rosenkrantz, G.; Mancera, O.; Sondheimer, F.; Djerassi, C. *J. Org. Chem.* **1956**, *21*, 520.
30. Ruzicka, L.; Goldberg, M.W.; Rosenberg, H.R. *Helv. Chim. Acta* **1935**, *18*, 1487.
31. Babcock, J.C.; Campbell, J.A. U. S. Patent **1967**, 3341557.
32. Ringold, H.J.; Batres, E.; Halpern, O.; Necoechea, E. *J. Am. Chem. Soc.* **1959**, *81*, 427.
33. Clinton, R.O.; Manson, A.J.; Stonner, F.W.; Beyler, A.L.; Potts, G.O.; Arnold, A. *J. Am. Chem. Soc.* **1959**, *81*, 1513.
34. Wiechert, R. German Offen. **1963**, 1152100; *Chem. Abstr.* **1963**, *59*, 11766.
35. Wiechert, R., Kaspar, E. *Chem. Ber.* **1960**, *93*, 1710.

36. Pappo, R.; Jung, C.J. *Tetrahedron Lett.* **1962**, 365.
37. Counsell, R.E.; Klimstra, P.D.; Ranney, R.E. *J. Med. Chem.* **1962**, *5*, 1224.
38. Herr, M.E.; Hogg, J.A.; Levin, R.H. *J. Am. Chem. Soc.* **1956**, *78*, 501.
39. Rasmusson, G.H.; Reynolds, G.F.; Steinberg, N.G.; Walton, E.; Patel, G.F.; Liang, T.; Cacieri, M.; Chueng, H.H.; Brooks, J.R.; Berman, C. *J. Med. Chem.* **1986**, *29*, 2298.
40. Battacharya, A.; De Michele, L.M.; Dolling, U.H.; Douglas, A.W.; Grabowski, E.J. *J. Am. Chem. Soc.* **1988**, *110*, 3318.
41. Cella, J.A.; Tweit, R.C. *J. Org. Chem.* **1961**, *83*, 4083.
42. Chinn, L.J. U. S. Patent **1974**, 3845041; *Chem. Abstr.* **1974**, *82*, 58011.
43. Weier, R.M.; Hoffman, L.M. *J. Med. Chem.* **1975**, *18*, 817.

CHAPTER 5

STEROIDS PART 2: COMPOUNDS RELATED TO PROGESTERONE AND CORTISONE

1. INTRODUCTION

The classical progestins as well as the corticosteroids are based on a steroid nucleus that is substituted at the 17-position by a two carbon side chain. This side chain most often includes a carbonyl group at the 21-position. The hypothetical parent hydrocarbon has been assigned the name ***pregnane*** for purposes of nomenclature. These two classes of compounds also share a common biogenetic pathway; their biological activities on the other hand differ markedly.

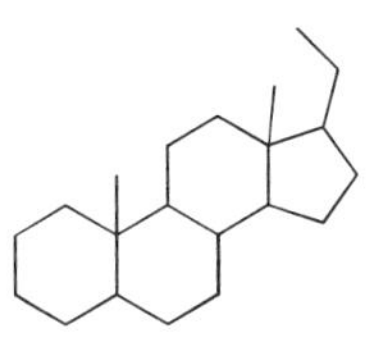

Pregnane

2. PROGESTINS

Progesterone ranks with estradiol as the key hormone in controlling the reproductive cycle in the mammalian female. This compound is synthesized by a structure on the ovary known as the corpus luteum. Progesterone is formed immediately postovulation and inhibits subsequent ovulation. Increased levels of progesterone are essential in permitting implantation of fertilized ova as well as providing a hospitable intrauterine milieu for the developing fetus. The most important application of the progestational 19-nor steroids (gonanes), discussed in Chapter 4, involves their use in oral contraceptives. The progestational agents more directly related structurally to progesterone itself are mainly used as drugs for treating various diseases related to progesterone deficiency. Several of these, however, had at one time been included as the progestational component in oral contraceptives; their discontinuation was probably due more to marketing than to medical reasons, since they had been preceded by their 19-nor counterparts by some years. Recall that progesterone itself is poorly bioavailable and readily metabolized. Most of the modifications seen below were aimed at overcoming these shortcomings.

Potency and oral activity of progestins in the pregnane class are enhanced by additional substitution at various positions; the presence of acyloxy groups at the 17α-position has proven to be a particularly important modification and is present in virtually all modified progestins. One of the original preparations for the prototype compounds starts by epoxidation of pregn-17-enolone, which is obtained from the saponification of the acetate (Chapter 4, compound **4**). The use of a peracid would preferentially lead to reaction at the 5,6 olefin because of its higher electron density over that at the 16,17-position. The initial step in the reaction with basic hydrogen peroxide is more akin to a Michael addition than a classic epoxidation; the reaction with this reagent thus occurs selectively at the conjugated double bond at 16,17 to give **1** as the observed product. Exposure to hydrogen bromide leads to the bromohydrin **2** from pseudodiaxial opening. Catalytic hydrogenation over palladium in the presence of ammonium acetate interestingly preferentially reduces the bromine over the remaining olefin to give **3**. Acetylation under standard conditions selectively acylates the hydroxyl at 3 over the very hindered, and tertiary, 17 alcohol.[1] The 17 alcohol can, however, be acylated under forcing conditions; thus reaction with hexanoic (caproic) anhydride in the presence of *p*-toluenesulfonic acid, leads the diester **4**. Methanolysis occurs selectively at the 3-position; the resulting alcohol is then oxidized with aluminum isopropoxide in the presence of cyclohexanone to give **hydroxyprogesterone caproate** (**5**). This very lipophilic ester is quite soluble in vegetable oils and is thus used for injection as depots that provide long-term progestin blood levels.

Addition of a methyl group at the 6α-position enhances activity as it does in the case of the androgens. Oppenauer oxidation of 17-hydroxypregenolone (**3**) leads in a straightforward manner to formation of the conjugated ketone and thus to 17α-hydroxyprogesterone (**6**). Reaction of **6** with an excess of ethylene glycol in the presence of acid gives the corresponding bis-acetal **7.** The very generally observed shift of the olefin to the 5,6-position on forming acetals of conjugated 3-one-4-enes has the fortunate consequence of providing a reaction center at the 6-position; the reasons for this rearrangement are, on the other hand, not immediately obvious. Reaction of the product with *m*-chloroperbenzoic acid (*m*CPBA) leads to formation of the α epoxide **8**. Compound **8** results from reaction at the more open face of the molecule. Condensation of **8** with methyl Grignard reagent leads to the product **9** from diaxial opening of the oxirane.

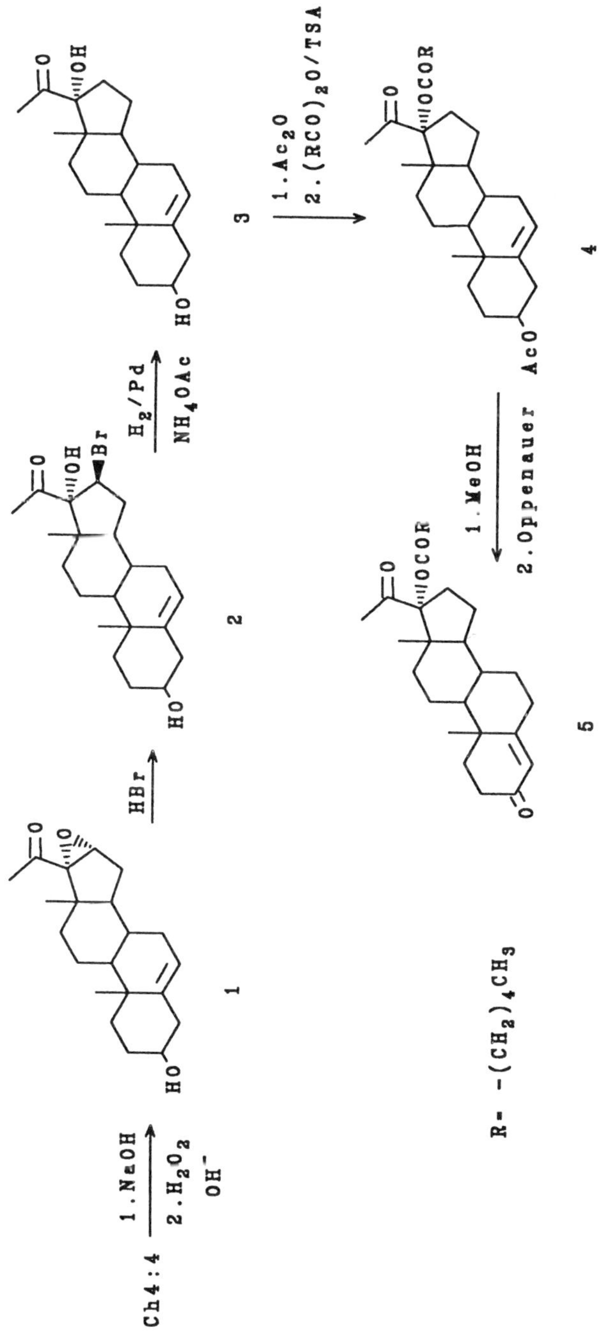

Ch4:4 → 1. NaOH; 2. H_2O_2, OH^- → 1

1 → HBr → 2

2 → H_2/Pd, NH_4OAc → 3

3 → 1. Ac_2O; 2. $(RCO)_2O$/TSA → 4

4 → 1. MeOH; 2. Oppenauer → 5

R- $-(CH_2)_4CH_3$

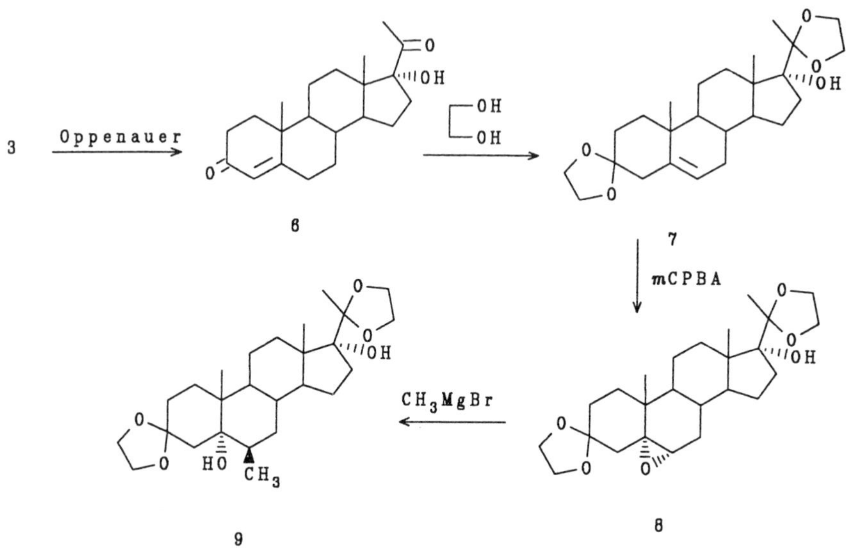

The acetal groups are then removed by hydrolysis of **9** in a mixture of acetone and dilute mineral acid to give the β-hydroxy ketone **10.** Compound **10** undergoes reverse Michael addition of water on treatment with base to give the conjugated ketone **11**. Acetylation with acetic anhydride under forcing conditions affords **medroxyprogesterone acetate** (**12**),[2] a quite potent orally active progestin. A far more important

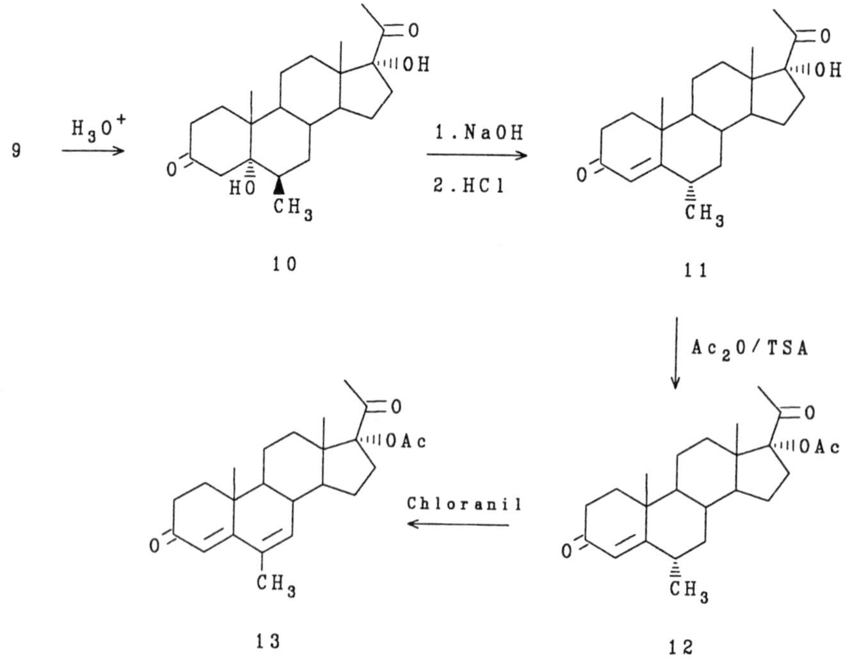

use for this drug hinges on its solubility in lipids; this drug can also be injected to form depots that provide long-term levels of progestin. Medroxyprogesterone acetate (**12**) is sufficiently potent so that these levels provide sufficient progestational activity to prevent ovulation and in effect provide an injectable contraceptive. This drug (Depo-Provera®) may incidentally have set a record in that it took close to two decades from application to final approval for this indication by the U.S. FDA; noncontraceptive uses, on the other hand, were approved in the early 1960s. Reaction of **12** with chloranil leads to the introduction of an additional double bond at the 6,7-position to afford the potent progestin **megestrol acetate** (**13**).[3] This compound has found some use in the treatment of breast cancer.

Biological activity is often retained when a methyl group is replaced by chlorine, suggesting that steric factors outweigh electronic considerations in these cases. The synthesis of 6-chloro progestins follows much the same route as the one used for the methylated counterparts. The scheme begins with the acetylation of 17α-hydroxyprogesterone (**6**) under forcing conditions; the 6,7 double bond is then introduced by dehydrogenation with chloranil to give dienone **14**. Reaction of **14** with basic hydrogen peroxide starts with initial conjugate addition of the hydroperoxy anion to the terminus of the conjugated system; the end product is the 6-α,7α epoxide **15**. Reaction of **15** with anhydrous hydrogen chloride proceeds via the normal diaxial opening to afford the 6-chloro-7-hydroxy derivative; this compound dehydrates under the reaction conditions to afford the chlorinated diene **chlormadinone acetate** (**16**).[4] The high potency and good bioavailability of this drug led to its use as a contraceptive. Note that it showed reasonable efficacy in the absence of coadministered estrogens; metabolic conversion to an estrogen, in contrast to the 19-nor progestins, is not a consideration with this compound. Use of the drug in women proved unattractive because it interfered with the menstrual cycle in a significant number of cases. It is, however, still

6 → (1. Ac_2O/TSA; 2. Chloranil) → 14 → (H_2O_2, NaOH) → 15 → (HCl) → 16 → (SeO_2) → 17

OAc, Cl

used to regulate estrus in some domestic animals. Dehydrogenation of this product with selenium dioxide leads to introduction of the double bond at the 1-position to afford the 1,4,6-trienone **delmadinone acetate** (**17**).[5]

One of the hallmarks of the medicinal chemistry of steroids is the observation that the effect of potentiating groups is additive. This phenomenon, illustrated time after time in the area of the corticosteroids, is also seen in the progestins. The drug **melengestrol acetate** (**29**), which includes activating alkyl groups at the 6- and 16-positions as well as extended conjugation, is one of the most potent progestins to have been described. One of the more interesting syntheses for this agent goes back to diosgenin (see Chapter 4) as starting material in order to take advantage of the fact that the 21-keto-16-ene functionality is present in this molecule in protected latent form. The first step after formation of a tosylate (**18**) at the 3-position involves solvolysis of this compound. These conditions lead the homoallylic tosylate to undergo a classic cyclopropylcarbinyl rearrangement (discussed in detail in Chapter 1) to afford the cyclopropyl carbinol **19**. Note that this rearrangement was in all probability first observed in the steroid series due to the ubiquity of the 3-hydroxy-5-ene functionality as well as the fact that the groups are ideally disposed sterically for the transformation. Oxidation of the alcohol with the chromium trioxide/pyridine complex leads to the 6 ketone. This intermediate is then condensed with methyl Grignard reagent to give the tertiary carbinol **20**. Solvolysis of **20** leads to the reverse rearrangement and formation of the homoallyl alcohol **21**, with the net addition of a methyl group at the 6-position.

Product **21** from this sequence, which is of course simply 6-methyldiosgenin, is then subjected to a series of degradation reactions which are similar to those used for the parent compound, to give the 6-methyl derivative of dehydropregnenolone (**22**). Reaction of the enone function of **22** with diazomethane leads to a 1,3-dipolar addition and formation of the pyrrazoline **23**. Intermediate **23** loses nitrogen on pyrolysis to

CH4:1 —TsCl→ 18 (TsO) —H_3O^+→ 19 (OH) —1. $CrO_3 \cdot Py$; 2. CH_3MgBr→ 20 (H_3C, OH) —H_3O^+→ 21 (HO, CH_3)

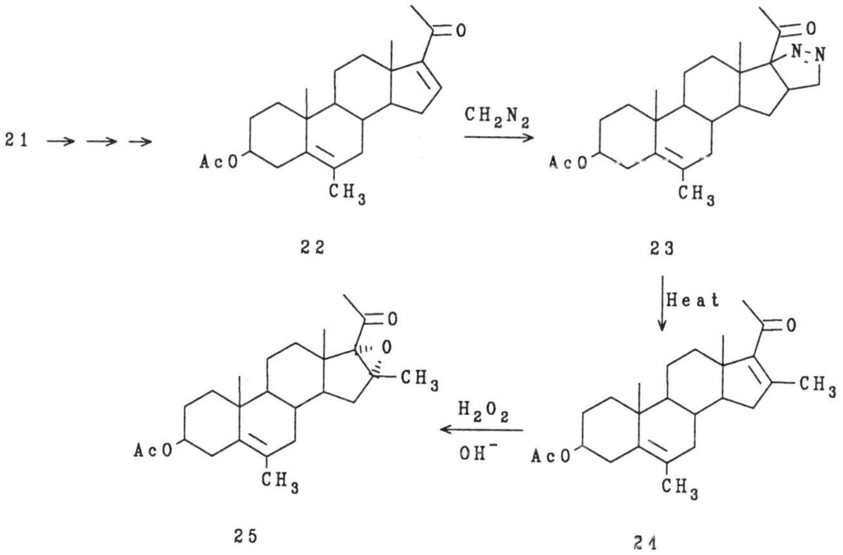

21 → → →

22

CH_2N_2

23

Heat

24

H_2O_2 / OH^-

25

form the 16 methyl derivative **24**, in keeping with the generalization that diazomethane adducts from enones form enone-methyl derivatives instead of cyclopropanes. Treatment of **24** with basic hydrogen peroxide leads to preferential formation of the epoxide **25** from the conjugated olefin at the 16-position.

The epoxyketone in **25** rearranges to an allyl alcohol when treated with acid. The reaction can be rationalized by assuming that the first step involves protonation of epoxide oxygen as in **26**. The subsequent step involves either concerted ring opening and proton loss as shown, or stepwise opening to a carbocation followed by loss of a proton from the methyl group. The allylic acetate in **27** is sequentially saponified and the alcohol is oxidized to an enone to give **28**. Dehydrogenation by means of chloranil completes the synthesis of **melengesterol** (**29**).[6]

The analogue of **medrogestone**, where a methyl group replaces the acetoxy function at the 17α-position, displays good oral progestational activity; this observation can be taken as evidence that mere bulk is required at this position. The first step in the preparation of this analogue involves reaction of the 16-dehydro-6-methylpregenolone acetate (**22**) with lithium in liquid ammonia in a modified Birch reduction to give the anion **30** as an intermediate. Quenching with methyl iodide results in alkylation of the anion at the 17-position; the usual steric course, that is, approach of the reagent from the more open α face, is followed to give **31** as the product; the basic conditions result in hydrolysis of the acetate at the 3-position. Oxidation of **31** under Oppenauer conditions leads to the conjugated enone; dehydrogenation of that product with chloranil establishes the dienone function. Thus, medrogestone (**32**)[7] is obtained.

An unspoken assumption in the medicinal chemistry of steroids involves the requirement of the "natural" configuration of the cyclopentaphenanthrene nucleus for biological activity. A notable departure from this assumption is the apparently equivalent antiprogestational activity of mifepristone and its analogue onapristone, with an

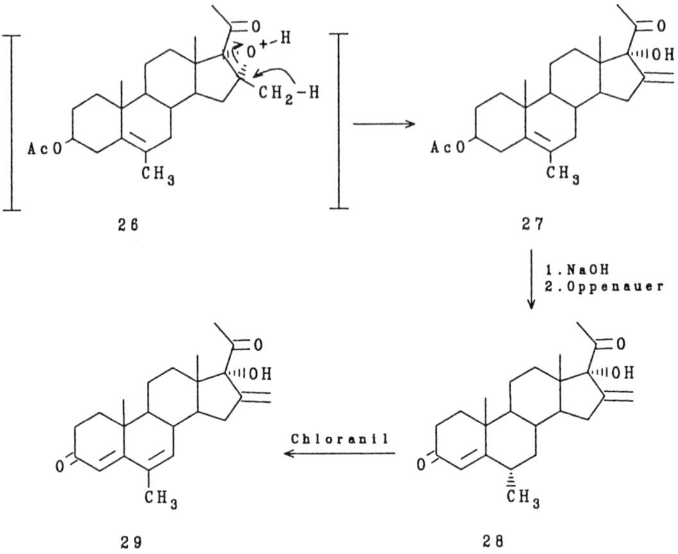

"unnatural" cis CD ring fusion and reversed stereochemistry at the 17-position (discussed in Chapter 4). A similar deviation from this assumption had been observed several decades earlier in the progestin series. The stereochemistry of the drug **didrogesterone** (**38**) is also determined by a photochemical reaction. The starting allylic bromide (**33**) for this drug is obtained by bromination of pregnenolone acetate with bromohydantoin. Dehydrobromination of the product with 2,4,6-collidine gives the conjugated diene **34**, which contains a functional array reminiscent of the Vitamin D precursor ergosterol. Like ergosterol, **34** ring opens on irradiation to the presumed intermediate triene **35**. Recyclization of the intermediate **35** results in a product (**36**)

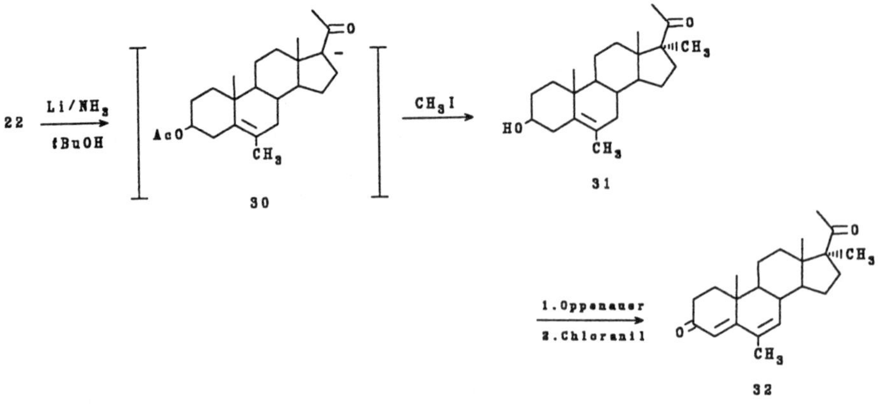

with reversed configuration at both C-9 and C-10. The stereochemistry of **36** can be rationalized by noting that the transition leading to this product is free from the nonbonding interaction between the angular methyl groups present in the transition state that leads back to the starting material. The same argument would apply if the cleavage and ring closure went via a 9,10 diradical. Oppenauer oxidation of **36** leads to the enone **37** in which only one of the double bonds migrates. Treatment of **37** with acid leads to the conjugated dienone, and thus to the progestin didrogestone (**38**).[8]

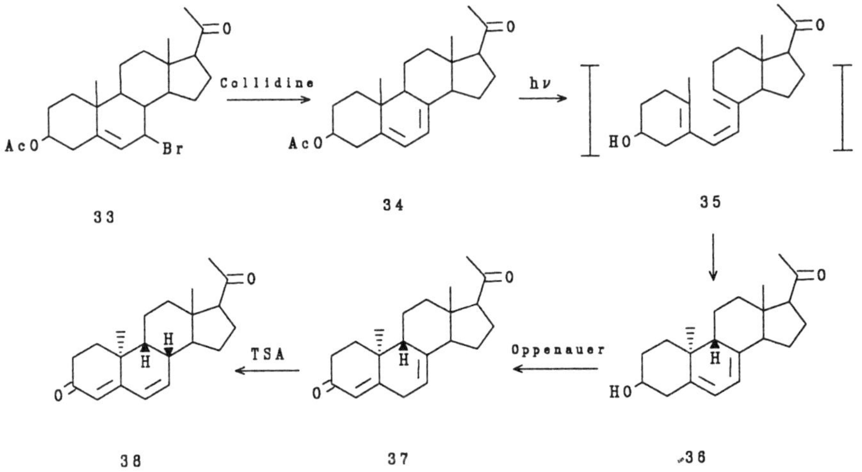

3. CORTICOSTEROIDS

The corticosteroids, a group of 11-oxygenated pregnanes secreted by the adrenal cortex, play a key role in the regulation of a sizeable number of homeostatic responses. It has been known since at least the mid-1920s that the pathology, which resulted in experimental animal models from adrenalectomy, could be reversed by administration of extracts from the adrenal cortex. Subsequent detailed investigation established that this activity was due mainly to two closely related steroids, cortisone and cortisol. The structures of these compounds differed from the progesterone in the presence of a ketone or hydroxyl group at the 11-position and further oxidation of the side chain at the 17-position to the dihydroxyacetone stage. Interest in this class of compounds might have been restricted to its use as a replacement drug in cases of adrenal insufficiency, but for the serendipitous discovery of their profound and widespread antiinflammatory activity. The observation that arthritis seemed to improve in the presence of elevated levels of endogenous cortisone together with the availability of adequate amounts of this drug from synthesis prompted clinical trials.[9] These trials confirmed the antiinflammatory activity of the compound, particularly when it was administered in supraphysiological doses. There followed several decades of research in many laboratories aimed at developing drugs that were more potent as well as agents that were antiinflammatory without the attendant corticosteroid activity. Although the first goal was eventually achieved, the development of nonhormonal analogues proved more elusive; as will be discussed, this split in activity was, however, partly achieved by changing the method of administration.

It is now known that corticoids owe their antiinflammatory activity to inhibition of phospholipase A_2, the enzyme that releases arachidonic acid. This class of compounds decreases levels of mediators from both legs of the arachidonic acid cascade; their widespread activity is thus attributed to decreased levels of both prostaglandins and leukotrienes. Recall that the NSAIDs affect only prostaglandins, since they specifically inhibit cyclooxygenase. The use of antiinflammatory corticoids came to be restricted as serious side effects directly traceable to their intrinsic hormonal activity became manifest. One of those effects results in a shutdown of endogenous cortisone production by a feedback mechanism at the level of the pituitary; this accounts for the fact that any course of treatment ends with gradual diminution of the dose rather than abrupt cessation so as to allow endogenous synthesis to resume.

The rarity of steroid natural products, which included an oxygen function at the 11-position combined with the chemical isolation of this position, proved a stumbling block to early steroid work. Although it had been shown that the mammalian biosynthesis of cortisone from progesterone could be replicated ex vivo by perfusing this compound through beef adrenal slices, this procedure was not practical for the scale required for preparing synthetic intermediates. The search for a counterpart to this reaction in a prokaryotic organism was rewarded at the Upjohn Company, by the finding that several molds of the *Rhizopus* family would hydroxylate progesterone to its 11α-hydroxyl derivative **39**. Compound **39** became the key to the preparation of synthetic corticoids; note, however, that the configuration of the new hydroxyl is reversed from that required for corticoid activity.

The next sequence represents one of the methods that had been used to convert the acetyl side chain to a dihydroxy acetone. In spite of its apparent complexity, this sequence was, and may still be, one of the methods for commercial large scale production of bulk corticoids. The first step consists in oxidation of the 11α alcohol to the somewhat more robust corresponding ketone **40**. In order to activate the 21-position to halogenation, it is converted to an oxalate. Condensation of the triketone with ethyl oxalate in the presence of alkoxide proceeds preferentially at the 21-position to give **41** due to the well-known high reactivity of methyl ketones. Reaction of the crude sodium enolate with bromine leads to the dibromide **42**, the oxalate moiety being cleaved under the reaction conditions. The Favorskii rearrangement is then used to in effect oxidize the 17-position so as to provide a site for the future hydroxyl group. Thus, treatment of **42** with an excess of sodium methoxide first provides an anion at

Rhizopus

CrO_3

Progesterone 39 40

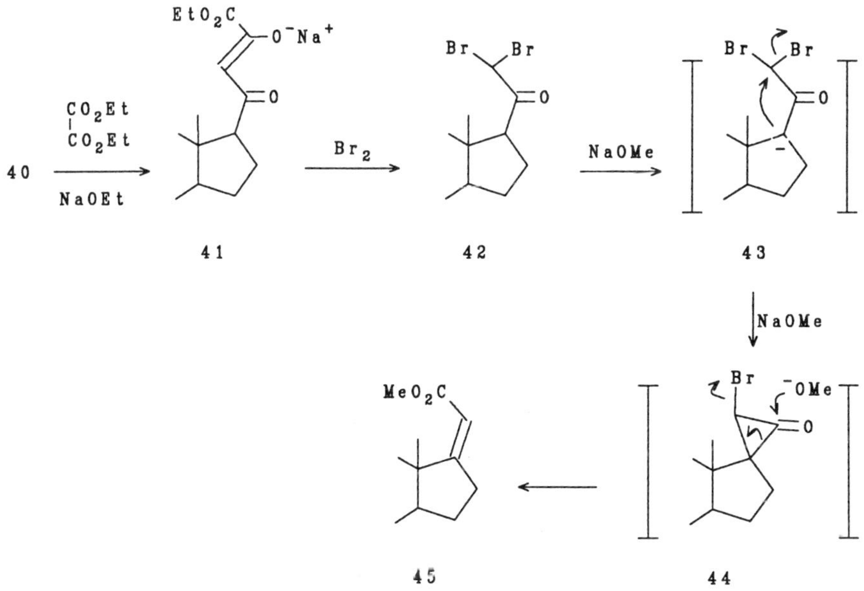

the 17-position. This compound (**43**) then cyclizes to the transient cyclopropanone **44**, which is characteristic of Favorskii rearrangements. Addition of the second mole of alkoxide to the carbonyl leads to acrylate ester **45** with simultaneous loss of bromide. Ring opening in the opposite sense is disfavored for several reasons, including the fact that it would lead to a bulky quaternary center at the 17-position. The cis stereochemistry about the double bond, which has been formally demonstrated, is probably due to the stereochemistry about the spirocyclic carbonyl group in **44**.

The ketone group at the 3-position is then protected as its acetal (**46**) by reaction with ethylene glycol, the carbonyl at 11 being totally resistant to these conditions; the enone double bond undergoes the customary shift. Treatment with lithium aluminum hydride reduces the acrylate ester at the 17-position to an allyl alcohol. The ketone at 11 is reduced to an alcohol in the same step; approach of hydride from the open α side results in a hydroxyl group that has the β configuration of active corticoids. Hydrolysis of the acetal to the 4-ene-3-one, followed by acetylation of the hydroxyl at the 21-position affords intermediate **47**, the 11β-hydroxyl group being resistant to acetylating conditions. The remaining task consists in conversion of the 17,20 double bond to a ketoalcohol. This conversion is accomplished in a single pot with a special combination of reagents. Osmium tetroxide is the reagent of choice for this transformation, since it is well known for its ability to convert olefins to trans 1,2-diols; it is, however, quite expensive and very toxic. The reaction is thus carried out in the presence of a cooxidant, *N*-methylmorpholine oxide, peroxide (NMOP). First, this compound regenerates osmium tetroxide from the initially formed cyclic osmate ester allowing this hydroxylating reagent to be used in catalytic amounts. The NMOP is also a sufficiently strong oxidant to then convert the hydroxyl at 20 to a ketone. Thus, reaction of **47** under these conditions completes the formation of the dihydroxyacetone side

45 → (HOCH$_2$CH$_2$OH) → 46 → (1. LAH; 2. Ac_2O; 3. H_3O^+) → 47 → (OsO_4, NMOP) → 48 → (CrO_3) → 49

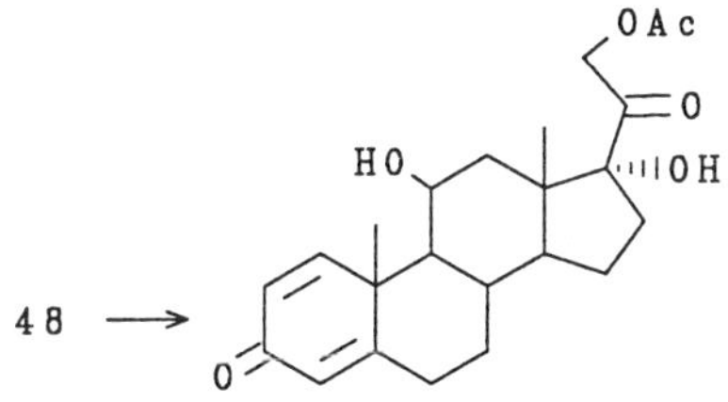

48 → 50

49 → 51

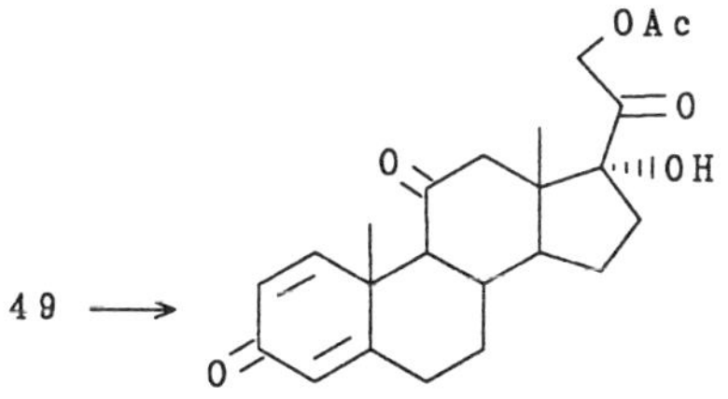

chain and **hydrocortisone acetate** (**48**). Further oxidation of the 11-hydroxyl group with, for example, Jones' reagent, give **cortisone acetate** (**49**).[10]

The increase in potency observed in the progestin series by incorporation of additional unsaturation in the A ring (see **delmadinone**) obtains in the corticoid series as well; in fact, the majority of commercial steroid antiinflammatories include this feature. The double bond at the 1-position may be formed by fermentation with an organism such as *Corynebacterium simplex*[11] or by reaction with selenium dioxide. **Dihydrocortisone acetate** (**48**) yields the widely used corticoid **prednisolone acetate** (**50**); in the same vein, cortisone acetate (**49**) goes to **prednisone acetate** (**51**). Mild saponification of either of these products yields the free alcohols **prednisolone** and **prednisone**, respectively.

The additive effect of potentiating groups is one hallmark of structure activity relations in steroids; while these cannot be simply summed arithmetically, cases where effects from two such groups cancel are very rare. Thus, substituents at the 6-, 9- and 16-positions generally lead to additive increases in potency. One the first potentiating groups to be recognized is a fluorine substituent at the 9α-position. The key intermediate to 9 fluorinated compounds consists of the corresponding corticoids, which have a 9,11 double bond. The 9α-fluoro-11β-hydroxy moiety is then constructed by the scheme that involves formation of a 9β-epoxide followed by its opening with the hydrogen fluoride–THF complex, which is described in detail in Chapter 4 in connection with the synthesis of fluoxymestrone. Dehydration of cortisol acetate (**48**) proceeds preferentially at the 11 hydroxyl to give the diene **52**; reactivity of the tertiary alcohol at the 17-position is seemingly decreased by adjacent ketone. The diene is then subjected to the standard hydroxy-fluorination scheme with the reaction proceeding at the more electron-rich double bonds. Thus **fludrocortisone acetate** (**53**),[12] a

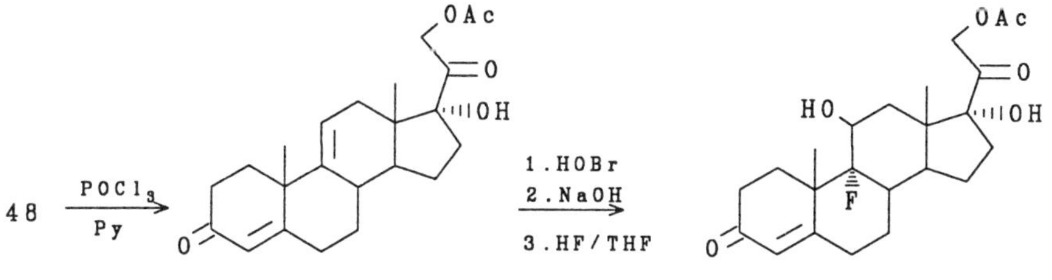

48 → ($POCl_3$, Py) → 52 → (1. HOBr 2. NaOH 3. HF/THF) → 53

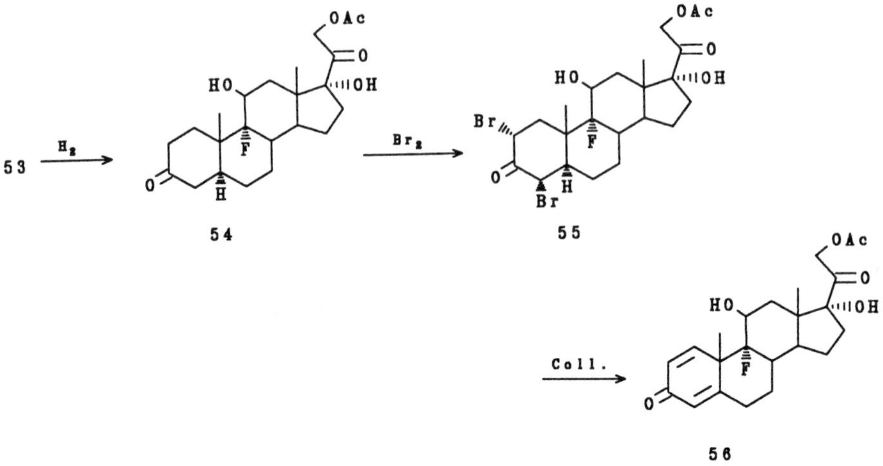

compound that is close to an order of magnitude more potent than cortisol acetate in some bioassays is obtained.

Addition of a second double bond in the A ring leads to a further increase in potency. One published synthesis of this compound begins with hydrogenation of fluorohydrin **53** to the saturated intermediate **54**. Treatment of **54** with bromine leads to the dibromide **55**; in fact, this reaction, as noted earlier, follows a quite complex pathway involving at one point a 2,2-dibromo derivative. Dehydrobromination with a base such as collidine establishes the 1,4-diene function to yield **fluoroprednisolone acetate** (**56**).[13]

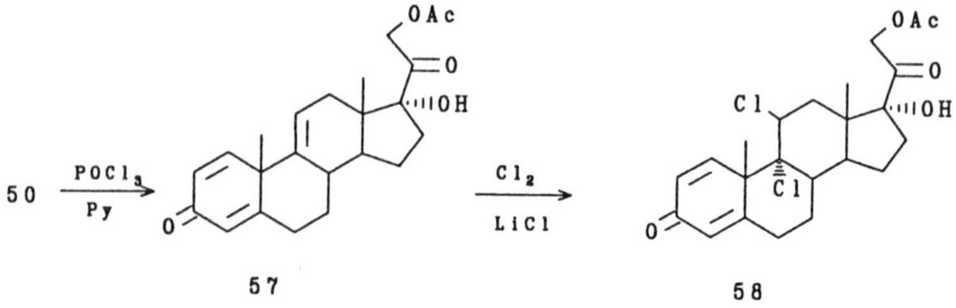

Corticoids that incorporate chlorine at the 9-position also show an increase in potency over the corresponding parent drugs bearing hydrogen at this position. The good activity of **dichlorisone** might at first sight seem to indicate that the 11-hydroxyl group is dispensable were it not for some evidence to indicate that this halogen is converted to hydroxyl in vivo. This compound is prepared in a straightforward manner by treatment of the triene **57** from dehydration of prednisolone acetate with chlorine. The 9α,11β stereochemistry of the product is that which would be predicted from diaxial opening of a 9α,11α-chloronium ion from addition of chlorine from the less hindered side; lithium chloride may provide the anion for starting the ring-opening sequence. Thus, the corticoid **dichlorisone acetate** (**58**)[14] is obtained.

Fluorine, at the equatorial 6α-position, also results in an increase in potency. One scheme for the production of such a compound relies on the shift of the double bond from the 4,5 to the 5,6-position, which follows formation of an acetal at the 3-position. The scheme starts with the selective reduction of the 11 ketone in the cortisone intermediate **46**. Epoxidation of **59** with a peracid takes place selectively at the unconjugated double bond to give the 5α,6α-oxirane **60**. Treatment of **60** with hydrogen fluoride in THF leads to fluorohydrin (**61**). (There is evidence to indicate that the reagent involves a HF–THF acid–base complex since omission of THF in at least some cases leads to a complex mixture of rearrangement products.[15]) The side chain in fluorohydrin (**61**) is then converted to the dihydroxyacetone derivative **62** by applying a set of transforms similar to those used in going from **46** to **49**. The ester is thus first reduced to an alcohol by means of lithium aluminum hydride. The resulting primary alcohol is converted to its acetate. Hydrolysis of the acetonide followed by oxidation of the olefin with a catalytic amount of osmium tetroxide in the presence of *N*-methylmorpholine oxide, then establishes the required corticoid side chain.

Treatment with base results in elimination of the alcohol at the 5-position by what is in effect a reverse Michael reaction to give the enone **63**. The additional double bond in ring A is then introduced by treating the last intermediate with selenium dioxide to give the 1,4-diene-3-one **64**. At this point, the 6-fluoro function still occupies the axial β configuration, which resulted from opening of the epoxide; treatment of **64** with strong acid leads the halogen to epimerize to the more stable equatorial 6α-position to give **fluprednisolone acetate** (**65**).[16]

46 → NaBH4 → 59 (MeO2C, HO, O, O) → mCPDA → 60 (MeO2C, HO, O, O, O) → HF / THF → 61 (MeO2C, HO, O, O, HO, F) → → → 62 (OAc, O, OH, HO, O, HO, F)

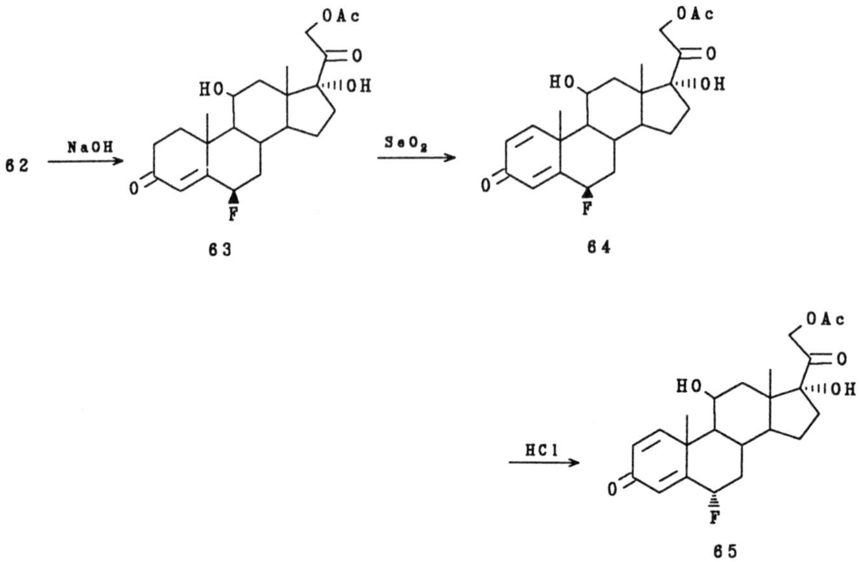

62 NaOH → 63 SeO$_2$ → 64 HCl → 65

A rather different scheme is used for the preparation of the analogue bearing a methyl group at the 6-position; since this group is most conveniently incorporated by means of a Grignard reaction, the carbonyl at the 20-position needs to be protected first. The protection group used to accomplish this takes advantage of the juxtaposition of two hydroxyl groups next to the ketone. Reaction of the side chain in **cortisone** (**66**) with formaldehyde to form a bismethylene-dioxy group (BMD) probably involves formation of a hemiacetal as the first step. The terminal hydroxyl group can then add to the adjacent ketone to give a cyclic hemiacetal; formation of a cyclic acetal from what is functionally a 1,2-glycol with the second mole of formaldehyde completes the formation of the BMD derivative, a particularly useful group since it protects both the ketone and the two adjacent hydroxyl groups.

The ketone group at the 3-position of product **67** is then converted to cyclic acetal **68**; the double bond undergoes the customary shift to the 5,6-position. Treatment of the olefin with a peracid such as *m*-perbenzoic acid gives the 5α,α6 epoxide admixed with a significant amount of its epimer (**69**); such a mixture could lead to regioisomers on direct reaction with an organometallic reagent. The alternate route chosen first involves rearrangement of the epoxide mixture with formic acid to give the 6 ketone **70**.

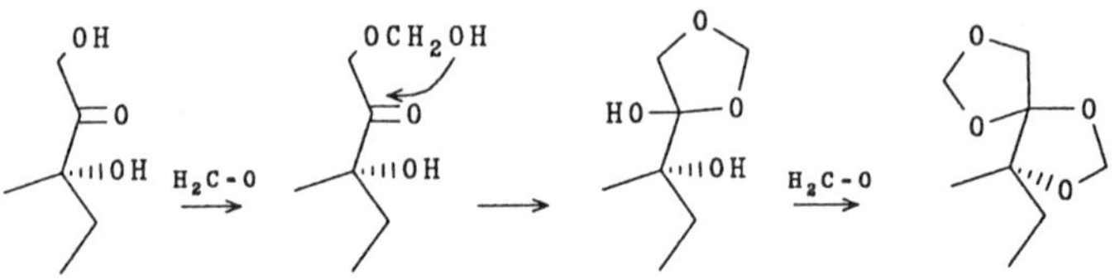

$H_2C{=}O$ → → $H_2C{=}O$ →

"BMD"

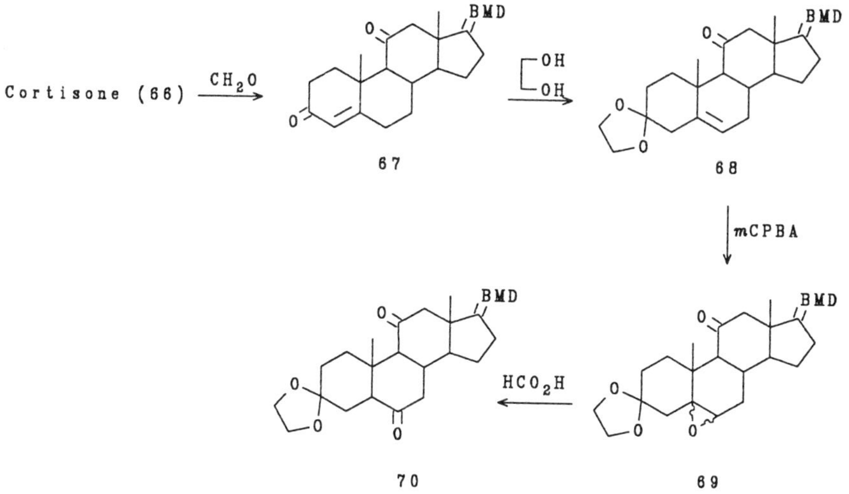

Cortisone (66) $\xrightarrow{CH_2O}$ 67 $\xrightarrow{HOCH_2CH_2OH}$ 68 $\xrightarrow{mCPBA}$ 69 $\xrightarrow{HCO_2H}$ 70

The resulting ketone (**70**) is then condensed with methylmagnesium bromide to give the corresponding methyl-carbinol; the highly unreactive nature of the carbonyl group at the 11-position is further illustrated by the selective reaction at the 6-position. The resulting tertiary alcohol is then dehydrated and the ketone at 11 is reduced to an alcohol by means of lithium aluminum hydride to give intermediate **71**. The acetal is then removed by exchange with acetone in the presence of dilute acid; the requisite 1,4-diene functionality is then put in place by dehydrogenation by means of selenium dioxide to give dienone **72**. Removal of the bismethylenedioxy protecting group by treatment with acetic acid completes the synthesis of the widely used corticoid **methylprednisolone 73**.[17]

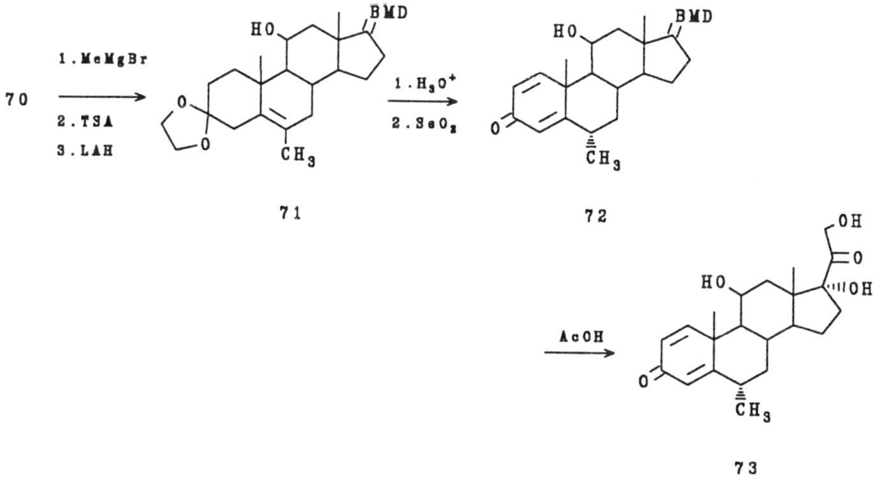

70 $\xrightarrow{1.\ MeMgBr;\ 2.\ TSA;\ 3.\ LAH}$ 71 $\xrightarrow{1.\ H_3O^+;\ 2.\ SeO_2}$ 72 $\xrightarrow{AcOH}$ 73

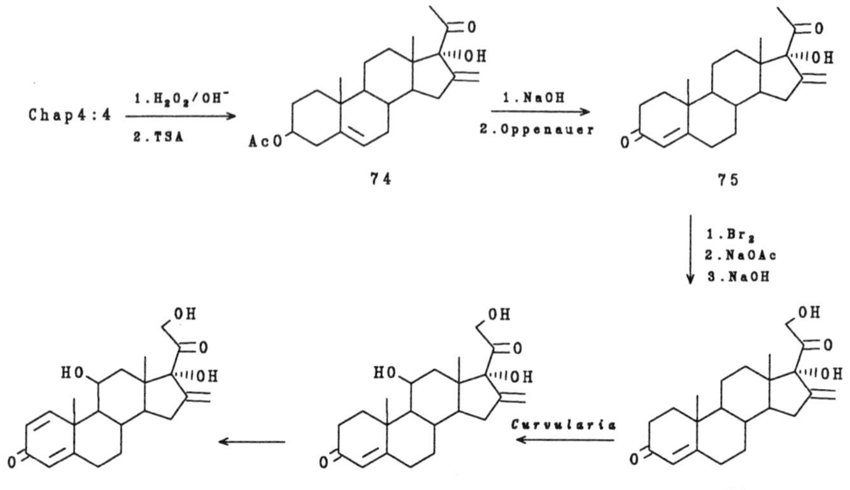

Chap4:4 → 1. H_2O_2/OH^- 2. TSA → 74 → 1. NaOH 2. Oppenauer → 75 → 1. Br_2 2. NaOAc 3. NaOH → 76 → *Curvularia* → 77 → 78

Substitution at the 16-position also leads to more potent compounds; the additional steric bulk introduced by such substituents also protects the dihydroxyacetone side chain against metabolic degradation. The synthesis used to prepare one compound bearing such substitution departs from those so far discussed by deferring the microbiological introduction of oxygen to a relatively late stage. The steps used to incorporate a 16 methylene group very closely parallel those used in the synthesis of melengestrol (**29**). The key intermediate **74** is obtained by the same sequence of reactions (cycloaddition of diazomethane, pyrolysis, epoxidation, and rearrangement) starting with 16-dehydropregnenolone instead of its 6-methyl derivative. Saponification of the acetate followed by Oppenauer oxidation of the resulting alcohol leads to conjugated ketone **75**. The requisite oxygen function is then established by a two-step sequence; reaction of **75** with bromine takes place selectively at the very reactive methyl group at the 21-position. Displacement of halogen from the thus formed bromoketone with acetate anion leads to the 21 acetoxy ketone and the intermediate **76**. This reaction in effect constitutes an alternative strategy for building the side chain from the one involving the Favorskii rearrangement. The 11-hydroxyl group is introduced by fermentation of intermediate **76** with a *Curvularia.* The newly introduced hydroxyl group in product **77** interestingly already has the 11β configuration, which is required for corticoid activity. Dehydrogenation of the product by further fermentation[18] yields **prednylidene** (**78**).

The additive effect of potentiating groups in the corticoid series is aptly illustrated by the fact that **dexamethasone** (**81**), which includes a substituent at 16, a fluorine at 9, and an additional double bond at the 1-position, is close to 100 times as potent as cortisone when used as an antiinflammatory agent in animal models. The 16β-epimer **betamethasone** retains much of this activity, showing about 30 times the potency of cortisone. Dehydration of 16β-methylprednisolone acetate (**79**)[19] leads to the 9,11-dehydro derivative **80.** The olefin is then converted to the corresponding 9α-bromo-11β-hydrin by means of a source of hypobromite such as basic NBS and is closed to an

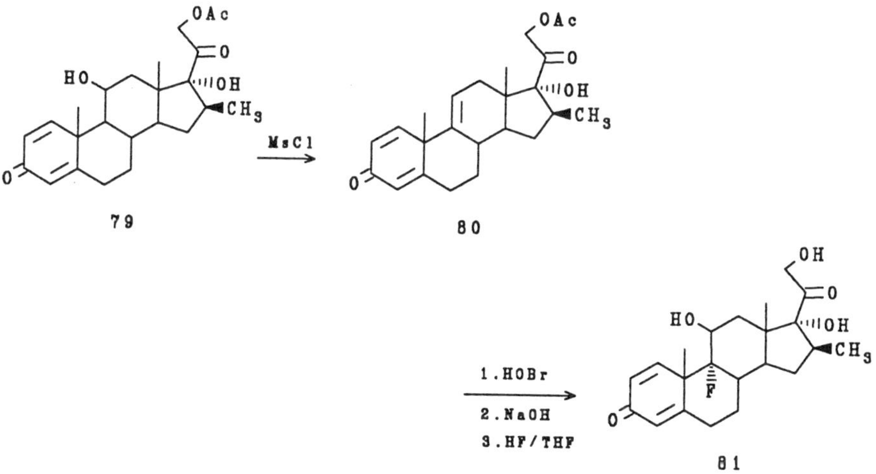

79 —MsCl→ 80

—1. HOBr, 2. NaOH, 3. HF/THF→ 81

epoxide. The standard ring-opening reaction with hydrogen fluoride in THF gives dexamethasone (**81**).[20]

The observation that 16α-hydroxy compounds and their derivatives also show increased potency is an indication that steric bulk at the 16-position is the more important factor in increases in potency of these drugs. The key intermediate in this case is a triene that includes olefins at both the 9,11- and 16,17-positions. The dehydration reactions described thus far involved only loss of the hydroxyl group at the 11-position; conversion of the ketone at 20 to its acetal apparently facilitates loss of the hydroxyl at the 17-position. The requisite intermediate **82** is obtained by reaction

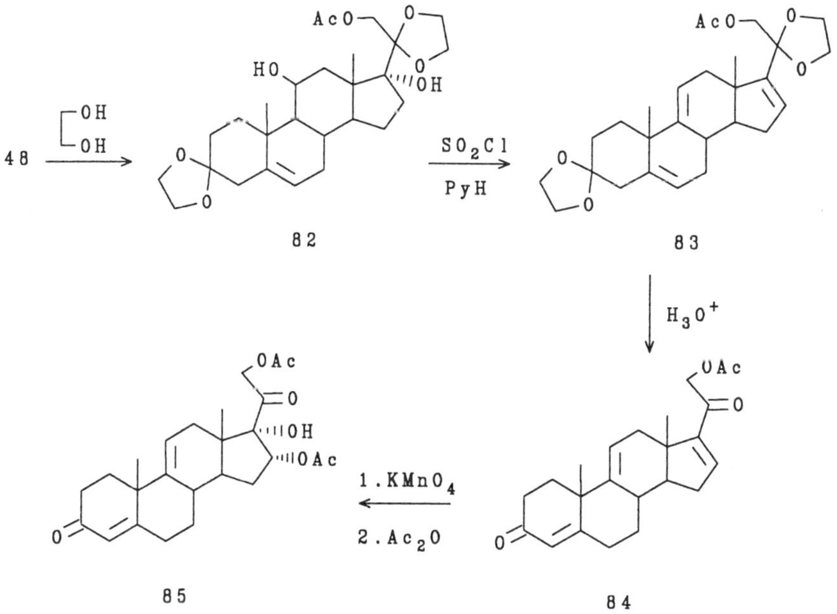

48 → 82 —SO_2Cl, PyH→ 83 —H_3O^+→ 84 —1. $KMnO_4$, 2. Ac_2O→ 85

of hydrocortisone acetate (**48**) with ethylene glycol under forcing conditions. Compound **83** goes on to the triene **83** on exposure to thionyl chloride in pyridine in the cold. The reaction probably proceeds by elimination of intermediate chlorosulfonate esters since the nature of the hydroxyl groups would seem to preclude those from going on to collapse to chloro compounds. The acetal groups are then removed by hydrolysis to give **84**. Oxidation of **84** with osmium tetroxide, or more practically, with potassium permanganate proceeds selectively at the olefin at 16,17 to give the corresponding 16α,17α-diol; reaction with acetic anhydride leads to the diacetate **85**; the stereochemistry is of course dictated by approach of the reagent from the open side of the molecule.

The 9,11 olefin in intermediate **85** is then converted to the 9α-fluoro-11β-hydrin by the customary scheme to afford **triamcinolone** (**87**). Reaction of **87** with acetone converts the 16,17-diol to an acetal to afford **triamcinolone acetonide** (**88**).[21] The latter is approximately 50 times more potent than cortisone in animal models; the free glycol interestingly shows only a 2.5-fold increase in potency over the cortisone in the same model.

The availability of fairly standardized procedures for the introduction of the various potentiating groups combined with the realization that their biological effects were often additive led to the synthesis of a plethora of highly substituted corticoids. The final steps in the synthesis of a corticoid, which combines four separate potentiating groups within the same molecule, leads to a compound, **flumethasone** (**92**), which is representative of this work. The synthesis starts with the 16β-methyl analogue **89** of the 6-fluoro corticoid **63** by using an analogous route for introducing the halogen substituent. The 11-hydroxyl function is then dehydrated to the corresponding olefin **90**; application of the by now familiar bromohydration, oxirane ring closure, and epoxide opening scheme leads to intermediate **91**, which now contains two fluorine atoms. Further dehydrogenation of ring A followed by saponification of the acetate at

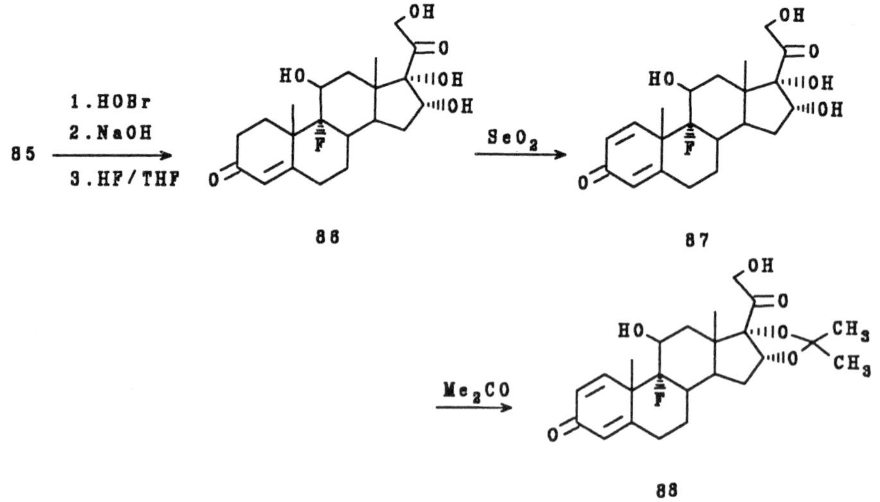

85 → (1. HOBr; 2. NaOH; 3. HF/THF) → 86 → (SeO_2) → 87 → (Me_2CO) → 88

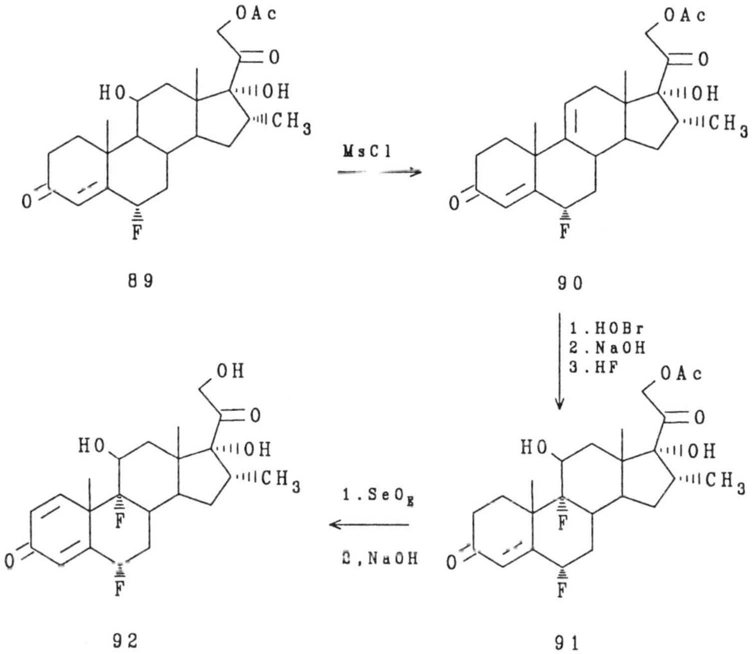

89 → (MsCl) → 90

90 → (1. HOBr; 2. NaOH; 3. HF) → 91

91 → (1. SeO_2; 2. NaOH) → 92

the 21 position affords flumethasone (**92**),[22] which shows fully 420 times the potency of cortisone in an animal model for antiinflammatory activity.

As in the case of the less highly substituted compounds, the acetonide of a 16α-hydroxyl corticoid shows about the same potency as its 16α-methyl counterpart. Thus, **fluocinonide** (**93**), which is obtained by an analogous set of transformations starting with a 6 fluorinated 16 dehydro intermediate,[23] is 440 times as potent as cortisone in the standard assay.

Although corticosteroids are significantly more effective as antiinflammatory agents than their nonsteroidal counterparts, they are, as a result of the potential for causing serious side effects, usually reserved for treating otherwise intractable inflammation. The activity of the former in relieving topical inflammation constitutes another notable difference between these compounds and the NSAIDs; the latter are virtually

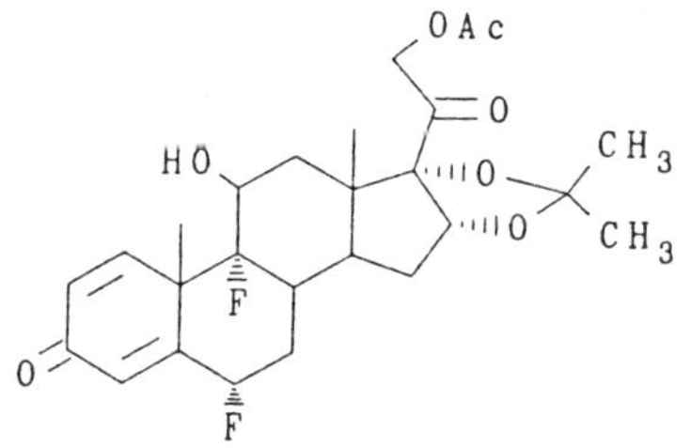

93

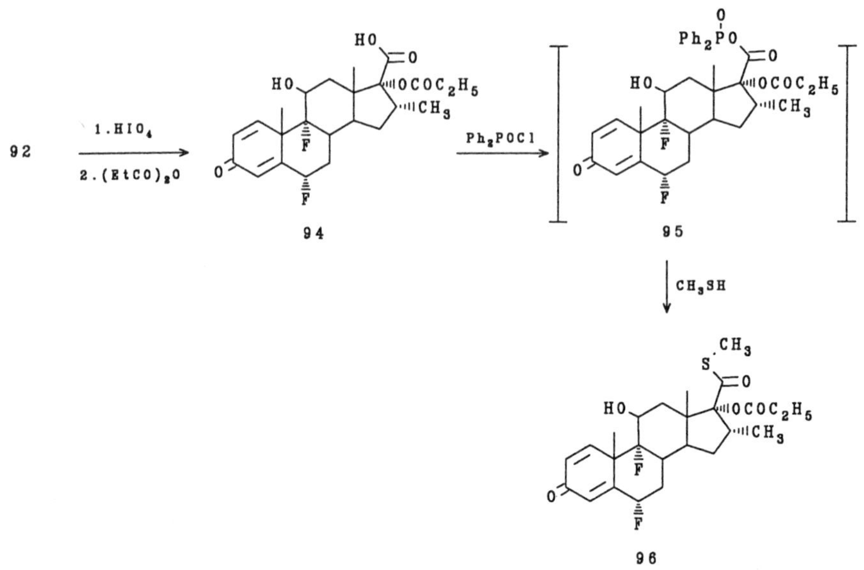

inactive by this route of administration. The availability of the extremely potent corticoids, such as those discussed above, has led to their widespread use for treating topical inflammations such as rashes and insect bites. One aspect of asthma involves inflammation of the bronchioles related to the release of leukotrienes. One important application of corticoids lies in the treatment of asthma by topical application to the lungs by inhalation. Although the use of high-potency compounds permits the use of very small doses, it carries with it the possibility of induction of side effects from the portion of the drug that enters the circulation via transdermal absorption. The observation that biological activity was retained in the face of considerable modification in the structure of the side chain permitted the design of corticoids with side chains that would be metabolized by plasma enzymes to inactive products. The cleavage of thioester moieties present in both **ticabesone** (**96**) and **fluticasone** (**98**) will lead to the corresponding biologically inactive C-17 free carboxylic acids.

Reaction of the hydroxyacetone side chain in flumethasone (**92**) with periodic acid leads to cleavage of this function to give carboxylic acid **94**, with the loss of the carbon atom at C-21. Further reaction of the very hindered acid group requires prior activation. Thus, acylation with diphenylchlorophosphate leads to the mixed anhydride; this compound (**95**) is not isolated, but treated immediately with methyl mercaptan. The product, ticabesone (**96**) is a quite effective topical antiinflammatory agent.[24]

The presence of the fluoromethyl thioester group in fluticasone (**98**) requires that the carboxyl group in **94** first be converted to a thioacid. This conversion can be accomplished most conveniently by taking advantage of the rearrangement of a mixed anhydride of **94** with an acid thione. This transient anhydride (**94a**) is obtained by reaction of the acid with *N,N*-dimethylformamidoyl chloride; this compound rearranges to its isomer **94b** under reaction conditions.

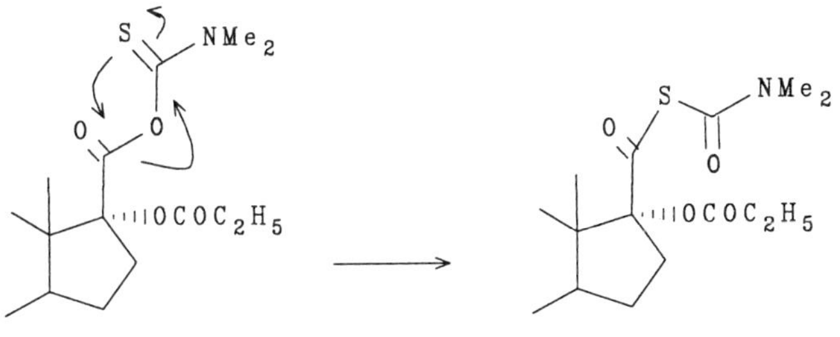

94A 94B

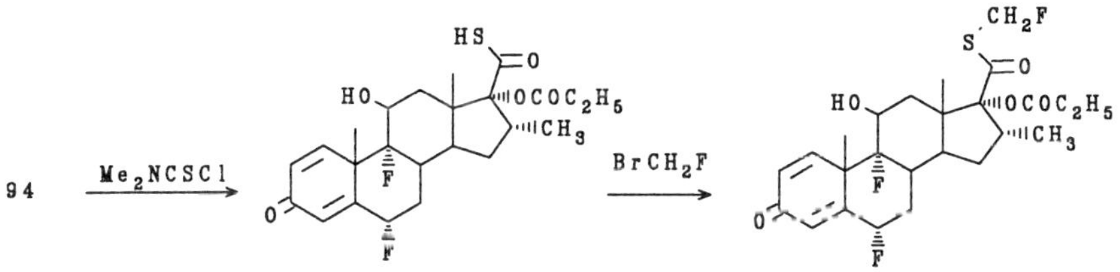

97 98

Saponification of the anhydride under mild conditions selectively cleaves the anhydride to give the thioacid **97**. Alkylation of **97** with bromofluoromethane in the presence of base affords fluticasone (**98**).[25]

Biological activity is interestingly maintained even in the face of the total omission of a carbonyl group from the side chain. Oxidation of the free alcohol from **fluoroprednisolone acetate (56)** under more drastic conditions with sodium bismuthate

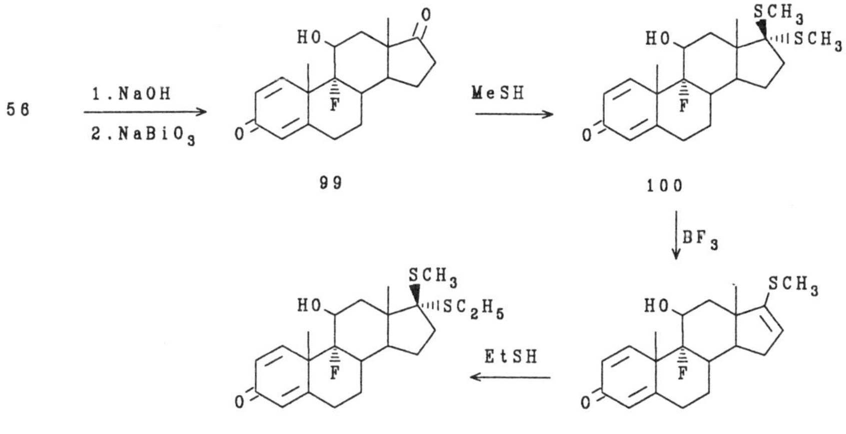

99 100 101 102

leads to complete loss of the dihydroxyacetone side chain to give the 17-keto derivative **99**. Reaction of Compound **99** with methyl mercaptan leads to selective formation of the dimethyl thioacetal at the 17-position, **100**. Reaction with strong acid leads to loss of one of the methyl mercapto groups and formation of the corresponding 17-thioenol ether **101**. Exposure of intermediate **101** to ethyl mercaptan results in formation of the unsymmetrical acetal **102** (**tiprednane**).[26] The stereochemistry follows from addition of ethyl mercaptan from the customary less hindered face of the steroid.

REFERENCES

1. Ringold, H.J.; Loken, B; Rosenkraz, G.; Sondheimer, F. *J. Am. Chem. Soc.* **1956**, *78*, 816.
2. Babcock, J.C.; Gutsell, E.S.; Herr, M.E.; Hogg, J.A.; Stucki, J.C.; Barnes, L.E.; Dulin, W.E. *J. Am. Chem. Soc.* **1958**, *80*, 2904.
3. Ringold, H.J.; Ruelas, J.P.; Batres, E.; Djerassi, C. *J. Am. Chem. Soc.* **1959**, *81*, 3712.
4. Bruckner, K.; Hampel, B.; Johnson, V. *Chem. Ber.* **1961**, *94*, 1225.
5. Ringold, H.J.; Batres, E.; Bowers, J.; Edwards, J.; Zderic, J. *J. Am. Chem. Soc.* **1959**, *81*, 3485.
6. Kirk, D.N.; Petrow, V.; Williamson, D.N. *J. Chem. Soc.* **1961**, 2821.
7. Degheghi, R.; Revesz, C.; Gaudri, R. *J. Med. Chem.* **1963**, *6*, 2821.
8. Rappolt, M.P.; Westerhoff, P. *Rec. Trav. Chim. Pays-Bas* **1961**, *80*, 43.
9. Hench, P.S.; Kendall, E.C.; Slocumb, C.H.; Polley, H.F. *Proc. Staff Meet. Mayo Clinic* **1949**, *24*, 181.
10. Hogg, J.A.; Beal, P.F.; Nathan, A.H.; Lincoln, F.H.; Schneider, W.P.; Magerlein, B.J.; Hanze, A.R.; Jackson, R.W. *J. Am. Chem. Soc.* **1955**, *77*, 4436.
11. Nobile, A.; Charney, W.; Perlman, P.L.; Herzog, H.L.; Payne, C.C.; Tully, M.E.; Jernik, M.A.; Hershberg, E.B. *J. Am. Chem. Soc.* **1955**, *77*, 4184.
12. Fried, J.; Sabo, E.F. *J. Am. Chem. Soc.* **1954**, *76*, 1455.
13. Fried, J.; Florey, K.; Sabo, E.F.; Herz, J.E.; Restivo, A.R.; Borman, A.; Singer, F.M. *J. Am. Chem. Soc.* **1955**, *77*, 4181.
14. Robinson, C.H.; Finckenor, L.; Olivetto, E.P.; Gould, D. *J. Am. Chem. Soc.* **1959**, *81*, 2191.
15. Author's personal, unpublished observations.
16. Hogg, J.A.; Spero, G.B. **1958**, U. S. Patent 2841600. *Chem. Abstr.* **1958**, *52*, 20274.
17. Fried, J.H.; Arth, G.E.; Sarett, L.H. *J. Am. Chem. Soc.* **1959**, *81*, 1235.
18. Manhardt, H.J.; van Werder, F.V.; Bork, K.H.; Metz, H.; Bruckner, K. *Tetrahedron Lett.* **1960**, 61.
19. Taub, D.; Hoffsomer, R.D.; Slates, H.L.; Wendler, N.L. *J. Am. Chem. Soc.* **1958**, *80*, 4435.
20. Arth, G.E.; Johnston, D.B.R.; Fried, J.; Spooner, W.; Hoff, D.; Sarett, L.H.; Silber, R.H.; Stoerk, H.C.; Winter, C.A. *J. Am. Chem. Soc.* **1958**, *80*, 3161.
21. Bernstein, S.; Lenhard, R.H.; Allen, W.S.; Heller, M.; Littel, R.; Stollar, S.M.; Feldman, L.I.; Blank, R.H. *J. Am. Chem. Soc.* **1959**, *81*, 1689.
22. Edwards, J.A.; Ringold, H.J.; Djerassi, C. *J. Am. Chem. Soc.* **1959**, *81*, 3156.
23. Mills, J.S.; Bowers, A.; Ringold, H.J.: Djerassi, C. *J. Am. Chem. Soc.* **1960**, *82*, 3399.

24. Kertesz, D.J.; Marx, M. *J. Org. Chem.* **1986**, *51*, 2315.
25. Phillips, G.H.; Bailey, E.J.; Bain, B.M.; Borella, R.A.; Buckton, J.B.; Clark, J.C.; Doherty, A.E.; English, A.F.; Fazakerley, H. *J. Med. Chem.* **1994**, *27*, 3717.
26. Varma, R.K. **1982**, U. S. Patent 4361559. *Chem. Abstr.* **1983**, *96*, 163044.

CHAPTER 6

NONSTEROIDAL ESTROGENS AND THEIR ANTAGONISTS

1. INTRODUCTION

Synthetic estrogens and their antagonists represent a small but important class of drugs, one which is finding extensive and increasing use in the treatment of breast cancer. Recently, another member of this group of drugs has been introduced in India as an oral contraceptive agent. The structures of these compounds at first sight comprise a much more heterogeneous collection than discussed in previous chapters. Careful examination will quickly disclose the fact that they all incorporate a common pharmacophore: a stilbene or a 1,2-diphenylethane moiety with one of the benzene rings bearing a dialkylaminoethoxy substituent in the case of the antagonists. Both agonists and antagonists have also been shown to bind to estrogen receptors.

2. SYNTHETIC ESTROGENS

As soon as the estrogens were isolated, it became apparent that these compounds were extremely potent and could have important, but then as yet unidentified, therapeutic applications. It was also realized that isolation of adequate supplies of these compounds from natural sources would be difficult because of the low levels at which they occur. Preparation of estrone or estradiol proper by total synthesis was ruled out because their full structure had yet to be established; the synthetic methods available in the early 1930s would probably have not been adequate to accomplish this task if the full detailed structure had been known. It had been observed that steroids yielded polycyclic hydrocarbons on destructive distillation such as the phenanthrene derivative

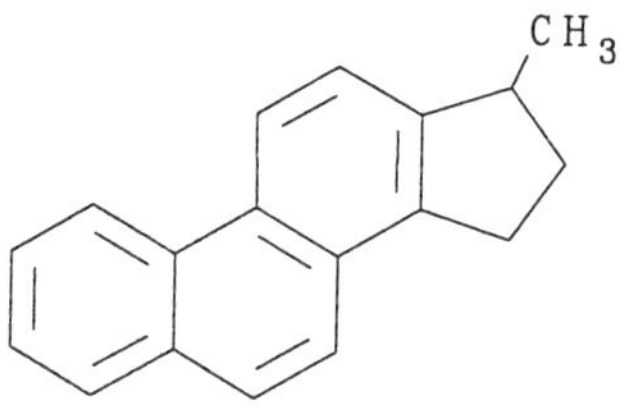

CH_3 O

1 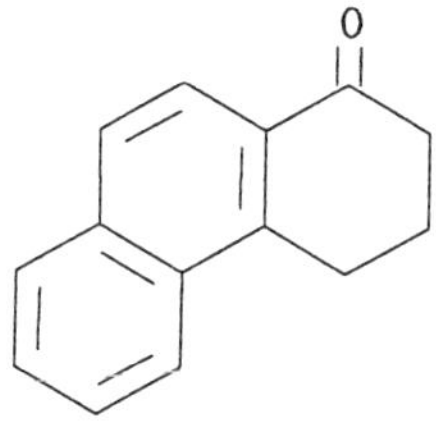 2

1 from degradation of cholesterol. The search for practical nonsteroid estrogens was given impetus by the finding that the structurally related phenanthrene derivative **2** showed this activity in animal models.[1]

Diethylstibestrol (**8**) also known as DES, was one of the first practical nonsteroidal estrogenic drugs. This agent was at one time widely used—and too often misused—for treatment of a number of hormone related disease states. The very belated discovery that DES elicited cancers in daughters of women who had been prescribed the drug during pregnancy virtually ended its use. The one time widespread use of DES led to the development of a large number of syntheses. One synthesis begins with chlorination of the benzylic position of *p*-ethylanisole (**3**) with *N*-chlorosuccinimide (NCS) to give **4**. Treatment of **4** with sodium amide in liquid ammonia initially leads to formation of a carbanion (**5**) at the benzylic position by abstraction of a proton. This anion then displaces chlorine on a second molecule of the halide; this displacement results in net alkylation and formation of the transient chlorinated dimer **6**. Dehydrochlorination of **6** by excess amide leads to formation of stilbene (**7**) as a mixture that consists predominantly of the desired trans isomer. The steric outcome of this reaction suggests that the alkylation product consists largely of the erythro isomer, indicating that the alkylation step leads to the least crowded intermediate. Demethylation of the stilbene (**7**) by means of hydrogen bromide in acetic acid leads to diethylstilbestrol (**8**).[2]

The estrogenic activity of diethylstilbestrol (8) was attributed to its very general structural mimicry of estradiol, where one of the phenolic rings acted as a surrogate A ring and the other filled in for the cyclopentano moiety; the relatively rigid stilbene framework would serve to hold those rings in place while supplying the required steric bulk; it was generally accepted that the compound interacted with a receptor for estradiol long before these entities were actually identified. More recent investigations showed that diethylstilbestrol and related synthetic estrogens show binding constants with estrogen receptors that, for selected compounds, exceed those of the endogenous hormones. This class is unique among steroid receptors because of its wide tolerance for structural modifications, as will be noted from the structures of the estrogen antagonists. It also stands out by the fact that none of the receptors for other classes of steroids show appreciable binding with nonsteroid compounds.

The conjugated diene system in the synthetic estrogen **dienestrol** (**11**) should provide even greater rigidity than the stilbene bond in **8**. The key intermediate **10** is obtained in a straightforward manner by pinacol coupling of the substituted propiophenone **9**. Dehydration of the glycol with a mixture of acetyl chloride and acetic

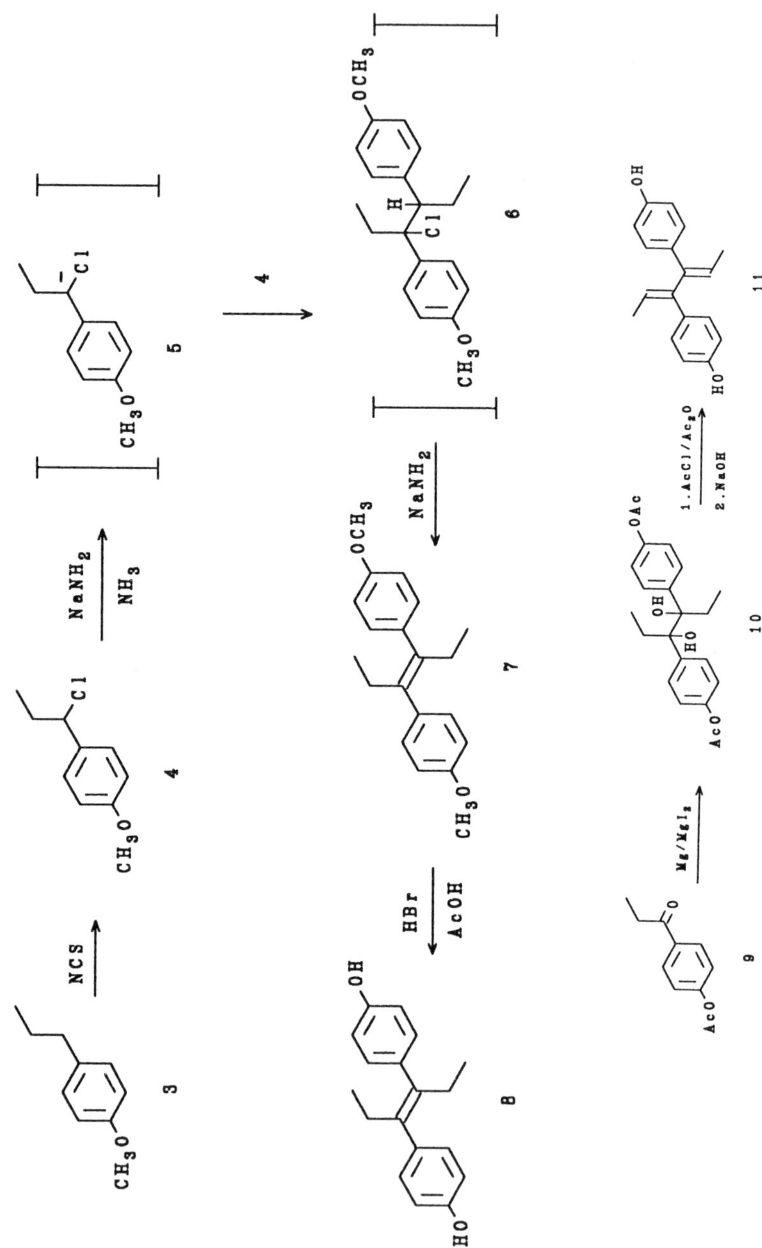
NCS
NaNH2
NH3
4
NaNH2
HBr
AcOH
Mg/MgI2
1. AcCl/Ac2O
2. NaOH
3
4
5
6
7
8
9
10
11

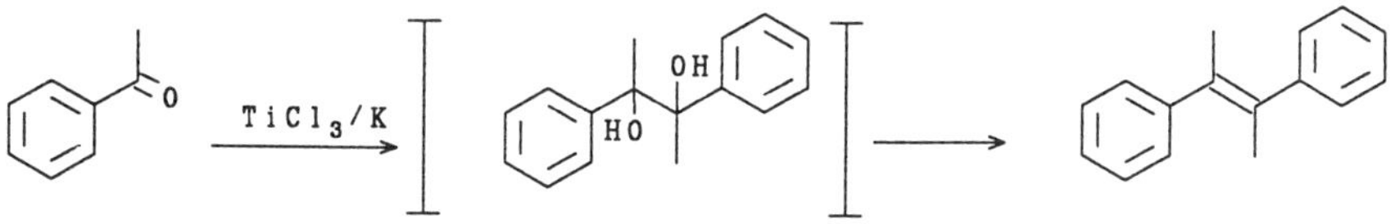

12 → ($TiCl_3/K$) → [13] → 14

anhydride leads to the diene transoid system; saponification removes the acetate groups and thus affords dienestrol (**11**).[3]

This sequence bears an interesting resemblance to a one-step coupling reaction developed many decades later, a transform that would seem applicable to the synthesis of DES if that were still an important drug. Thus, treatment of acetophenone (**12**) with titanium trichloride has been demonstrated to yield the pinacol product **13** initially; if a reducing agent such as potassium metal is present, the glycol eliminates to form a double bond. Product **14** consists predominantly (~90%) of the trans isomer.[4]

The following two examples owe their significance mainly to the fact that they led to the development of the much more important estrogen antagonists. The finding that the unsubstituted hydrocarbon triphenylethylene showed some activity as an estrogen in animal models aptly illustrates the nonspecific nature of the estrogen receptor. Adding methoxyl groups to the para positions of the hydrocarbon increases lipophilicity, thus leading to an increase in potency. Condensation of the Grignard reagent from 4-bromomethylanisole with anisophenone (**15**) gives the alcohol **16**; treatment of the tertiary alcohol with acid leads to formation of the triarylethylene **trianisene** (**17**).[5]

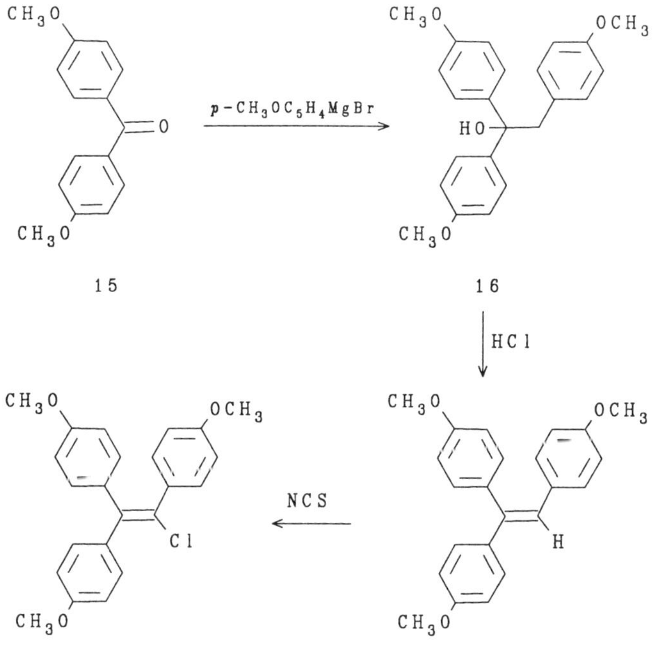

15 → (p-$CH_3OC_5H_4MgBr$) → 16 → (HCl) → 17 → (NCS) → 18

Replacement of the remaining hydrogen on the ethylene by chlorine from the reaction of **17** with *N*-chlorosuccinimide leads to **chlorotrianisene** (**18**).[6] This replacement leads to another increase in potency, perhaps by providing additional bulk.

3. NONSTEROID ESTROGEN ANTAGONISTS

Incorporation of a side chain bearing basic nitrogen in a molecule closely related to chlorotrianisene (**18**) markedly changes biological properties. While the compound still interacts with the estrogen receptor, it largely blocks the effects of endogenous or exogenous estrogens. This activity of the prototype **clomiphene** (**24/25**) is less clear than that of later congeners, because it in fact consists of a mixture of geometric isomers, only one of which is an antagonist. The observation that this compound showed contraceptive activity in rats by inhibiting estrogen-dependent implantation of fertilized ova led to the first concentrated research effort in this area. The known dependence on estrogen of the majority of breast cancers provided additional motivation for one research group; this research led to **tamoxifen** (**34**), an important drug for adjuvant therapy of mammary cancer. The original goal, development of a nonsteroid antifertility compound, was fulfilled in India by the recent approval of the use of the anti-estrogen **centchroman** (**71**) as an oral contraceptive.

The synthetic route to **clomiphene (24/25**) is in fact very close to the one used for its nonbasic parent. The basic side chain, usually referred to as a basic ether, is incorporated in the first step by alkylation of the phenol in 4-hydroxybenzophenone (**19**) with 2-chlorotriethylamine. Addition of benzylmagnesium bromide to **20** affords the tertiary alcohol **21**.

The known preference for transoid elimination of the elements of water from alcohols such as **21** controls the stereochemistry of the product. The arrangement in the starting material of the groups about the incipient olefin actually determines the steric identity of the product. The two rotamers of alcohol **21** that have the trans hydrogen and hydroxyl, shown as their Newman projections **21a** and **21b**, are equally probable, since they differ only in the placement of the remote basic ether. The dehydration in fact gives a mixture of the trans isomer **22** and the cis isomer **23**, presumably from rotamers **21a** and **21b**, respectively. Reaction of the mixture with NCS gives a mixture of the chlorinated derivatives;[7] in fact, commercial clomiphene

HO — **19** → ($Et_2NCH_2CH_2Cl$, NaOH) → Et_2N–CH₂CH₂–O — **20** → ($C_6H_5CH_2Br$, Mg) → Et_2N–CH₂CH₂–O, OH, CH_2 — **21**

consists of an approximately 6:4 mixture of the isomers **24** and **25**. It was determined later that the isomer **25** acts as an estrogen antagonist while the isomer **24** retains considerable agonist activity. The mixture is, however, quite effective in inducing ovulation in cases of infertility, the approved indication for this drug.

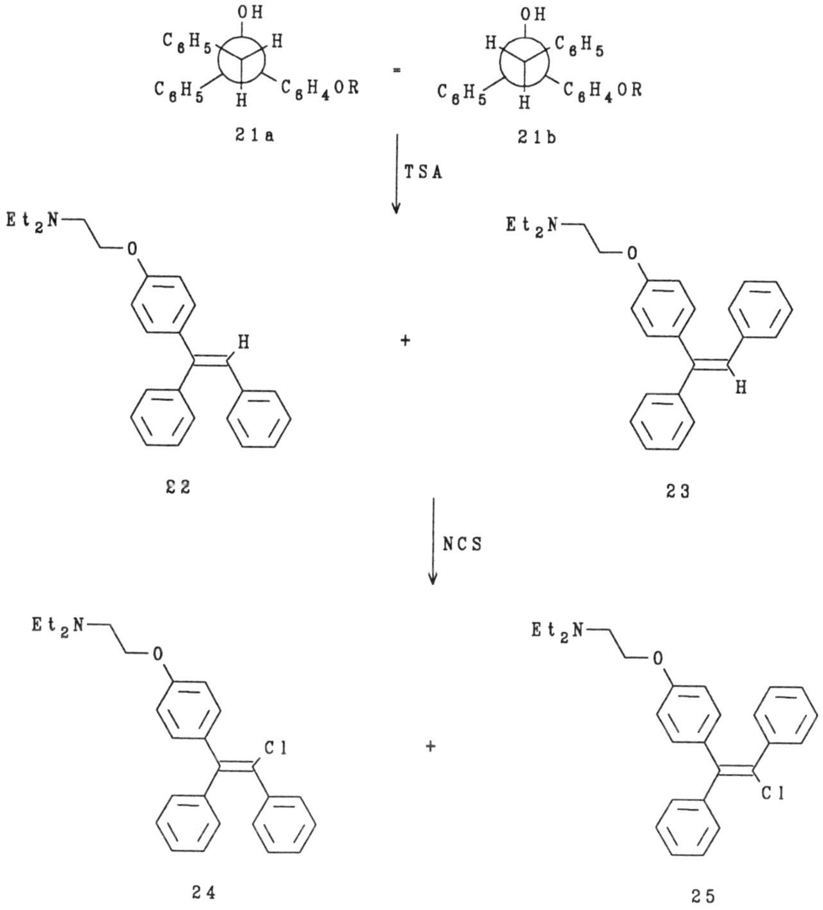

Essentially the same route is followed for the synthesis of the triphenylethylene **nitromifene** (**31**). The sequence begins with Friedel–Crafts acylation of the alkylation product **26** from phenol and 1,2-dibromoethane with the acid chloride from anisic acid (**27**). Displacement of bromine in product **28** with pyrrolidine leads to formation of the basic ether and thus to **29**. Condensation of **29** with benzylmagnesium bromide gives the tertiary alcohol **30**. Compound **30** is then treated with a mixture of nitric and acetic acid. The dehydration products from the first step almost certainly consist of a mixture of the *(E)* and *(Z)* isomers for the same reasons as those advanced above. The olefin undergoes nitration under reaction conditions to lead to nitromifene (**31**) as a mixture of isomers;[8] the separated compounds are reported to show surprisingly equivalent agonist–antagonist activities.

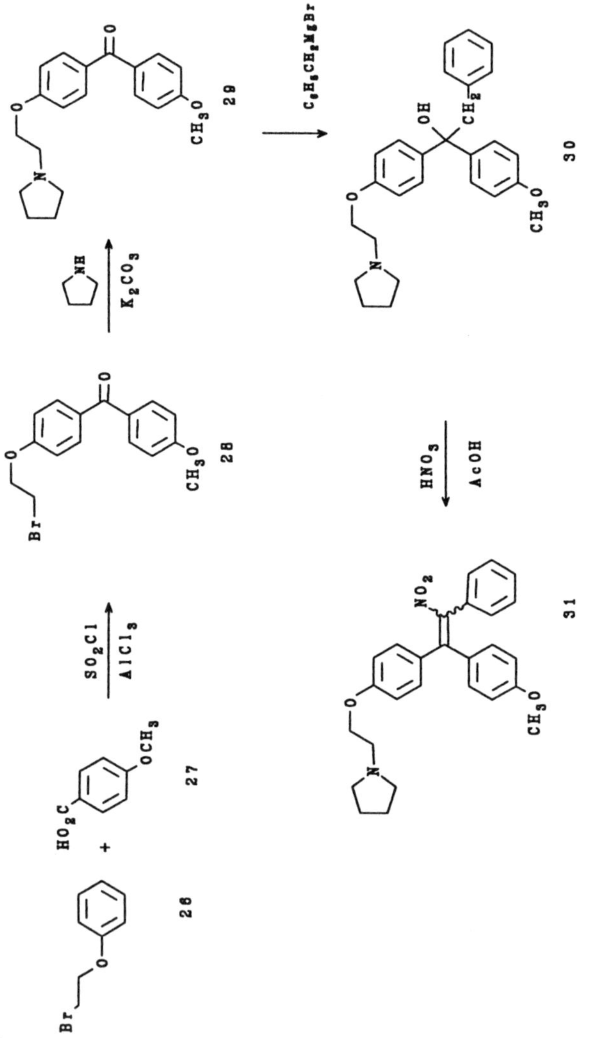
Br
HO2C
OCH3
26
27
SO2Cl
AlCl3
CH3O
28
NH
K2CO3
O
29
C6H5CH2MgBr
OH
CH2
30
HNO3
AcOH
NO2
31

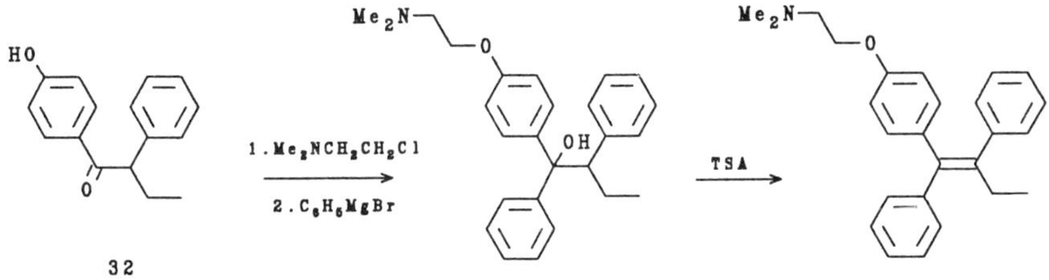

Although the preparation of the estrogen antagonist tamoxifen (**34**) appears to be comparable to the foregoing, the presence of the ethyl substituent on the ethylene has important stereochemical consequences. This sequence also differs from the foregoing by starting with an intermediate (**32**), which already includes both carbon atoms of the future ethylene moiety. The substituted desoxybenzoin (**32**) is first alkylated to its basic ether with 2-chlorotriethylamine; reaction with phenylmagnesium bromide leads to the corresponding tertiary alcohol **33**. Dehydration by means of *p*-toluenesulfonic acid gives the triphenylethylene **34**, this time as predominantly the (*Z*) isomer, tamoxifen (**34**).[9] Isomerically pure tamoxifen is a potent estrogen antagonist with little intrinsic agonist activity; the (*E*) isomer, by contrast, displays largely estrogen agonist activity.

The stereochemical outcome of this sequence traces back to the Grignard addition step. The rotamer of the starting ketone **32**, which involves the fewest nonbonding interactions, is that which opposes the two aromatic rings as shown in the Newman projection **32**. Isomer **33a** will be formed if it is assumed that the organometallic reagent adds from this direction, which will lead to the fewest interactions in the product giving **33a**. The rotamer **33b** from this product, which has the trans hydroxyl–proton relation for dehydration, leads to the observed product **34**.

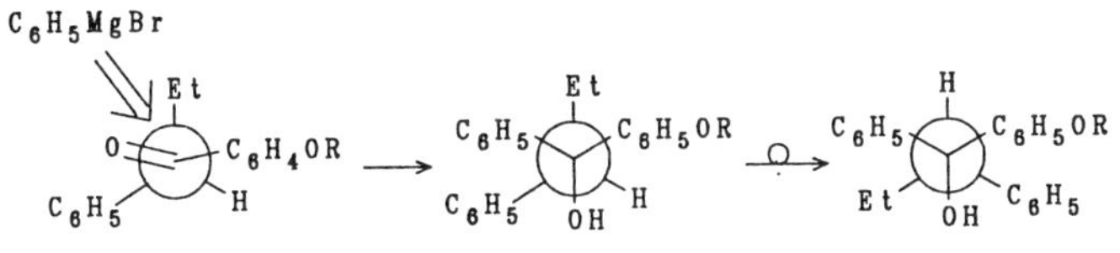

Some recent work on estrogen antagonists takes advantage of the stereochemical control afforded by starting with preformed ethylene fragments. Alkylation of desoxybenzoin proper (**35**) with the benzyl ether from 2-bromoethanol affords intermediate **36**. Compound **36** is then condensed with the Grignard reagent from the tetrahydropyranyl ether of 4-bromophenol to give the corresponding tertiary alcohol. Since this reaction is quite analogous with the one leading to **33**, this product would be expected to consist predominantly of a single diastereomer. The benzyl ether protecting group is then removed by hydrogenation over palladium to yield the diol **37**. Strong acid leads to formation of a tetrahydrofuran ring from the tertiary alcohol and that on the ethyl side chain as well as hydrolysis of the terahydropyranyl protecting group to form **38**. Ring formation almost certainly starts by loss of the protonated benzhydryl hydroxyl group; the fact that the final olefin consists predominantly of the (*Z*) isomer argues that this stereochemical identity must be retained during ring formation. This finding can be rationalized by assuming that loss of the benzhydryl hydroxyl is simultaneous with attack by the terminal primary alcohol. The requisite basic ether is then added by alkyation of the free phenol with 2-chlorotriethylamine to form **39**. Treatment of **39** with hydrogen chloride leads to ring opening of the tetrahydrofuran ring with simultaneous formation of the (*Z*) olefin; the terminal oxygen is converted to a chloro substituent under the reaction conditions. Thus, the estrogen antagonist **toremifene** (**40**)[10] is obtained.

Compounds designed as rigid or conformationally constrained analogues have provided several classes of very potent estrogen antagonists. Note that the fused bicyclic moieties chosen to lock these compounds more closely resemble the steroid AB ring moiety than do the styrene fragment of the open chain analogues discussed above.

Synthesis of the anti-estrogen **nafoxidine** (**45**) starts with reduction of 3-methoxyphenylacetic acid (**41**) to the corresponding phenethyl alcohol by means of lithium aluminum hydride; esterification of the alcohol with methanesulfonyl chloride gives the mesylate **42**. Displacement of the newly introduced leaving group with the carbanion obtained from reaction of phenylacetonitrile with sodium amide in liquid ammonia gives the diaryl butyronitrile derivative **43**. The cyano group is then hydrolyzed to the corresponding carboxylic acid with sulfuric acid. This compound is in turn converted to its acid chloride with phosphorus pentachloride and is cyclized by treatment with stannous chloride to afford the key tetralone **44**. Compound **44** is then condensed with the Grignard reagent from *N*-(*p*-bromophenoxy-ethyl)-pyrrolidine; the first obtained benzylic tertiary alcohol dehydrates on workup to give a 1,2-diaryldihydronaphthalene. Thus, the very potent estrogen antagonist nafoxidine (**45**)[11] is obtained.

An alternative synthesis for **45** involves assembly of the full carbon skeleton prior to formation of the bicyclic nucleus. The first product isolated from the reaction of deoxybenzoin (**46**) with ethyl formate and sodium ethoxide consists of the sodium salt **47** of the enol from the β-formyl derivative. This salt (**47**) is then reacted without prior purification with the phosphonium salt from 3-methoxybenzyl bromide and triphenylphosphine. These reaction conditions result in net Wittig condensation to afford an undefined mixture of isomeric enones. The small amount of enolate **47** present in the essentially heterogeneous mixture probably reacts on the surface of the phosphonium salt to form a small amount of phosphorane; this phosphorane then condenses with the minor fraction of neutral **47** present as its keto–aldehyde tautomer. (This reaction gives intractable mixtures when carried out in the traditional manner by preforming the ylide.) Selective demethylation of the methyl ether in the hydrogenation product **48** on the future 1-aryl ring to give phenol **49** is achieved by reaction of the compound with 3 equivalents (one for each oxygen) of aluminum chloride; the ether directly para to the carbonyl is sufficiently more electron deficient so as to cleave preferentially. The ketone is then cyclized by heating in the presence of *p*-toluenesulfonic acid to give dihydronaphthalene **50**. Alkylation with *N*-chloroethylpyrrolidine of the salt obtained by treating the phenol with sodium hydride affords nafoxidine (**45**).[12]

35 $Bz\text{-}CH_2C_6H_5$

1. EtONa
2. $BrCH_2CH_2OBz$

36

1. $THPOC_6H_4MgBr$
2. H_2/Pd

37

H_3O^+

38

$Me_2NCH_2CH_2Cl$

39

HCl

40

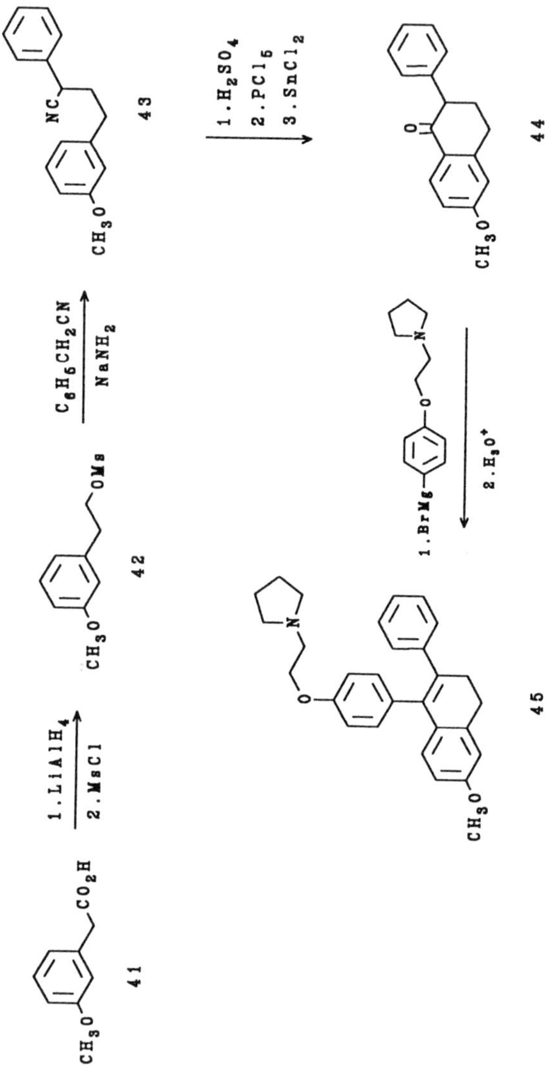
CH3O
CO2H
41
1. LiAlH4
2. MsCl
CH3O
OMs
42
C6H5CH2CN
NaNH2
CH3O
NC
43
1. H2SO4
2. PCl5
3. SnCl2
CH3O
O
44
1. BrMg
O
N
2. H3O+
CH3O
O
N
45

OCH3 HCO2Et NaOEt O 46 OCH3 O CHO⁻Na⁺ 47 1. mMeOC6H4CH2P⁺Ph3 2. H2 OCH3 O CH3O 48 AlCl3 OH O CH3O 49 TSA OH CH3O 50 1. NaH 2. Cl N 45

The wide tolerance for structural modification in agonists of the estrogen receptor is applied to antagonists as well. The dihydronaphthalene **trioxifene** (**55**), which contains an extra carbonyl group between the ring bearing the basic ether and the bicyclic moiety, retains antiestrogenic activity. Condensation of the enolate from β-tetralone (**51**) with ethyl anisoate leads to the corresponding acylation product **52**, shown here as the more highly conjugated of the two possible enolates. Reaction of **52** with the Grignard reagent from 4-anisoyl chloride leads to attack on the ketone that is expected to show more carbonyl character, that is, the ketone at the 2-position on the ring. The first formed product dehydrates on workup to give the conjugated product **53**. Demethylation by means of the sodium salt of ethyl mercaptan proceeds selectively at the ether para to the carbonyl (**54**) that is the more electron deficient of the two. The newly generated free phenol group is then alkylated as its phenolate with *N*-(2-chloroethyl)pyrrolidine to afford trioxifene (**55**).[13]

O 51 p-MeOC6H4CO2Et NaOEt HO OCH3 O 52 p-MeOC6H4MgCl CH3O O OCH3 53 EtSNa HO O OCH3 54 Cl(CH2)2N N O O OCH3 55

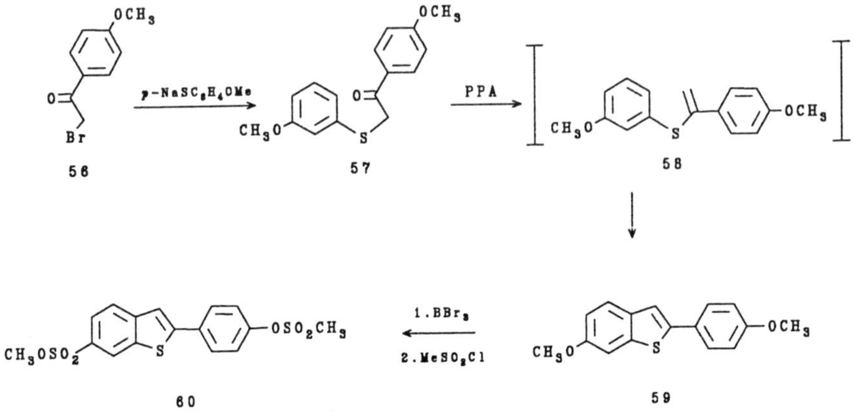

Replacement of the dihydronaphthalene nucleus by benzothiophene invokes one of the classic bioisosteric equivalences, which involves the fact that sulfur is approximately the same size as an ethyl group. The synthesis of this compound starts with displacement of halogen on phenacyl bromide (**56**) by *m*-methoxythiophenoxide to give the alkylated product **57**. Treatment of **57** with polyphosphoric acid (PPA) results in formation of the 2-anisoyl substituted benzothiophene **59** rather than the expected 1-anisoyl derivative. The unexpected product can be rationalized by invoking initial rearrangement of intermediate **57** to a thioenol ether such as **58**; Friedel–Crafts cyclization of this compound will give **59** directly. The alternative possibility requires 1,2 migration of the anisoyl group from the expected 1-anisoyl cyclization product. The methyl ethers are then cleaved by a reagent such as bromine tribromide; acylation of the thus freed phenol groups with methanesulfonyl chloride gives the corresponding bismesylate **60**; those protecting groups can be removed under relatively mild conditions.

Alkylation of the sodium salt from methyl 4-hydroxybenzoate (**61**) with the ubiquitous chloroethylpyrrolidine gives the acid **62**, after removal of the methyl ester with aqueous base. The carboxyl group is then converted to its acid chloride; the acid chloride is reacted with benzothiophene (**60**) in the presence of aluminum chloride to give the 2-acylated derivative. The mesylate groups are then removed by means of mild base to give the anti-estrogen **raloxifene** (**63**).[14]

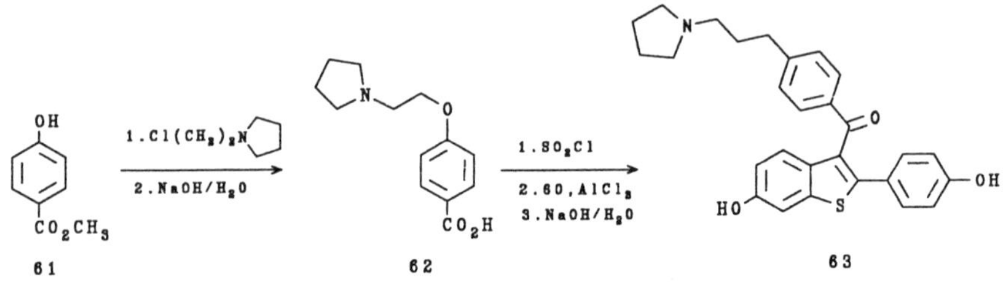

A benzopyran, also known as a chroman, forms the nucleus of the anti-estrogen oral contraceptive drug **centchroman** (**71**). The preparation begins with straightforward Friedel–Crafts acylation of the monomethyl ether of resorcinol (**64**), with the acid chloride from 4-hydroxybenzoic acid (**65**) to give the benzophenone **66**. Condensation of **66** with phenylacetic acid in acetic anhydride in the presence of triethylamine leads to formation of coumarin **67**; the overall reaction, of course, involves aldol condensation of phenylacetate with the benzophenone carbonyl and esterification of the ortho phenol group; the order in which those steps occur is not clear.[15]

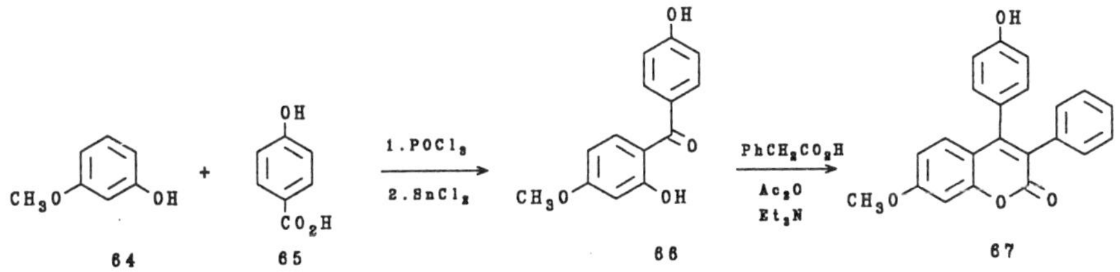

Reaction of the coumarin with excess methylmagnesium bromide leads to addition of two methyl groups to the ester carbonyl with consequent ring opening and formation of tertiary carbinol **68**. Compound **68** cyclizes to chromene **69** on treatment with ethanolic hydrogen chloride in all likelihood via the carbocation from loss of the tertiary hydroxyl group. Alkylation of the free phenol group with the familiar chloroethylpyrrolidine gives the corresponding basic ether **70**. High-pressure hydrogenation of **70** initially leads to the expected 3,4-*cis*-dihydro derivative; this compound is equilibrated to the more stable trans derivative by treatment with butyllithium. Thus centchroman (**71**)[16] is obtained.

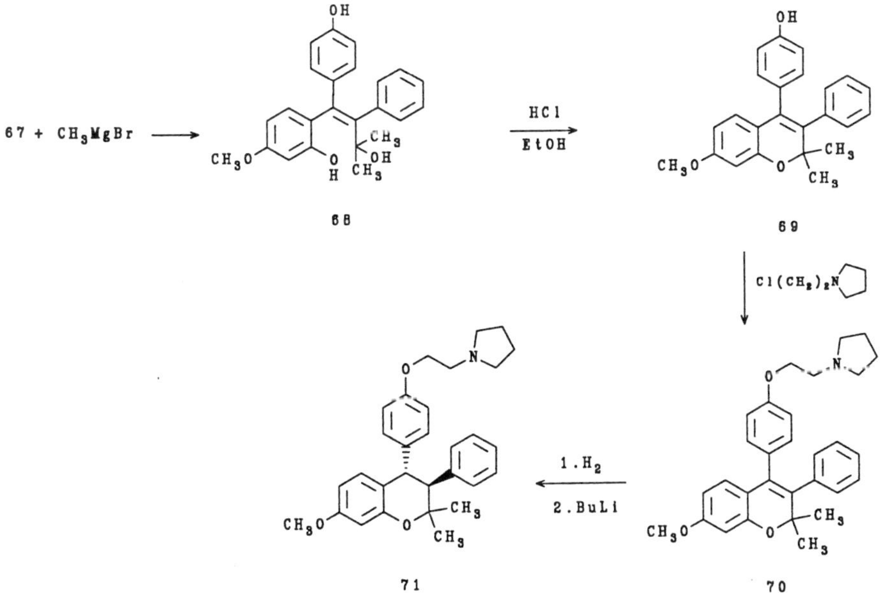

REFERENCES

1. Cook, J.W.; Dodds, E.C.; Hewett, C.L. *Nature (London)* **1933**, *131*, 56.
2. Kharash, M.S.; Kleinman, M. *J. Am. Chem. Soc.* **1943**, *65*, 11.
3. Dodds, E.C.; Goldberg, L.; Lawson, W.; Robinson, R. *Proc. R. Soc. (London)* **1939**, *B127*, 148.
4. McMurry, J.E.; Fleming, M.P.; Kees, K.L.; Krespi, L.R. *J. Org. Chem.* **1978**, *43*, 3255.
5. Brasford, F.R. Br. Patent **1944**, 561508.
6. Shelton, S.R.; Vancampen, M.G. U. S. Patent **1947**, 2430991.
7. Allen, R.E.; Palapoli, F.P.; Schumann, E.L.; Vancampen, M.G. U. S. Patent **1959**, 2914563. *Chem. Abstr.* **1960**, *54*, 5581.
8. DeWald, H.A.; Bird, O.D.; Rodney, G.; Kaump, D.H.; Black, M.L. *Nature (London)* **1966**, *211*, 538.
9. Bedford, G.R.; Richardson, D.N. *Nature (London)* **1966**, *212*, 733.
10. Toivola, R.J.; Karjlaainen, A.J.; Kurkelka, K.O.A.; Sodervall, M.J.; Kangas, L.V.M.; Blanco, L.G.; Sundquist, H.K. Eur. Patent Appl. **1983**, 95875; *Chem. Abstr.* **1985**, *102*, 166452.
11. Lednicer, D.; Lyster, D.C.; Aspergren, B.D.; Duncan, G.W. *J. Med. Chem.* **1966**, *9*, 172.
12. Lednicer, D.; Emmert, E.E.; Lyster, S.C.; Duncan, G.W. *J. Med. Chem.* **1969**, *12*, 881.
13. Jones, C.D.; Suarez, T.; Massey, E.H.; Black, L.J.; Tinsley, F.C. *J. Med. Chem.* **1979**, *22*, 962.
14. Jones, C.D.; Jevnikar, M.D.; Pike, A.J.; Peters, M.K.; Black, L.J.; Thompson, A.R.; Falcone, J.F.; Clemens, J.A. *J. Med. Chem.* **1984**, *27*, 1057.
15. Ray, S.; Grover, P.K.; Anand, N. *Indian J. Chem.* **1973**, *9*, 619.
16. Ray, S.; Grover, P.K.; Kamboj, V.P.; Sethy, B.S.; Kar, A.B.; Anand, N. *J. Med. Chem.* **1976**, *19*, 276.

CHAPTER 7

OPIOID ANALGESICS

1. INTRODUCTION

The adventitious discovery, in prehistory, of the analgesic, soporific, and euphoriant properties of the dried sap from the flower bulb of the poppy, *papaver somniferum*, has been treated too often elsewhere to merit repetition. By the nineteenth century,

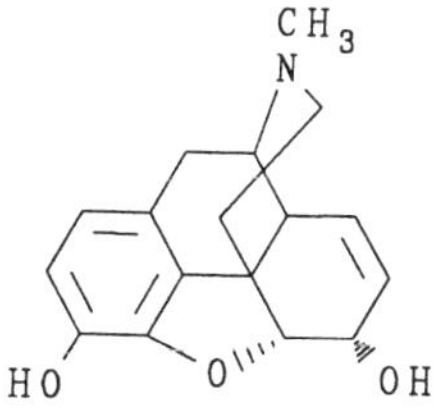

1 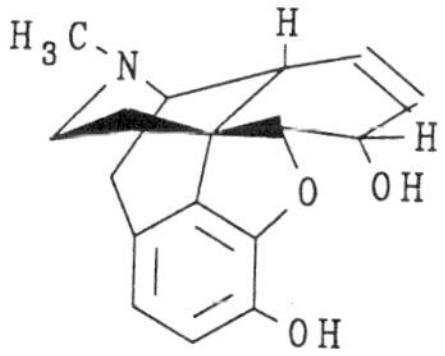 1

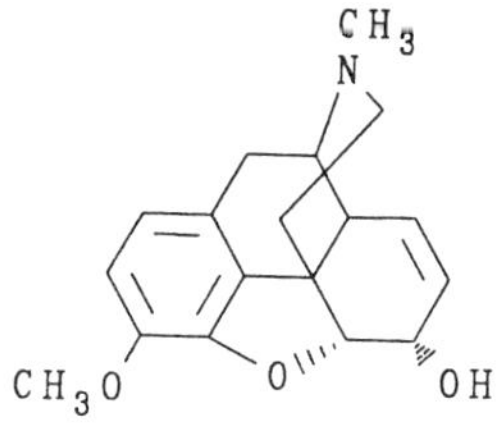

2 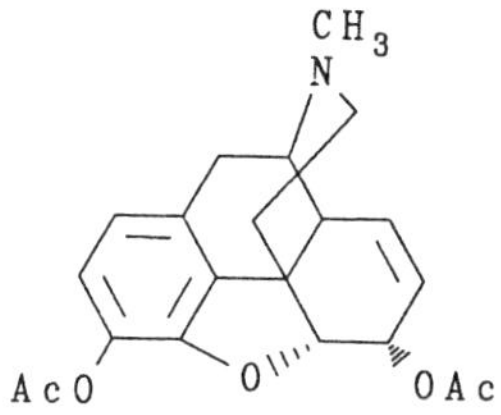 3

organic chemistry had advanced far enough so that the active principle from opium had been isolated, purified, and crystallized. Increasing clinical use of this compound, **morphine** (1) and its naturally occurring methyl ether **codeine** (**2**) disclosed a host of side effects, the most daunting of which was, and still is, these compounds' propensity for inducing physical dependence. A modest effort was launched at modifying this compound, even though its detailed structure was still unknown and was to remain so for close to another one-half of a century. The product from treatment of morphine with acetic anhydride is the corresponding diacetate **heroin** (**3**), a drug, somewhat optimistically, originally thought to be less addicting than its parent.

2. DRUGS DERIVED FROM MORPHINE

The fact that the structure of morphine was not solved until 1925,[1] severely limited early synthetic work on its derivatives. The relatively early availability of the *N*-demethyl derivative **4**, via von Braun degradation, did lead to an investigation of the effect on biological activity of substituents on nitrogen. The classical version of this degradation involves reaction of a tertiary *N*-methyl compound with cyanogen bromide. The initial product consists of a quaternary *N*-cyano betaine; internal displacement of the methyl group by the bromide counterion leads to scission of the *N*-methyl bond and loss of methyl bromide with consequent formation of a cyanamide. This functionality readily hydrolyzes to the secondary amine with loss of cyanide. A more recent version of this reaction replaces cyanogen bromide with ethyl chloroformate. The reaction follows a very similar sequence with the product of the alkylation–internal displacement sequence, in this case consisting of a urethane. The same secondary amine is obtained on hydrolysis of the initial product.

The *N*-allyl derivative, **nalorphine** (**5**), is prepared from *N*-demethylmorphine (**4**) by alkylation with allyl bromide.[2] The discovery that this compound proved to antagonize the activity of morphine in experimental animals led to the synthesis of the potent opioid antagonist **naloxone (19)**, which is discussed below. Nalorphine (**5**), in marked contrast to the latter shows some analgesic activity in humans.

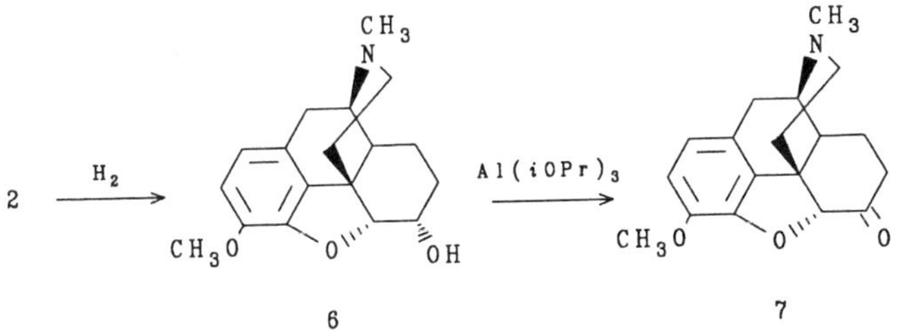

One of the more benign ancillary properties of morphine is its activity in suppressing the cough reflex. Catalytic reduction of codeine (**2**) leads to the dihydro derivative **6**. Oppenauer oxidation of the hydroxyl group leads to **hydrocodone** (**7**),[3] a compound used extensively in cough remedies; note that this drug retains considerable opioid activity.

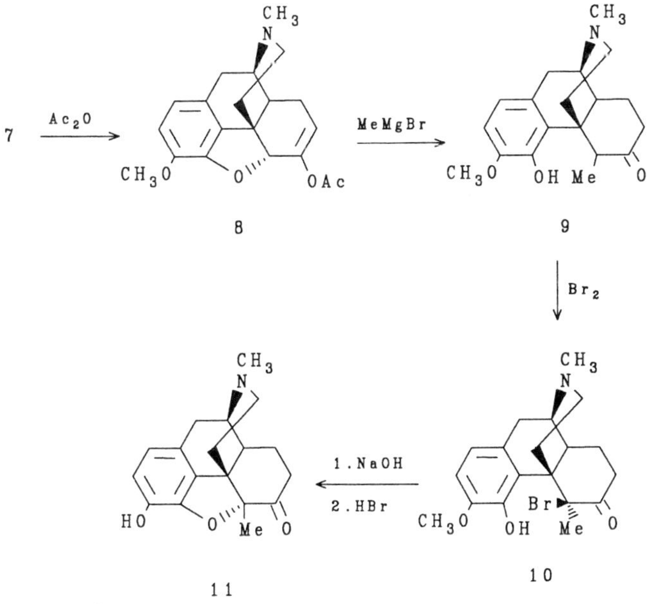

In addition to the shortcomings of morphine (**1**) noted above, its poor absorption from the GI tract means that it must be administered parenterally. Although its methyl ether, codeine (**2**), does show oral activity, this derivative does not afford the same measure of analgesia as the parent compound. An analogue methylated on the carbon bearing the furan ring combines increased oral absorption and analgesic activity. The key starting material for this derivative, enol acetate **8**, is obtained by treatment of codeine with acetic anhydride. Reaction of **8** with methylmagnesium bromide leads to the somewhat unusual displacement of the allylic oxygen of the fused furan and formation of the alkylated product **9**; the stability of the phenoxide anion leaving group may provide some of the driving force for this reaction. Excess reagent presumably reacts with the acyl group without, however, freeing the latent carbonyl group. Bromination of the ketone (**9**) proceeds, as expected, on the more highly alkylated carbon atom to afford the bromide (**10**). Treatment of **10** with base initially leads to formation of a phenoxide; the phenoxide displaces the adjacent halide to lead to reclosure of the furan ring; cleavage of the methyl ether by means of hydrogen bromide completes the synthesis of **metopon** (**11**).[4]

3. COMPOUNDS PREPARED FROM THEBAINE

As is often the case with complex natural products, opium, the dried sap from *papaver sominferum*, contains a number of structurally closely related compounds. One of these minor constituents, thebaine (**12**), although itself devoid of significant analgesic activity, incorporates a reactive diene system that has been exploited for preparing several important opioids. This discovery in turn led to the search for strains of the poppy in which this compound predominated; one such strain, *papaver bracteum*, has been reported to yield as much as 26% thebaine from the dried sap of its seed pod.

The product (**13**) from reaction of thebaine with hydrogen peroxide results formally from 1,4 addition of two hydroxyl groups across the diene. The perspective depiction of thebaine shown below reveals that the addition in fact occurs at the far more open face of the molecule. The product from this oxidation incorporates at new hydroxyl group at the 14β-position and a hemiacetal at the 6-position. Treatment with mild acid leads to hydrolysis of this last function and formation of enone **14**.

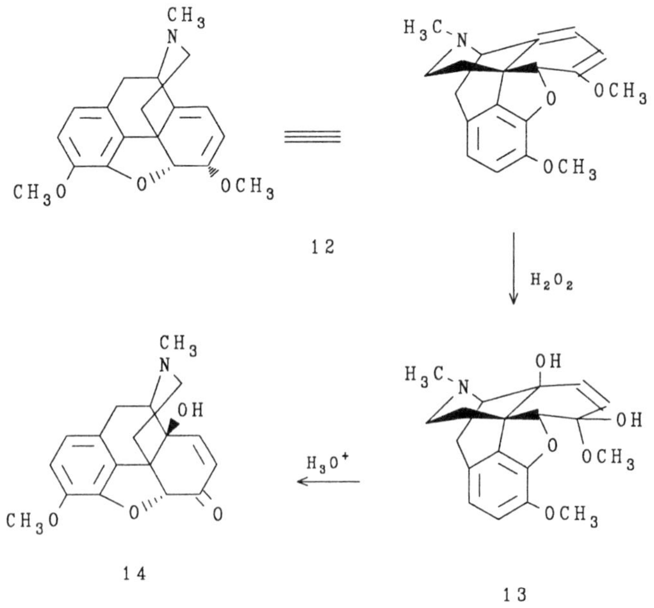

CH3 N CH3O O OCH3 ≡ H3C N O OCH3 OCH3

12

↓ H_2O_2

CH3 N OH CH3O O O ← H_3O^+ H3C N OH OH O OCH3 OCH3

14 13

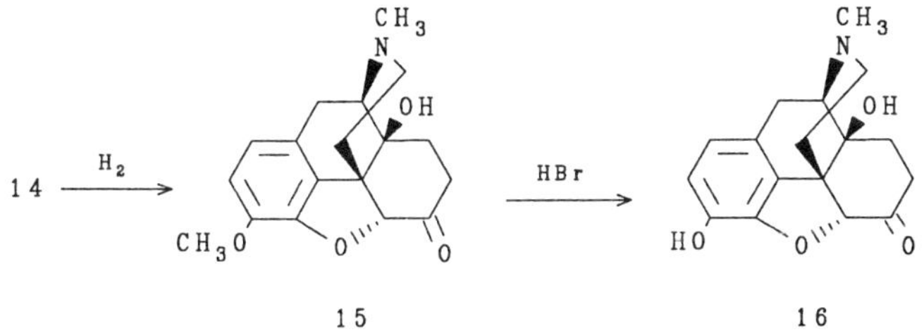

Catalytic hydrogenation of the hydrolysis product leads to the orally active compound **oxycodone (15)**, which is used in a number of analgesic drugs. Cleavage of the methyl ether to its free phenol leads to one of the most potent close analogues of morphine, **oxymorphone (16)**.[5] Note that both of these compounds carry the hazard of classical opiate dependence liability.

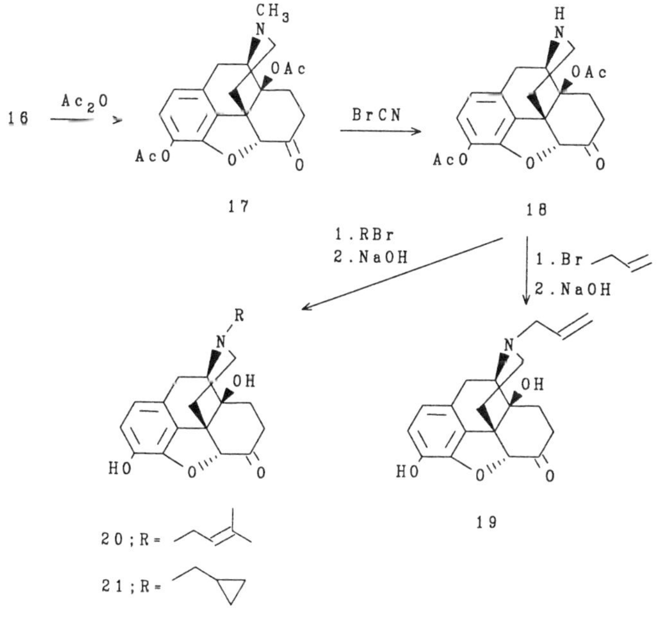

Replacement of the *N*-methyl group in oxymorphone (**16**) by allyl affords a potent opiate antagonist. This compound, **naloxone (19)**, in contrast to its desoxy analogue, nalorphine (**5**), is quite devoid of any analgesic or other opiate activity in humans. The drug will in fact precipitate withdrawal symptoms in opiate-dependent individuals. Preparation of the compound begins with protection of **16** as its diacetate **17**. Removal of the methyl group on nitrogen by reaction with cyanogen bromide, followed by hydrolysis of the first formed cyanamide, gives the corresponding secondary amine **18**. Alkylation with allyl bromide and subsequent saponification of the acetates affords naloxone (**19**).[6] Very small changes in the structure of the side chain substituents interestingly lead to compounds that show a mixture of opiate agonist and antagonist activity; such mixed agonist–antagonists have on the whole shown somewhat reduced dependence liability while retaining a good measure of analgesic activity. Thus, the alkylation–saponification sequence on **18**, using 3,3-dimethylallyl bromide, gives **nalmexone** (**20**);[7] the use of cyclopropylmethyl bromide leads to **naltrexone** (**21**).[8]

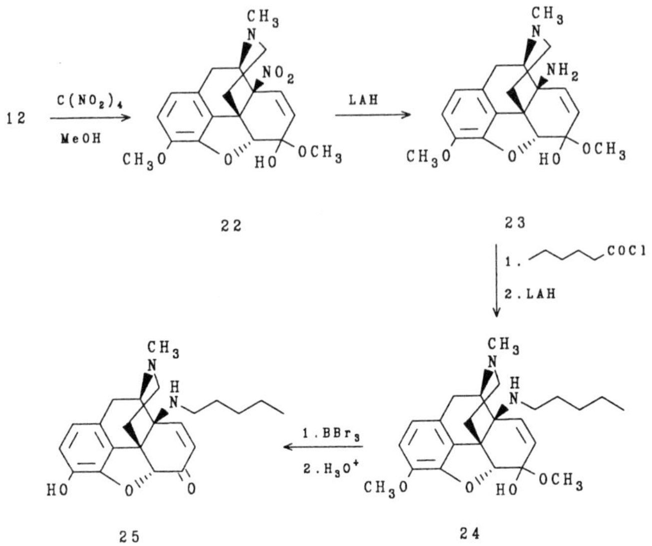

An interesting variation on this theme includes basic nitrogen rather than oxygen at the quaternary 14β-position. The first step in this sequence involves reaction of thebaine (**12**) with tetranitromethane in methanol with the overall result of addition of a nitro and a methoxyl group across the termini on the diene to form **22**. The reaction can be rationalized by assuming initial addition of a nitro free radical from decomposition of the reagent. The newly introduced nitro group is then reduced with lithium aluminum hydride to the primary amine **23**; the hemiacetal produced at the 6-position in the addition reaction protects the future carbonyl in the next couple of steps. The amine is then acylated with valeroyl chloride and the resulting amide is reduced to an alkyl group, again by treatment with lithium aluminum hydride to give **24**. The *O*-demethylation with boron tribromide followed by mild acid hydrolysis to remove the hemiacetal affords the opiate analgesic **pentamorphone** (**25**).[9]

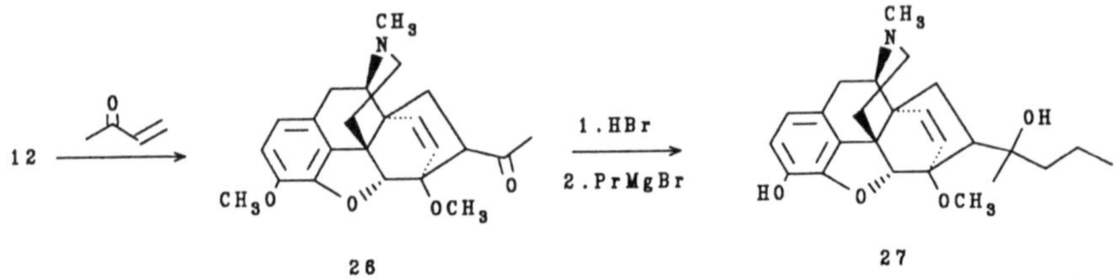

The diene function in thebaine is of course ideally set for Diels–Alder addition. Condensation of **12** with methyl vinyl ketone proceeds to give the cycloadduct **26**, addition proceeding from the more open face of the molecule; the acetyl side chain occupies the customary endo position on the 2,2,2-bicyclooctyl moiety. Addition of propylmagnesium bromide to the side chain ketone of this already potent analgesic leads to the tertiary carbinol and **etorphine** (**27**).[10] This very active compound is between 1000 and 10,000 times as potent as morphine as an analgesic in vivo in a variety of mammalian species, including humans. The drug is commonly used in flechettes to bring down large game for various studies; a subsequent single injection of naloxone (**19**) counteracts the agent and restores consciousness within a short time.

4. MORPHINANS

Detailed inspection of the structure of morphine reveals that the backbone of the molecule consists of a reduced phenanthrene ring system.[11] Initial efforts at producing this compound thus relied on the known chemistry used to prepare polynuclear aromatic hydrocarbons; acid catalyzed Friedel–Crafts cyclization played a key role in those approaches. This research culminated in the early 1940s with the successful synthesis by Grewe and his associate Mondon of the morphinan **29**,[12] which consists of the bare morphine skelcton. Thus, treatment of benzyl octahydroisoquinoline (**28**) with strong acid leads to formation of the bis-cation by successive protonation of the basic nitrogen and the isolated olefin. Attack of the pendant benzene ring by the resulting carbocation leads to cyclization and formation of the morphinan. The fact that **29** showed some analgesic activity in animal models demonstrated that the fused furan ring could be dispensed with; this finding led as well to analogue programs based on this simplified structure.

One of the syntheses for a morphinan that has been approved for sale starts by Knoevnagel reaction of cyclohexanone with ethyl cyanoacetate to give condensation product **30**. Hydrolysis of **30** leads to the corresponding cyanoacid. The latter loses

H_3C H_3C

28

H_3PO_4

H_3C H_3C H

29

carbon dioxide under reaction conditions to give **31**; the out-of-conjugation shift of the olefin is a direct consequence of the mechanism of the decarboxylation reaction. Treatment of the nitrile with lithium aluminum hydride then leads to the corresponding primary amine **32**. Acylation of **32** with *p*-methoxyphenylacethyl chloride adds the aromatic ring required for the morphinan. The use of the Bischler–Napieralski cyclodehydration of phenylacetamides of arylethylamines constitutes one of the standard methods for synthesis of dihydroisoquinolines; the reaction interestingly works equally well when the benzene ring involved in the cyclization is replaced by cyclohexene. Thus, reaction of **33** with phosphoric acid gives the hexahydroisoquinoline **34** via the enol form of the amide. Exposure of **34** to sodium borohydride leads to selective reduction of the imine bond and formation of the morphinan precursor **35**.[13]

Treatment of **35** with strong acid proceeds to morphinan **36** in a manner quite analogous to the unsubstituted compound. This specific product, which was actually produced some years after the drugs discussed below, has been used to prepare a number of analogues bearing differing substituents on nitrogen. Acylation of **36** with ethyl chloroformate followed by reduction of the intermediate urethane leads to the *N*-methyl derivative **37**. Cleavage of the methyl ether leads to the very potent analgesic drug **racemorphan** (**37**).[14] **Levorphanol**, the levorotatory isomer obtained either by resolution of the racemate[15] or by a scheme that involves prior resolution of an intermediate, is about 6–8 times as potent in humans as morphine itself. The dextrorotatory enantiomer of the O-methyl precursor **37**, **dextromethorphan**, is interestingly quite devoid of analgesic activity; the compound is, however, used extensively in cough remedies, since it shows antitussive activity comparable to that of morphine. Alkylation of resolved levorotatory **35** with allyl bromide followed by cleavage of the methyl ether by means of hydrogen bromide leads to the *N*-allyl analogue **levallorphan** (**39**), a compound that acts largely as an opiate antagonist.

Hydroxylation at the equivalent of the 14-position in morphine has much the same effect in the morphinan series in that it increases potency and oral activity. The somewhat involved synthesis of such a compound begins with the spirocycloalkylation

CO_2Et CN — NaOH → CN — LAH → NH_2 — MeO–C₆H₄–CH_2COCl → O NH OCH_3

30 31 32 33

↓ H_3PO_4

NH ← $NaBH_4$ — N

OCH_3 OCH_3

35 34

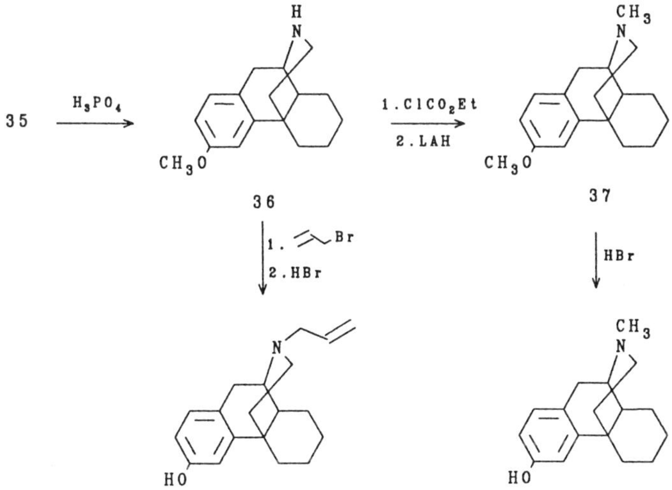

of 7-methoxy-1-tetralone with 1,4-dibromobutane to give **40**. Addition of the anion from acetonitrile to the ketone in **40** gives the tertiary alcohol adduct **41**; the nitrile is then reduced to the corresponding primary amine by means of lithium aluminum hydride to give **42**. The benzylic tertiary hydroxyl group eliminates to give a carbocation when treated with strong acid; this ion then rearranges to the hydrophenanthrene **43**.

The first step in forming the nitrogen-containing bridging ring involves reaction of the olefin **43** with bromine. The reaction probably starts by formation of a bromonium ion; ring opening by the adjacent primary amine leads to **44** as the hydrobromide salt in which nitrogen occupies the 14-position. Neutralization in the cold frees the still quite basic amine; this compound displaces the remaining bromine to form aziridine

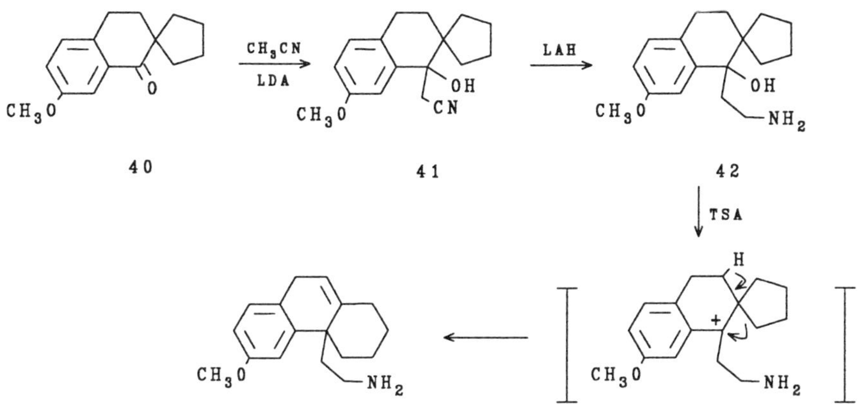

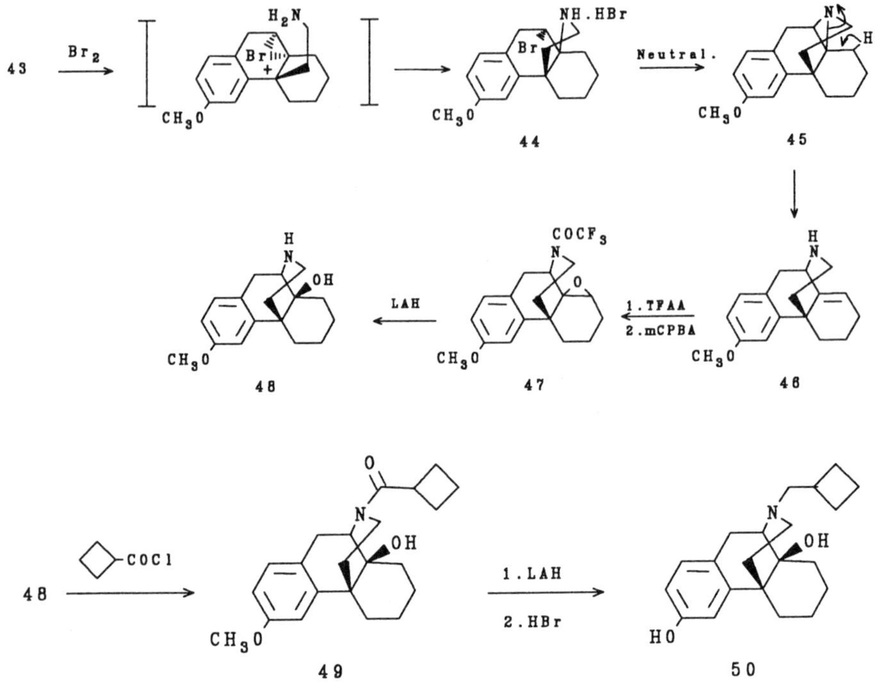

45. This ring opens on warming to **46** with concomitant formation of an olefin at the ring junction; the reaction from **44** goes directly to this intermediate when carried out warm. The amine is then protected as its trifluoroacetamide by reaction with trifluoroacetic anhydride (TFAA) and the olefin is converted to an epoxide with peracid to give **47**; the stereochemistry of epoxidation results from the fact that these morphinans closely resemble thebaine sterically. Reduction of the oxirane by means of lithium aluminum hydride leads the intermediate **48.**

It is often more advantageous to attach substituents on nitrogen by a two-stage acylation–reduction scheme than by direct alkylation, particularly in those cases where alkyl halogen is unreactive or where steric hindrance may interfere. Thus, acylation of **48** with cyclobutylcarbonyl chloride leads to the amide **49**; reduction using lithium aluminum hydride gives the cyclobutylmethyl derivative **butorphanol** (**50**),[16] a mixed agonist–antagonist analgesic.

5. BENZOMORPHANS

The use of tetrahydropyridines instead of tertahydroisoquinolines in the Grewe synthesis leads to benzomorphans, which may be viewed as morphine missing both a furan and one alicyclic ring. The fact that the original benzomorphan, which was actually produced by a different synthetic route, showed reasonable analgesic potency, combined with the ready access provided by the Grewe approach, prompted the preparation of a host of analogues.[17] The observation that these compounds seemed to offer few advantages over the accepted central analgesics probably accounts for the fact that a surprisingly small number of these compounds were investigated clinically. The little known addition of Grignard reagents to ternary imminium salts provides a flexible method for building the two-ring intermediate required for the Grewe based route. Thus, reaction of *p*-methoxybenzylmagnesium chloride with the quaternary salt **51** from reaction of 3,4-lutidine with methyl iodide gives the product **52** from addition of the organometallic to the ternary imminium function; the reason for reaction at the more highly hindered position is not immediately apparent. Treatment of **52** with sodium borohydride leads to selective reduction of the enamine bond to lead to the tertrahydropyridine **53**. Compound **53** undergoes ring closure with strong acid to give the benzomorphan **54** in direct analogy to the more complex morphinans. This product consists predominantly of the isomer that bears the equatorial secondary methyl group.[18]

Most of the compounds that have been biologically investigated in greater detail bear more complex side chains on nitrogen. Starting materials are available by the use of an alternative synthesis that leads to intermediates that include unsubstituted piperidine nitrogen. Acylation of the aliphatic amine **55** with *p*-methoxyphenylacetyl chloride gives the corresponding amide **56**.[19] Compound **56** cyclizes to the dihydropyridine **57** when treated under the conditions of the Bischler–Napieralski reaction. Reaction with sodium borohydride results in reduction of the enamine double bond exactly as above. Cyclization of **58** with strong acid proceeds to the benzomorphan as in the case of the *N*-methyl analogue. The methyl ether is then cleaved with hydrogen bromide to afford the key intermediate **59**.

Alkylation of **59** with phenyethyl bromide leads to **phenazocine** (**60**), a quite potent analgesic that, however, exhibits the same profile of side effects as the classical opiates.[20] The 2,2-dimethylallyl analogue **pentazocine** (**61**),[21] on the other hand shows mixed agonist and antagonist properties even in humans; the drug also seems to have relatively low dependence potential. The cyclopropylmethyl derivative **cyclazocine** (**62**) has very similar pharmacological properties; this agent does, however, tend to induce hallucinations.

Central analgesic activity is retained when the ring methylene group adjacent to the benzene ring occurs as a carbonyl group. Preparation of **ketazocine** (**63**) starts by protection of the amine on cyclization product **63**, as its acetamide; oxidation with chromium trioxide followed by deprotection leads to intermediate **64**. Alkylation on nitrogen with bromomethylcyclopropane and subsequent cleavage of the phenolic ether gives **65**.[22] Ketazocine (**63**) also exhibits mixed opiate agonist and antagonist activity.

A compound that presents functionality very similar to that of the benzomorphans but with different connectivity is also a central analgesic; the presence of basic nitrogen as a primary amine is, however, most unusual for an opiate. In this case, the bridging ring is constructed by alkylation. Reaction of tetralone (**66**) with 1,5-dibromopentane in the presence of two equivalents of sodium hydride leads to the bridged bicylic ketone **67**; the first alkylation step almost certainly occurs at the more acidic benzylic position, eliminating the possibility of formation of a spiran at the 3-position. The now highly hindered ketone is converted to its oxime (**68**) by reaction under forcing conditions

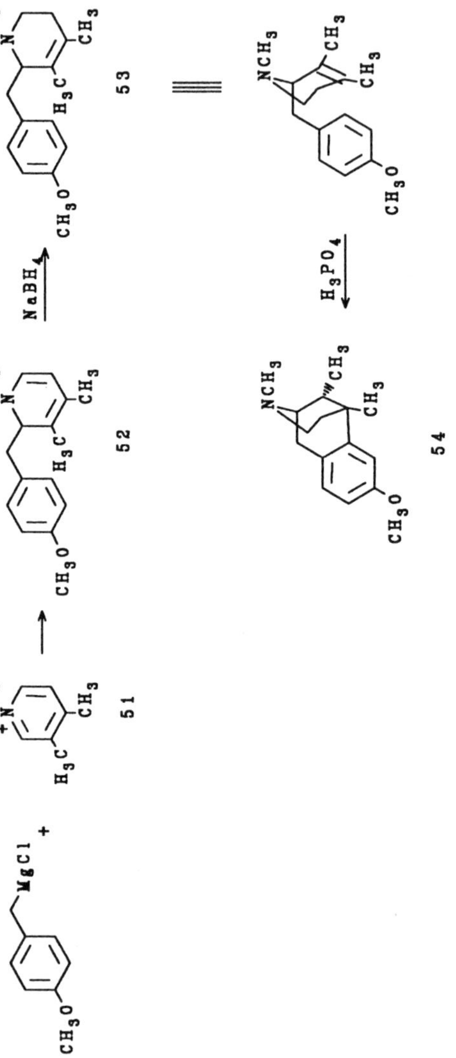

CH_3O — MgCl + 51 → 52 $\xrightarrow{NaBH_4}$ 53 ≡ $\xrightarrow{H_3PO_4}$ 54

51

52

53

54

CH_3O — $COCl$ + H_2N, H_3C, CH_3

55

→

56

H_3PO_4 →

57

$NaBH_4$ ↓

58

1. H_3PO_4
2. HBr

↓

59

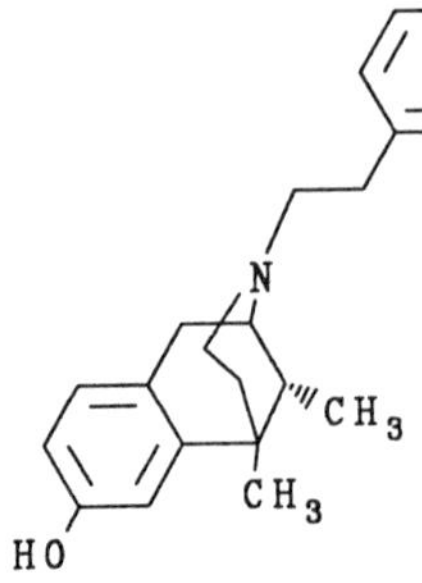

60 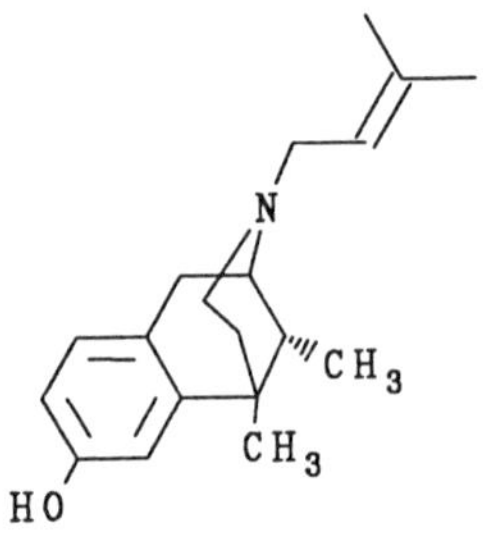 61 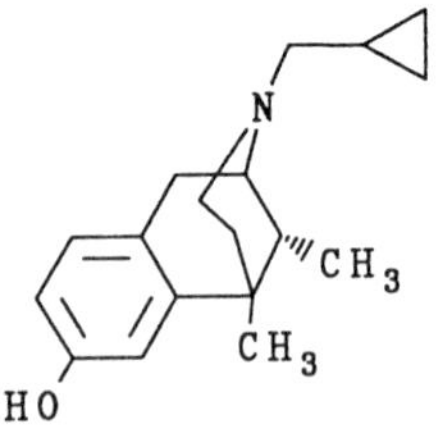 62

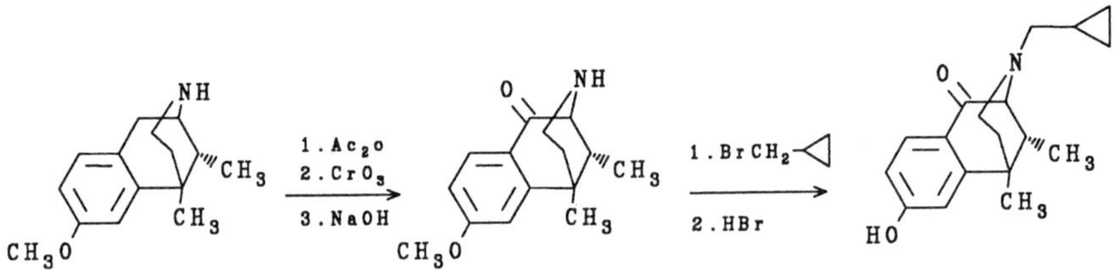

63 64 65

1. Ac_2O
2. CrO_3
3. NaOH

1. $BrCH_2$–cyclopropyl
2. HBr

with hydroxylamine. Reduction with hydrogen over Raney nickel takes place largely by approach of the reagent from the more open side away from the bridgehead with a consequent formation of predominantly the all-cis isomer **69**. Cleavage of the methyl ether to a free phenol in the usual manner, followed by resolution, affords the central analgesic **dezocine (70)**.[23]

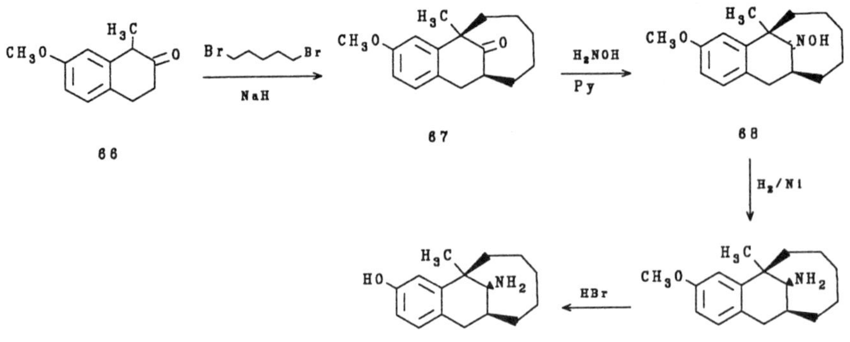

66 67 68 69 70

Br(CH2)5Br / NaH

H_2NOH / Py

H_2/Ni

HBr

6. ANALGESICS BASED ON NONFUSED PIPERIDINES

A. 4-Arylpipridines: Meperidine and Its Analogues

Screening of compounds in a broad battery of biological assays without regard to the intended activities of these agents, although currently in disfavor, has in fact led to the discovery of a disproportionate number of classes of therapeutic agents. The discovery of yet a further simplification of the minimal opiate analgesic traces back to such an adventitious finding.[24] **Meperidine (74)**, also known as pethidine, retains only two of the rings in morphine yet shows full analgesic activity. This compound is still used extensively in the clinic. The key step in one established synthesis of **74** involves spiroalkylation of the carbanion from phenylacetonitrile with the nitrogen mustard **71**, to afford the piperidine **72**; the intramolecular reaction is favored over polymerization because of the enhanced reactivity of the first monoalkylated product. The nitrile group is then hydrolyzed with strong base, and the resulting acid **73** is esterified with ethanolic hydrogen chloride to afford meperidine (**74**).[25]

The secondary amine (**77**) required for preparation of compounds bearing other substituents on nitrogen could in principle be prepared by *N*-demethylation of **meperidine** (**74**) itself. A more versatile approach employs the *p*-toluenesulfonyl (Ts) analogue **75** of the nitrogen mustard; this compound leads to production of the corresponding piperidine intermediate **76**. Base hydrolysis as shown above leads to the free acid. Reaction of **76** with ethanolic hydrogen chloride results in formation of the ethyl ester with concomitant cleavage of the tosylamide amide bond to afford the so-called **normeperidine** (**77**). Alkylation on nitrogen with chloroethylaniline (**78**) leads to **anileridine** (**79**),[26] an analgesic considerably more potent than its *N*-methyl analogue. The effect on potency of such arylethyl substituents will be revisited below.

Morphine, as noted previously, exhibits a large number of side effects which are not directly related to its analgesic activity; among these effects is inhibition of intestinal peristalsis. The finding that meperidine (**74**) shares this activity led to the development of a highly substituted derivative, **dlphenoxylate** (**82**), which inhibits intestinal motility, acts as an antidiarrheal agent. The side chain for **82** is prepared by alkylation of the carbanion from diphenylacetonitrile (**80**) with 1,2-dibromoethane to give the bromoethyl derivative **81**. Reaction of **81** with normeperidine (**77**), lead to alkylation of the secondary amine and formation of diphenoxylate (**82**),[27] familiar to world travelers under one of its trade names, Lomotil®.

It will become clear from the next several examples that the carboxyl group in meperidine (**74**) acts largely as a bulky group, since activity is maintained when it is replaced by other space-filling functions. The starting material **83** for one of these agents can be obtained by spiroalkyaltion of *m*-methoxyphenylacetonitrile with the nitrogen mustard **71**. The nitrile, in this case, is then condensed with ethylmagnesium bromide; hydrolysis of the first formed imine with mild aqueous acid gives the corresponding ketone, **84**. Standard demethylation of the ether group using hydrogen bromide yields the analgesic agent **ketobemidone** (**85**).[28]

Extensive studies on the SAR of the meperidine (**74**) series revealed that incorporation of a methyl group on the ring position adjacent to the carboxyl markedly

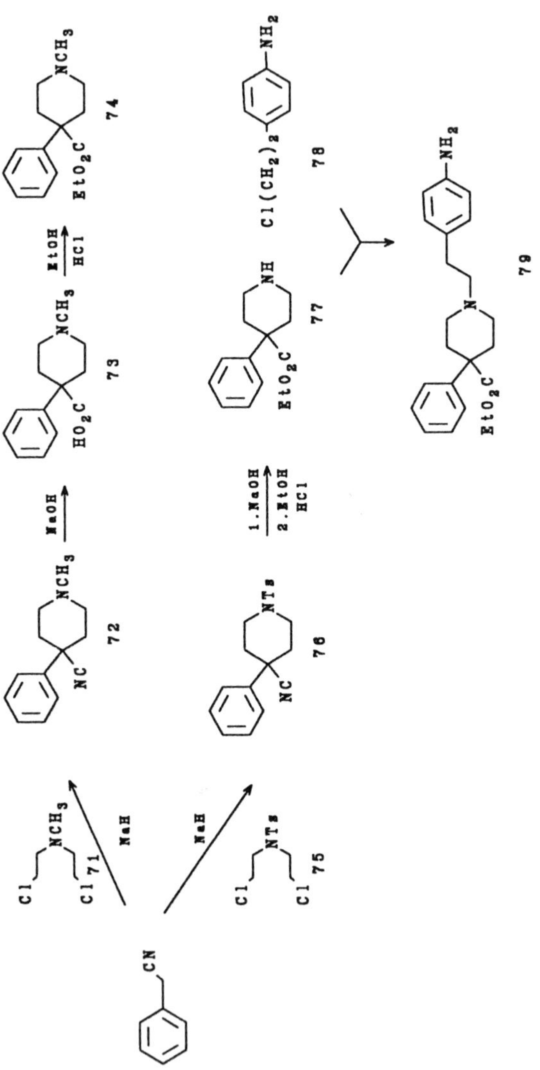
NaH
NaOH
EtOH
HCl
1. NaOH
2. EtOH
HCl
71
72
73
74
75
76
77
78
79

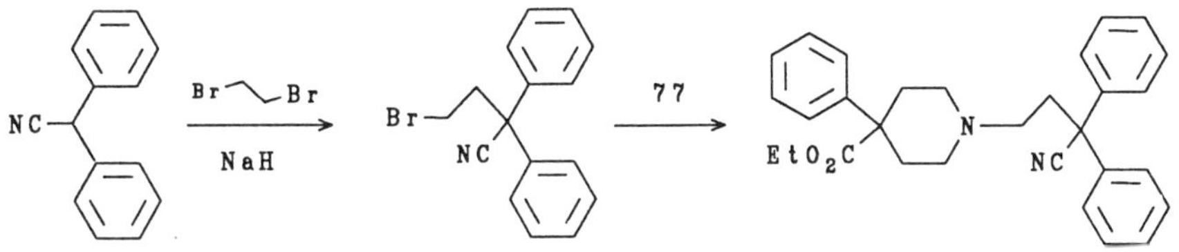

80 81 82

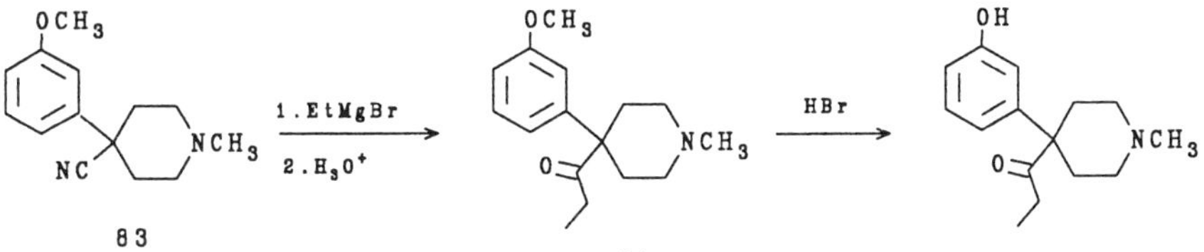

83 84 85

increased potency. Both this finding and the fact that the carbethoxy group could be replaced by another function of similar bulk are combined in the design of the analgesic agent **picenadol** (**93**). The synthesis of this compound departs significantly from those discussed above. Thus, in the original route, the required two-ring intermediate **87** is obtained by condensation of *N*-methyl-4-piperidone (**86**) with the Grignard reagent

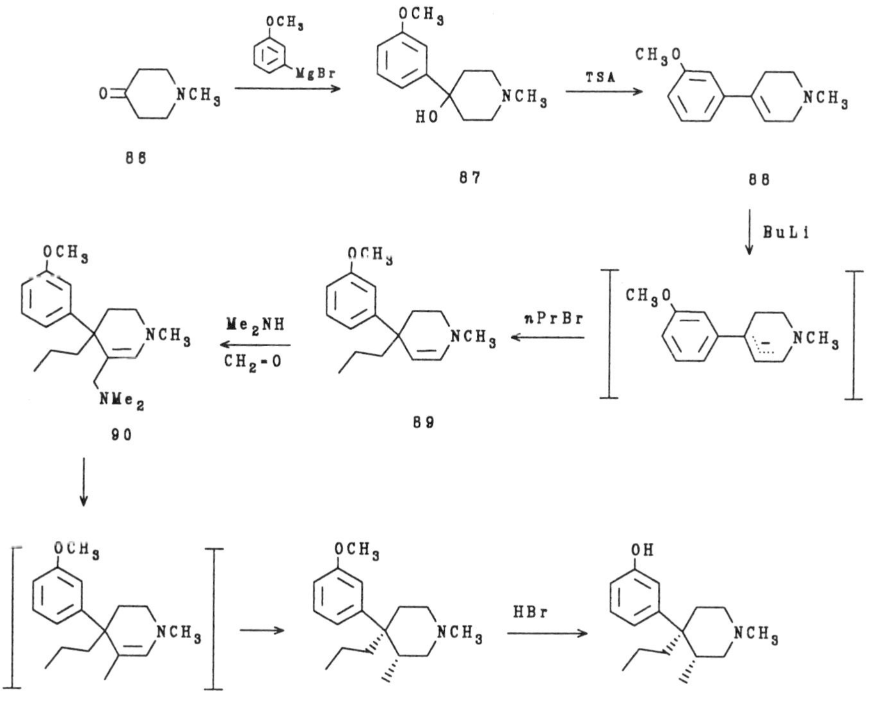

86 87 88 89 90 91 92 93

from *m*-methoxybromobenzene. The resulting tertiary alcohol is then dehydrated under acidic conditions to give **88**. Deprotonation of **88** with butyllithium, leads to an ambident anion; treatment with propyl bromide occurs at the more electrophilic terminus of the anion to give the 4-propyl derivative **89**, which, it should be noted, is now an enamine. This functional group undergoes the Mannich reaction with dimethylamine and formalin to give the corresponding aminomethylated derivative **90**. Catalytic reduction probably proceeds initially to give **91** from hydrogenolysis of the allylic dimethylamino group. Reduction of the enamine double bond gives a mixture of isomers that consists predominantly of the isomer containing cis alkyl groups. The methyl ether is then removed from **92** by reaction with hydrogen bromide. Fractional crystallization gives the pure cis isomer picenadol (**93**).[29]

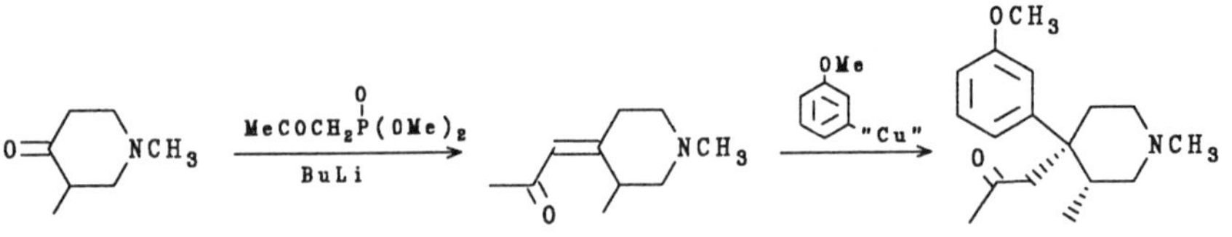

A significantly shorter method for preparing **93** hinges on a cuprate based conjugate addition. Condensation of the ylide from the dimethylphosphonate obtained from bromoacetone with substituted piperidone (**94**) yields the conjugated ketone **95**. Reaction of **95** with the cuprate reagent from *m*-methoxybromobenzene proceeds at the terminus of the enone to give intermediate **96**, which contains the required carbon skeleton; **picenadol** (**93**) is obtained on reduction of the ketone to a methylene group followed by cleavage of the methyl ether.[30]

B. 4-Amidopiperidines: Compounds Related to Fentanyl

All the opiate analgesics discussed thus far incorporated a quaternary carbon removed by two carbon atoms from the basic nitrogen, which at one time was considered an absolute requirement for activity. Several compounds in which this center consists of a simple secondary amine in fact exhibit considerably enhanced potency over the classical analgesics; the prototype **fentanyl** (**99**) is some 50 times more potent in humans than morphine and as much as 300 times more potent in experimental animals. This high potency is mirrored by higher in vitro affinity to the opioid μ receptor. There is evidence from the SAR in this and related series that the potentiating effect is due largely to the presence of the arylethyl side chain. Preparation of the parent molecule begins by alkylation of 4-piperidone with 2-phenethyl chloride to give the intermediate **97**. Reaction of **97** with aniline gives the corresponding Schiff base **98**. Catalytic reduction of the imine bond followed by acylation of the newly formed secondary amine with propionic anhydride gives fentanyl (**99**).[31] Condensation of the piperidone **97** with *o*-fluroaniline gives the more highly substituted derivative **100**. This compound is reduced as above to give a secondary amine; acylation of this intermediate with 2-methoxyacetyl chloride affords the potent analgesic **ocfentanil** (**101**).[32]

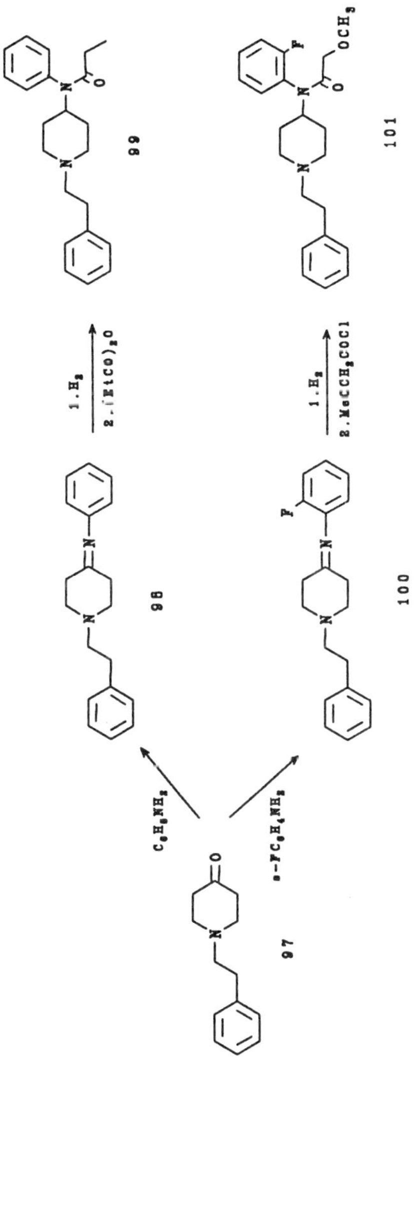

$C_6H_5NH_2$

$o\text{-}FC_6H_4NH_2$

97

98

100

1. H_2
2. $(EtCO)_2O$

1. H_2
2. $MeOCH_2COCl$

99

101

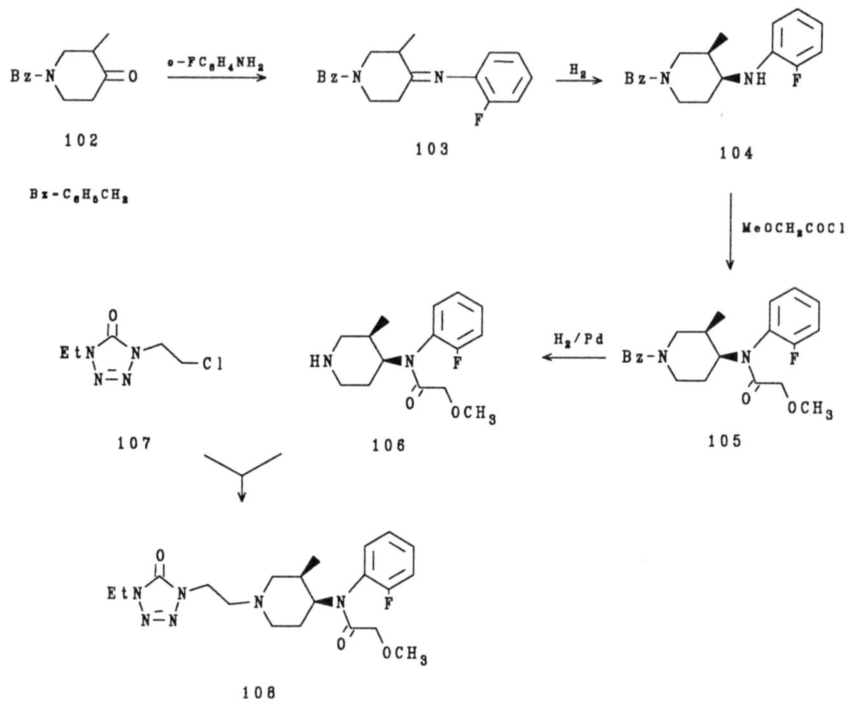

Bz = $C_6H_5CH_2$

The enhancement of potency due to the presence of an equatorial methyl group noted in the meperidine series also applies to fentanyl analogues. The ring nitrogen of the starting material (**102**) for this particular derivative is protected with a benzyl group since it will later be replaced by a somewhat more complex side chain. Condensation of **102** with *o*-fluroaniline gives the Schiff base **103**. This compound gives a mixture of isomeric amines on catalytic reduction; the cis isomer is then separated and acylated with 2-methoxyactyl chloride to give the amide **105**. Hydrogenolyis of the benzyl group over palladium then leads to the secondary amine **106**. Reaction with the chloroethyl tetrazolone **107** leads to the alkylated product and **brifentanil** (**108**);[33] the heterocycle in this case acts as a surrogate potentiating aryl group.

Restoring the quaternary center in the fentanyl series results in yet further enhancement in potency resulting in compounds that show analgesic activity at one-ten thousandth the dose of morphine in some animal models; side effects, however, show a parallel increase in potency since these agents are classical opiates. The key reaction in the syntheses of these compounds consists in the formation of α-aminonitriles, a functional group related to cyanohydrins. Thus reaction of piperidone (**109**) with potassium cyanide and aniline hydrochloride leads to the α-aminonitrile **110**. Hydrolysis of the cyano group by means of sulfuric acid affords the corresponding amide **111**; the benzyl protecting group is then removed by hydrogenation over palladium to give the secondary amine **112**. The all important phenethyl group is then incorporated by alkylation at the more basic ring nitrogen with 2-phenethyl chloride to give **113**. Interchange with ethanolic hydrogen chloride then converts the amide to an ethyl ester; acylation of the remaining secondary amine with propionic anhydride affords **carfen-**

109

Bz-$C_6H_5CH_2$

KCN

$C_6H_5NH_3^+Cl^-$

110

H_2SO_4

111

H_2

112

$C_6H_5CH_2CH_2Cl$

113

1. MeOH/HCl
2. $(EtCO)_2O$

114

115

tanil (**114**).[34] The potentiating effect of a ring methyl group obtains here as well; the analogous sequence starting from the 3-methyl piperidone **102** leads to **lofentanil** (**115**).[35]

The exact nature of the carbon substituent at the quaternary center is apparently not critical for good potency as in the meperidine series. Replacement of the ester by a methoxymethyl group removes one site for potential metabolic inactivation. Alkylation on the ring nitrogen of the secondary amine **112** with 2-chloroethylthiophene (**116**) gives intermediate **117**; the amide is then converted to an ester as shown above

116 + 112 → 117

EtOH
HCl

118

LAH

119

CH_3I
NaOEt

120

$(EtCO)_2O$

121

111 → (1. EtOH/HCl; 2. $(EtCO)_2O$) → 122 → (H_2) → 123 → (CH_2=$CHCO_2Et$) → 124

by an interchange reaction. Reduction of the **118** by means of lithium aluminum hydride converts the carbethoxy group to the corresponding carbinol, resulting in **119**. The alkoxide from the free hydroxyl is then alkylated with methyl to give the methyl ether **120**. Acylation of the aniline nitrogen with propionic anhydride completes the synthesis of **sufentanil** (**121**).[36]

As alluded to in the previous paragraph, the design of drugs that would be protected from metabolic inactivation has been a long-term preoccupation in medicinal chemistry. The reversed concept, drugs that contain deliberate metabolic weak links so that blood levels of parenteral agents can be quickly dropped, is of much more recent origin. The β-blocker **esmolol** represents one of the pioneer efforts in this area (see Chapter 2). Lately, this concept has been applied to the opiates, where a terminal ester on the side chain of ring nitrogen is compatible with analgesic activity. The carboxylic acid product from interaction with serum esterases on the other hand will presumably not cross the blood–brain barrier and consequently not reach the opiate receptors. Preparation of the compound based on this rationale begins with the amide–ester interchange of the *N*-benzylpiperidone derived intermediate **111**; acylation with propionic anhydride gives **122**; the benzyl protecting group is then removed by hydrogenolysis over palladium to give **123**. Michael addition of the secondary piperidine nitrogen to ethyl acrylate provides the weak-link containing side chain. Thus, **remifentanil** (**124**)[37] is obtained.

REFERENCES

1. Gulland, J.M.; Robinson, R. *Mem. Proc. Manch. Lit. Philos. Soc.* **1926**, *69*, 79. It is startling to note that communication of this landmark development was not followed by

a report in one of the standard chemical journals. Though duly recorded by Chemical Abstracts. (*Chem. Abstr.* **1926**, *20*, 765) the abstract does not include a structural diagram.

2. Weijlard, J.; Erickson, A.E. *J. Am. Chem. Soc.* **1942**, *69*, 869.
3. Pfister, K.; Tischler, M. U. S. Patent **1955**, 2715626. *Chem. Abstr.* **1956**, *50*, 7886.
4. Small, L.; Turnbull, S.G.; Fitch, H.M. *J. Org. Chem.* **1939**, *3*, 204.
5. Weiss, U. *J. Am. Chem. Soc.* **1955**, *77*, 5891.
6. Lowenstein, M.J. Br. Patent **1967,** 955493.
7. Lowenstein, M.J.; Fishman, J. U. S. Patent **1967,** 3320262; *Chem. Abstr.* **1967**, 67, 90989.
8. Blumberg, H.; Pachter, I.J.; Metossian, Z. U. S. Patent **1967,** 3320950; *Chem. Abstr.* **1967**, *67*, 100301.
9. Kobylecki, R.J.; Guest, I.G.; Lewis, J.W.; Kirby, G.W. German Offen. **1978**, 2812581; *Chem. Abstr.* **1979**, *90*, 39100.
10. Lewis, J.W.; Redhead, M.J. *J. Med. Chem.* **1970**, *13*, 525.
11. The presence of this tricyclic system in both morphine and the steroids prompted a book by Louis Fieser, entitled *Natural Products Related to Phenanthrene*. This tome, long out of print, reviewed the chemistry of both these two classes.
12. See Lednicer, D. *J. Chem. Ed.* **1989**, 718, for a discussion of the evolution of the Grewe synthesis.
13. Hellerbach, J.; Grussner, A.; Schnider, O. *Helv. Chim. Acta* **1956**, *39*, 429.
14. Schnider, O.; Grussner, A. *Helv. Chim. Acta* **1949**, *32*, 821.
15. Schnider, O.; Grussner, A. *Helv. Chim. Acta* **1951**, *34*, 2211.
16. Monkovic, I.; Wong, H.; Pircio, A.W.; Peron, Y.G.; Pachter, I.J.; Belleau, B. *Can. J. Chem.* **1975**, *53*, 3094.
17. See Palmer, D.C.; Strauss, M.J. *Chem. Rev.* **1977**, *77*, 1, for an extensive review on benzomorphan chemistry.
18. Eddy, N.B.; Murphy, J.G.; May, E.L. *J. Org. Chem.* **1957**, *22*, 2592.
19. Kametano, T.; Kisagawa, K.; Hiiraga, M.; Satoh, F.; Sugi, H.; Uryu, T. *J. Heterocycl. Chem.* **1972**, *9*, 1065.
20. May, E.L.; Eddy, N.B. *J. Org. Chem.* **1959**, *24*, 295.
21. Archer, S.; Albertson, N.F.; Harris, L.S.; Pierson, A.K.; Bird, J.G. *J. Med. Chem.* **1964**, *7*, 123.
22. Michne, W.F.; Albertson, N.F. *J. Med. Chem.* **1972**, *15*, 1278.
23. Freed, M.E.; Potoski, J.R.; Freed, E.H.; Conklin, G.L.; Malis, J.L. *J. Med. Chem.* **1973**, *16*, 595.
24. Eisleb, O.; Schaumann, O. *Deut. Med. Wochenschr.* **1939**, *65*, 967.
25. Eisleb, O. *Chem. Ber.* **1941**, *74*, 1433.
26. Weijlard, J.; Dorahovats, A.P.; Sullivan, G.; Purdue, G.; Heath, F.K; Pfister, K. *J. Am. Chem. Soc.* **1956**, *78*, 2342.
27. Janssen, P.A.J.; Jagenau, A.H.; Huygens, J. *J. Med. Chem.* **1959**, *1*, 299.
28. Aridon, A.W.D.; Morison, A.L. *J. Chem. Soc.* **1950**, 1471.
29. Feth, G.; Mills, J.E. U. S. Patent, **1985**, 4499274. *Chem. Abstr.* **1985**, *102*, 166624.
30. Martinelli, M.J.; Peterson, B.C. *Tetrahedron Lett.* **1990**, *31*, 5401.
31. Janssen, P.A.J.; Niemegeers, C.J.E.; Dony, J.G.H. *Arzneim.-Forsch.* **1963**, *13*, 502.

32. Huang, B.S.; Deutsche, K.H.; Lalinde, N.L.; Terrell, R.C.; Kudzma, L.V. Eur. Patent Appl. **1985**, 160422; *Chem. Abstr.* **1986**, *104*, 186308.
33. Lalinde, N.; Moliterni, J.; Wright, D.; Spencer, H.K.; Ossipov, M.H.; Spauling, T.C.; Rudo, F.G. *J. Med. Chem.* **1990**, *33*, 2876.
34. Vandaele, P.G.H.; DeBruyn, M.L.F.; Boey, J.M.; Sanczuk, S.; Agten, J.T.M.; Janssen, P.A.J. *Arzneim.-Forsch.* **1976**, *26*, 1521.
35. VanBever, W.F.M.; Niemegeers, C.J.E.; Janssen, P.A.J. *J. Med. Chem.* **1974**, *17*, 1047.
36. VanBever, W.F.M.; Niemegeers, C.J.E.; Schellekens, K.H.; Janssen, P.A.J. *Arzneim.-Forsch.* **1976**, *26*, 1548.
37. Feldman, P.L.; James, M.K.; Brackeen, M.C.; Marcus, F.; Billota, J.M.; Schuster, S.V.; Lahey, A.P.; Johnson, M.R.; Leighton, H.J. *J. Med. Chem.* **1991**, *34*, 2202.

CHAPTER 8

DRUGS BASED ON FIVE-MEMBERED HETEROCYCLES

1. INTRODUCTION

It has been estimated that over one-half of all therapeutic agents consist of heterocyclic compounds. The heterocyclic ring system in many cases comprises the very core of the active moiety or pharmacophore; the antibiotic activity of the cephalosporin antibiotics is, for example, clearly attributable to the presence of the fused azetidone ring, while the anxiolytic activity of the benzodiazepines can be traced to the aryl fused diazepine present in these drugs. Examples discussed in this chapter and those that follow have been chosen because either their heterocyclic component is believed to form part of a pharmacophore or, alternatively, they illustrate aspects of the chemistry of particular heterocyclic rings. Many drugs do exist in which the heterocyclic component is a surrogate for an open-chain amine, as illustrated by those drugs bearing piperidine or pyrrolidine rings in lieu of open-chain tertiary amines. Discussions of these compounds, if present, will generally be found in earlier chapters.

2. RINGS WITH ONE HETEROATOM

A. Furans

Virtually all drugs that contain a furan ring involve starting materials in which that ring is pre-formed. A sizeable number of antibacterial agents were at one time available that were based on relatively simple derivatives of 5-nitrofuran. These compounds have diminished greatly in importance due to the discovery of side effects and the availability of safer and more effective antibacterials. **Nitrofurazone** (**2**) and its more hydrophilic congener **nidroxyzone** (**3**) are typical of some of these drugs. Nitrofurazone (**2**) is prepared by reaction of 5-nitrofuran (**1**) with semicarbazone, while its analogue results from condensation of the aldehyde with the carbamide from 2- hydrazinoethanol.[1]

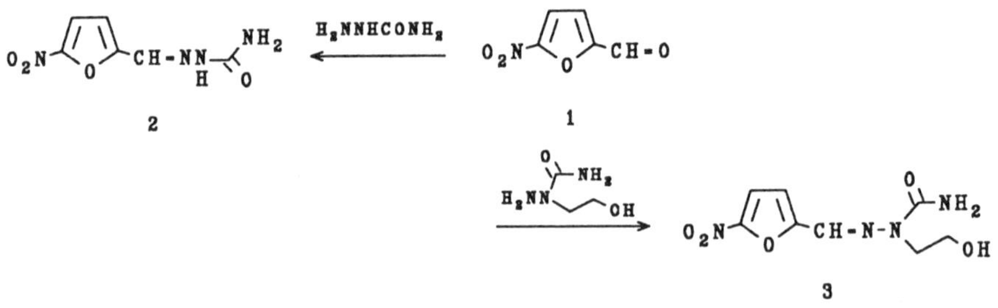

The imine function is replaced by a carbon–carbon double bond in a somewhat more complex nitrofuran antibacterial agent. Condensation of 5-nitrofurfuraldehyde with the carbanion obtained by treating 2,6-lutidine with strong base affords the diarylethylene **4** of unstated stereochemistry. Oxidation of the remaining pyridil methyl group starts by reaction of the product **4** with a peracid to give the corresponding *N*-oxide **5.** Treatment of **5** with acetic anhydride initially leads to the formation of a transient dehydro *O*-acetate; the acyl group then undergoes O to C migration with consequent bond reorganization; the overall result of this version of the Polonovski reaction is formation of acetate **6**. Saponification yields the final product **nifurpirinol** (**7**).[2]

The side chain for furan derivatives yet further structurally removed from the nitrofurans is obtained by internal amide–ester interchange of the *N*-aminoglycine derivative **8** with resulting cyclization to the imidazolinedione derivative **9**. Cuprous chloride catalyzed coupling of the diazonium salt from *p*-nitroaniline (**10**) and furfural leads to the 5-nitrophenyl furan derivative **11**. Reaction of the aldehyde group in **11** with hydrazine **9** leads to **dantrolene** (**12**), a compound that now exhibits muscle relaxant rather than antibacterial activity.[3] Replacement of the 4-nitro group in **12**, a group that may have been traced conceptually to the antibacterials by 3,4-dichloro, retains muscle relaxant activity. This compound, **clodanolene** (**15**), is prepared by the same sequence except it is started with 3,4-dichloroaniline (**13**).[3]

An imidazole ring provided the nucleus for the first of the antiulcer drugs, which act by blocking histamine H_2 receptors, **cimetidine** (**212**). The presence of this heterocyclic ring, as discussed later in this chapter, traces back the occurrence of this moiety in the normal agonist for the receptor histamine. Replacement of the ring by furan, in conjunction with some rearrangement of the functionality, leads to the very widely used histamine H_2 receptor blocker **ranitidine** (**21**).[4] The first step in the synthesis consists of addition of a dimethylaminomethyl group to furfuryl alcohol (**16**) by Mannich reaction with dimethylamine and formalin to give **17**. Treatment with thio-2-ethanolamine in the presence of hydrogen chloride leads to displacement of hydroxyl by the thiol and formation of thioether **18**, a reaction that probably proceeds via initial formation of a furfuryl carbocation. The key reagent **19**, may be viewed

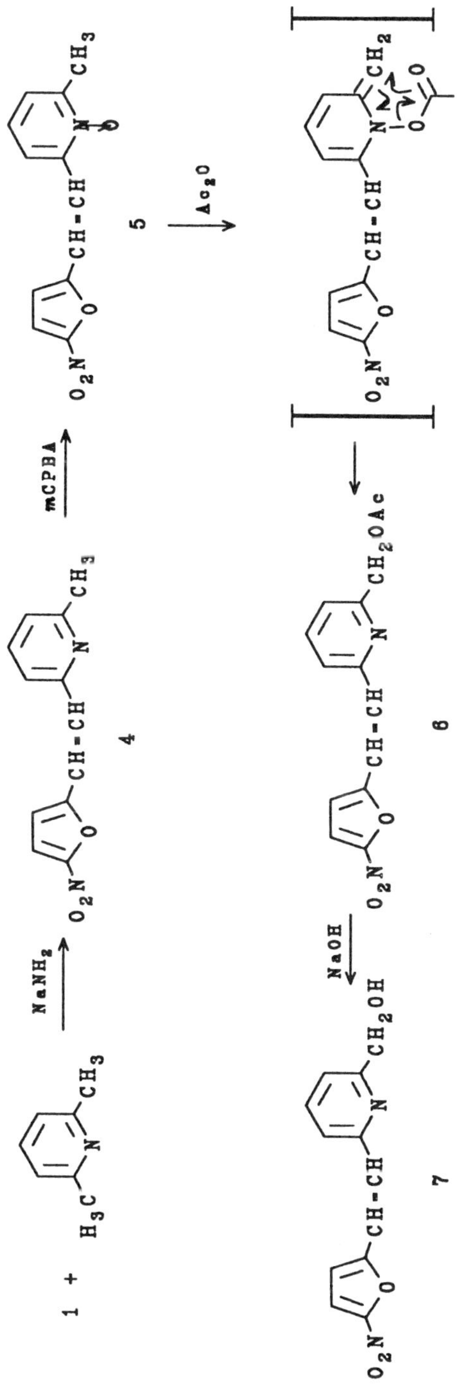
1 +
NaNH2
mCPBA
Ac2O
NaOH
4
5
6
7
CH2OAc
CH2OH

H_2N-N(CONH$_2$)-CH$_2$-CO_2Et (8) → 9

O_2N-C$_6$H$_4$-NH_2 (10) —1. HONO/CuCl; 2. Furfural→ 11 —9→ 12

Cl$_2$C$_6$H$_3$-NH_2 (13) —1. HONO/CuCl; 2. Furfural→ 14 —9→ 15

16 $\xrightarrow[CH_2\text{-}O]{Me_2NH}$ 17 $\xrightarrow{HSCH_2CH_2NH_2}$ 18

$(CH_3)_2N$ — 17 — OH

$(CH_3)_2N$ — 18 — S — NH_2

18 + 19 (MeS)(MeS)C=CH—NO_2 → 20

20: $(CH_3)_2N$ … S … N(H) — C(SCH_3)=CH—NO_2

20 $\xrightarrow{CH_3NH_2}$ 21

21: $(CH_3)_2N$ … S … N(H) — C($NHCH_3$)=CH—NO_2

formally as the bisthiomethyl acetal of nitroacetic acid. Reaction of **19** with **18** leads to displacement of one of the thiomethyl groups by the terminal amine on the side chain to give **20**, probably by an addition–elimination sequence. Displacement of the remaining thiomethyl group by methylamine leads to formation of the nitrovinyl function, a group that is bioisosteric with a cyanoguanidine. Thus, ranitidine (**21**) is obtained.

B. Pyrrole and Its Derivatives

Benzene rings provide the aromatic nucleus for the majority of the NSAIDs. A propionic acid attached at its 2-position provides the side chain for most of these compounds, as noted in Chapter 2. Inhibition of the cyclooxygenase enzyme and, consequently, NSAID activity is retained when the role of an aromatic ring is filled by a pyrrole. Note that the side chain for all but one of these compounds consists of an otherwise unsubstituted acetic acid.

The starting material for the simplest of these compounds, **tolmetin** (**26**), consists of the Mannich product **22**, from reaction of pyrrole itself with dimethylamine and formalin. Alkylation of **22** with methyl iodide gives the quaternary salt **23.** Displacement of the now highly reactive benzyl-like salt with cyanide ion leads to formation of the corresponding acetonitrile **24** with loss of the excellent leaving group trimethylamine. Friedel–Crafts acylation of **24** with the acid chloride from *p*-toluic acid leads to the ketone **25**. Hydrolysis of the nitrile group with aqueous base affords the corresponding carboxylic acid and thus tolmetin (**26**).[5]

The preparation of the NSAID **clopirac** (**31**) starts with the construction of the *N*-phenylpyrrole moiety. Thus, reaction of *p*-chloroaniline with the 2,5-dimethoxy furan **27**, a latent form of hexane-2,5-dione, leads to the desired 2,5-dimethylarylpyrrole **28**. A Mannich reaction on this intermediate adds the first carbon of the side chain at the only free position on the pyrrole ring yielding **29**. Compound **29** is then converted to the nitrile **30**, which is hydrolyzed as shown above to afford clopirac (**31**).[6]

A modification of the classical Hantzsch synthesis provides a pyrrole ring that bears a prebuilt acetic acid chain. The reaction in this case involves treatment of a mixture of ethyl acetonedicarboxylate (**32**), shown below as its enolate, and chloroacetone with

$N(CH_3)_2$ $\xrightarrow{CH_3I}$ $N^+(CH_3)_3\,I^-$ $\xrightarrow{KCN}$ CN

N, CH_3 — 22, 23, 24

MeC_6H_5COCl, $AlCl_3$

H_3C … CO_2H $\xleftarrow{NaOH}$ H_3C … CN

O, N, CH_3 — 26, 25

Cl–C₆H₄–NH₂ + CH₃O, OCH₃, H₃C, CH₃ (27) → 28

28 → (CH₃)₂NH, CH₂=O → 29 (N(CH₃)₂)

29 → 1. CH₃I 2. KCN → 30 (CN)

30 → NaOH → 31 (CO₂H)

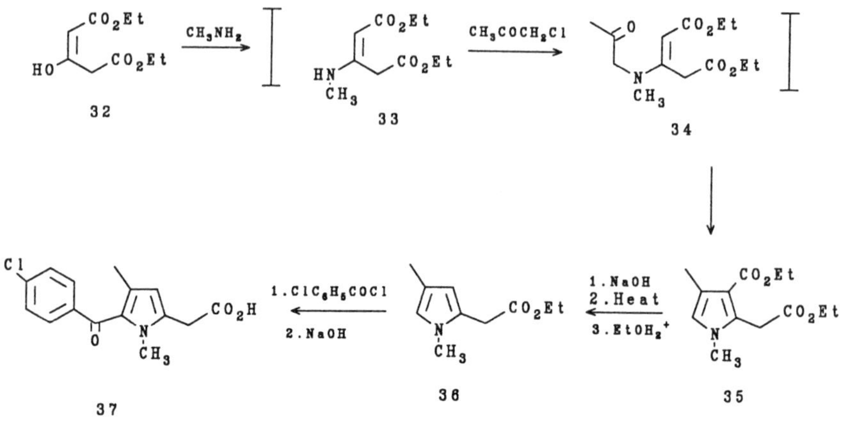

aqueous methylamine to give pyrrole **35** as the first observable product. The reaction can be rationalized by assuming formation of enamine **33** as the first step; alkylation with chloroacetone will then give ketoester **34**. Internal aldol condensation will then lead to the observed product **35**. This compound is then saponified to a dicarboxylic acid; heating of this product leads to loss of the ring carboxyl group. The remaining acid is then reesterified (**36**) for protection in the next step. Friedel–Crafts acylation with the acid chloride from *p*-chlorobenzoic acid occurs at the more reactive 2-position of the pyrrole ring. Saponification then gives the free acid **zomepirac** (**37**)[7].

The carboxylic acid in the most active of these pyrrole NSAIDs, **ketorolac** (**44**) is actually disposed on an alicyclic ring fused onto the pyrrole; the connectivity of the carbon bearing this acid is in fact closer to the better known arylpropionic acid NSAIDs. The starting material (**38**) for the preparation of this agent is obtained by electrophilic substitution on pyrrole proper by the sulfonium chloride obtained from reaction of dimethyl sulfide with *N*-chlorosuccinimide. This intermediate loses a methyl group on heating, probably as chloromethane to give the mercapto ether **39**. The presence of the free pyrrole proton precludes normal Friedel–Crafts acylation. Instead, the benzoyl group is added by a variation of the Villsmeier reaction where **39** is treated with *N,N*-dimethylbenzamide and phosphorus oxychloride to give **40** on workup. Reaction of **40** with the spirocyclopropyl substituted Meldrum's acid derivative in the presence of base leads to attack on the three-membered ring by the anion from the pyrrolidine nitrogen to give the ring-opened intermediate **41**. The sulfur is then oxidized to the sulfoxide by means of peracid to increase its propensity to act as a leaving group; methanolysis opens the dioxolane ring and leads to the key diester **42**. Treatment with strong base initially forms a carbanion at the carbon bearing the two carboxylates; internal attack on the pyrrole ring leads to displacement of the sulfoxide group, possibly by an addition–elimination sequence and formation of the fused ring to give **43**. The sequence toward ketorolac (**44**) is completed by saponification of the ester groups and monodecarboxylation of the resulting diacid.[8]

The renin–angiotensin system plays a pivotal role in the maintenance of the circulatory system, exerting control over blood volume and pressure as well as levels of tissue sodium and potassium. As noted in the discussion of protease inhibitors in

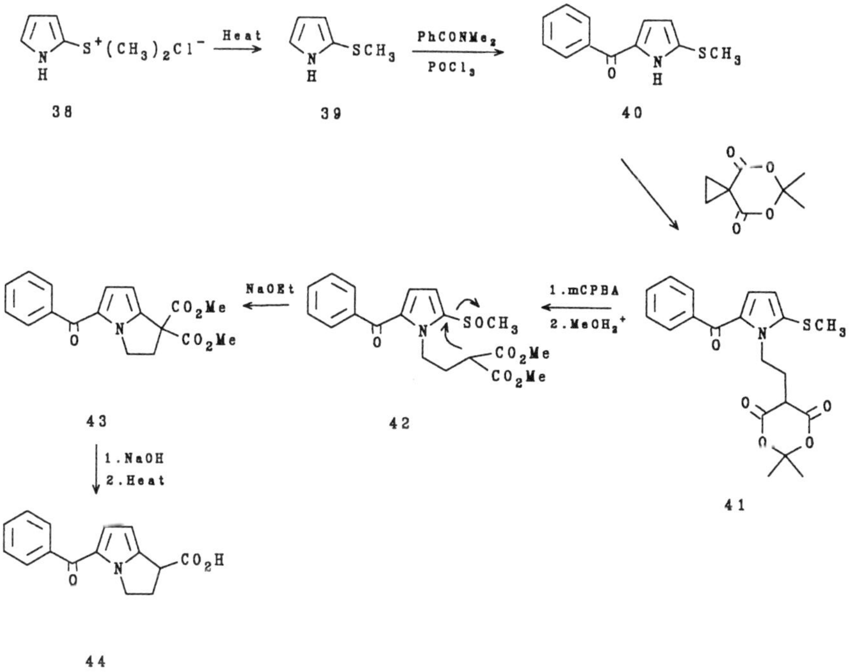

Chapter 1, angiotensin II, the controlling octapeptide is actually the product of a degradative scheme that starts from the large peptide angiotensinogen. Reaction of angiotensin I, the decapeptide product, from renin catalyzed cleavage of the latter with ACE leads to the vasoconstrictor angiotensin II. The development of ACE inhibitors as antihypertensive drugs was at least partly spurred by the observation of the hypotensive effect and ACE inhibiting activity of a nonapeptide from the snake venom from *Bothropos jararaca*. This compound and its successors all inhibit ACE by acting as false substrates at a cleavage site on the enzyme. A synthesis program aimed at finding an orally effective non-peptide counterpart culminated in the identification of **captopril** (**48**).[9] The mechanism of action is reflected in the fact that both this compound and its successors retain some peptide-like features as well as the fact that these drugs are also all single enantiomers.

In one synthesis of **48**, L-proline (**46**) is acylated with the acid chloride **45**, obtained from addition of hydrogen chloride to the double bond in methacrylic acid followed by reaction with thionyl chloride, to give the amide **47** as a mixture of diastereomers. The pure (2*S*) isomer is then isolated from the mixture by fractionation as the dicyclohexylamine salt. Treatment of this compound with ammonium hydrosulfite leads to displacement of chlorine by a thiol group and formation of captopril (**48**).[10]

The effectiveness of ACE inhibitors as antihypertensive agents combined with the low incidence of side effects led to intensive investigation of the SAR of this class of drugs. Replacement of the methylthiol fragment by an alkylated amino group formally changes that fragment to an alanine derivative; the core of the structure thus becomes a dipeptide. Compounds in this series have proven very active ACE inhibitors when the alkylating function is provided with an additional carboxyl group. Reductive alkylation of the dipeptide alanylproline (**50**), prepared by standard coupling methods, with ketoester **49** and hydrogen, leads directly to the secondary amine **51** as a mixture of diastereomers in which the desired product predominates. This compound is isolated by a crystallization as its maleate salt to give the antihypertensive drug **enalapril** (**51**).[11] It has been established that the active agent is in fact the dicarboxylic acid **enalaprilat**, which results from in vivo saponification. This product is, however, not active by the oral route, probably as a result of its highly polar nature.

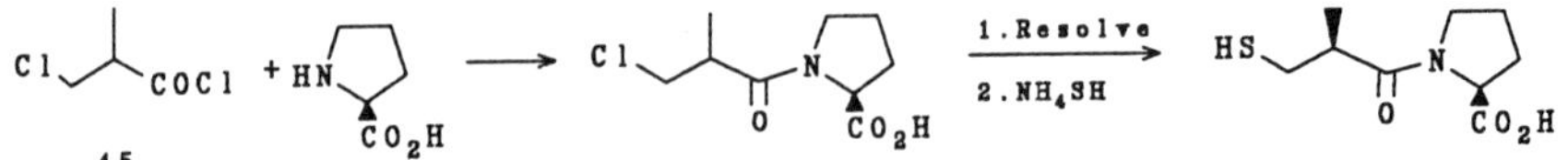

Replacement of the alanine residue by a lysine moiety leads to **lisinopril** (**53**), which is administered in this case as a free dicarboxylic acid. This compound is quite actively orally in spite of its polarity. The synthesis is quite analogous to the one shown above, which involves reductive alkylation of the *tert*-butoxycarbonyl protected lysylproline **52** with ketoester **49**. Separation of the desired diastereomer followed by removal of the *t*-BOC group with trifluoroacetic acid and then saponification gives lisinopril (**53**).[11]

Oxidized phosphorus, included as part of the alanyl residue, can also serve as the second acidic group. The required side chain **55** is prepared by addition of the free radical from treatment of the sodium salt of phosphinic acid with azobisisobutyronitrile (AIBN) to 4-phenylbut-1-ene (**54**).

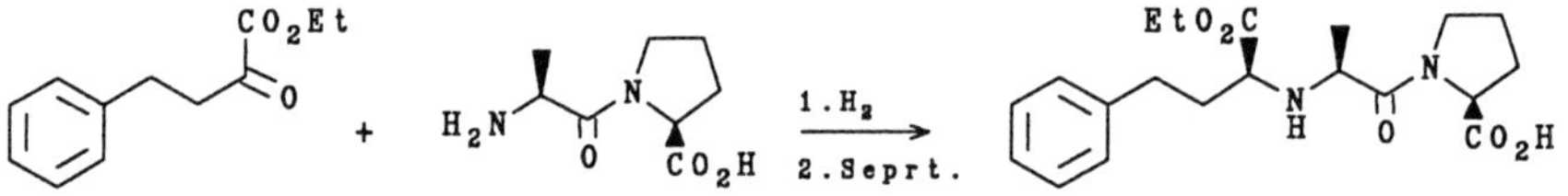

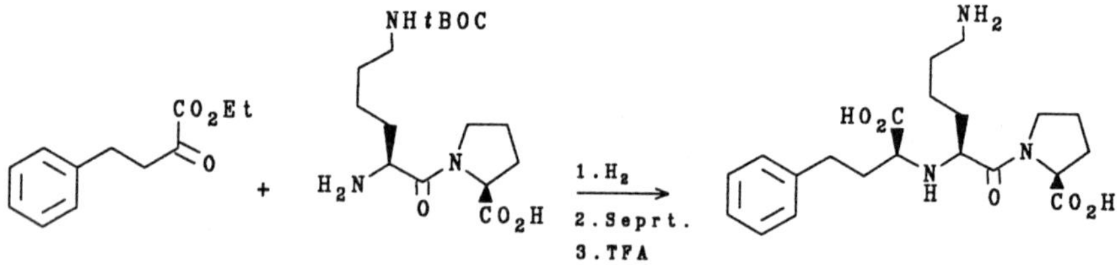

tBOC= OCOtBu

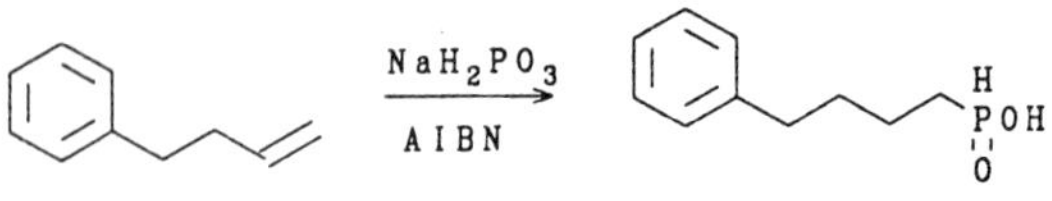

NHCbz HO CO_2H **56** + HN CO_2Bz **57** —DCC→ NHCbz HO CO_2Bz **58** —59/DCC→ H P O NCbz CO_2Bz **59** —1. mCPBA 2. H_2/Pd→ HO P O NH_2 CO_2H **60**

Cbz - OCOBz
Bz - $CH_2C_6H_5$

Dicyclohehylcarbodiimide mediated amide formation between the carbobenzyloxy (Cbz) derivative of the lysine derivative **56**, where the hydroxyl group replaces the amino and the benzyl ester of L-proline (**57**), affords the amide **58**. Reaction of the hydroxyl group in **58** with the phosphonic acid **55** in the presence of DCC leads to the formation of an ether bond to give intermediate **59**. Oxidation of the phosphine P–H bond with peracid leads to a phosphinic acid. Catalytic reduction of this product leads to cleavage of both of the benzyl based protecting groups. Thus, **ceronapril (60)** is obtained.[12]

Considerable latitude also seems to exist for structural modification of the proline moiety. This ring may form part of an indole, an indoline, or even a tetrahydroisoquinoline. A less profound change involves fusion of a spiro-thioacetal onto the pyrrolidine ring. Thus, reaction of the Cbz derivative of 3-ketoproline methyl ester (**61**) with ethane-1,2-thiol affords the corresponding thiocacetal, **62**. The Cbz group is then removed by hydrogenolysis over palladium and the resulting free amine is coupled with an appropriately protected alanine derivative to afford the dipeptide-like intermediate **63**. This compound is then subjected to reductive alkylation with the same ketoester used above. Separation of the desired isomer gives the ACE inhibitor **spiralpril (64)**.[13]

The ACE inhibiting activity is retained when the pyrrolidine ring of the proline moiety is imbedded in a fused bicyclic system; this modification leads to several quite potent antihypertensive drugs. The first step in the preparation of the alanine based side chain involves conjugate addition of the benzyl ester of alanine to the unsaturated ketoester **65**, with addition occurring at the terminus of the double bond conjugated with the ketone to give intermediate **66** as a mixture of diastereomers; the compound whose stereochemistry corresponding to **67** is then isolated. Exhaustive catalytic hydrogenation over palladium removes the benzyl protecting group and at the same

CbzN CO_2CH_3 61 → [SH SH] → CbzN S S CO_2CH_3 62 → 1.H_2/Pd 2.Gly → H_2N N O S S CO_2CH_3 63 → 1.45 2.Sepr. → EtO_2C N H O N S S CO_2CH_3 64

Cbz = $OCOCH_2C_6H_5$

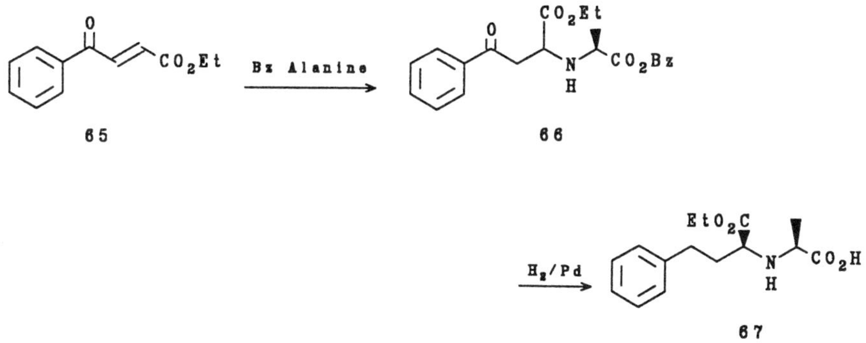

time reduces the ketone to a methylene group via hydrogenolysis of the initially formed carbinol. Thus, the completed side chain **67** is obtained. This intermediate could in principle be obtained in a single step by reductive amination of the ketoester **49** with alanine.

Construction of the heterobicyclic fragment start with alkylation of the enamine **69** from cyclopentanone with the chloro derivative **68**, which is obtained in several steps from *N*-acetyl serine, to give amidoketone **70** as the product. Treatment of **70** with aqueous acid can be viewed as involving initial hydrolysis of the acetamide; imine formation between the thus freed primary amine and the adjacent carbonyl group will lead to formation of the dihydropyrrolidine ring; hydrolysis of the ester group in the acid medium leads to formation of iminoester **71** as a mixture of stereoisomers. The acid is then converted to its benzyl ester and resolved; hydrogenolysis affords the desired isomer **72**. The DCC mediated coupling between **72** and the side chain acid **67** results in formation of the ACE inhibitor **ramipril** (**73**).[14]

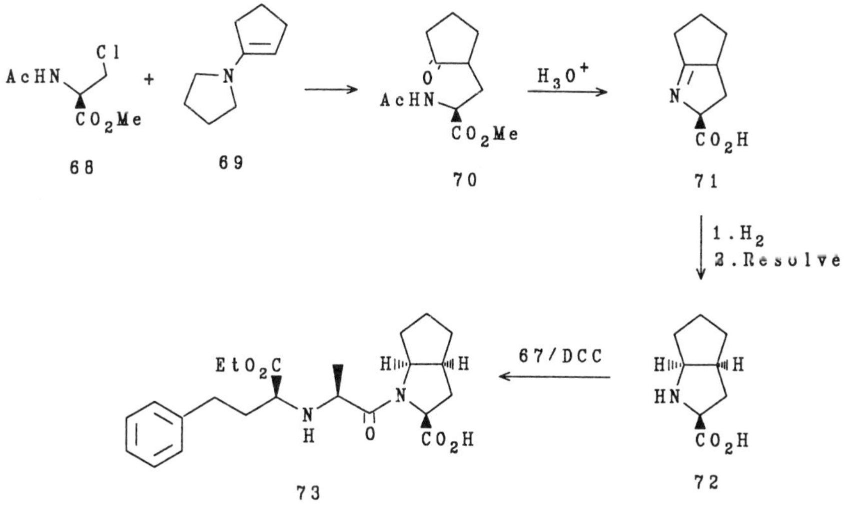

The nucleus of the ACE inhibitor **perindopril** (**77**) consists, as the name suggests, of a perhydroindole; this compound also differs from the foregoing by omitting the benzene ring at the side chain terminus. The side chain **74** for this compound is obtained in a single step by sodium cyanoborohydride mediated reductive amination of pyruvic acid with (*S*)-norvaline. The reaction can be visualized as first involving the formation of an imine between the primary amine and the ketone; the proximity of this newly formed bond to the chiral carbon on the amino acid results in stereoselective reduction and consequent formation of **74** as a single diastereomer. Exhaustive hydrogenation of the ester (**75**) of chiral indoline 2-acetic acid leads to the totally reduced derivative with the cis ring fusion. This compound is saponified and the acid is protected as its *tert*-butyl ester **76** by treatment with isobutylene. Coupling of **76** with the side chain **74**, followed by removal of the protecting group, affords perindopril (**77**).[15]

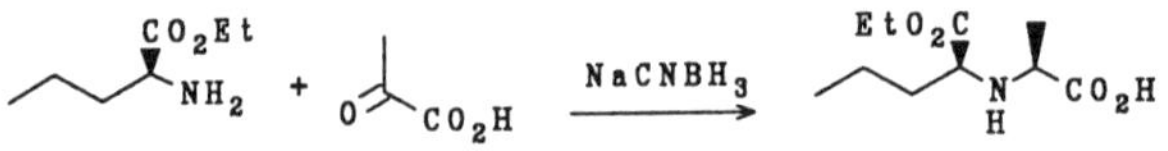

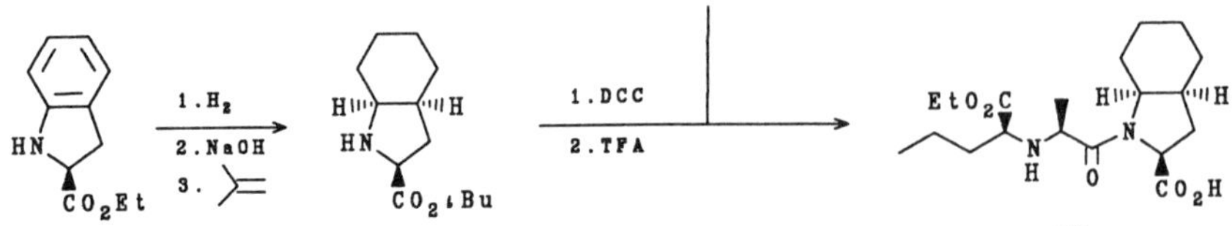

A counterpart of **77**, in which the proline is incorporated in an indoline, has a markedly altered peptidomimetic side chain similar to the statine-based protease inhibitors discussed in greater detail in Chapter 1. Chiral indoline carboxylic acid (**78**) is obtained by first resolving the corresponding acetamide and saponifying the resulting product. The key half-ester **79** can be obtained by careful ethanolysis of *meso*-2,4-dimethylglutaric anhydride, a compound in which both carboxyl groups are sterically identical. Coupling the two synthons by means of DCC gives **pentopril** (**80**).[16]

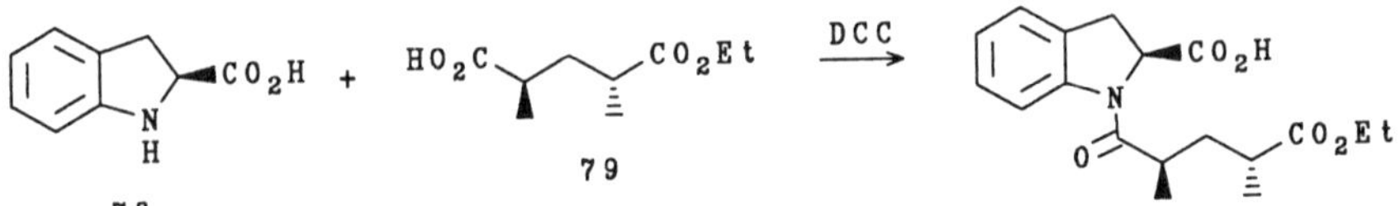

Quite analogous methodology is used to prepare a ring enlarged version of the foregoing; the resulting tetrahydroisoquinoline derivative **quinapril** (**82**) does, however, retain a peptide-like side chain. The heterocyclic nucleus for this ACE inhibitor is obtained by Pictet–Spengler condensation of *S*-phenylalanine with formaldehyde in the presence of sulfuric acid. The reaction probably proceeds via an intermediate imine to afford the heterocycle as predominantly a single enantiomer. The carboxylic acid then converted to its *tert*-butyl ester by reaction with isobutylene to give the aminoester **81.** The final product (**82**) is obtained by acylation of **81** with the acid **67**, whose preparation is described above, followed by acid catalyzed removal of the *tert*-butyl ester protecting group.[17]

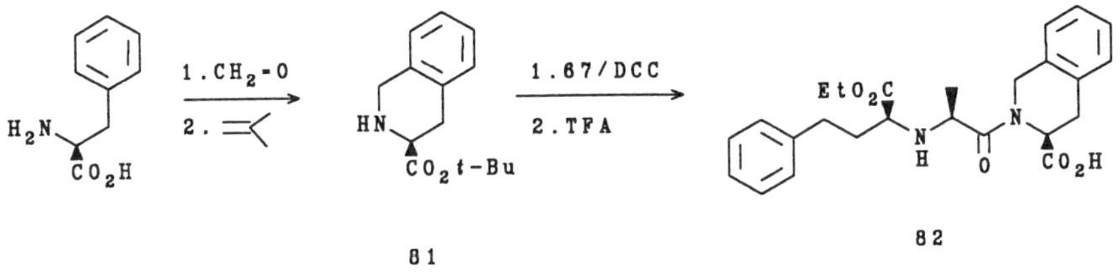

An unusual rearrangement provides the key to the preparation of a highly substituted pyrrolidone, **doxapram** (**89**), which is used as a respiratory stimulant. The synthesis starts with displacement of chlorine on pyrrolidine (**83**) by the carbanion from diphenylacetonitrile (**84**) to give **85** as the product. The quite hindered nitrile is then hydrolyzed to the corresponding carboxylic acid **86** by basic hydrolysis. Reaction of **86** with thionyl chloride presumably initially proceeds to formation of the corresponding acid chloride. The close proximity of this group to basic nitrogen, which is not readily apparent from planar drawings, leads to internal reaction of these functions to form the reactive, bicyclic quaternary acylation product **87**. The chloride counterion liberated in this reaction then displaces one of the more electrophilic branches of the caged structure. This reaction leads to ring opening to form a pyrrolidone (**88**) with simultaneous formation of a halogenated side chain. Displacement of chlorine by nitrogen on morpholine leads to the alkylation product doxapram (**89**).[18]

Hydantoins, that is, imidazo-2,4-diones, constitute one of the oldest and most effective classes of anticonvulant drugs for the treatment of epilepsy, as noted in a later discussion of imidazole derivatives. Several somewhat simpler imides have also found some use for this indication, although these compounds are said to act mainly on petit

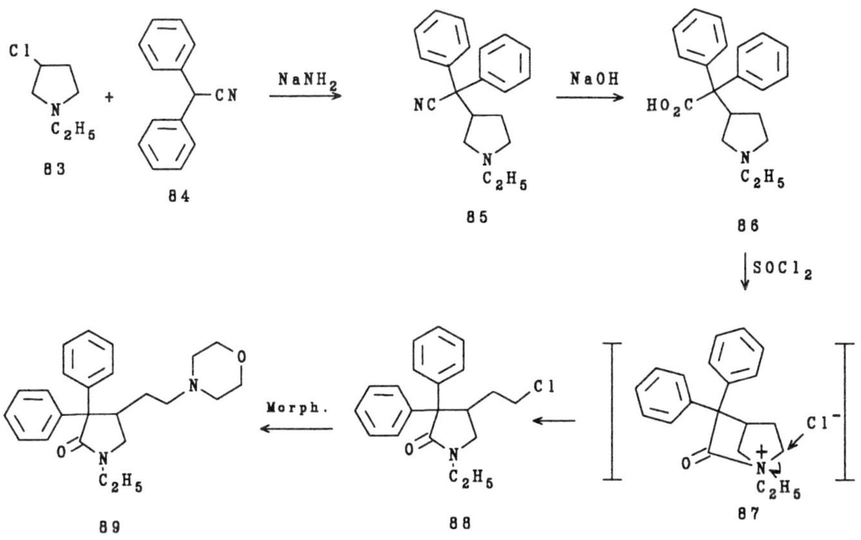

CN KCN CN NaOH CH_3NH_2
CO_2Et NC CO_2Et HO_2C CO_2H O N O CH_3
90 91 92 93

CN H_3C O N O CH_3
H_3C CO_2Et
94 95

C_2H_5 CN 1.KCN 2.H_3O^+ CH_3 C_2H_5 HO_2C CO_2H NH_3 CH_3 C_2H_5 O N H O
CH_3 CO_2Et
96 97 98

mal seizures. The synthesis of a typical agent, **phensuximide (93)**, starts with the conjugate addition of cyanide ion to the product **90** from Knoevnagel condensation of benzaldehyde with ethyl cyanoacetate. Acid hydrolysis initially proceeds to a tricarboxylic acid; decarboxylation of the β-dicarboxylic acid in the intermediate leads to the observed product, the succinic acid **92**. Reaction of **92** with methylamine leads to formation of the imide ring and thus to **93**.[19] The same sequence starting from the Knoevnagel product **94** from acetophenone affords the anticonvulsant agent **methsuximide (95)**.[19]

The first report of the preparation of the dialkyl succinimide **98** dates back to early in the twentieth century. It is consequently surprising to note that it was introduced as an anticonvulsant, under the name **ethosuximide**, well after its more recently synthesized congeners. The synthetic route starting from the methyl ethyl ketone generally follows the one above except for the use of ammonia in the last step. The compound thus differs as well by possessing a somewhat acidic imide proton.[20]

3. RINGS WITH TWO HETEROATOMS

A. Oxazoles, Isoxazoles, and Their Derivatives

The wide diversity of biological activities displayed by therapeutic agents based on unfused oxazole rings indicate that moiety plays a role similar to the one of a simple aromatic ring. Thus, it is not surprising to note that the small class of agents based on this ring system includes NSAIDs whose activity in all probability hinges on the presence of a pendant propionic acid. The synthesis of the first of these begins with the acylation of the benzoin derivative **99** with the half-ester acid chloride from succinic acid. Treatment of the resulting 1,4-dicarbonyl compound **100** with phosphorus oxychloride leads to formation of the oxazole ring by a cyclodehydration reaction; this reaction is a well-precedented standard method for forming five-membered heterocycles, which in this case may well proceed via the dienol form **101**. Saponification of the ester affords **oxaprozin** (**102**).[21]

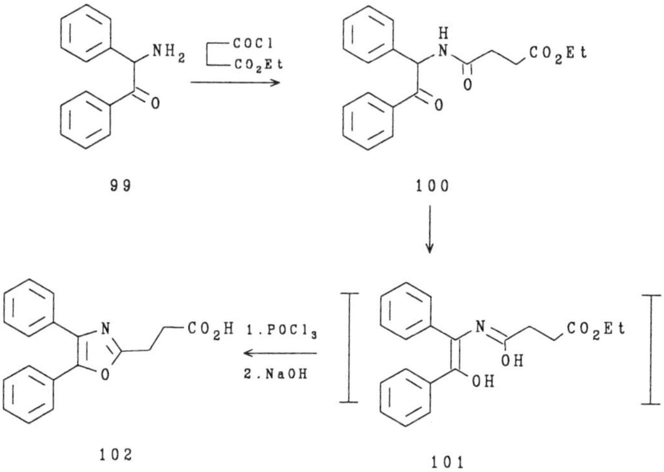

The strategy used for building the oxazole ring for the NSAID **romazerit** (**107**) relies on closing the ring about the nitrogen rather than on the oxygen atom. The intermediate that incorporates the requisite oxygen atom is obtained by acylation of ethyl 2-hydroxyacetoacetate with 4-chlorobenzoyl chloride to give the diester **103**, which is shown below in its enol form. Heating **103** with formamide in the presence of acid probably initially leads to replacement of the enol hydroxyl group by ammonia, perhaps by a conjugate addition–elimination sequence. Imine formation between this amine and the benzoyl carbonyl group results in formation of an oxazole ring and thus affords **105**. The ester group is then reduced to a carbinol with lithium aluminum hydride and the resulting alcohol **106** is replaced with chlorine by means of thionyl chloride. Displacement of this activated halogen with the enolate obtained from ethyl 2,2-dimethylglycolate and sodium ethoxide leads to displacement and formation of an ether bond. The ester on the newly connected side chain is then saponified to afford the corresponding acid and romazerit (**107**).[22]

A related approach is used to prepare a relatively simple 2-aminooxazole derivative. Reaction of hydroxyacetone with the alkylated cyanamide **108** leads to addition of the hydroxyl group to the cyano function and formation of the hypothetical iminoether **109**. This compound cyclizes to an oxazole again by internal imine formation under the reaction conditions to give the intermediate **110**. Acylation with isobutyryl chloride gives **isamoxole** (**111**), a compound that exhibits antiasthmatic activity.[23]

An oxazole substituted with a complex aminohydantoin side chain is described as a muscle relaxant. Imine formation between glyoxylic acid and aminohydantoin (**112**) results in the imino acid **113**. Use of **113** to acylate the amine on 4-chloro-2′-aminoace-

103

$HCONH_2$

104

105

1. LAH
2. SO_2Cl

106

1. HO–C(Me)₂–CO_2Et
2. NaOH

107

108

109

110

$Me_2CHCOCl$

111

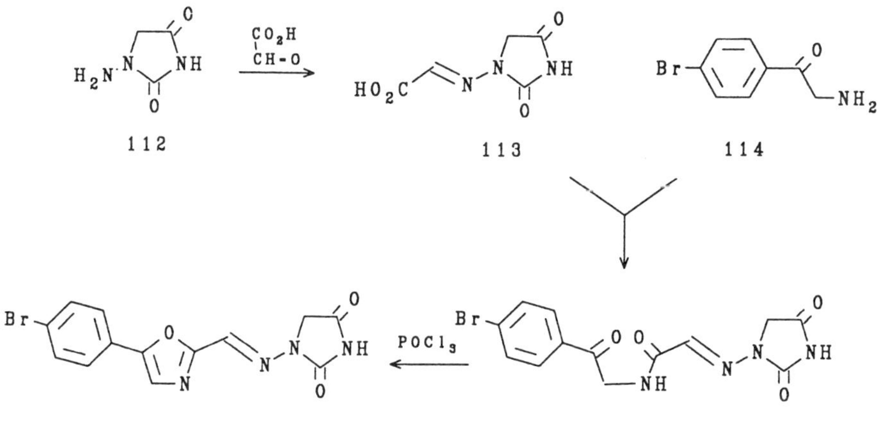

tophenone (**114**) leads to the amide **115**, which now includes a 1,4-dicarbonyl array. Treatment of the keto amide with phosphorus oxychloride leads to cyclodehydration and the consequent formation of an oxazole ring. Thus, **azumolene** (**116**)[24] is obtained.

Of course, the fully saturated version of an oxazole ring is the cyclic equivalent of a carbinolamine. Aside from providing a protecting group for aminoalcohol intermediates, these derivatives sometimes prove useful as latentiated forms of drugs. As an example, reaction of the *N*-demethyl analogue **117** of the α-adrenergic agonist **phenylephrine** with cycloheptanone affords **ciclafrine** (**118**), a drug that shows the same activity as the parent.[25]

The simple phenethylamine derivative amphetamine has well-recognized appetite suppressant and euphoriant activity; the latter has led to significant street use of the agent as an "upper." The SAR for this activity is sufficiently broad so that similar activity is shown by many compounds that show comparable spacing of an aromatic ring and an amino group. Reaction of the phenylethanolamine **119** with cyanogen bromide can be envisaged as proceeding first to the corresponding cyanamide **120**; internal attack by transient alkoxide oxygen on the nitrile triple bond leads to cyclization to the iminooxazolidine **121**. Tautomerization to the endocyclic isomer gives the appetite suppressant agent **aminorex** (**122**).[26]

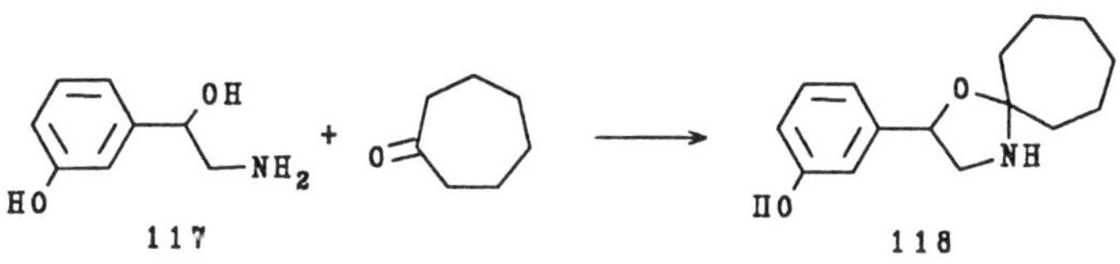

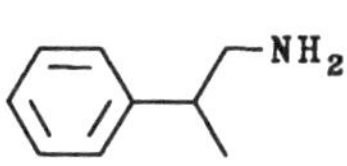

Amphetamine

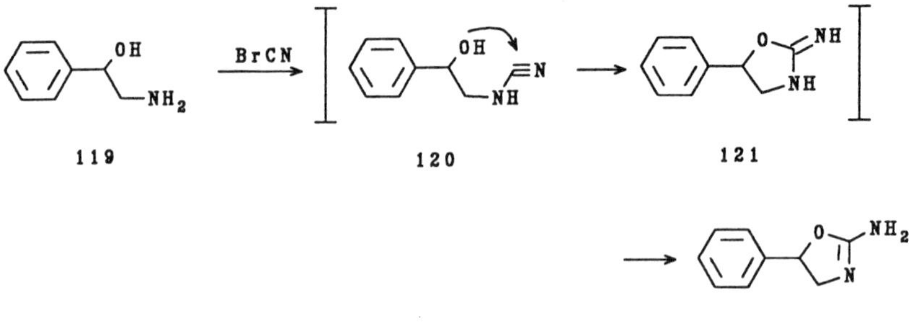

BrCN

119 120 121

122

The dimethylaminooxazolidone derivative **thozalinone** (**125**) is described as an antidepressant. Once again, the synthesis of this agent uses a cyanamide, provided in this case as a preformed reagent. Thus, reaction of the alkoxide from ethyl mandelate (**123**) with *N*,*N*-dimethylcyanamide leads to the amidine **124** by addition to the nitrile. Internal displacement of the ester ethoxide group closes the ring to an oxazolidinone forming the product **125**.[27]

The structural requirements for anticonvulsant activity are also quite broad. As noted in Section 3,B, hydantoins are the best known antiepileptic drugs: selected succinimides, as noted earlier, and some oxazolidinediones show the same activity. The parent nucleus for the latter, **129**, was first prepared over four score years ago.[28] The original route involves the reaction of ethyl lactate (**126**) with guanidine; the first step probably involves interchange of the ester to an acylated guanidine derivative such as **127**. Addition of alkoxide to the imine followed by loss of ammonia leads to formation of the iminooxazolidone **128**. The imino group is then hydrolyzed to a carbonyl and the resulting imide is methylated by means of base and methyl iodide to give the oxazolidinedione **129.** Further alkylation of this intermediate leads to a number of anticonvulsants. In a typical example, treatment of the carbanion from **129** and sodium ethoxide with methyl iodide gives **trimethadione** (**130**);[29] use of allyl bromide gives **aloxidone**.

The classical method for preparing isoxazole involves condensation of a 1,3-dicarbonyl compound with hydroxylamine, a reagent that contains the preformed N–O bond. The regiochemistry of the reactions can usually be rationalized by assuming that the first step involves imine bond formation at the more reactive carbonyl group. Thus, reaction of formyl ketone **131** with hydroxylamine gives isoxazole **132**, the product of the initial reaction at the aldehyde. Preparation of the second part of the compound in question, **disoxaril** (**135**), illustrates an alternative method for preparing oxazolines. Ester–amide interchange between ethyl 4-hydroxybenzoate and ethanolamine leads to the amide **133**. Treatment of **133** with thionyl chloride probably first proceeds to

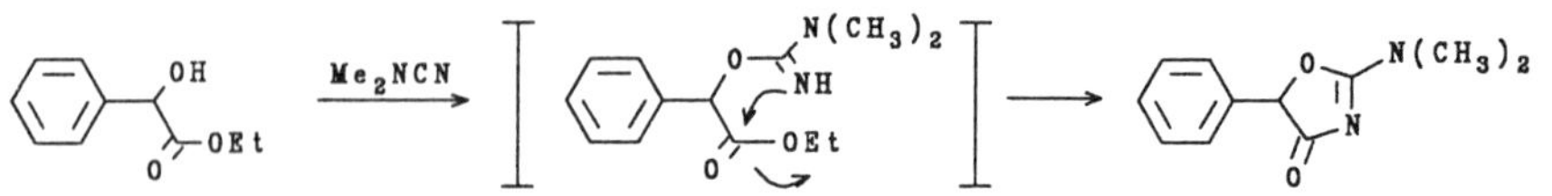

Me_2NCN

123 124 125

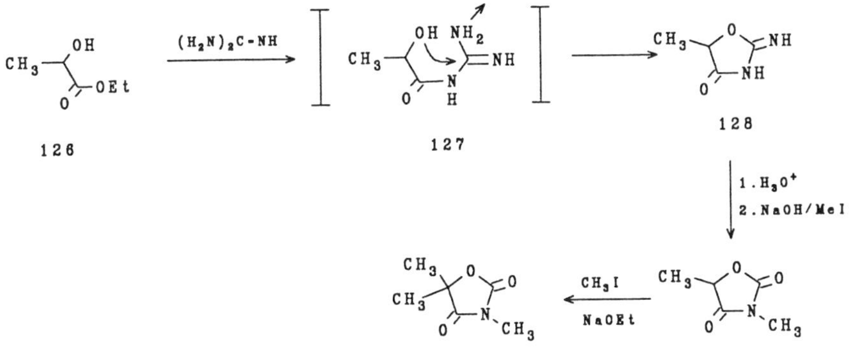

the imino chloride; addition–elimination of the terminal side chain hydroxyl group leads to formation of the oxazoline ring. Alkylation of the phenolate from **134** with bromide **132** then affords the antiviral agent disoxaril (**135**).[30]

Well over one-half a century after its discovery, **isoniazide**, the hydrazide of pyridine-4-carboxylic acid, is still one of the mainstays for the treatment of tuberculosis. Widespread use led to the serendipitous discovery of its antidepressant activity. This latter activity is retained when pyridine is replaced by isoxazole. The requisite ester **139** is obtained in a single step by condensation of diketoester **136**, which is obtained by aldol condensation of acetone with diethyl oxalate, and hydroxylamine. One explanation of the outcome for this reaction assumes the first step to consist of conjugate addition–elimination of hydroxylamine to the enolized diketone to afford **137**. Ester amide interchange of **139** with hydrazine then affords the corresponding hydrazide; reductive alkylation with benzaldehyde completes the synthesis of **isocarboxazid** (**140**).[31]

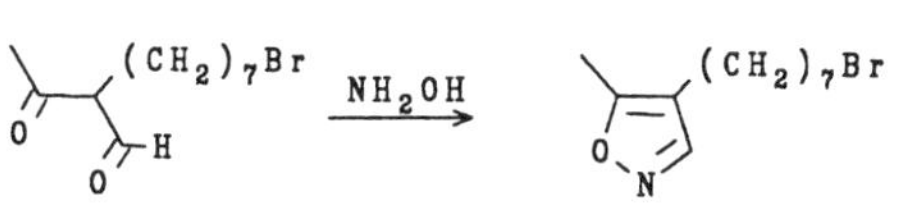

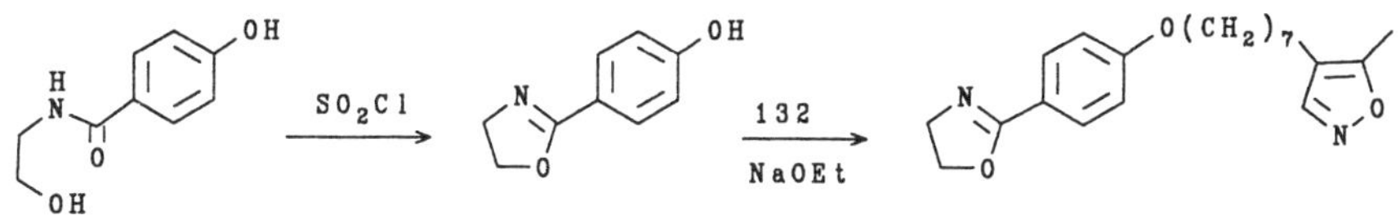

Fermentation products have proven an unusually rich source for antibiotics and antineoplastic agents due, at least in part, to the fact that they are elaborated by microorganisms as potential toxins toward other potentially threatening life forms. The antitumor activity of the amino acid derivative **acivicin** (**147**) elaborated by a *Streptomyces svicens*, is due to its interference with glutamate metabolism; the quite reactive

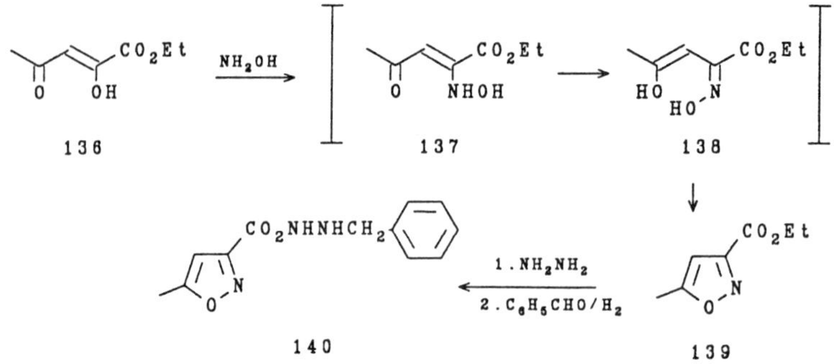

imino chloro isoxazole moiety can be viewed as a surrogate for the second glutamate carboxylic acid. One total synthesis for this compound relies on a 1,3-dipolar cycloaddition reaction involving a chiral auxiliary for construction of the isoxazoline ring. The sequence starts by formation of the oxime **141**, shown in cyclic form, from the D-ribose in which the hydroxyl group at 3,4, and 5 are protected. Reaction with paraformaldehyde leads to the transient nitrone **142**. In a convergent synthesis, vinylglycine[32] is protected as its cyclic formaldehyde carbinolamine derivative **143**. Treatment of **143** with **142** formed in situ leads to the 1,3-dipolar cycloaddition product **144** as a single chiral enantiomer. Exposure to formic acid cleaves the bond to the anomeric carbon to afford the bicyclic product **145**. The all important enol chloride is then introduced by treatment of this last product with *N*-chlorosuccinimide. Thus, compound **146** is obtained. Reaction of **146** with boron trichloride removes the remaining protecting

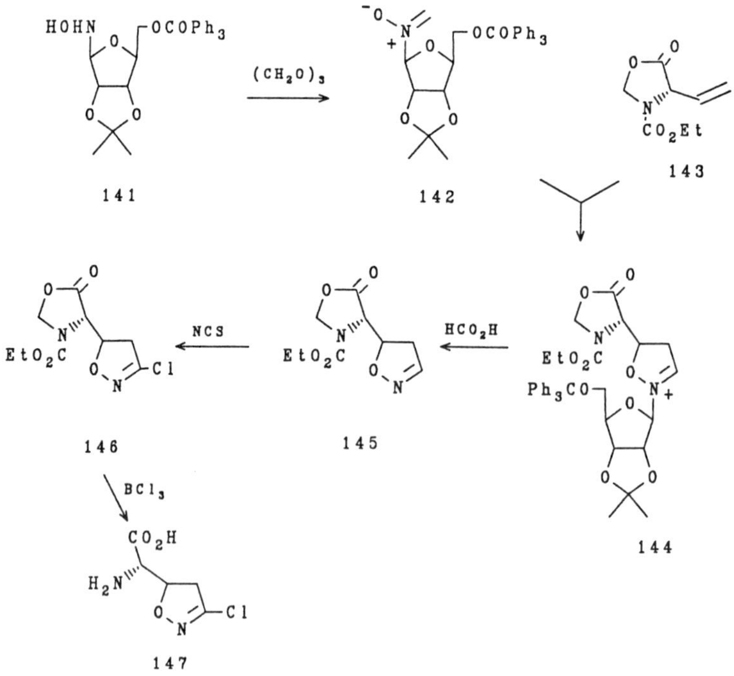

groups to afford finally the free amino acid side chain. Thus, **acivicin** (**147**)[33] is obtained.

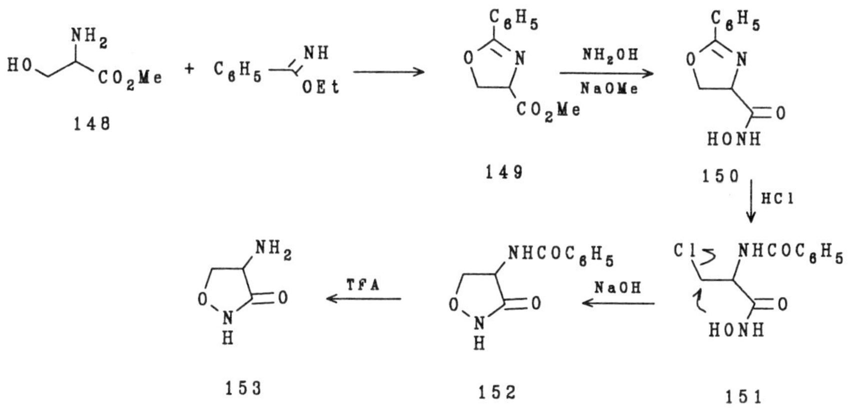

The antibiotic **cycloserine** (**153**), which was also originally isolated from fermentation broths, is still one of the important drugs used to control tuberculosis; it too, interestingly, dates back five decades. The quite concise original synthesis for the compound involved racemic material possibly as a result of the difficulty of obtaining resolved starting materials in those days. The first step consists in protection of serine (**149**) as its isoxazoline **149** by reaction with the iminoether obtained from benzonitrile. Ester–amide interchange of the intermediate with hydroxylamine in the presence of sodium methoxide gives hydroxamic acid **150** on neutralization of the product. Reaction of **150** with anhydrous hydrogen chloride leads to ring opening of the isoxazoline with concurrent conversion of the latent hydroxyl group to a chloride (**151**), a reaction that is probably facilitated by the good leaving group properties of the amide function. Sodium hydroxide converts the hydroxamic acid to its anion; the negatively charged oxygen then displaces the chlorine internally to form the isoxazoline and thus afford the benzamide **152**.[34] Several steps were initially required to remove the amide; it was subsequently shown that cycloserine (**153**) could be obtained directly from the benzamide by treatment with trifluoroacetic acid.

B. Imidazoles

Fully unsaturated imidazole rings feature prominently in several classes of therapeutic agents. The specific substitution pattern associated with each class of drugs suggests that these moieties play an important pharmacophoric role.

The antiprotozoal activity exhibited by nitroimidazoles bearing a substituent on the adjacent nitrogen prompted an enormous amount of work in this series. The extensive record of efficacy and safety of one of these compounds, **metronidazole** (**158**) has led to its being made available on a nonprescription basis for the treatment of vaginal trichomonal infections. The first step in the synthesis of this agent points up a structural ambiguity inherent in imidazoles. The starting material, 2-methylimidazole (**154**), exists as two freely equilibrating tautomers; the symmetry of the molecule in this case results in their exact equivalence. Introduction of a second substituent, for example, a nitro group, leads to the production of two nonequivalent isomers; the product in fact consists of the pair of nitroimidazoles, **155** and **156**, which freely equilibrate. Treatment with base under aprotic conditions has been found empirically to favor the formation of anion **157**, which bears the charge adjacent to the nitro group. Treatment with ethylene chlorohydrin leads to alkylation and formation of metronidazole (**158**).[35]

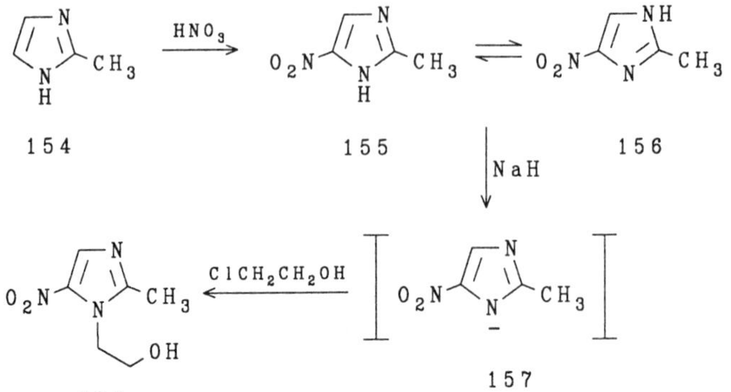

The same regiochemistry is observed when nitroimidazole (**155/156**) acts as a nucleophile in its un-ionized form. Thus, reaction of **155/156** with benzoylaziridine in the presence of boron trifluoride probably involves initial salt formation with this amide; attack by the imidazole results in ring opening and formation of the alkylated product **159**; the free primary amine **160** is obtained on basic hydrolysis. Acylation of **160** with methyl thiochloroformate gives the corresponding thiourethane **carnidazole** (**161**).[36]

The *N*-methyl group in the starting material (**162**) for the antiprotozoal compound **ronidazole** (**164**) in essence locks this compound into a single isomeric form. Nitration occurs predominantly at the presumably more electron-rich position to give the 5 isomer; saponification of the acetate protecting group then leads to the free alcohol **163**. Reaction of **163** with potassium isocyanate in hydrofluoric acid gives the product from reaction of the alcohol with isocyanic acid generated in situ, giving ronidazole (**164**).[37]

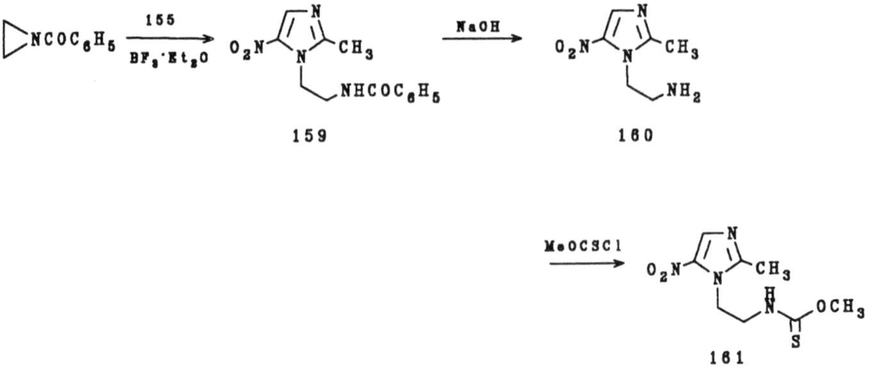

162 → (1. HNO_3, 2. NaOH) → 163 → (KNCO, HF) → 164

The preparation of the antitrichomonal agent substituted with an aromatic ring on the imidazole starts with the condensation of the iminoether from 4-fluorobenzonitrile with the dimethylacetal from 2-aminoacetaldehyde; the reaction can be envisaged to involve initial formation of an amidine by exchange of the ether methoxyl with the primary amine; hydrolysis of the acetal groups under reaction conditions would give the imine–aldehyde that then cyclizes to **165** by Schiff base formation. Nitration gives the usual mixture of isomers, only one of which, **166**, is shown. A special stratagem is required to make up for the tendency of this imidazole pair to give the undesired alkylation product, which essentially consists of initially introducing an easily removable alkyl group. Thus reaction with methylchloromethyl ether gives the imidazole **167**. The future hydroxyethyl side chain is then incorporated by alkylation of the second nitrogen with the highly reactive fluoroborate from the cyclic ethylene acetal of acetic acid; this reaction yields the unstable quaternary salt **168**. Heating in pyridine leads to cleavage of the methoxymethyl group, which results in net removal of one of the nitrogen alkyl groups leaving the hydroxyethyl at the desired position. Hydrolysis of the acetate completes the synthesis of **flunidazole** (**169**).[38]

The very simple nitroimidazole **azomycin** (**171**) is one of the very early compounds uncovered by the extensive search for antibacterial agents produced by *Streptomyces* fermentation. Its development as an antibiotic drug was probably precluded by its toxicity. Incidentally, the **mycin** ending for the name indicates that it was isolated from a *Streptomyces*. A slightly circuitous route was used for its total synthesis, since it is not available by direct nitration of imidazole. The key 2-aminoimidazole **170** is obtained by condensation of cyanamid with the acetal from 2-aminoacetaldehyde. The amine is then diazotized and the diazonium salt is replaced by a nitro group with a Sandmeyer reaction using sodium nitrite in the presence of cuprous chloride[39] to yield **171**.

Ionizing radiation, for example, X-rays, is one of the oldest, and still extensively used methods for treating cancer. The generalized organ and tissue toxicity of this form of treatment puts a premium on delivering the radiation preferentially to the cancerous tumors. It was found adventitiously that 2-nitroimidazoles such as **metronidazole** increased the sensitivity of solid tumors to the cell killing effect of radiation. These agents seem particularly effective in the oxygen deficient environment found in solid tumors. Azomycin (**171**) provides the starting material for one of these agents. Thus, alkylation of **171** with ethyl chloroacetate in the presence of sodium hydroxide gives the ester **172**. Ester interchange with ethanolamine gives the corresponding amide **etanidazole** (**173**).[40]

The synthesis of the antifungal agent **ethonam** (**179**) illustrates yet one more method for building imidazole rings. Reaction of 5-aminotetralin (**174**) with ethyl chloroacetate under carefully controlled conditions leads to the monoalkyl product;

NH
OMe
+ $H_2NCH_2CH(OMe)_2$ →
F
165
HNO_3
H_2SO_4
166
$ClCH_2OCH_3$
CH_2OCH_3
NO_2
167
$\cdot BF_3$
168
OAc
1. Py
2. H_3O^+
169
NO_2
OH

H_2NCN + $H_2NCH_2CH(OMe)_2$ →
170
NH
NH_2
HONO
$NaNO_2$
$CuCl_2$
171
NH
NO_2

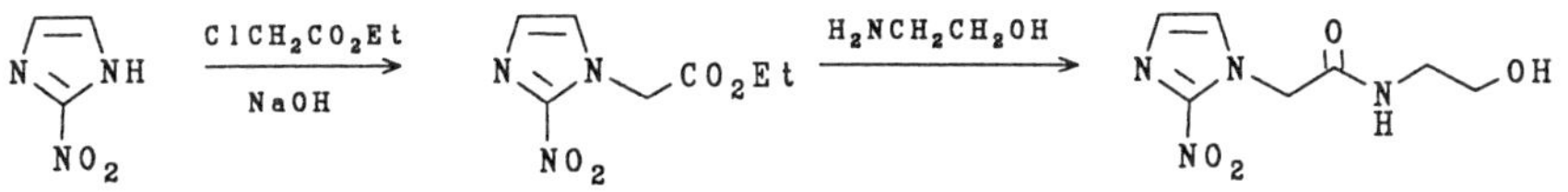

171 → ($ClCH_2CO_2Et$, NaOH) → 172 → ($H_2NCH_2CH_2OH$) → 173

this product is then converted to thc formamide **175** with formic acid. Condensation of **175** with ethyl formate in the presence of sodium ethoxide gives the hydroxymethylene derivative **176**. Treatment with isothiocyanic acid, from sodium isothiocyanate and mineral acid, can be envisaged as involving initial replacement of the hydroxyl group in **176** by a conjugate addition–elimination sequence to give an intermediate such as **177**. Internal exchange of the formyl group with the adjacent thiocarbonyl group will lead to a thioimidazoline; this compound can then tautomerize to the imidazothiol **178**. Desulfurization with Raney nickel then gives ethonam (**179**).[41]

Otherwise unsubstituted imidazole rings and to a lesser extent 1,2,4-triazole rings, which are discussed in a subsequent section, form an essential part of the most important class of antifungal agents, the so-called conazoles. The selective toxicity of these drugs depends on the fact that they inhibit fungal enzymes responsible for the biosynthesis of ergosterol, a steroid that plays an important structural role in the organism's cell membrane. Most conazoles possess, in addition to the heterocyclic ring, one or more chlorinated aromatic rings; very wide latitude seems to exist concerning the skeleton that connects these moieties. The agents discussed below constitute a very small sample of the enormous number of antifungal compounds that have been described.

Bromination of 2,4-dichloroacetophenone provides the α-bromoketone **180**, which is the key to the synthesis of a number of conazoles. The corresponding α-imidazo

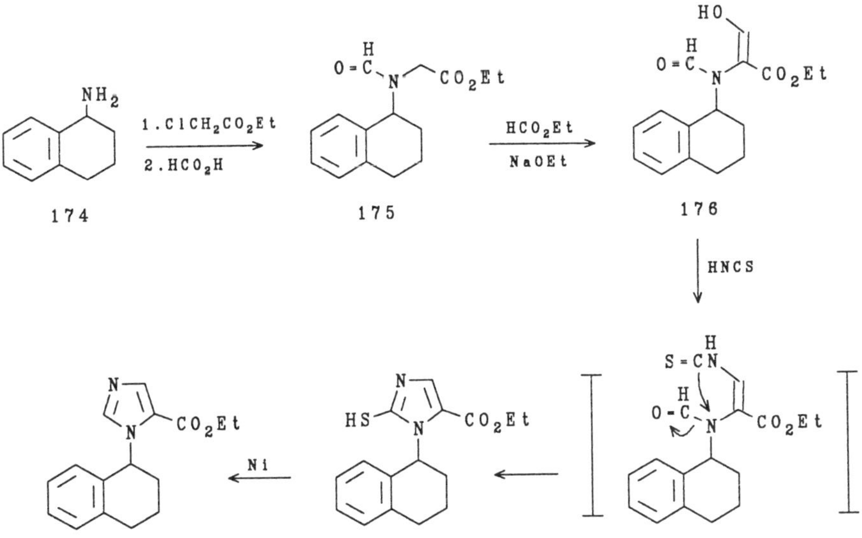

174 → (1. $ClCH_2CO_2Et$, 2. HCO_2H) → 175 → (HCO_2Et, NaOEt) → 176 → (HNCS) → [177] → 178 → (Ni) → 179

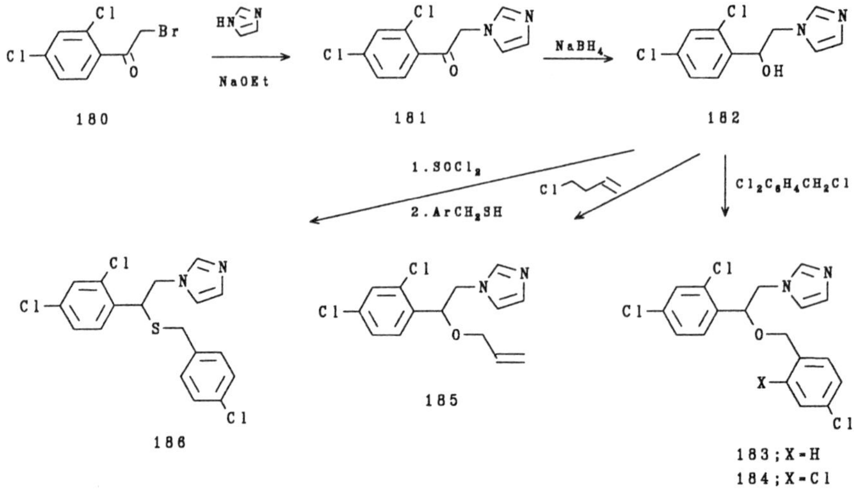

derivative **181** is obtained by displacement of halogen by imidazole; the carbonyl function is then typically reduced to the alcohol **182** with sodium borohydride. Displacement of chlorine from α,4-dichlorotoluene with the alkoxide from **182** affords the antifungal agent **econazole** (**183**);[42] the analogous reaction using α,2,4-trichlorotoluene gives **miconazole** (**184**).[42] The bioisosteric relation of oxygen and sulfur apparently maintains in the case of the conazoles; thus, conversion of the hydroxyl in **182** to chlorine followed by displacement with 4-chlorobenzylthiol as its anion gives the antifungal agent **sulconazole** (**186**).[43] The second halogenated ring is apparently not absolutely required for activity; thus reaction of the alkoxide from **182** with allyl chloride gives **enilconazole** (**185**).[42]

Interestingly, an imine moiety can be interposed in the ether linkage to connect the two halogenated aromatic rings. The requisite oxime **187** is obtained in straightforward fashion by reaction of imidazo acetophenone **181** with hydroxylamine; alkylation with α,2,3-trichlorotoluene leads to **oxiconazole** (**188**).[44]

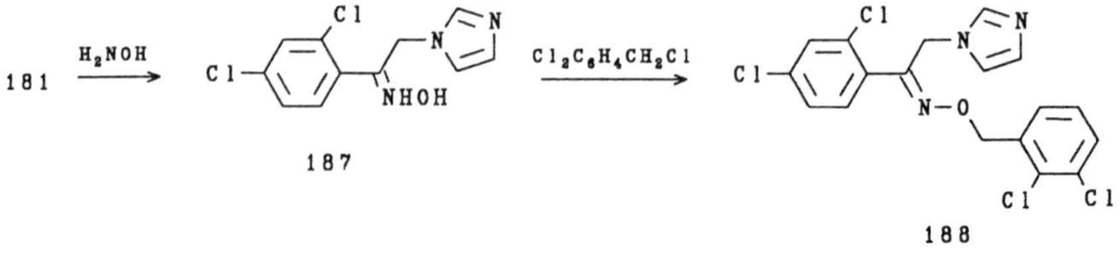

The orally active antifungal agent **ketoconazole** (**193**) was widely used in the treatment of opportunistic fungal infections in AIDS patients until the advent of newer, more effective drugs. The synthesis of this agent begins by formation of the acetal from 2,4-dichloroacetophenone (**189**) and glycerol. Reaction with bromine, which may well proceed via a small amount of ketone in equilibrium with the acetal, gives the intermediate **190**. The aliphatic halogen is then displaced with imidazole and the free hydroxyl group is converted to a mesylate leaving group by reaction with

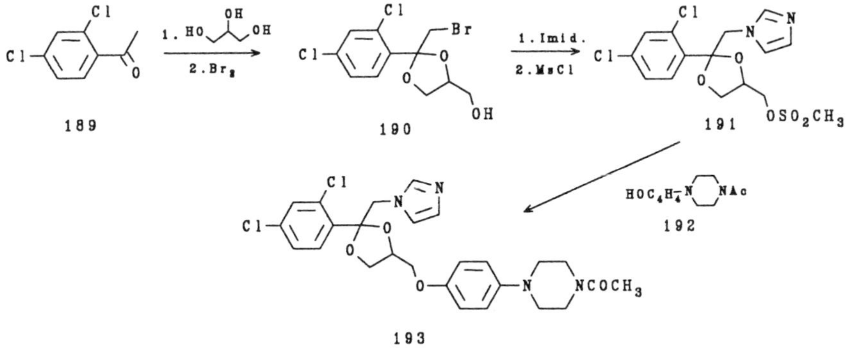

189 190 191 192 193

methanesulfonyl chloride. Displacement of the mesylate with the phenoxide from treatment of the side chain **192** with base affords ketoconazole (**193**).[45]

The structural latitude that exists in this series is further illustrated because the activity is retained when the other ring is replaced by thiophene and a hydrazide provides the linkage. The key intermediate **195** is obtained by displacement of halogen in **194** with imidazole; condensation with 2,6-dichlorophenylhydrazine gives the hydrazone **zinoconazole** (**196**).[46]

The topical antifungal **bifonazole** (**198**) dispenses with virtually all but the imidazole ring; the intermediate **197** is obtained by sequential reduction of 4-phenylbenzophenone and reaction of the alcohol with thionyl chloride. Displacement of chlorine by imidazole gives **198**.[47]

The pivotal role of histamine (**199**) in allergic reactions was recognized as early as 1911. Further pharmacological studies uncovered the role of that compound in the release of gastric acid. The first antiallergic drugs, such as **diphenhydramine** (**200**), were available some three decades later. However, neither this class of drugs nor its more effective tricyclic successors, which are discussed in Chapter 13, had any effect on histamine mediated gastric acid secretion. Detailed studies, which involved the study of a large number of analogues of histamine, led to the recognition of two

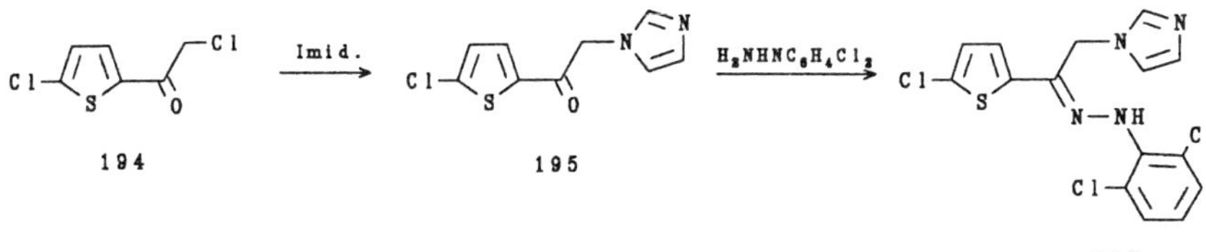

194 195 196

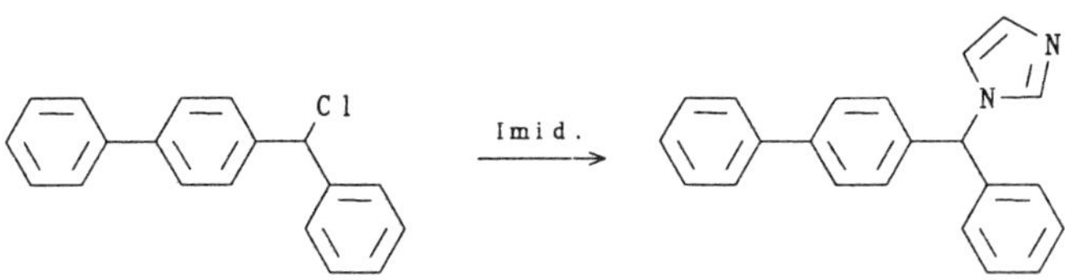

197 198

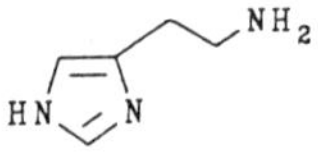

NH2
HN N
199

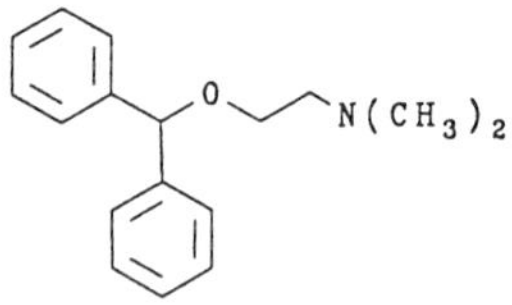

O N(CH3)2
200

separate populations of receptors for this agent: interaction with the so-called H_1 receptors leads to allergic manifestation; gastric acid secretion on the other hand involves activation of the H_2 receptors. The lack of effect of classical then-available antihistamines on acid secretion could thus be attributed to their selective binding to H_1 receptors. A direct consequence of this research was the development of a series of H_2 receptor antagonists that inhibit acid secretion and have found widespread use in treating gastric ulcers. **Cimetidine** (**212**), the first of these drugs to reach the market, was developed by the laboratory responsible for the work that resulted in recognition of two separate receptors; this agent is now, because of a long record of freedom from side effects, available as a nonprescription drug. The follow-on furan competitor **ranitidine** (**21**), described earlier in this chapter, was approved for over-the-counter sales about a year later.

The structure of the first H_2 antagonist, **burimamide** (**204**), bears a close structural resemblance to histamine, differing only by the presence of an extended side chain that bears a slightly acidic thiourea group instead of a basic nitrogen. The synthesis starts with the reaction of the diaminoalcohol **201** (itself obtained by reduction of the corresponding ester) with ammonium isothiocyanate. The initial step probably consists in formation of a thiourea at the amino group adjacent to the alcohol; the hypothetical aldehyde from oxidation of this function could then form an imine with ammonia; cyclization would then lead to the observed thioimidazoline **202**. Treatment with iron powder reductively removes the sulfur so as to form the imidazole **203**. Reaction with methyl isothiocyanate gives the corresponding thiourea burimamide (**204**).

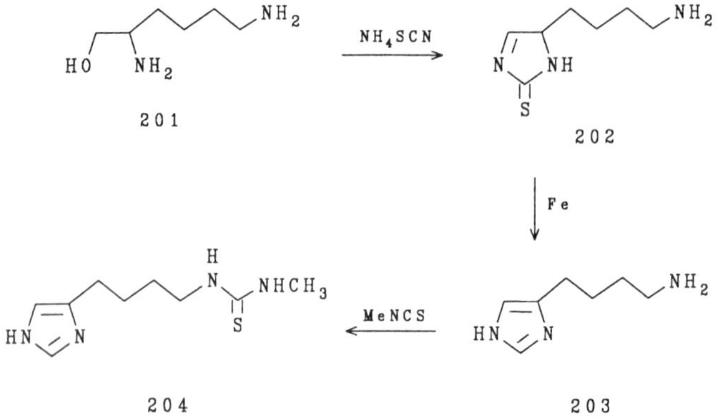

NH2
HO NH2
201

NH_4SCN

NH2
N NH
S
202

Fe

H
N NHCH3
S
HN N
204

MeNCS

NH2
HN N
203

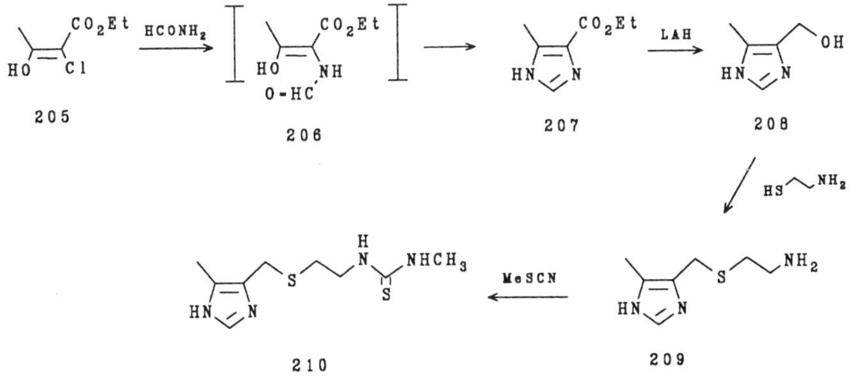

Side effects noted in the clinic with this agent led to its abandonment. The preparation of the second compound illustrates another method for forming imidazole rings. Thus, the first step in the reaction of 2-chloroacetoacetate, shown as its enolate **205**, with formamide can be envisaged to involve formation of intermediate **206** by an addition–elimination sequence; the enolic β hydroxyl can then undergo a similar displacement. Loss of one formyl group followed by internal imine formation will then lead to a imidazole ring to afford **207**. The carbethoxy group in this intermediate is then reduced to the corresponding alcohol **208** by means of lithium aluminum hydride. Treatment of **208** with the cysteamine in the presence of hydrogen hydrochloride leads to displacement of the benzylic-type alcohol by sulfur to give the thioether **209**. Condensation of **209** with methyl isothiocyanate gives **metiamide** (**210**).

The fact that this agent caused similar side effects as its predecessor, led to the search of a replacement for the thiourea group. The cyanoguanidine function that was eventually found equally suitable is bioisosteric with thiourea in that the cyano group reduces the basicity of the guanidine; the nitrovinyl amidine in **ranitidine** serves the same function. Reaction of the intermediate **209** with dimethylthiocyanoimidocarbonate leads to replacement of one of the methylthio groups by the side chain amino group to afford intermediate **211**. Treatment of **211** with methylamine under somewhat more strenuous conditions replaces the remaining methylthio group to afford **cimetidine** (**212**).[48,49]

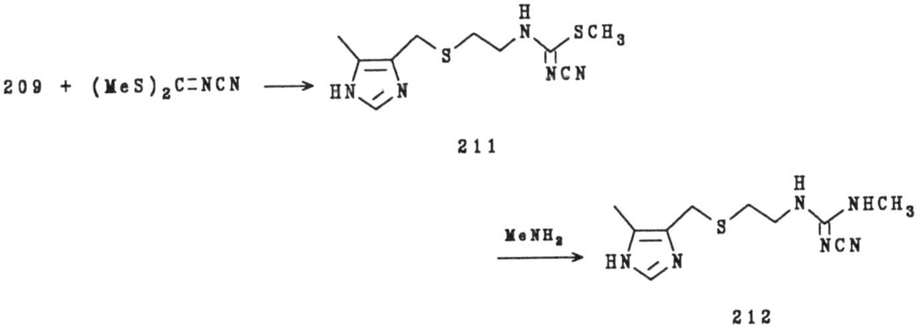

209 + $(MeS)_2C{=}NCN$ → 211 $\xrightarrow{MeNH_2}$ 212

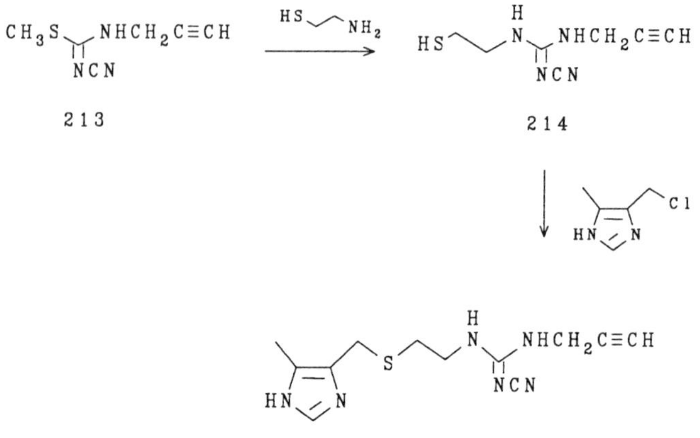

213

214

215

The surprising fact that biological activity is sometimes maintained when a methyl group is replaced by propargyl provides a fast route to analogues that are distinct for the purposes of patents. The preparation of the propargyl derivative of **cimetidine** (**212**) also changes the order of the steps in the synthesis. Thus, reaction of cyanoamidine (**213**) (obtained by treatment of the above thiocarbonate with propargylamine) with cysteamine leads to replacement of the second methylsulfide group and formation of the cyanoguanidine **214**. The reaction of chloromethyl imidazole, which is obtained from **208**, with the side chain as its mercaptide leads to replacement of halogen and formation of the histamine H_2 **etintidine** (**215**).[50]

Work from other laboratories, some of which is discussed in Chapter 9, later showed that 2-aminopyrimidones could fulfill the same function in histamine H_2 blockers as thioureas and modified guanidines; many of these compounds are devoid of histamine-like nuclei. **Oxmetidine** (**217**) represents a hybrid in that it includes moieties from both series. The pyrimidone-2-thiol (**216**) is prepared in a manner analogous to the one described in Chapter 9 for **lupitidine**; condensation of **216** with intermediate **209** leads to replacement of the thiol by the terminal amino group. Thus, oxmetidine (**217**)[51] is obtained.

The great majority of cyclooxygenase inhibitors include strongly acidic functions that range from carboxylic acids in the various "profens" and "enacs" discussed in earlier chapters to acidic protons on the dicarbonyl functions characteristic of the "oxicam" fused heterocycles discussed in Chapter 11. That several imidazoles seem to show this same activity is unexpected because of the very weak acidity of this heterocyclic ring (pK_a ~ 6). The simplest of these agents, **lofemizole** (**220**), is prepared by reaction of the hydroxypropiophenone **218** with formamide.[52] This transformation,

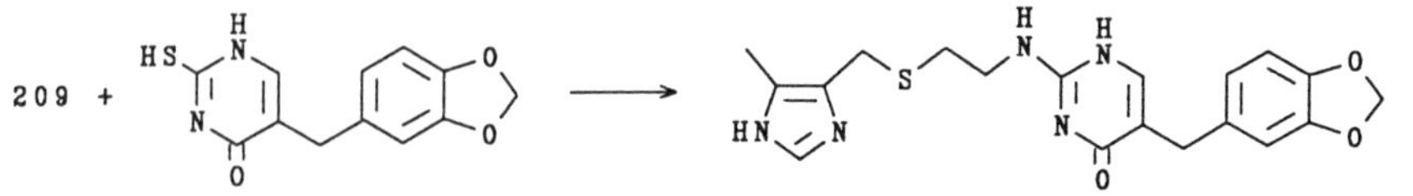

209 +

216

217

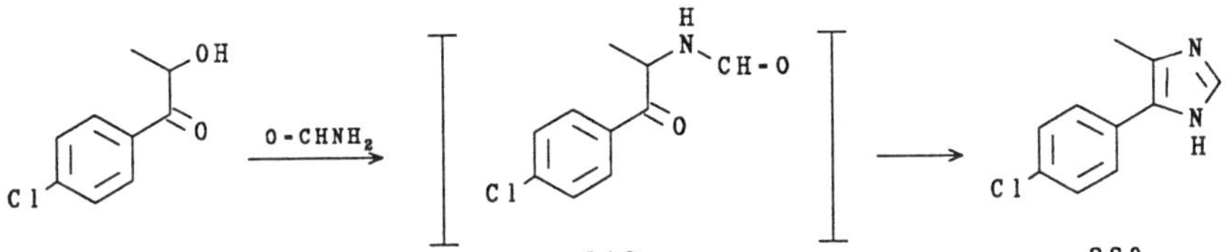

218 219 220

which is reminiscent of the one described above for the **metiamide** intermediate (**207**), can be rationalized by assuming displacement of hydroxyl by the amide to form **219** as the first step. Imine formation from the ketone with a second mole of formamide, followed by cyclization with loss of formate, would then give the observed product.[52]

An alternative approach to imidazoles involves adding nitrogen in the form of an ammonium salt and the carbonyl component that will form the carbon atom at the 2-position as separate components. The reaction of dibenzyl **223** with ammonium acetate and trifluroacetaldehyde ethyl hemiacetal (**221**) is probably preceded by equilibration of the latter two reagents to form **222**. Condensation of **222** with one of the carbonyl groups will lead to the intermediate **224**; the remaining nitrogen can be added either by further acetal exchange or by imine formation. Cyclization then affords the antiinflammatory agent **flumizole** (**225**).[53]

A similar approach is used for **trifenagrel** (**228**), an antithrombotic agent. Alkylation of the phenoxide from salicylaldehyde with 1,2-dibromoethane affords the bromoethyl ether **226**; displacement of halogen with dimethylamine gives the corresponding basic ether **227**. This intermediate is then condensed with benzyl in the presence of ammonium acetate; this results in the formation of an imidazole ring by a process directly analogous to the one shown above, giving trifenagrel (**228**).[54]

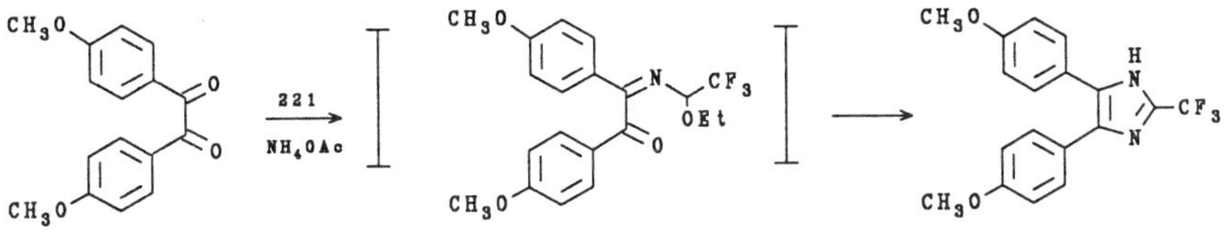

223 224 225

221 222

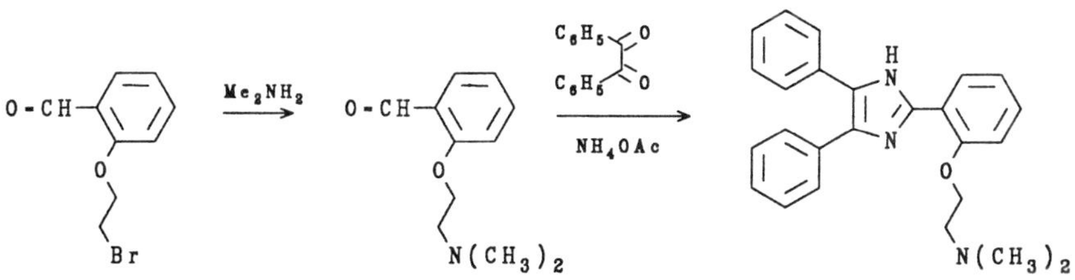

226 227 228

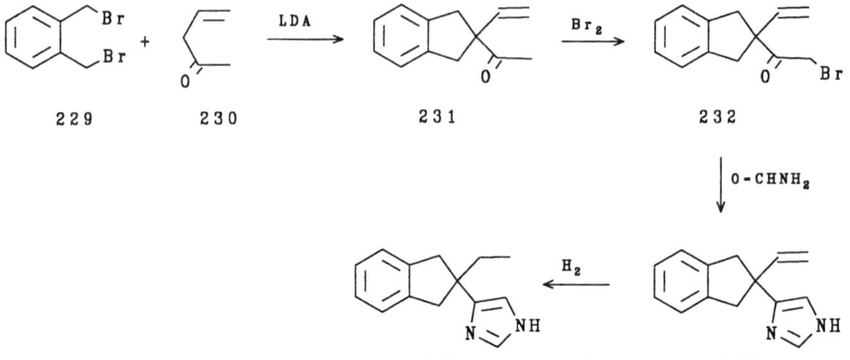

Compounds that display either agonist or antagonist activity of the β-adrenergic system consist of the arylethanolamines or aryloxypropanolamines discussed in Chapter 2; few if any other structural types show this activity. The situation is quite different for the α-adrenergic system; although epinephrine is the endogenous agonist, some of the most potent synthetic agonists for this receptor consist of imidazolines. A few imidazoles also show α-adrenergic activity; note that the recent identification of α-receptor subtypes has led to work aimed at developing compounds with selectivity for such discrete receptor subclasses. The imidazole **atipamezole** (**234**), for example, has been found to display α_2 antagonist and non-opioid analgesic activity. Successive alkylation by dibromide **229** on the anion obtained from allylmethyl ketone (**230**) and lithium diisopropylamide (LDA) leads to spirocyclization and formation of the indan **231.** Reaction with bromine then gives the bromoketone **232**. Treatment of **232** with formamide proceeds to form the imidazole **233**, similar to the case of the 2-hydroxyketone shown above. Catalytic reduction of the pendant vinyl group affords atipamezole (**234**).[55]

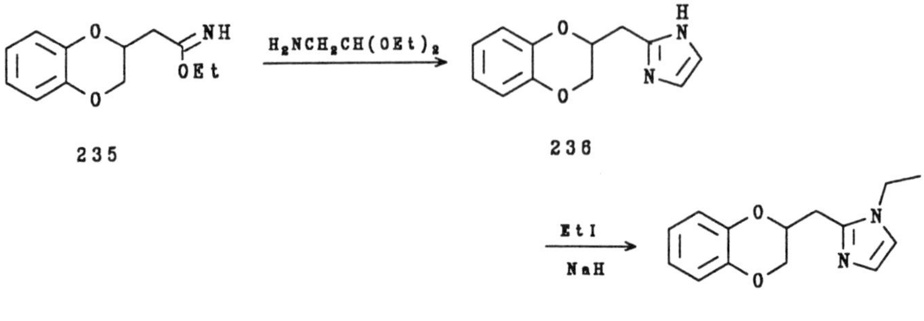

The benzodioxan nucleus has itself been associated with α-adrenergic agents; it is of interest that combining this moiety with an imidazole leads to an α_2-agonist. Reaction of iminoether **235**, which is obtained by treatment of the corresponding nitrile with ethanolic hydrogen chloride with the diethylacetal of aminoacetaldehyde, probably proceeds to give the amidine replacement product as a transient intermediate. Internal imine formation with the aldehyde or alternatively replacement of acetal groups by amino, leads to cyclization and formation of the imidazole intermediate **236**. Alkylation of the salt obtained by means of sodium hydride with ethyl iodide gives **imiloxan** (**237**),[56] a compound that displays antidepressant activity.

Generally, activation of α-adrenergic receptors leads to contraction of muscles in the innervated tissues. The net result of this action on the vasculature will be vasoconstriction and a consequent increase in blood pressure. In fact, some of the first antihypertensive agents consisted of imidazolines that blocked the response to endogenous α-adrenergic agonists; the extensive menu of side effects elicited by these drugs, possibly because they blocked all α-receptor subtypes, led to their eventual replacement by more selective agents. Subtle structural differences lead to imidazolines that act as α-adrenergic agonists. These compounds are used to this day to induce vasoconstriction in peripheral tissue; they comprise the over-the-counter nasal decongestants and drops for treating eye irritation

The schemes for preparing imidazolines take advantage of the fact that this heterocyclic system in effect simply consists of a cyclic amidine of ethylenediamine. Thus, treatment of the iminoether **238** from phenylacetonitrile with ethylenediamine can be envisaged to result in initial displacement of ethoxide to give a transient intermediate such as **239**; internal addition–elimination by the remaining side chain amino group leads to formation of the imidazoline ring. In this case, this reaction leads to the formation of the α-adrenergic blocker **tolazoline** (**240**),[57] a one time vasodilator. The same sequence starting with the cyanotetralin **241** affords **tetrahydrolozine** (**242**),[58] this compound, in contrast to **240**, is an α-agonist that is used extensively as a vasoconstrictor for treating eye inflammation.

A very lipophilic highly substituted benzene ring provides the side chain for the nasal decongestant **xylometazoline** (**246**). The synthesis of this compound starts with the chloromethylation of the mesitylene derivative **243** followed by displacement of chlorine by cyanide. Thus, the arylacetonitrile **244** is obtained. This compound is then converted to the corresponding imidazoline **246** by the sequence outlined above.[59]

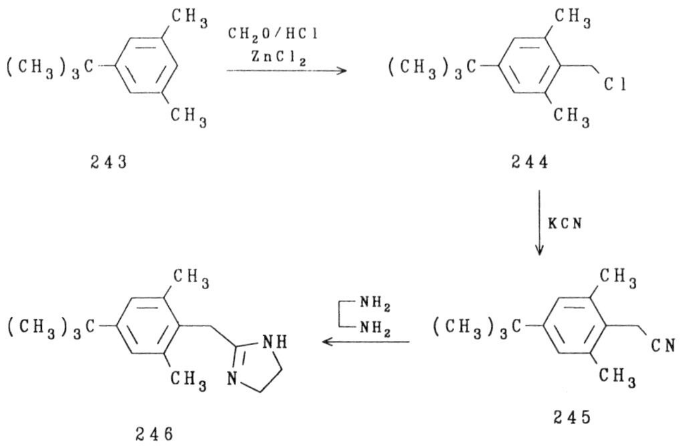

The α-blocker **phentolamine** (**249**) was used fairly extensively as an antihypertensive agent prior to the availability of more effective and far better tolerated drugs. Some of the most important side effects from use of this drug, such as fluid retention and tachycardia, were at that time attributed to compensatory mechanisms casting a cloud upon α-blockers; the use of this class of agents for controlling elevated blood pressure awaited the discovery of receptor subtype-specific agents. One synthesis for this agent depends on the fact that the key intermediate **248** is in fact an α-aminonitrile derivative of formaldehyde. Thus, the intermediate is obtained by reaction of the diarylamine **247** with formalin and potassium cyanide in the presence of mineral acid. The product **248** is converted to phentolamine (**249**) by the usual sequence involving ethylene diamine.[60]

One approach in work aimed at developing structurally novel nasal decongestants involved replacement of the methylene link present in most α-blockers by an amine.[61] Initial pharmacology on one of these analogues, **clonidine** (**253**), indicated that the α-blocking activity had in fact been retained; on oral administration, this compound was, however, surprisingly found to be hypotensive in humans. This activity was subsequently traced to the fact that the drug readily crosses the blood–brain barrier and that stimulation of α-adrenergic receptors in the brain resulted in lowering of blood pressure. The drug soon found widespread use as an antihypertensive agent because of its good tolerability compared to the then available alternatives. One of the early syntheses of this compound starts by formation of the methyl ether **251** from the cyclic thiocarbonate **250** from ethylene diamine by means of methyl iodide and base; this transformation in essence converts the sulfur to a better (if odiferous) leaving group.

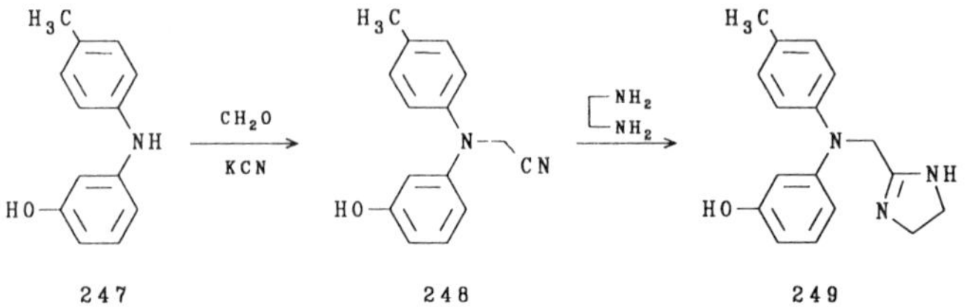

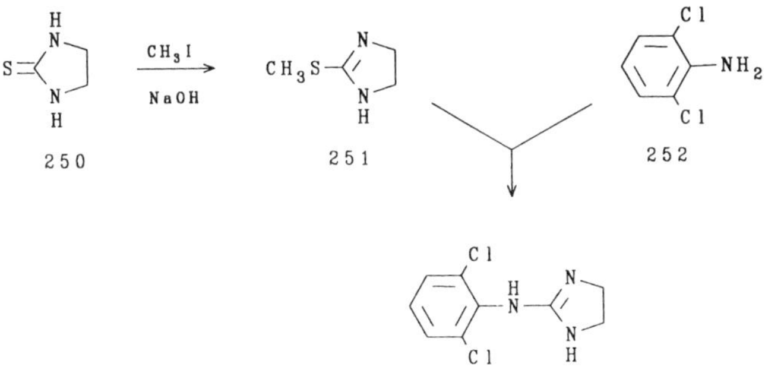

Reaction of the thiomethyl ether with a 2,6-disubstituted aniline (**252**) results in displacement and formation of the cyclic guanidine; the ortho substituents are crucial for activity. Thus, clonidine (**253**),[62] shown as the so-called "amino" tautomer, is obtained.

The frequently observed bioisosteric relation of benzene and thiophene applies to the **clonidine** series as well. Reaction of the thiophenyl thiourea **254**, in which the amine group is also flanked by substituents, gives the corresponding methyl thioether **255**. Reaction of **255** with ethylenediamine leads to formation of an imidazoline ring and the antihypertensive agent **tiamenidine** (**256**);[63] shown as its "imino" tautomer.

Antihypertensive agents, and α-blockers in particular, have been found to lower the intraocular pressure that marks glaucoma. The use of such drugs for treatment of this disease is limited by their propensity to lower blood pressure even in normotensive individuals. Sufficient drug is often absorbed even on topical application to make this a problem. Increasing the polarity of clonidine by addition of an amino group keeps the drug out of its locus of action in the brain while retaining its effect on intraocular pressure when applied locally. Nitration of the clonidine intermediate **252** leads to the nitroaniline **257**; this compound is then converted to the corresponding formamide **258**, for example, by ester interchange with ethyl formate. Treatment of **258** with a mixture of sulfuryl chloride and thionyl chloride results in formation of the bischloroimino derivative **259**. Nitrogen then replaces the halogens by reaction of this intermediate with ethylenediamine to form the iminoimidazolidine **260** speculatively by sequential addition–elimination steps. Reduction of the nitro group with iron powder and hydrochloric acid yields **apraclonidine** (**261**).[64]

A compound that includes a carbonyl group on the imidazoline ring is described as a sedative. Treatment of the guanidyl substituted amino acid creatine (**262**) with hydrochloric acid results in cyclization to the iminoimidazolinone creatinine **263**.

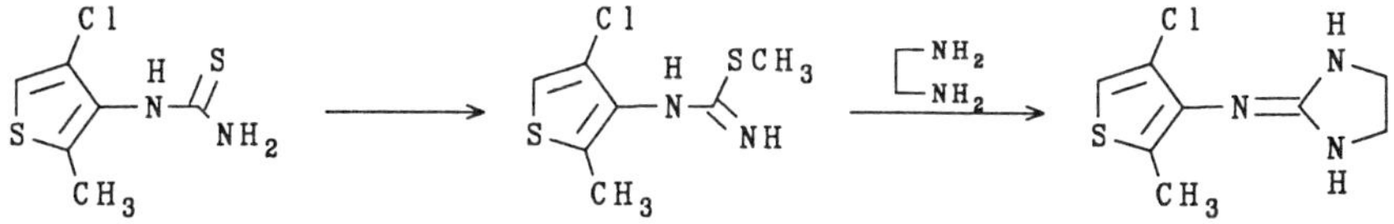

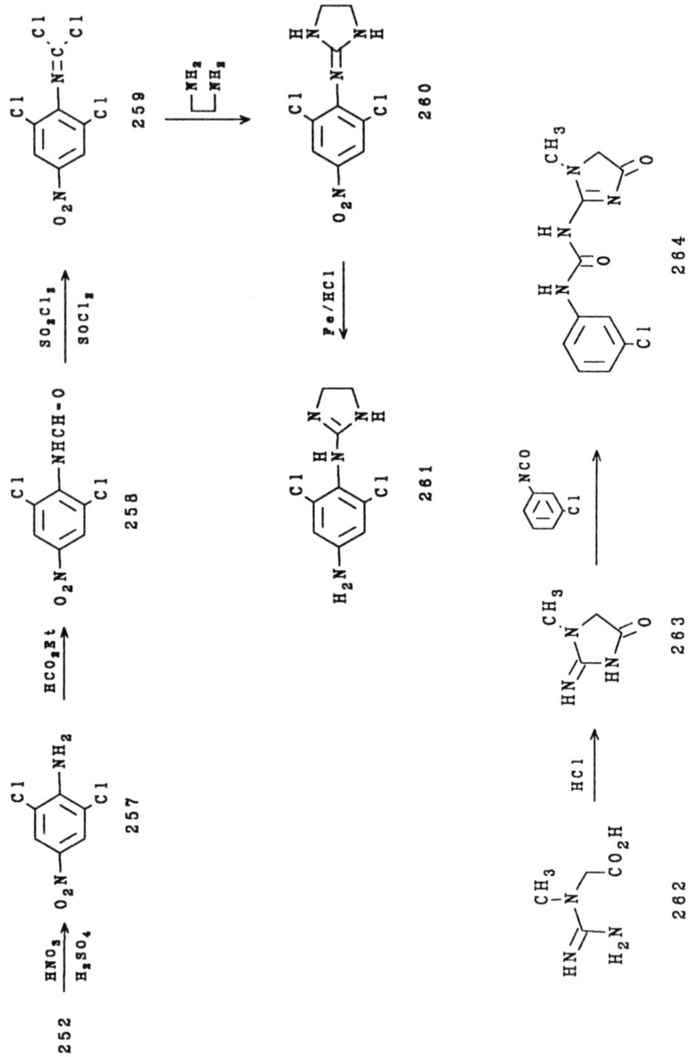
252
257
258
259
260
261
262
263
264
HNO3
H2SO4
HCO2Et
SO2Cl2
SOCl2
Fe/HCl
HCl
NCO

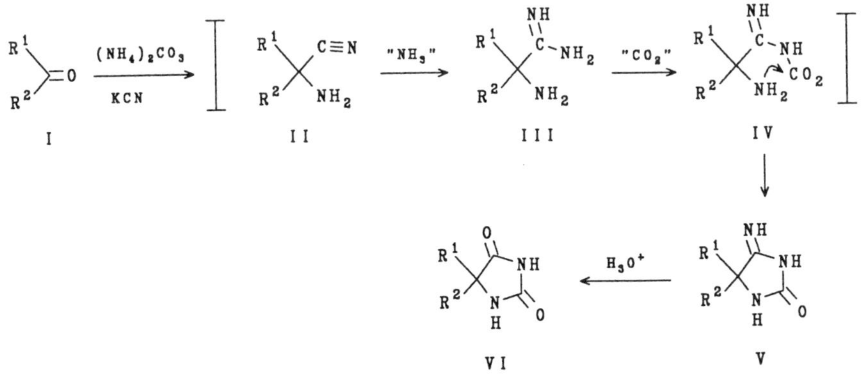

Condensation of **263** with *m*-chlorophenylisocyanate leads to formation of a urea by condensation of this reactive function with the imidazole as its amino tautomer. Thus, **fenobam (264)**[65] is obtained.

The apparently quite broad structural requirements for anticonvulsant activity, which was noted earlier in this chapter, extends to yet another class of five-membered heterocycles, which include an imide function. Imidazo-2,4-diones, better known as hydantoins, provide some of the most widely used drugs for treating severe motor and psychomotor epileptic seizures. The general reaction used to prepare this heterocyclic system involves treatment of a carbonyl compound with ammonium carbonate and potassium cyanide. The first step in this complex sequence can be visualized as addition of the elements of ammonia and hydrogen cyanide to give an α-aminonitrile (**II**). Addition of ammonia to the cyano group would then lead to an amidine (**III**). Carbon dioxide or carbonate ion present in the reaction mixture can then add to the quite basic amidine to afford a carbamic acid-like intermediate (**IV**); attack by the adjacent amino group will then close the ring and afford the isolable imino derivative **V**. This compound is then hydrolyzed to a hydantoin (**VI**) by treatment with aqueous acid.

Alkylation of the hydantoin **265** prepared by that scheme from benzaldehyde with ethyl iodide takes place at the imide nitrogen to afford **ethotoin (266)**.[66] In much the same vein, treatment of the hydantoin **267** from propiophenone with methyl iodide in the presence of base affords **mephenytoin (268)**.[67] Replacement of the quite acidic imide proton by an alkyl group is not required for activity; the well-known anticonvulsant **phenytoin (269)** consists of simply the hydantoin obtained from benzophenone;[68] this compound is often formulated as its sodium salt.

A somewhat different scheme is used for the preparation of an all-aliphatic thiohydantoin. Thus, reaction of racemic leucine (**270**) with allylisothiocyanate leads to the thiourea **271**. Attack of the anion from treatment of **271** with strong base leads to ring closure and formation of the imidazoline ring. Thus, the anticonvulsant agent **albutoin (272)**[69] is obtained.

A pair of iminohydantoins both depend on addition of amide nitrogen to a cyano group for formation of the imidazole ring; both agents further exhibit unexpected biological activities. Reaction of the cyanamide **273** from *p*-chloroaniline and cyano-

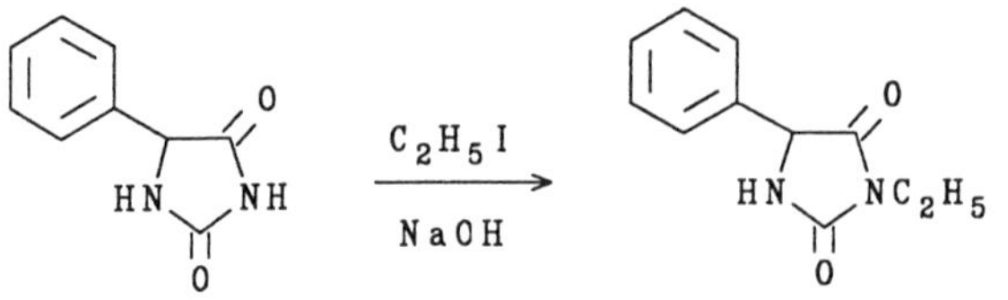

C_2H_5I, NaOH

265 → 266

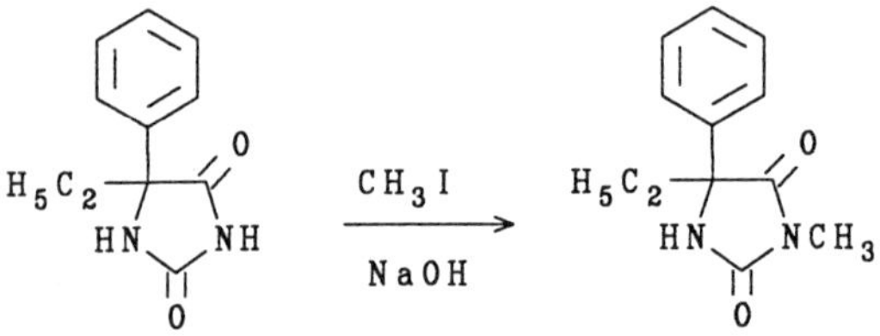

CH_3I, NaOH

267 → 268

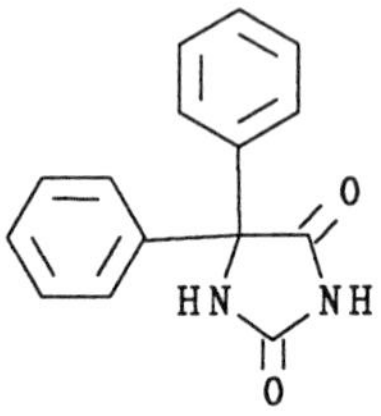

269

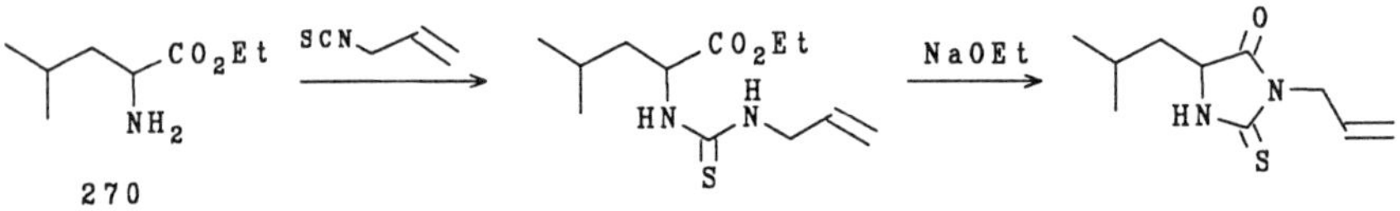

SCN, NaOEt

270 → 271 → 272

gen bromide with *N*-methylchloroacetamide can be visualized to lead initially to the alkylation product **274**. Cyclization by addition to the nitrile group then affords **clazolamine** (**275**),[70] a compound described as a diuretic.

In a similar vein, reaction of bis-cyanoethylamine (**277**) with *p*-chlorophenylisocyanate (**276**) gives the urea **278**, in this case, as an isolable product. Heating **278** leads to an analogous addition reaction to form the imidazole ring. The product **nimazone** (**279**)[71] displays antiinflammatory activity.

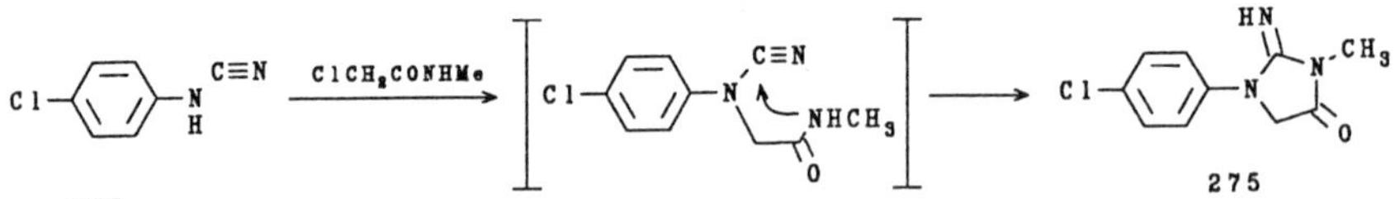

ClCH₂CONHMe

273 → 274 → 275

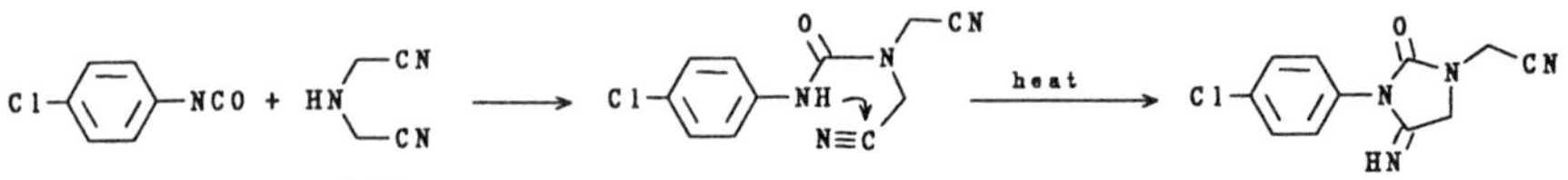

heat

276 + 277 → 278 → 279

C. Pyrrazolones and Pyrrazolodiones

Some of the first therapeutic agents to show documented activity in humans consisted of substituted pyrrazoles. The pyrrazolone based antipyretic and antiinflammatory agent **antipyrine** (**281**) first came into use at about the same time as aspirin. The large number of side effects due to the use of **281** and some of its analogues led to their eventual abandonment. The compounds are discussed here for the illustrative value of their chemistry; note that these compounds do not possess an acidic center, unlike the diones that follow. The parent pyrrazolone **280** is obtainable in straightforward fashion from the reaction of ethyl acetoacetate with phenylhydrazine; the regiochemistry is dictated by formation of an enamide by reaction of the acetoacetate enol with the more basic, less substituted, hydrazine amino group; ester–amine interchange involving the anilide nitrogen leads to ring formation. Alkylation of **280** with methyl iodide completes the preparation of antipyrine (**281**). Reaction of **281** with nitrous acid interestingly leads to reaction on the heterocycle rather than the benzene ring, forming the nitroso derivative **282**. Reduction of the newly introduced function gives the intermediate **283**. Alkylation with isopropyl bromide gives **isopyrine** (**284**),[72] while acylation with nicotinoyl chloride affords **nifemazone** (**285**).[73]

The diarylpyrrazolodione class of antiinflammatory agents are of more recent origin, having been developed in the late 1950s. Only representative examples are discussed below, because this class also fell into disuse as a result of toxicity problems. The first and most widely used drug from this class is **phenylbutazone** (**288**). This

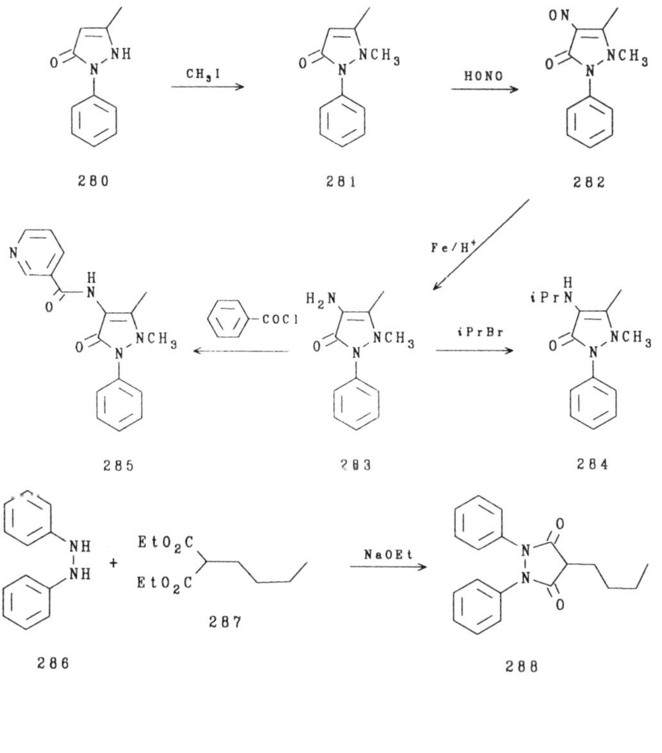

CH_3I HONO Fe/H^+ iPrBr NaOEt

280 281 282 285 283 284 286 287 288

$C_6H_5NH_2$ + C_6H_5OH →(1. HONO; 2. BzBr) 289 →(H_2) 290

Bz-$C_6H_5CH_2$-

290 →($BuCH(CO_2Et)_2$) 291 →(H_2/Pd) 292

293 + $C_6H_5NHNHC_6H_5$ → 294 →(H_2O_2) 295

compound, which in effect is a double lactam of *sym*-diphenylhydrazine (**286**), can be prepared by reaction of the latter with diethyl butylmalonate (**287**) in the presence of base;[74] note that this compound has the acidic proton that marks the classic heterocyclic NSAIDs.

Detailed pharmacokinetic studies on **phenylbutazone** (**288**) revealed that the phenol **292** from hydroxylation of the aromatic ring constituted one of the main metabolites. The finding that this compound was more potent than the parent led to its introduction as a drug in its own right. Coupling of the benzenediazonium chloride with phenol gives the corresponding azo derivative (**289**); the free hydroxyl is then protected as a benzyl ether (**290**) by reaction of the phenoxide with benzyl bromide. Condensation of this compound with diethyl butylmalonate leads to the pyrrazolodione **291**. Removal of the protecting group by hydrogenolysis over palladium affords the NSAID **oxyphenbutazone** (**292**).[75]

Addition of a phenylsulfoxide moiety to the end of the side chain markedly changes the activity of this class of compounds. The product **sulfinpyrazone** (**295**) stimulates uric acid excretion, making it a valuable drug for dealing with the high serum uric acid levels associated with gout. This compound is still one of the more important uricosuric agents available today The starting ester **293** is available by alkylation of the dianion from ethyl malonate with 2-chloroethylphenyl thioether. Condensation with diphenylhydrazine in the presence of base then affords the pyrrazolodione **294**. Oxidation of sulfur with a controlled amount of hydrogen peroxide leads to the sulfoxide and thus sulfinpyrazone (**295**).[76]

D. Thiazoles and Related Sulfur–Nitrogen-Containing Heterocycles

The very broad structural requirements for NSAIDs, which allows replacement of a benzene ring by a pyrrole or an oxazole ring, have been noted earlier in this chapter. It has been found that thiazoles also can substitute for the central benzene ring.

One of the two classic schemes for construction of the thiazole ring involves condensation of a thioamide or its equivalent with an α-haloketone. The first step of this reaction can be visualized as involving the displacement of halogen by sulfur from the enol form of the amide; imine formation will then close the ring. Thus, reaction of the bromoketone (**297**), which is obtained from bromination of the corresponding ketoacid with thioamide **296**, affords the thiazole **298** in a single step. Thus, the NSAID **fentiazac**[77] (**298**) is obtained.

In a similar vein reaction of the benzamide **299** with ethyl 4-bromoacetoacetate in the presence of base starts by displacement of bromine by sulfur to afford a transitory addition product such as **300**. This compound can then undergo internal imine

S NH2 + Br CO2H O Cl → S N CO2H Cl

296 297 298

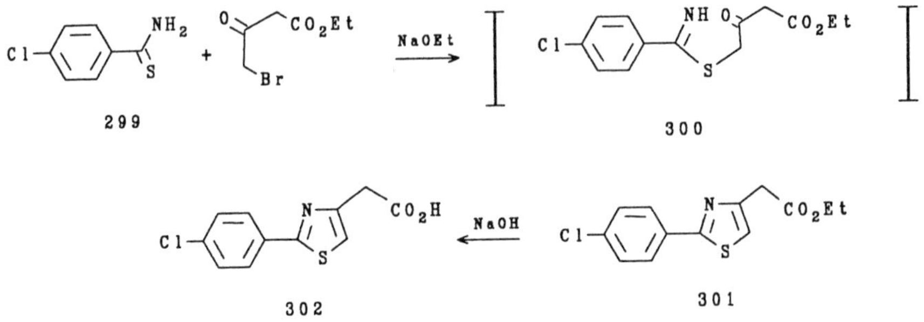

formation to afford the thiazole **301**. Saponification then leads to the NSAID **fenclozic acid** (**302**).[78]

Replacement of the imidazole ring by thiazole in the antiinflammatory drug flumizole (**225**) retains the cyclooxygenase inhibiting activity of that compound; the resulting compound **itazigrel** (**305**) was, however, investigated largely as a platelet aggregation inhibitor. It is of interest that the analogue in which the trifluoromethyl group is replaced by methyl shows very similar activity.[79] Itazigrel (**305**) is obtained in straightforward fashion by condensation of trifluoroacetamide (**303**) with bromodesoxyanisoin (**304**).[80]

A 1,2,5-thiadiazole ring replaces the benzene ring in one of the more widely used β-blockers, **timolol** (**372**), which is discussed later in this chapter. A simpler analogue based on a thiazole ring also inhibits the β-adrenergic system. The first step in its synthesis consists in displacement of the enol bromide like halogen in bromothialzole (**306**) with the alkoxide from racemic glycerol acetonide (**307**) to give the ether (**308**). The diol obtained on hydrolysis of the acetonide function is then preferentially

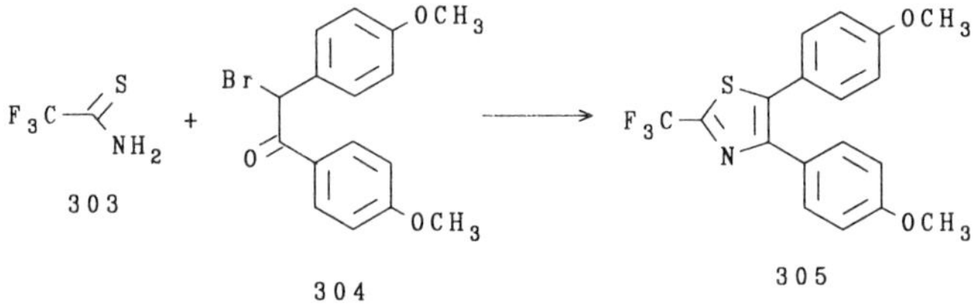

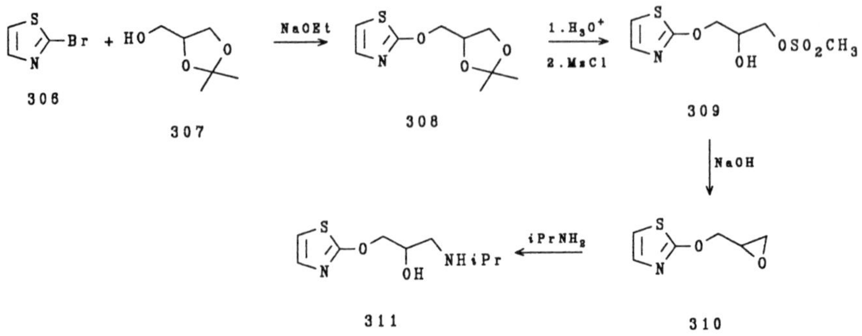

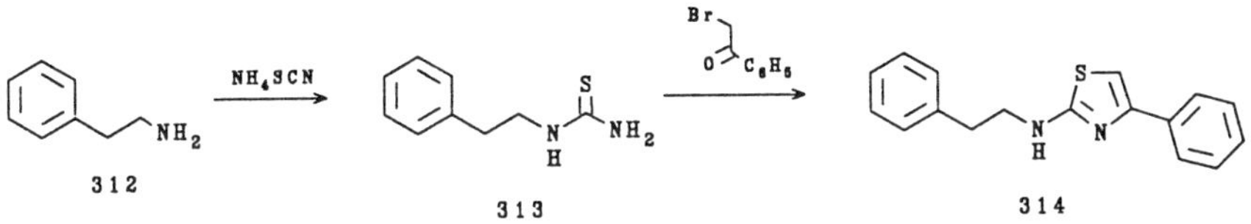

NH_4SCN

312 313 314

converted to the mesylate (**309**) from the primary alcohol by reaction with methanesulfonyl chloride. The mesylate is then converted to the aminoalcohol by the standard β-blocker sequence: Conversion to the epoxide (**310**) with sodium hydroxide, followed by ring opening with isopropylamine. Thus, **tazolol** (**311**)[81] is obtained.

Thioureas serve as a convenient starting material for 2-aminothiazoles. Reaction of β-phenethylamine (**312**) with ammonium isothiocyanate gives the thiourea **313**. Treatment of **313** with phenacyl bromide leads to the corresponding thiourea. This compound, **fanetizole** (**314**),[82] shows immunoregulating activity.

The interchangeabilty of five-membered heterocyclic rings extends to histamine H_2 blockers as well. Replacement of the imidazole in cimetidine by furan to give ranitidine includes some additional changes in functionality. The furan ring in the latter can, however, be replaced directly by thiazole. The starting thiazole **316** is prepared by the standard route, that is, by condensation of thioamide (**315**) with ethyl bromoacetate. Reduction of the ester group in **316** by means of lithium aluminum hydride leads to the corresponding methyl carbinol; this compound is then converted to the bromide **317**. Displacement of bromine by cysteamine incorporates the required side chain (**318**). Reagent **319**, can be prepared by reaction of the ranitidine bis-methylthionitrovinyl intermediate with methylamine. Treatment of primary amine **318** with **319** leads to addition of the nitrovinyl group and formation of **nizatidine** (**320**).[83]

Addition of a mercaptide to a nitrile provides the key reaction to the preparation of a thiazolidinone that shows diuretic activity. The first step in the reaction of ethyl 2-mercaptoacetate with ethyl cyanoacetate thus probably results in the formation of the addition product **321**. The imino group, or its precursor anion, then attacks the adjacent ester group to form the lactam, thus forming the thiazolidinone **322**. Reaction of **322** with base apparently results in an ambident anion; this anion alkylates on nitrogen on treatment with methyl iodide to afford the product with the exocyclic olefin **323**, as the single Z isomer. Bromination proceeds at the position adjacent to the carbonyl group to give **324**. Displacement of halogen with piperidine followed by saponification then gives **ozolinone** (**325**).[84]

The enzyme xanthine oxidase mediates the metabolic disposition of xanthine and hypoxathine, which results in the formation of uric acid. The pathology of gout, as discussed earlier, is due in most part to accumulation of this metabolite. One of the earliest drugs for treating gout is **allopurinol** (see Chapter 14), which in fact inhibits accumulation of uric acid by inhibiting xanthine oxidase. One of the rare isothiazoles, whose generic name is **amflutizole** (**330**), shares this activity. The key intermediate for the synthesis of **330** consists of the toluenesulonyl derivative (**326**) of the oxime of the acid cyanide of *m*-trifluromethylbenzoic acid. Reaction with ethyl 2-mercaptoacetate in the presence of base results in displacement of the tosylate by mercaptide

F_3C — NHOTs, CN

326

Ts = $OSO_2C_6H_4Me$

$HSCH_2CO_2Et$, NaOEt

F_3C — N–S, NC, CO_2Et

327

F_3C — N–S, HN, CO_2Et

328

F_3C — N–S, H_2N, CO_2Et

329

NaOH

F_3C — N–S, H_2N, CO_2H

330

and formation of the N–S bond. This compound is then converted to the carbanion **327** by a second equivalent of base. The anion in **327** then adds to the cyano group; protonation then goes on to form the imine **328**, which tautomerizes to the amino form **329**. Saponification of the ester then affords amflutizole (**330**).[85]

4. RINGS WITH THREE OR MORE HETEROATOMS

A. 1,2,4-Oxadiazoles

The syntheses of the 1,2,4-oxadiazole system described below all rely on hydroxylamine for providing a preformed N–O linkage. The preparation of the respiratory antiinflammatory and antitussive agent **oxolamine** (**334**) starts by acylation of the alkoxide from the *N*-hydroxyamidine (**331**) with 3-chloropropionyl chloride; the presence of the negative charge on oxygen results in formation of the O-acylated product **332**. Treatment of **332** with triethylamine leads to cyclization via imine formation to afford the 1,2,4-oxadiazole **333**. Displacement of the terminal chlorine with diethylamine gives the corresponding amine oxolamine (**334**).[86]

Acylation of *N*-hydroxy-2-phenylbutyramidine with 3-chloropropionyl chloride in the absence of added base proceeds as might be expected to give the product **335** from acylation on the more basic nitrogen. Heating **335** leads to formation of the oxadiazole **336** almost certainly via the enol tautomer of the amide. Displacement of the terminal chlorine with diethylamine leads to the tertiary amine and thus to **proxazole** (**337**),[87] which exhibits antispasmodic activity.

Reaction of the urea **338** derived from the hydroxyamidine **331** with phosphorus oxychloride represents an alternative method for forming the oxadiazole ring. The first step in this sequence can be visualized as formation of the imino chloride **339**. Internal displacement by oxime oxygen gives the cyclization product **340**, which is in tautomeric equilibrium with the imino form **341**. Alkylation with 2-chlorotriethylamine interestingly proceeds via the latter unconjugated tautomer. Thus, **imolamine** (**342**),[88] a compound that has been described as an antianginal agent, is obtained.

NH2
N—OH
331
Cl–C(O)–CH2CH2–Cl
NaOEt
NH2 O
N—O
Cl
332
Et3N
N
N—O
Cl
333
Et2NH
N
N—O
NEt2
334

Heat

335 336

Et_2NH

337

$POCl_3$

338 339 340

Cl⁀NEt₂

NaOH

342 341

B. Triazoles

The great majority of "conazole" antifungal agents discussed above share an *N*-substituted imidazole ring as well as a dichlorphenyl ring and some form of ether linkage. Doubling the heterocyclic ring, now replaced by 1,2,4-triazole, as well as replacement of chlorine by fluorine leads to the very effective orally active antifungal agent **fluconazole** (**346**); this drug has proved particularly useful in treating the opportunistic fungal infections contracted by AIDS patients. The starting material (**343**) for the synthesis is available by acylation of *m*-difluorbenzene with chloroacetyl chloride. Displacement of chlorine by triazole affords the intermediate **344**. Condensation of the carbonyl group with the ylide from trimethylsulfonium iodide leads initially to an addition product. The anion formed on the carbonyl oxygen then internally displaces dimethyl sulfide to give an oxirane yielding epoxide **345**. Reaction of **345** with 1,2,4-triazine leads to epoxide ring opening with consequent incorporation of the second heterocyclic moiety. Thus, fluconazole (**346**) is obtained.[89]

Recall that the first histamine H_2 antagonist retained the imidazole of histamine and achieved its blocking activity by the presence of a thiourea function. Both functions,

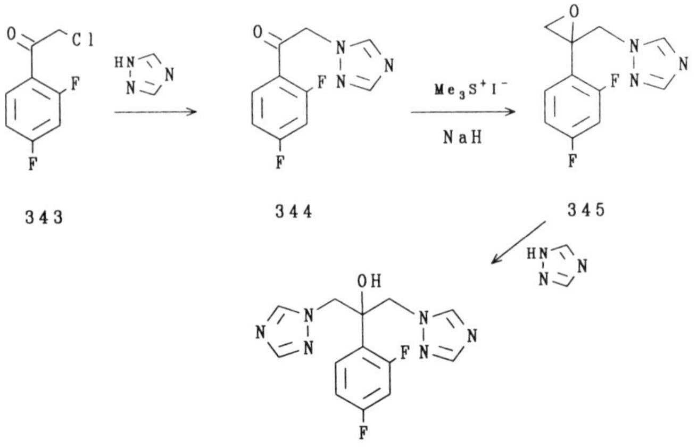

343 344 345 346

it was later found, could be extensively modified; the thiourea group, as noted above, can be replaced by appropriately substituted guanidines, amidine, or amidines. These compounds form part of a heterocyclic ring [see oxmetidine (**273**)] and the imidazole replaced by other heterocycles. In the H_2 blocker **lavoltidine** (**352**), the aminoimidazole is replaced by a phenoxy benzylamine while aminotriazole serves as surrogate guanidine. The synthesis starts with acylation of imine **347** with acetylglycolyl chloride to give amide **348**. Reaction of **348** with the *N*-methylhydrazone of benzaldehyde leads to the amidine **349** by displacement of one of the thiomethyl groups by means of an addition–elimination sequence. Treatment of this intermediate with the primary amine **350** leads to replacement of the remaining thiomethyl group and formation of guanidine **351**. Acid hydrolysis of the hydrazone leads to the corresponding hydrazine; the terminal amino group cyclizes with the carbonyl group to form a triazole ring. Saponification of the acetyl protecting group gives the corresponding free alcohol and thus lavoltidine (**352**).[90]

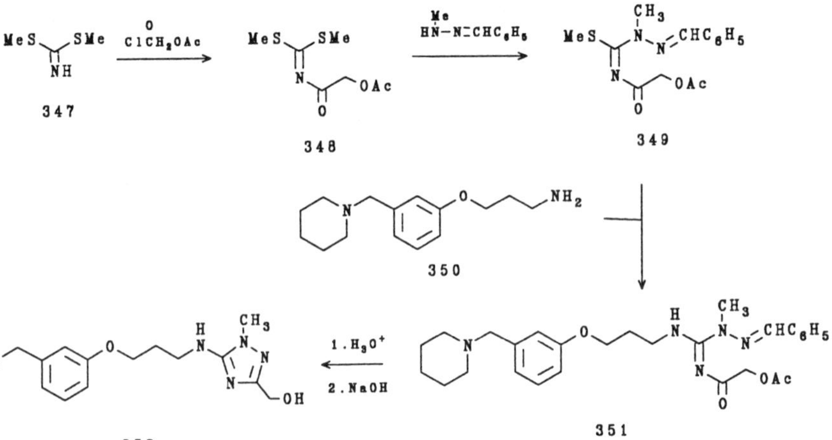

347 348 349 350 351 352

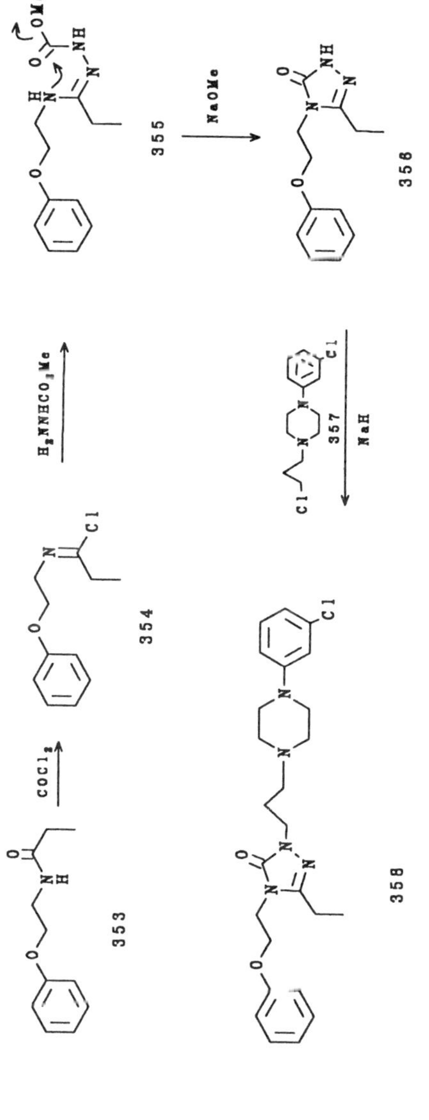

353 $\xrightarrow{COCl_2}$ 354 $\xrightarrow{H_2NNHCO_2Me}$ 355 $\xrightarrow{NaOMe}$ 356 $\xrightarrow[NaH]{357}$ 358

Arylpiperazines have a venerable history as psychotropic agents; one of these compounds, which also incorporates a 1,2,4-triazol-3-one ring, shows significant antidepressant activity. Reaction of propionamide **353** with phosgene leads to the corresponding imino chloride **354**. Condensation of **354** with hydrazine methylurethane gives the corresponding guanidine **355** from displacement of chlorine by the basic hydrazine nitrogen. Reaction with sodium methoxide leads first to ionization of the amine on the side chain; this intermediate then cyclizes to triazolone (**356**) by displacing the urethane methoxyl. Alkylation of the anion from the reaction of **356** with strong base with chloride **357** affords **nefazodone** (**358**).[91]

C. Thiadiazoles

The sequence of events that led from the observation of excess urine excretion on the administration of high doses of sulfonamide antibacterial agents to the development of diuretic drugs was touched upon in Chapter 2. It is of interest that one of the first sulfonamide diuretic agents consisted of a compound in which the benzene ring of the antibacterial sulfonamides was replaced by a thiadiazine. Treatment of commercially available 1,3,4-thiadiazine (**359**) with acetic anhydride gives the corresponding acetamide **360**. The mercapto group is then oxidized to a sulfonyl chloride **361** by reaction with aqueous chlorine. Ammonolysis of **361** affords **acetozolamide** (**362**).[92]

359 —Ac_2O→ 360 —Cl_2, H_2O→ 361 —NH_4OH→ 362

ClOC —$H_2NNHCSNH_2$→ [363] → 364 —p-$MeOC_6H_4SO_2Cl$→ 365

HO, OH, CHO + $H_2NC(CH_3)_3$ —H_2→ HO, OH, $NHC(CH_3)_3$

366 367

H_2NCOCN —S_2Cl_2→ [Cl, O, N, NH_2, ClS] → Cl, OH, N, N, S —TsCl / Py→ Cl, OTs, N, N, S

368 369 370

367/NaH

O, N, O, OH, $NHC(CH_3)_3$, N, N, S ← O NH — Cl, O, OH, $NHC(CH_3)_3$, N, N, S

372 371

The sulfonylurea hypoglycemic agents, as noted in Chapter 2, can also trace their ancestry to the sulfonamides. It is interesting that activity is retained when a substituted 2-amino-1,3,4-thiadiazole replaces the urea function. Reaction of isobutyryl chloride with thiosemicarbazone leads initially to the transient 1,2-diacylhydrazine **363**. This compound apparently spontaneously cyclizes to thiadiazine (**364**) under the reaction conditions. Acylation with *p*-methoxybenzenesulfonyl chloride affords the oral hypoglycemic agent **isobuzole** (**365**).[93]

The increasing availability of practical enantioselective synthetic methods have combined with the growing awareness of the stereospecific nature of drug action to place new regulatory emphasis on providing drugs in chiral form. The synthesis of the β-blocker **timolol** (**372**) illustrates such an entantiospecific synthesis. The key to the synthesis involves the preparation of the propyl aminoalcohol side chain by use of a chiral starting material. The required side chain glycol (*R*)-glyceraldehyde (**366**) is available in chiral form as a degradation product from the sugar mannitol, which constitutes a tonnage chemical. Thus, reductive amination of the aldehyde with *tert*-butylamine gives the ethanolamine side chain **367** in chiral form

Preparation of the 1,3,5 tiadiazine intermediate **370** involves the reaction of cyanoamide with sulfur monochloride. The reaction can be rationalized by invoking addition of the reagent across the nitrile group to form an intermediate such as **368**. Displacement of the chlorine on sulfur by amide nitrogen serves to close the ring; tautomerization to the all-enol form then gives the observed product **369**. The hydroxyl group is then converted to its *p*-toluenesulfonate **370**. Displacement of this good leaving group with the terminal alkoxide from diol **367** incorporates the β-blocker side chain to give **371**. Replacement of the remaining leaving group on the heterocycle by morpholine affords timolol (**372**).[94]

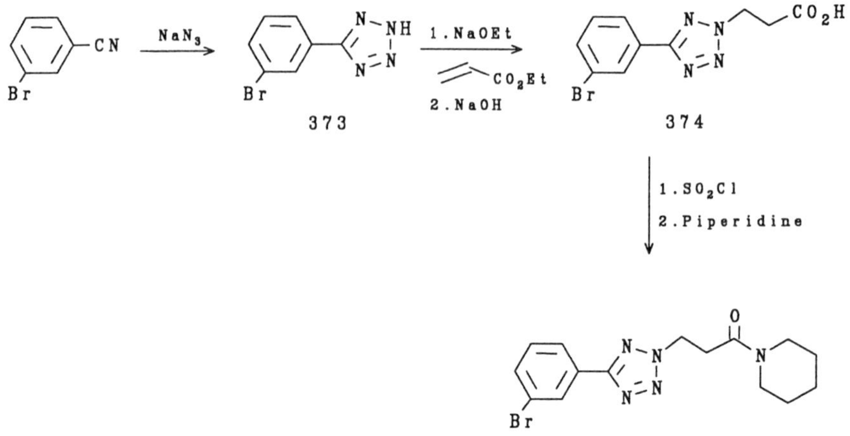

D. Tetrazole

The acidity of carboxylic acids is markedly enhanced by the fact that the negative charge in its anion can be dispersed over two atoms. Much the same holds true for a tetrazole devoid of nitrogen substituents where the charge in the anion can be dispersed over all four nitrogen atoms. When this ring system occurs in therapeutic agents it usually replaces a carboxylic acid in the parent drug; these terazoles are found elsewhere in this book.

The structure of the antiinflammatory agent **broperamole** (**375**) is reminiscent of the heterocycle based NSAID propionic acids. The activity may in this case be due to the acid that would result on hydrolysis of the amide. Tetrazoles are virtually always prepared by the reaction of a nitrile with hydrazoic acid, or more commonly sodium azide in the presence of acid, in a reaction very analogous to a 1,3-dipolar cycloaddition. Reaction of the anion from tetrazole **373**, which is made from *m*-bromobenzonitrile, with ethyl acrylate leads to the product from Michael addition; saponification gives the corresponding carboxylic acid **374**. This compound is then converted to the acid chloride; reaction with piperidine affords broperamole (**375**).[95]

REFERENCES

1. Stillman, W.B.; Scott, A.B. U. S. Patent **1947**, 241664.
2. Fujita, A.; Nakata, M.; Minami, S.; Takamatsu, H. *J. Pharm. Soc. Jpn.* **1966**, 1014.
3. Snyder, H.R.; Davis, C.S.; Bickerton, R.K.; Halliday, K.P. *J. Med. Chem.* **1967**, *10*, 891.
4. Bradshaw, J. In *Chronicles of Drug Discovery*, Lednicer, D., Ed.; ACS Books: Washington DC, 1993, Vol. 3, p. 45.

5. Carson, J.R.; McKinstry, D.N.; Wong, J. *J. Med. Chem.* **1971**, *14*, 646.
6. Lambelin, G., Roba, J.; Gillet, C.; Buu-Hoi, N.P. German Offen. **1973**, 2261965; *Chem. Abstr.* **1973**, *79*, 78604.
7. Carson, J.R.; Wong, J. *J. Med. Chem.* **1973**, *16*, 172.
8. Franco, F.; Greenhouse, R.; Muchowski, J.M. *J. Org. Chem.* **1982**, *47*, 1682.
9. Ondetti, M.A.; Rubin, B.; Cushman, D.W. *Science* **1977**, *196*, 441.
10. Nam, D.H.; Lee, C.S.; Ryu, D.D. *J. Pharm. Sci.* **1984**, *73*, 1843.
11. Patchett, A.A. In *Chronicles of Drug Discovery*, Lednicer, D., Ed.; ACS Books: Washington DC, 1993, Vol. 3, p. 125.
12. Karanewsky, D.S.; Badia, M.C.; Cushman, D.W.; DeForrest, J.M.; Dejneka, T.; Loots, M.L.; Perri, M.G.; Petrillo, E.W.; Powell, J.R. *J. Med. Chem.* **1988**, *31*, 204.
13. Gold, E.H.; Neustadt, B.R.; Smith, E.M. U. S. Patent **1984**, 4470972, *Chem. Abstr.* **1984**, *102*, 96083.
14. Teetz, V.; Geiger, R., Henning, R.; Urbach, H. *Arzneim.-Forsch.* **1984**, *34*, 1399; Teetz, V.; Geiger, R.; Gaul, H. *Tetrahedron Lett.* **1984**, *25*, 4479.
15. Vincent, M.; Remond, G.; Portevin, B.; Serkiz, B.: Laubie, M. *Tetrahedron Lett.* **1982**, *23*, 1677.
16. Gruenfeld, N.; Stanton, J.L.; Yuan, A.M.; Ebetino, F.H.; Browne, L.J.; Gude, C.; Huebner, C.F. *J. Med. Chem.* **1983**, *26*, 1277.
17. Klutchko, S.; Blankley, C.J.; Fleming, R.W.; Hinkley, J.M.; Werner, A.E.; Nordin, I.; Holmes, A.; Hoefle, M.L.; Cohen, D.M.; Essenburg, A.D.; Kaplan, H.R. *J. Med. Chem.* **1986**, *29*, 1953.
18. Lunsford, C.D.; Cale, A.D.; Ward, J.W.; Franko, B.V.; Jenkins, H. *J. Med. Chem.* **1964**, *7*, 302.
19. Miller, C.A.; Long, L.M. *J. Am. Chem. Soc.* **1951**, *73*, 4895.
20. Sircar, S.S.G. *J. Chem. Soc.* **1927**, 600.
21. Brown, K. U. S. Patent **1971**, 3578671, *Chem. Abstr.* **1972**, *75*, 36005.
22. Self, C.R.; Barber, W.E., Machin, P.J.; Osbond., J.M.; Smithen., C.E.; Tong, B.P.; Wickens, J.C.; Bloxham, D.P.; Bradshaw, D.; Cashin, C.H.; Dodge, B.B.; Ford, H.; Lewis, E.S.; Westmacott, D. *J. Med. Chem.* **1991**, *34*, 772.
23. Ross, W.J.; Harrison, R.G.; Jolley, M.J.R.; Neville, M.C.; Todd, A.; Berge, J.P.; Dawson, W.; Sweatman, W.J.F. *J. Med. Chem.* **1979**, *22*, 412.
24. White, R.L.; Wessels, F.L; Swan, T.J.; Ellis, K.O. *J. Med. Chem.* **1987**, *30*, 236.
25. Satzinger, G.; Herrman, M. German Offen. **1975**, 2336746; *Chem. Abstr.* **1975**, *83*, 10047.
26. Poos, G.I.; Carson, J.R.; Roseneau, J.D.; Roszowski, A.P.; Kelley, N.M.; McGowin, J. *J. Med. Chem.* **1963**, *6*, 266.
27. Hardy, R.A.; Howell, C.F.; Quinones, C.Q. U. S. Patent **1962**, 3037990.
28. Aschar, R. *Chem. Ber.* **1913**, *46*, 2077.
29. Spielman, M.A. *J. Am. Chem. Soc.* **1944**, *66*, 1244.
30. Diana, G.D.; Ogelsby, R.C.; Akullian, V.; Carabateas, P.M.; Cutliffe, D.; Mallamo, J.P.; Otto, M.,J.; McKinley, M.A.; Maliski, E.G.; Michalek, S.J. *J. Med. Chem.* **1987**, *30*, 383.
31. Gardner, T.S.; Wenis, E.; Lee, J. *J. Med. Chem.* **1960**, *2*, 133.
32. Friis, P.; Helboe, P.; Larsen, P.O. *Acta Chem. Scan.* **1974**, *28*, 317.

33. Mzengeza, S., Yang, C.M.; Whitney, R.A. *J. Am. Chem. Soc.* **1987**, *109*, 276.
34. Stammer, C.H.; Wilson, A.N.; Spencer, C.F.; Bachelor, F.W.; Holly, F.W.; Folkers, K. *J. Am. Chem. Soc.* **1957**, *79*, 3236.
35. Cosar, C.; Crisan, C.; Horclois, R.; Jacob, R.M.; Robert, J.; Tchelitcheff, S.; Vaupre, R. *Arzneim.-Forsch.* **1966**, *16*, 23.
36. Heeres, J.; Mostmans, J.H.; Maes, R. German Offen. **1975**, 2429755. *Chem. Abstr.* **1975**, *82*, 156309.
37. Verdi, V.F. U. S. Patent **1969**, 3450710. *Chem. Abstr.* **1970**, *71*, 70597.
38. Grabowski, E.J.; Liu, T.M; Salce, M.; Schoenewaldt, E.F. *J. Med. Chem.* **1974**, *17*, 547.
39. Lancini, G.C.; Lazari, E. *Experientia* **1965**, *21*, 83.
40. Beaman, A.G.; Tautz, W.; Duchinsky, R. *Antimicrob. Agents Chemother.* **1986**, *1967*, 520.
41. Godefroi, E.F.; Van Cutsem, J.; Van der Eycken, C.A.M.; Janssen, P.A.J. *J. Med. Chem.* **1967**, *10*, 1160.
42. Godefroi, E.F.; Heeres J.; Van Cutsem, J.; Janssen, P.A.J. *J. Med. Chem.* **1969**, *12*, 784.
43. Walker, K.A.M.; Marx, M. U. S. Patent, **1976**, 4038409. *Chem. Abstr.* **1977**, *87*, 152210.
44. Mixich, G.; Thiele, K. **1986**, 4550175. *Chem. Abstr.* **1986**, *105*, 6508.
45. Heeres J.; Backx, L.J.J.; Mostmans J.H.; Van Cutsem, J. *J. Med. Chem.* **1979**, *22*, 1003.
46. Dyer, R.L.; Ellames, G.J.; Hamill, B.J.; Manley, P.W.; Pope, A.M.S. *J. Med. Chem.* **1983**, *26*, 442.
47. Regel, E.; Draber, W.; Buechel, K.H.; Plempel, L. German Offen. **1978**. *Chem. Abstr.* **1978**, *89*, 24307.
48. Durant, G.J.; Emmett, J.C.; Ganellin, C.R.; Miles, P.D.; Prain, H.D.; Parsons, M.E.; White, G.R. *J. Med. Chem.* **1977**, *20*, 901.
49. For a succinct account of the development of these drugs see, Ganellin, C.R. In *Chronicles of Drug Discovery*, Bindra, J.S.; Lednicer, D., Eds.; Vol. 1. Wiley: New York, 1982, p. 1.
50. Crenshaw, R.R.; Kavadias, G.; Santonge, R.F. U. S. Patent **1979**, 4157340; *Chem. Abstr.* **1979**, *90*, 157312.
51. Brown, T.H.; Durant, G.J.; Emmett, J.C.; Ganellin, C.R. U. S. Patent **1979**, 4145546; *Chem. Abstr.* **1979**, *91*, 204137.
52. Scalesciani, J. German Offen. **1977**, 2721835; *Chem. Abstr.* **1978**, *88*, 89671.
53. Lombardino, J.G. German Offen. **1972**, 2155558; *Chem. Abstr.* **1972**, *77*, 101607.
54. Phillips, A.P.; White, H.L.; Rosen, S. Eur. Patent Appl. **1982**, 58890; *Chem. Abstr.* **1982**, *98*, 53894.
55. Anonymous *Drugs Future* **1950**, 15, 448.
56. Caroon, J.M.; Clark, R.D.; Kluge, A.F.; Olah, R.; Repke, D.B.; Unger, S.H.; Michel, A.D.; Whiting, R.L. *J. Med. Chem.* **1982**, *25*, 666.
57. Sohn, A. U. S. Patent, **1939**, 2161938. *Chem. Abstr.* **1939**, *33*, 7316.
58. Synerholm, F.E.; Jules, L.H.; Sayhun, M. U. S. Patent **1956**, 2731471.
59. Hueni, A. U. S. Patent, **1959**, 2868802. *Chem. Abstr.* **1959**, *53*, 10253.
60. Miescher, K.; Marxer, A.; Urech, E. U. S. Patent **1950**, 2503509.
61. For an informal account of the discovery and SAR of **clonidine**, see Stahl, H. In *Chronicles of Drug Discovery*, Bindra, J.S.; Lednicer, D.; Eds.; Vol. 1. Wiley: New York, **1982**, p. 87.

62. Berg, A. German Offen. **1963**, 1191381. *Chem. Abstr.* **1965**, *63*, 8373.
63. Rippel, H.; Ruschig, H.; Linder, E.; Schorr, M. German Offen. **1971**, 1941761. *Chem. Abstr.* **1971**, *74*, 100054.
64. Ruout, B.; Leclerc, G. *Bull. Soc. Chim. Fr.* **1979**, *9–10(Pt.2)*, 520.
65. Rasmussen, C.R. U. S. Patent **1976**, 3983135. *Chem. Abstr.* **1977**, *86*, 55441.
66. Pinner, A. *Chem. Ber.* **1988**, *21*, 2320.
67. Anonymous Swiss Patent, **1934**, 166004.
68. Henze, H.R. U. S. Patent, **1946**, 2409754. *Chem. Abstr.* **1947**, *41*, 1250.
69. Oba, S.; Koseki, Y.; Fukawa, K. *J. Soc. Sci. Phot. Jpn.* **1951**, *13*, 33. *Chem. Abstr.* **1952**, *46*, 3885.
70. Hanifin, J.W.; Gussin, R.Z.; Cohen, E. German Offen. **1973**, 2251354. *Chem. Abstr.* **1973**, *79*, 18717.
71. Perronet, J.; Demonte, J.P. *Bull. Soc. Chim. Fr.* **1970**, 1168.
72. Skita, E.; Stummer, W. German Offen. **1955**, 930328. *Chem. Abstr.* **1958**, *52*, 163721.
73. Ponngrantz, A.; Zirm, K.L. *Monatsh. Chem.* **1957**, *88*, 330.
74. Stenzl, H. U. S. Patent **1951**, 2562830. *Chem. Abstr.* **1953**, *47*, 1191.
75. Pfister, R.; Hafliger, F. *Helv. Chim. Acta* **1957**, *40*, 395.
76. Pfister, R.; Hafliger, F. *Helv. Chim. Acta* **1961**, *44*, 232.
77. Brown, K., Cater, D.P.; Cavalla, J.F.; Green, D.; Newberry, R.A.; Wilson, A.B. *J. Med. Chem.* **1974**, *17*, 1177.
78. Anonymous Neth. Patent **1967**, 6614130. *Chem. Abstr.* **1968**, *68*, 68976.
79. Lednicer, D. U. S. Patent **1971**, 3558644.
80. Rynbrandt, R.H.; Nishizawa, E.E.; Balgoyen, D.P.; Mendoza, A.R.; Annis, K.A. *J. Med. Chem.* **1981**, *24*, 1507.
81. Roszowski, A.P.; Strosberg, A.M.; Miller, L.M.; Edwards, J.A.; Berkoz, B.; Lewis, G.S.; Halpern, O.; Fried, J.H. *Experientia* **1972**, *28*, 1336.
82. Lombardino, J.G. U. S. Patent **1981**, 4307106; *Chem. Abstr.* **1981**, *96*, 122784.
83. Pioch, R.P. U. S. Patent **1983**, 4382090; *Chem. Abstr.* **1983**, *99*, 43548.
84. Hayao, S.; Havera, H.J.; Stryker, W.G. U. S. Patent **1983**, 4006232; *Chem. Abstr.* **1977**, *87*, 5968.
85. Beck, J.R.; Gajewski, R.P.; Hackler, R.E. Eur. Patent Appl., **1982**, 48615; *Chem. Abstr.* **1982**, *97*, 55798.
86. Anonymous German Offen. **1961**, 1097998; *Chem. Abstr.* **1962**, *56*, 11598.
87. Palazzo, G.; Silvestrini, B. U. S. Patent **1963**, 3141019; *Chem. Abstr.* **1963**, *59*, 6415.
88. Aron-Samuel, M.D.; Sterne, J.J. Fr. Patent **1963**, M2023; *Chem. Abstr.* **1964**, *60*, 2952.
89. Richardson, R.K. Br. Patent **1982**, 2099818; *Chem. Abstr.* **1983**, *99*, 38467.
90. Cliterow, J.W.; Bradshaw, J.; Price, B.J.; Martin-Smith, M.; Mackinon, J.W.M.; Judd, D.B.; Hayes, R.; Carey, L. Eur. Patent Appl. **1980**, 16565; *Chem. Abstr.* **1981**, *94*, 192345.
91. Madding, G.D.; Smith, D.W.; Sheldon, R.I.; Lee, B. *J. Heterocycl. Chem.* **1985**, *22*, 1121.
92. Roblin, R.O.; Clapp, J.W. *J. Am. Chem. Soc.* **1950**, *72*, 4890.
93. Chubb, F.L.; Nissenbaum, J. *Can. J. Chem.* **1959**, *37*, 1121.
94. Weinstock, L.M.; Mulvey, D.M.; Tull, R. *J. Org. Chem.* **1976**, *41*, 3121.
95. Anonymous Br. Patent **1973**, 1319357; *Chem. Abstr.* **1973**, *79*, 92231.

CHAPTER 9

DRUGS BASED ON SIX-MEMBERED HETEROCYCLES

1. RINGS WITH ONE HETEROATOM

A. Pyridines

The unmodified pyridine ring, in contrast to the imidazole or imidazoline ring, is not associated with any specific pharmacophoric activity. It does, however, appear in a number of therapeutic agents because benzene rings in biologically active compounds can often be replaced by pyridines despite the presence of the basic nitrogen atom. Therefore, the relatively small number of pyridines discussed below do not reflect the frequency of their occurrence in therapeutic agents; instead, examples have been selected that illustrate some facet of pyridine chemistry.

The tubercule bacillus, causative agent of tuberculosis, is extremely resistant to most antibacterial agents. The discovery that a very simple derivative of pyridine, **isoniazid** (pyridine-4-carboxylic acid hydrazide) showed antibacterial activity against this microbe permitted major inroads against tuberculosis in Western society. The structurally related thioamide **ethionamide** (**6**) has also proven to be active against tuberculosis in humans. One synthesis of this agent begins with the aldol condensation of diethyl oxalate and 2-butanone to give the diketoester **1**. Condensation of **1** with cyanoacetamide leads to the pyridone (**2**), which is depicted as its unconjugated tautomer. This reaction can be visualized as involving initial conjugate addition of the cyanoacetamide anion to **1**, followed by elimination of hydroxide; internal imine formation then closes the ring. Hydrolysis of the intermediate leads to a β-ketoacid that loses carbon dioxide under the reaction conditions to afford the pyridone acid **3**. Treatment of **3** with phosphorus oxychloride converts the amide to its imino chloride;

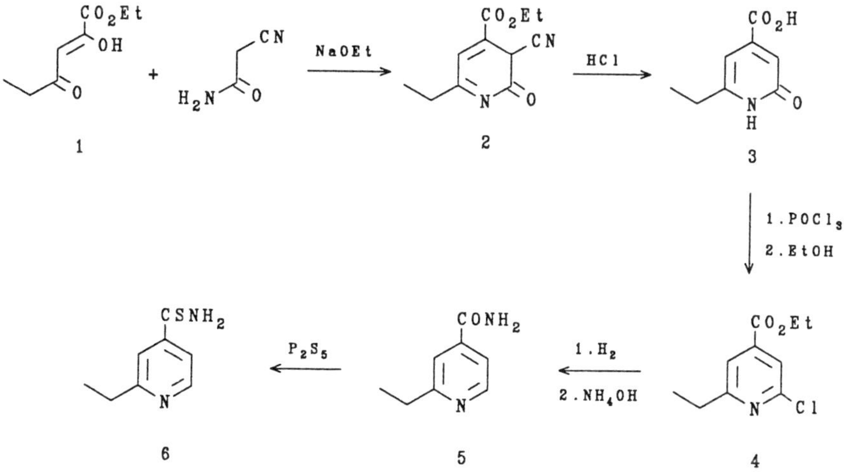

the carboxyl group is converted to the acid chloride under reaction conditions. Exposure of the first formed product to ethanol then gives the ester **4**. The ring chlorine is then removed by catalytic hydrogenation and the ester is exchanged to an amide with ammonia to give **5**. The two-step sequence used originally[1] to convert this function to a thioamide would now be accomplished directly with phosphorus pentasulfide to give ethionamide (**6**).

One broad category of NSAIDs, which was discussed in Chapter 2, consists of *N*-arylanthranilic acids and their "fenac" homologues. As an example of the equivalence noted above, cyclooxygenase inhibiting activity is retained when the ring bearing the carboxylic acid is replaced by pyridine. The key reaction to the series involves treatment of nicotinic acid *N*-oxide (**7**) with phosphorus oxychloride. This reaction results in chlorination of the adjacent ring carbon with loss of oxygen in a Polonovski-like rearrangement; the regiochemistry is probably dictated by migration to the lower electron density center. Nucleophilic aromatic displacement of this chlorine in **8** with *m*-trifluoromethylaniline leads to the diaryl amine and thus to **niflumic acid** (**9**).[2]

The closely related compound, **flunixin** (**11**) has much the same activity. Its preparation is analogous to the one shown above replacing the aniline by its analogue **10**.[3]

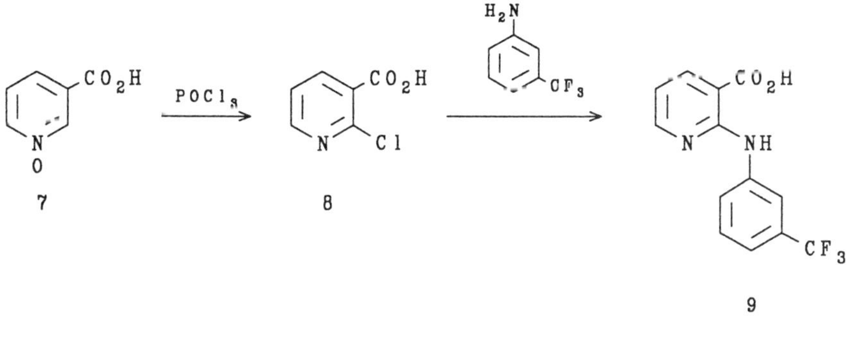

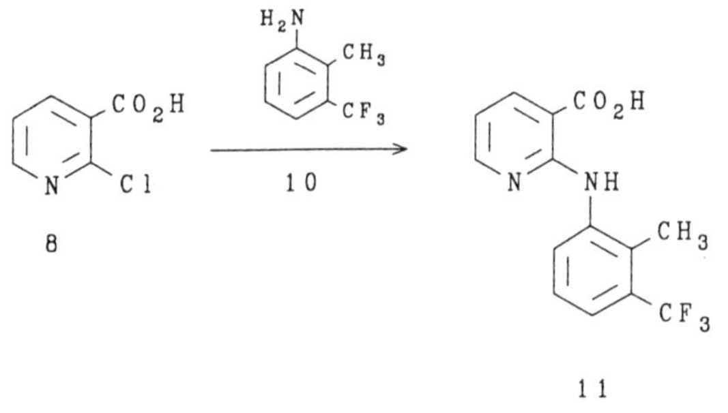

Chapter 2 noted that interposition of a methylene group between the benzene ring and the meta hydroxyl in β-adrenergic agonists led to very potent compounds exemplified by the very successful bronchodilator albuterol. In this case also, activity is retained when benzene is replaced by pyridine. Reaction of 3-pyridinol benzyl ether (**12**) with formaldehyde and strong acid leads to the bis-hydroxymethylation product **13**. In the key reaction, treatment of this diol with manganese dioxide can be controlled to give the product **14** from oxidation of only the more sterically accessible hydroxyl group. Addition of the anion from nitromethane to the carbonyl group results in formation of intermediate **15**, which now has the requisite side chain carbon atoms. The nitro group is then reduced by means of Raney nickel and the resulting primary amine is alkylated with *tert*-butyl bromide. Hydrogenolysis over palladium removes benzyl protecting group to afford the β-adrenergic agonist **pirbuterol** (**16**).[4]

A simplified derivative of **quinine** in which substituted pyridine replaces quinoline retains the antimalarial activity of the parent drug (see Chapter 11). The highly substituted pyridine ring is formed in this case by reaction of acetyl acrylic acid **17**

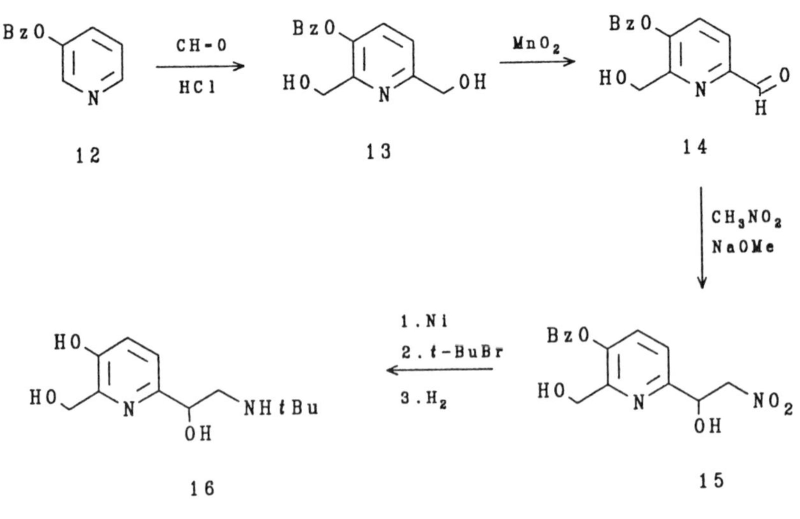

This reaction can be visualized as starting with formation of an anion (or ylide) at the highly activated acyl methylene group on **18**. This compound then adds to the enone function in **17**; loss of the excellent leaving group, pyridine, leads to formation of the hypothetical intermediate **19**. The ammonia in the reaction medium then converts the 1,5-diketone to a pyridine ring yielding intermediate **20**. Condensation of the carboxyl group in **20** with 2-lithiopyridine from halogen exchange of 2-bromopyridine and butyllithium then affords the diaryl ketone **21**. Basic nitrogen in the terminal ring of the final product mimics nitrogen on the quinuclidine ring in quinine. The discrimination between the two pyridines in **21** required to selectively reduce one of these compounds depends on the relatively higher basicity of the terminal ring, since it contains fewer electron-withdrawing substituents than the central pyridine ring. Thus, catalytic hydrogenation of **21** in the presence of acid leads to reduction of the protonated ring; the ketone is reduced to an alcohol in the same reaction. Isolation of the (*R*, *R*) isomer by fractional crystallization affords **enpiroline** (**22**).[5]

Virtually all of the older antiallergic H_1 antihistamines show significant sedation as a side effect; in fact, many nonprescription sleep remedies simply consist of these antihistamines. The finding that this class of drugs can act locally on the allergic end target organs has led to the development of a spate of polar antihistamine drugs that do not cross the blood–brain barrier and consequently do not cause central nervous system (CNS) mediated sedation. The majority of these drugs achieve this by incorporating polar carboxylic acid moieties. The synthesis of one of these compounds, **acrivastine** (**28**), starts by addition to 4-toluonitrile (**24**) of the monolithio derivative from 2,5-dibromopyridine (**23**). Hydrolysis of the first-formed imine with dilute acid leads to the ketone **25**. This compound is then condensed with the ylide from triphenyl-2-*N*-pyrrolidylethylphosphonium bromide to give the olefin **26** as a mixture of isomers. The remaining bromine on the pyridine ring is then converted to its lithio reagent by halogen exchange; reaction with DMF leads to incorporation of a formyl group and formation of intermediate **27**. Horner–Wadsworth condensation of the carbonyl group leads to formation of an acrylic acid side chain; the product from the last reaction probably consists largely of the trans isomer. The (*E*,*E*) isomer is then separated to afford acrivastine (**28**).[6]

B. 1,4-Dihydropyridines

Calcium ions (Ca^{2+}) play a pivotal role in a host of biological responses and particularly those involved in muscle contraction. Excitation induced movement of calcium ions into the cell can lead to inappropriate contractions of smooth muscle in the cardiovascular system; this condition is manifested in the pathology that leads to spasms of coronary vessels manifested as anginal pain; contracted vascular musculature can also be expressed as hypertension. Three structurally very diverse classes of drugs have all been shown to relax vascular smooth skeletal muscle by acting specifically on calcium movement in these tissues. These drugs, variously known as calcium antagonists, calcium blockers, or calcium channel blockers comprise verapamil and its analogues (see Chapter 2), the benzo-1,5-thiazepine related to diltiazem (see Chapter 12), and

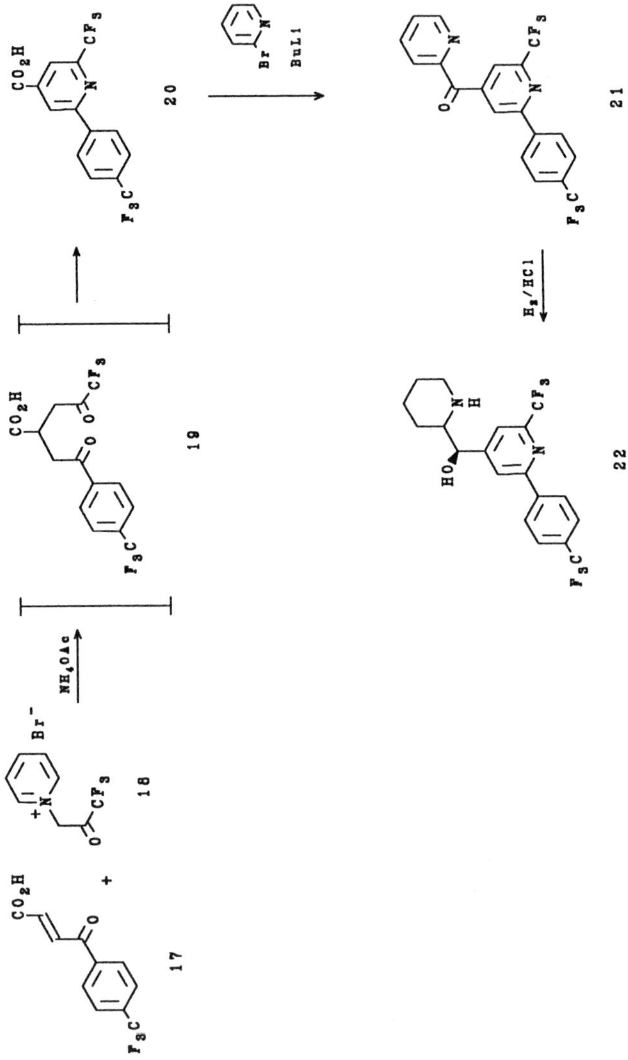
NH4OAc
BuLi
H2/HCl
17
18
19
20
21
22

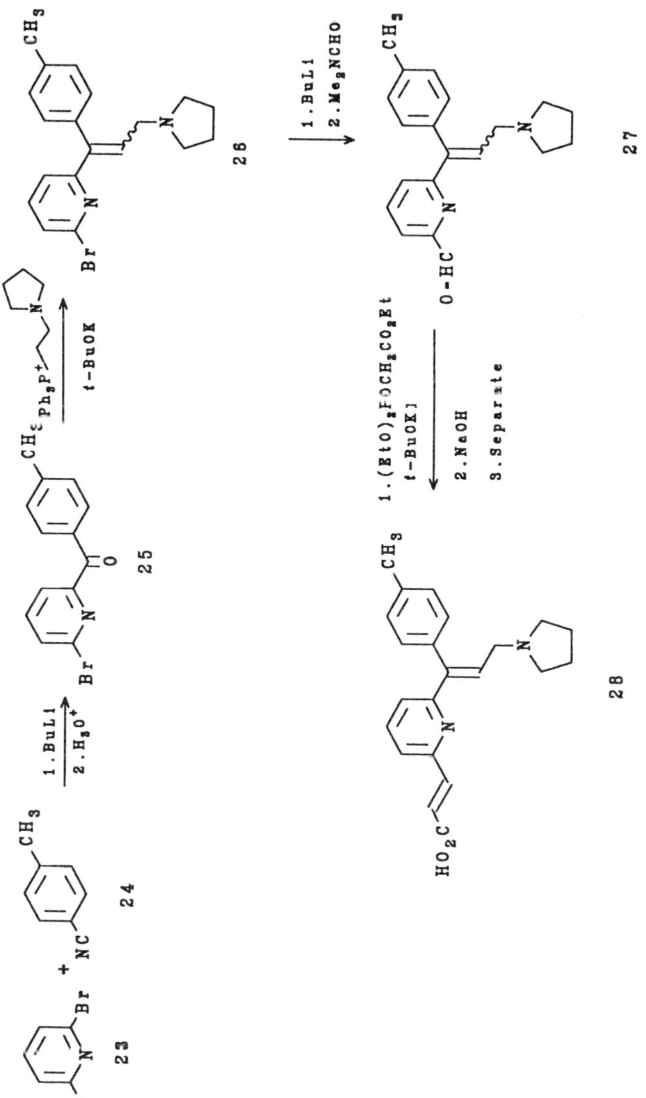

23
24
1. BuLi
2. H3O+
25
t-BuOK
26
1. BuLi
2. Me2NCHO
27
1. (EtO)2POCH2CO2Et
t-BuOK
2. NaOH
3. Separate
28

the 1,4-dihydropyridines. By far the largest groups of calcium blocker analogue series fall into this last category, possibly because of their accessibility.

The great majority of 1,4-dihydropyridines are prepared by using the classical Hantzsch pyridine synthesis or one of its variants. The first dihydropyridine was in fact isolated back in 1882 as a stable intermediate from this method. In its simplest form, the synthesis involves heating an aldehyde such as a *o*-nitrobenzaldehyde (**29**) with methyl acetoacetate and ammonia. The reaction almost certainly involves aldol condensation to form the benzylidene derivative **31** as the first step. Conjugate addition of a second mole of acetoacetate would then afford the 1,5-diketone **32**. Reaction of the carbonyl groups with ammonia will lead to the formation of the dihydropyridine ring to give **nifedipine** (**33**),[7] a drug that has been used extensively for treatment of angina and hypertension.

An alternate approach, useful mainly for the preparation of unsymmetrical compounds, consists in reaction of the aldehyde with a single mole of acetoacetate to give **31** as an isolable product. The appropriate acetoacetate is then converted to its ammonia enamine, such as **30**. Condensation with **30** probably involves Michael addition of the enamine to the conjugated double bond in **31** as the initial step. Internal imine formation completes the sequence to **33.** Extensive SAR studies in this series revealed that unsymmetrical substitution on the heterocyclic ring and, hence, introduction of chirality on the tetrahedral carbon, led to increased potency. The stepwise variation of the Hantzsch synthesis is essential for preparing such compounds. Thus, aldol condensation of methyl acetoacetate with 2,3-dichlorobenzaldehyde (**34**) gives the cinnamyl ketone **35**. Reaction of **35** with the enamine **30** from ethyl acetoacetate gives the calcium channel blocker **felodipine** (**36**).[8] The key acetoacetate (**38**) for the

NO2 CH=O 29 + CO2CH3 NH3 → [NO2 CO2CH3 O] 31

↓ NH3 NH2 CO2CH3 30

H3CO2C NO2 CO2CH3 N H 33 ← [H3CO2C NO2 CO2CH3 O O] 32

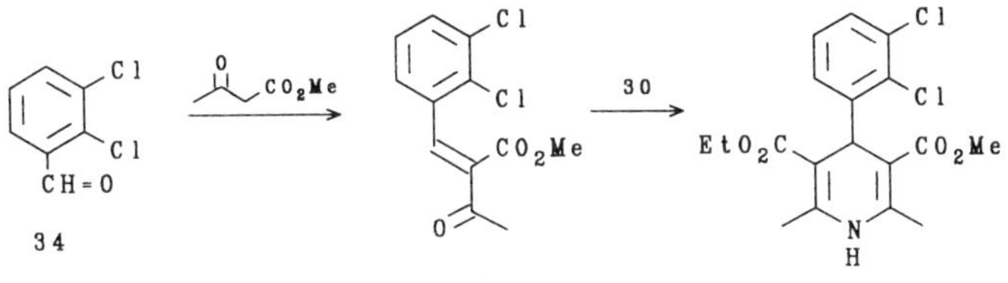

synthesis of **nimodipine** (**39**) is obtained by alkylation of the sodium acetoacetate with 2-methoxyethyl chloride. Aldol condensation of *m*-nitrobenzene (**37**) and subsequent reaction of the intermediate with enamine (**30**) gives nimodipine (**39**).[9]

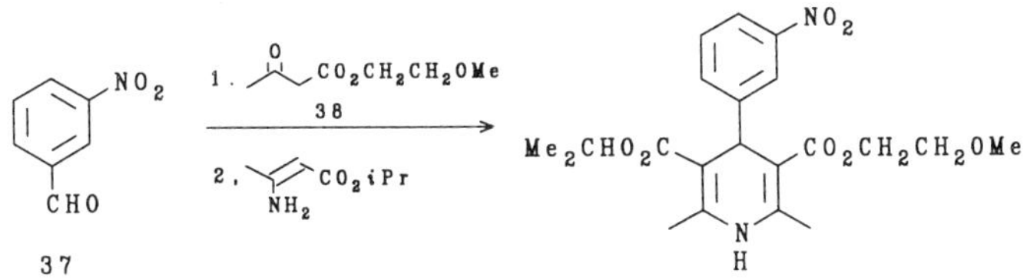

The product **40** from aldol condensation of *m*-nitrobenzaldehyde (**37**) with the dimethyl acetal from isopropyl 4-formylacetoacetate provides the starting material for a dihydropyridine in which one of the methyl groups is replaced by a nitrile. Reaction of **40** with the enamine from isopropyl acetoacetate gives the corresponding dihydropyridine; hydrolysis of the acetal function with aqueous acid affords the aldehyde **41**. This function is then converted to its oxime by reaction with hydroxylamine; exposure to hot acetic acid leads to dehydration of the oxime and formation of a nitrile. Thus, **nilvadipine** (**42**)[10] is obtained.

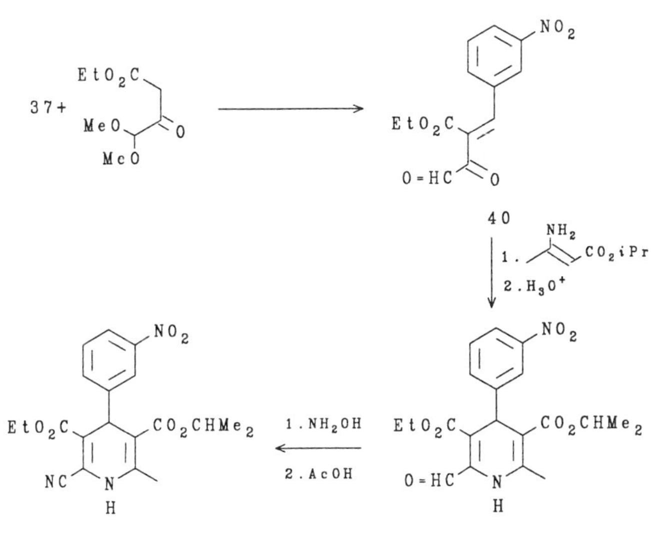

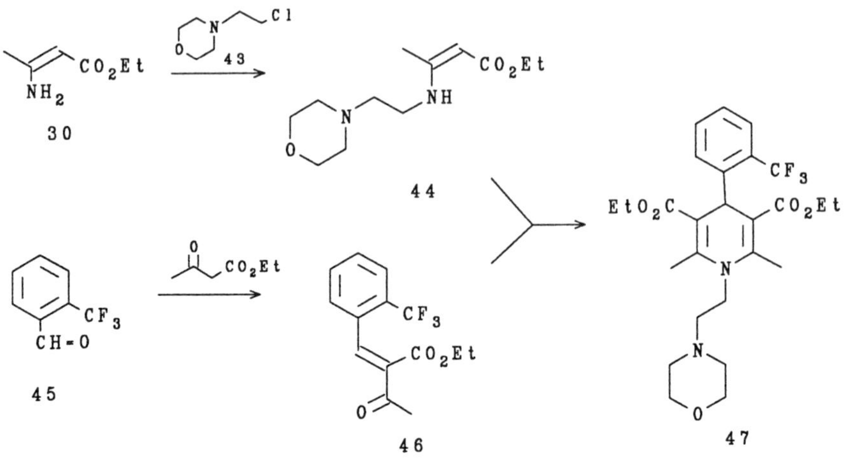

One approach to dihydropyridines substituted on ring nitrogen consists in simply alkylating the corresponding secondary amine; reaction of the dihydropyridine from the Hantzsch synthesis of *m*-trifluoromethylbenzaldehyde (**45**) with halide **43** in the presence of sodium hydride leads directly to **flordipine** (**47**). An alternate approach relies on the stepwise formation of the dihydropyridine ring. The convergent scheme involves alkylation of the enamine **30** from ethyl acetoacetate with the chloroethylmorpholine (**43**) to give intermediate **44**. Reaction of **44** with the aldol condensation product **46** from aldehyde **45** and acetoacetate again leads to flordipine (**47**).[11]

The synthesis of two rather complex dihydropyridines begins with Wittig condensation of phthalaldehyde (**48**) with the ylide from the triphenylphosphonium salt **49**. The trans stereochemistry of **50** prevails because the reaction is not carried out under salt free conditions; selectivity for monoalkylidation is probably due to steric hindrance from the newly introduced adjacent side chain at the remaining formyl group. Reaction of this intermediate with ethyl acetoacetate and ammonia gives the dihydropyridine **lacidipine** (**51**).[12] Further modification of this compound depends on the allylic nature of the ring methyl groups. Thus, reaction of **51** with pyridinium perbromide leads to bromination of one of these groups and the formation of **52**. Displacement of halogen by dimethylamine leads the tertiary amine **taludipine** (**53**).[13]

C. Piperidines

As noted earlier, the piperidines and pyrrolidines, which are so frequently found in side chains of therapeutic agents, are not usually associated with pharmacophores; they simply serve as surrogates for open-chain tertiary amines. The series of 4-aryl or 4-heteroaryl piperidines, which are discussed below, however, constitute an important group of clinically useful psychotropic drugs. The dopamine antagonism which forms the basis for the activity of all these compounds, suggests that the piperidine ring may be part of a pharmacophore in this series.

Interestingly, the first agent of this class to be introduced in the clinic, **haloperidol** (**61**), shares the 4-phenylpiperidine structural fragment found in the central analgesic

48 + $Ph_3P^+CH_2CO_2tBu\ Br^-$ 49 — LDA → 50

50 — $CH_3COCH_2CO_2Et$, NH_3 → 51

51 — $PyH^+Br_3^-$ → 52

52 — $HNMe_2$ → 53

54 — 1. $HOCH_2CH_2OH$; 2. $SOCl_2$; 3. NH_3 → 55

55 — LAH → 56

agent meperidine and its derivatives (see Chapter 7). The former compound may well have been discovered in the course of further SAR studies on the opiate.[14] An unusual synthesis of haloperidol (**61**) starts with the product **54** from Friedel–Crafts acylation of fluorobenzene with succinic anhydride. Successive protection of the ketone as its acetal by reaction with ethylene glycol, conversion of the carboxyl group to acid chloride, and finally ammonolysis gives the amide **55**. This compound is then reduced with lithium aluminum hydride to the primary amine **56**.

Construction of the piperidine begins with conjugate addition of methyl acrylate to the primary amino group to give the diester **57**. Dieckmann condensation of **57** leads to formation of the β-ketoester **58** and thus to cyclization. Heating **58** in aqueous methanol leads to hydrolysis to the ketoacid; this compound quickly decarboxylates to give the piperidone **59**. Condensation with the Grignard reagent from *p*-bromochlorobenzene gives the tertiary carbinol **60**. Hydrolysis with aqueous acid leads to removal of the acetal protecting group and formation of **haloperidol** (**61**).[15] It has been found that this drug, which is widely used for the treatment of schizophrenia, is a competitive inhibitor at dopamine receptor sites in the brain, diminishing the effect of excess neurotransmitter, or receptor supersensitivity, associated with the disease. The series of often serious side effects caused by this agent seem to be a consequence of its dopamine antagonism.

A benzimidazolone moiety replaces the aryl group on the piperidine ring in a somewhat more complex antipsychotic drug. The starting piperidone (**62**) for this sequence is available by a sequence analogous to the one outlined for preparing **58**; that is, Michael addition of benzylamine with ethyl acrylate followed by Dieckmann cyclization. Reaction of this product with *o*-phenylenediamine (**63**) leads directly to the imidazolone **65**. This seemingly complex transform can be rationalized by assuming that the first step consists in formation of the imine **63**. The carboxylate group then migrates to nitrogen by a process analogous to decarboxylation of β-ketoacids; the accompanying double-bond rearrangement leads to formation of the endocyclic olefin (**64**). Attack on the carboxylate by the adjacent amine leads to the cyclic urea with the net effect of formation of unsaturated benzimidazolone (**65**). Catalytic reduction over palladium leads to the loss of the benzyl protecting group and formation of **66**. Somewhat more strenuous condition lead directly to the corresponding saturated piperidine **67**.

Alkylation of the basic amino group in **66** with butyrophenone (**68**), available from acylation of fluorobenzene with 4-chlorobutyryl chloride (**66**), affords the antipsychotic drug **droperidol** (**69**).[16] Alkylation of the reduced intermediate **67** with side chain **70** yields **pimozide** (**71**).[17]

The presence of a quaternary carbon atom at the 4-position of the piperidine in the form of a spiro substituent seems to enhance potency. The starting piperidine **72**, whose preparation is described in Chapter 7, is the product from formal addition of cyanide and aniline to the 4-position of *N*-benzyl-4-piperidone. Reaction of **72** with formamide serves to form the spiroimidazoline ring. The benzyl protecting group in **73** is removed by hydrogenolysis over palladium to give the secondary amine **74**. Alkylation of the *p*-fluoro analogue of **74** with 4-chlorobutyrophenone affords **fluspiperone** (**75**). Reaction with the bisfluorophenyl side chain **70** gives **fluspiriline** (**76**).[18]

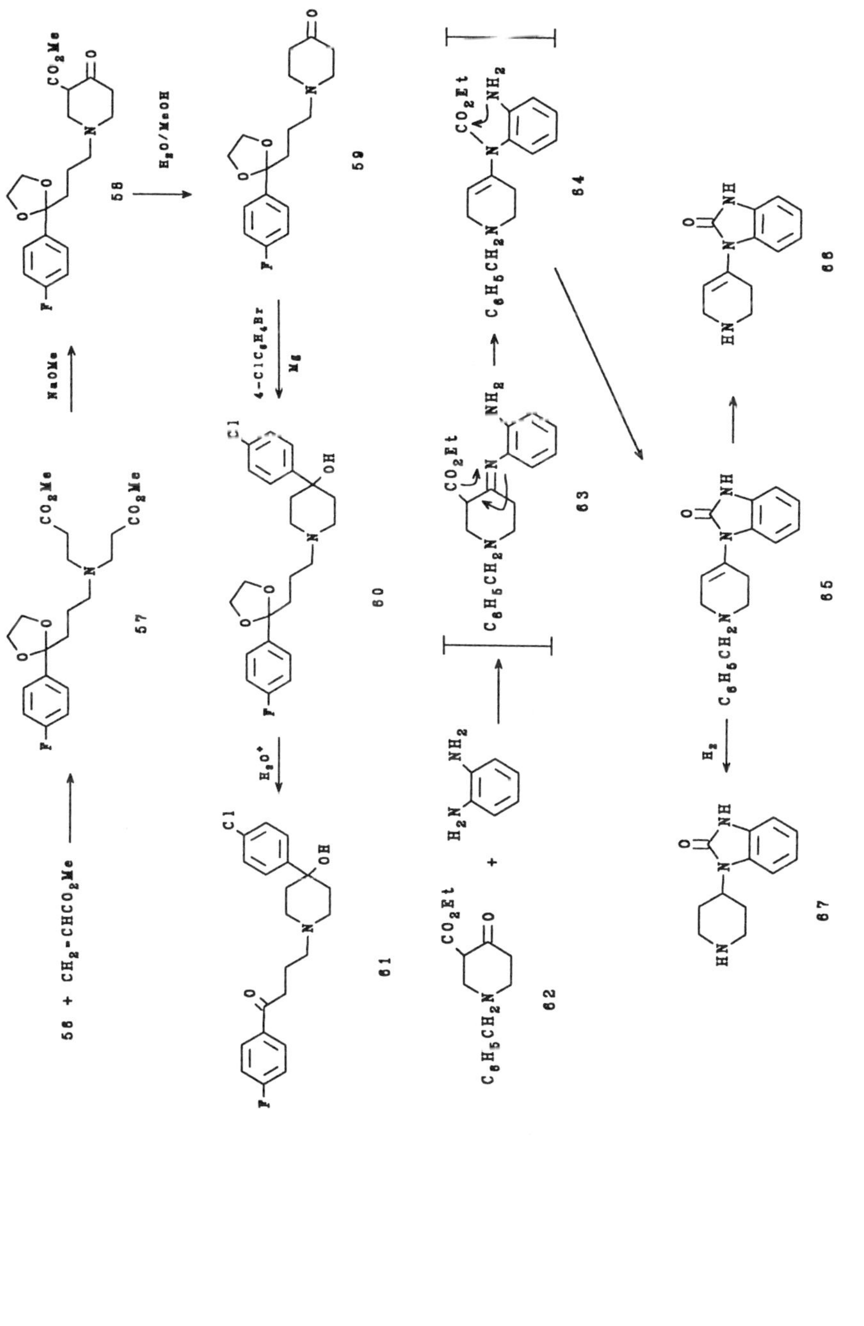

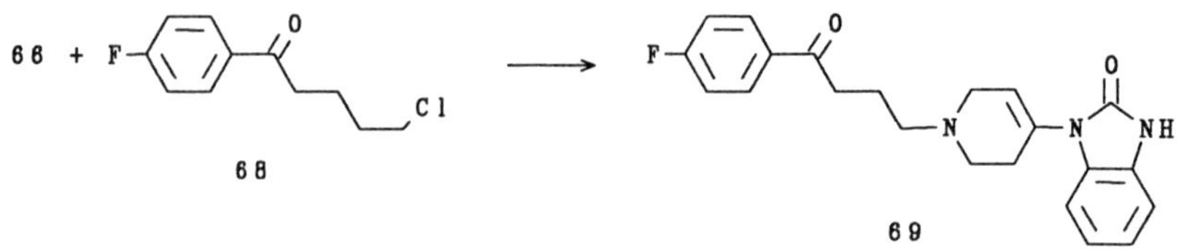

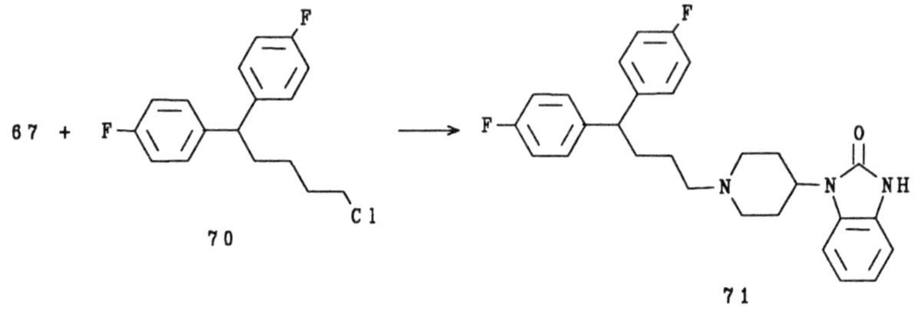

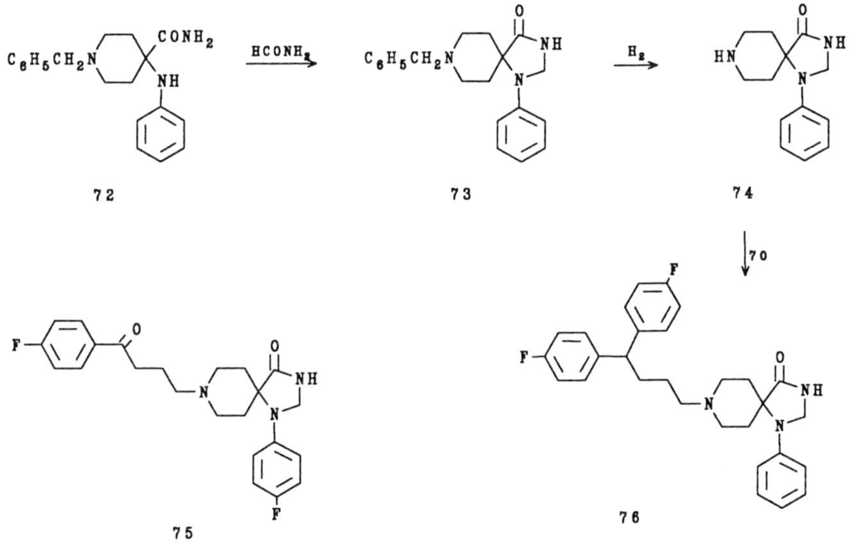

66 + F ... Cl → 69

68

67 + 70 → 71

72 —HCONH₂→ 73 —H₂→ 74 —70→ 76

75

D. Pyridones and Glutarimides

Decrease in the force of contraction of the heart muscle is one of the symptoms that marks heart failure. Digitoxin and related steroid glycosides constitute one of the oldest and, until quite recently, most effective means of increasing cardiac contractile force. These drugs do, however, show a very narrow therapeutic index and thus require careful patient monitoring. The science of pharmacokinetics may have owed its origin at least in part to need for monitoring blood levels of these drugs. The two pyridones that follow show clinical useful cardiotonic activity on oral administration without the narrow therapeutic activity of the so-called cardiac glycosides; a much larger series of pyrazinones that show the same activity are discussed later in this chapter.

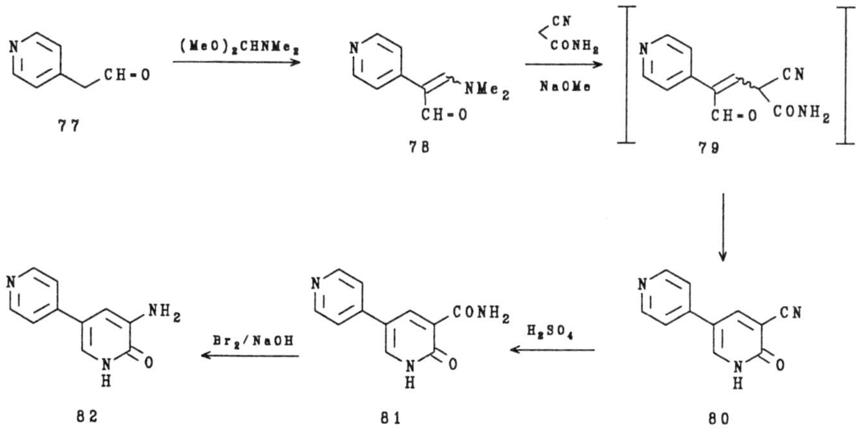

Reaction of pyridyl-4-acetaldehyde (**77**) with DMF acetal gives the corresponding aminoformyl derivative **78**. The first step in the reaction of this intermediate with the anion from cyanoacetamide can be envisaged as involving replacement of the dimethylamino group to form **79**, probably by an addition–elimination sequence. Imine formation of amide nitrogen with the formyl group will then lead to cyclization and formation of the pyridone ring in **80**. Hydrolysis with strong acid then gives the corresponding amide (**81**). Treatment of **81** with bromine in the presence of base leads to a classical Hofmann rearrangement of the amide to an amine with loss of the carbonyl carbon. Thus, the cardiotonic drug **amrinone** (**82**)[19] is obtained.

An analogous scheme is used for the synthesis of the cardiotonic agent **milrinone** (**85**).[20] The key aminoformyl derivative **84** is obtained in this case by condensation of 4-pyridylacetone (**83**) with DMF acetal. Reaction with cyanoacetamide in the presence of strong base proceeds exactly as above to give the pyridone **85** by a parallel set of transformations.

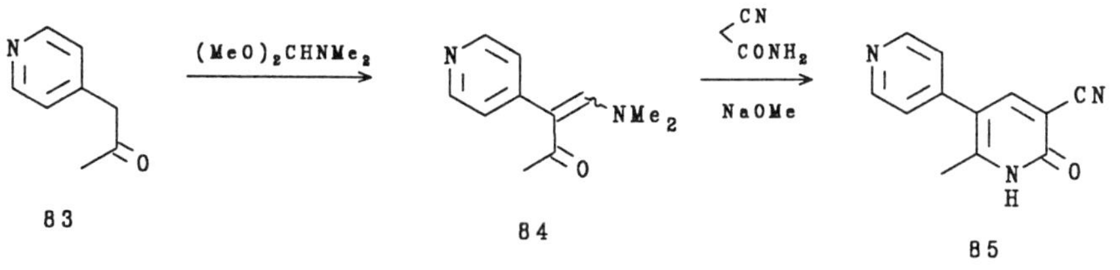

The utility of barbituric acid derivatives as sedative–hypnotic agents is discussed later in this chapter. Studies on chemical simplification of these pyrimidinetrione derivatives led to the discovery that pyridinediones, or glutarimides, also acted as sedative–hypnotic agents. The synthesis of the parent compound in this series, **gluthethimide** (**88**), starts by conjugate addition of the product **86** from ethylation of phenylacetonitrile with ethyl acrylate to give the cyanoester **87**. Heating **87** in acetic acid leads to formation of the glutarimide, glutethimide (**88**),[21] probably via initial hydrolysis of the nitrile to an amide. Prior nitration of the acetonitrile gives the analogue **89**. This compound yields the nitro glutarimide **90** when subjected to the two-step sequence. Reduction of the nitro group by catalytic reduction affords **aminoglutethimide (91)**.[22]

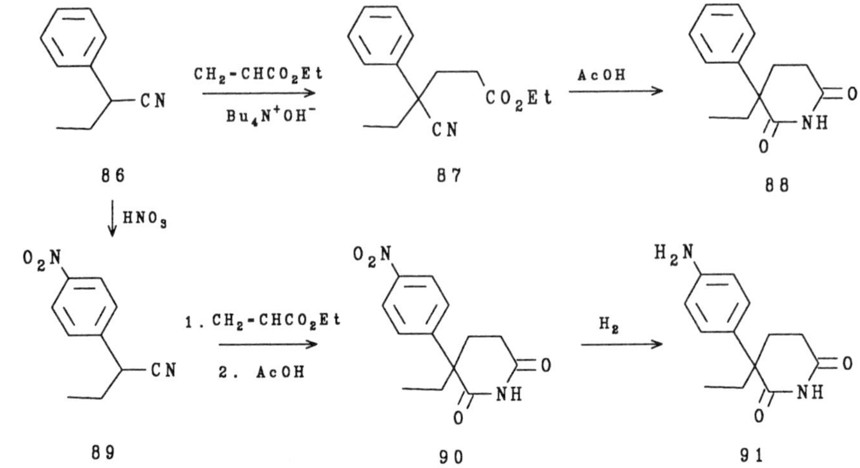

The steroid aromatase inhibiting activity of aminoglutethimide (**91**), which was discovered quite adventitiously in the clinic, led to some use of this drug in the treatment of breast cancer. The very modest enzyme inhibiting activity displayed by the drug led to a search for more potent analogues. The derivative in which a pyridyl group replaces the aniline moiety was designed specifically as an aromatase inhibitor. The synthesis of this agent, which is analogous to those described above involves conjugate addition of the ester **92** to acrylamide to give the ester–amide **93**. Treatment of **93** with sodium ethoxide leads to imide formation to give **rogletimide (94)**.[23]

No volume that deals with therapeutic agents would be complete without at least mentioning the ill-fated notorious sedative–hypnotic agent **thalidomide (96)**. It may be noted that the finding of this drug's surprising teratogenic activity when administered to pregnant women has led to many of today's regulations for drug testing and licensing. The compound was in fact prepared in quite straightforward manner by

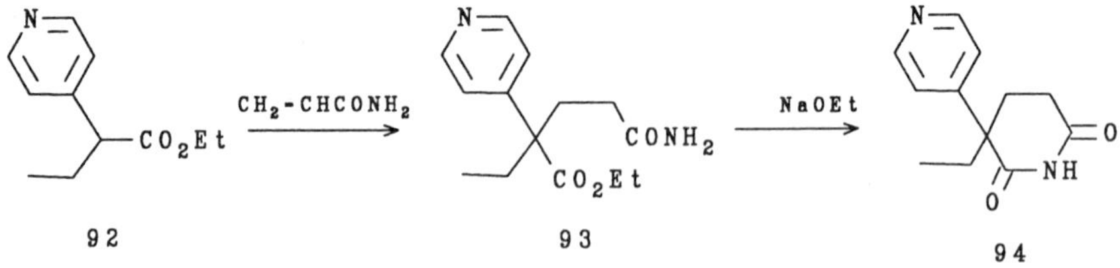

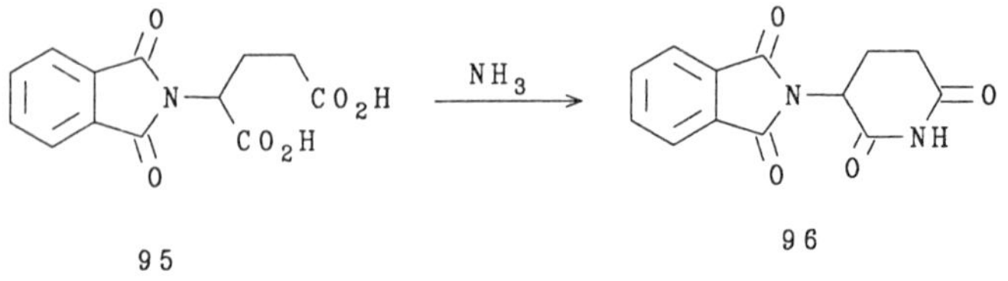

97 $\xrightarrow[\text{NaOEt}]{\text{HCO}_2\text{Et}}$ 98 $\xrightarrow{\text{NH}_3}$ 99 $\xrightarrow{\text{heat}}$ 100 $\xrightarrow[\text{NaHSO}_3]{\text{CH}_2\text{=O}}$ 101 $\xrightarrow{\text{H}_2}$ 102

reaction of the phthalimide (**95**) of (*RS*)-glutamic acid with ammonia. More recent work suggests that while both isomers are a sedative, the teratogenic effect is due solely to a single enantiomer.

Moving one of the carbonyl groups away from the amide function markedly reduces the acidity of the proton on nitrogen. It is interesting that sedative–hypnotic activity is retained in the face of this change. The synthesis of this agent, **methyprylon** (**102**), starts with the formylation of the doubly ethylated acetoacetate **97** by means of ethyl formate and sodium ethoxide to give **98**. Reaction of the formyl derivative, shown as the hydroxymethyl tautomer with ammonia, converts the product to its enamine **99**; this compound cyclizes to the pyrideinedione **100** on heating. Treatment of **100** with formaldehyde in the presence of sodium bisulfite leads to the carbinol **101** by aldol condensation. Catalytic reduction first leads to loss of the allylic hydroxyl by hydrogenolysis; reduction of the ring double bond then gives methyprylon (**102**).[24]

2. RINGS WITH TWO HETEROATOMS

A. Pyridazines

One approach to the treatment of hypertension consists of overcoming the vascular resistance due to contracted arterioles by use of vasodilators. The structures of such agents quite frequently incorporate the elements of an amidine or guanidine function. The preparation of the antihypertensive agent **hydracarbazine** (**108**), which may be viewed as containing an embedded amidine, begins with the reaction of ketoacid **103** with hydrazine. Sequential formation of a hydrazide and hydrazone, although not necessarily in that order, leads to the pyridazinone **104**. Treatment of **104** with bromine in acetic acid results in the introduction of a double bond and formation of the derivative **105**, which has aromatic character. The chemically inert nature of the ring is illustrated by the fact that exposure to potassium dichromate leads to oxidation of the ring methyl group to a carboxylic acid and thus formation of **106**; compound **106**

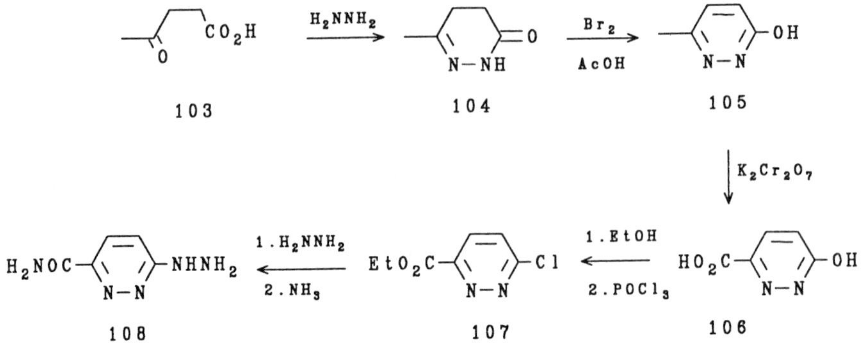

H_2NNH_2 Br_2 / AcOH $K_2Cr_2O_7$ 1. EtOH 2. $POCl_3$ 1. H_2NNH_2 2. NH_3

103 104 105 106 107 108

is converted to the corresponding ester by treatment with ethanolic acid. A standard reaction of heterocyclic enols consists in their conversion to chlorides by means of phosphorus oxychloride. Displacement of chlorine from the thus obtained intermediate **107** by hydrazine, followed by treatment with ammonia to form the amide, yields **hydracarbazine** (**108**).[25]

The side effects that accompany the use of vasodilators, such as fluid retention and increase in heart rate, may be due to sympathetic reflex effects; these effects can be diminished by coadministration of β-blockers. The antihypertensive agent **prizidilol** (**113**) may be viewed as an analogue of hydracarbazine (**108**) with a built-in β-blocker moiety. The key compound chloropyridazine (**111**) is prepared by a sequence quite analogous to the one shown above. This reaction involves formation of the pyridazinone ring from **109** followed by dehydrogenation to **110**, which is shown as its keto tautomer; treatment with phosphorus oxychloride then gives **111**. The oxypropanolamine side chain is then built onto the phenol function in **111** by the standard scheme (epichlorohydrin followed by *tert*-butylamine), which is discussed in detail in Chapter 2 to give **112**. Displacement of chlorine by hydrazine then gives prizidilol (**113**).[26]

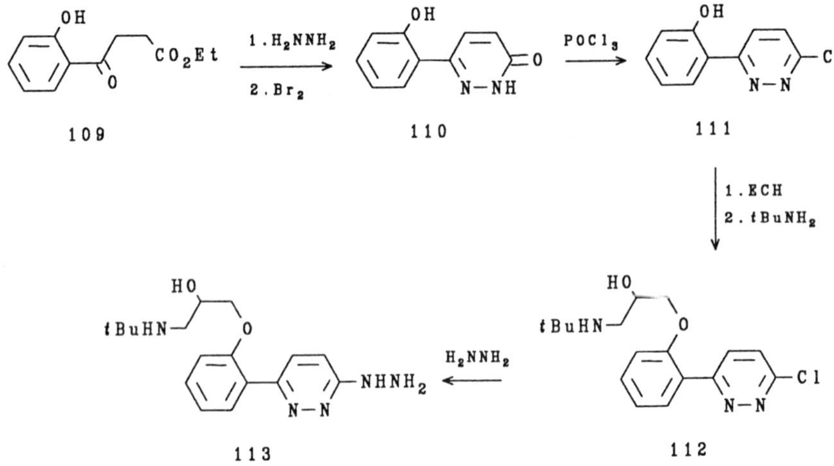

1. H_2NNH_2 2. Br_2 $POCl_3$ 1. ECH 2. tBuNH$_2$ H_2NNH_2

109 110 111 112 113

HO—N—N—OH POCl3 → Cl—N—N—Cl

114 115 116

117

The pyridazine ring provides the nucleus for a structurally atypical analgesic that acts by an unspecified mechanism. The starting material (**114**) is in fact simply the dienol tautomer of maleic hydrazide; treatment with phosphorus oxychloride leads to the dihalide **115**. Displacement of a single chlorine with cinnamyl piperazine (**116**) leads directly to **lorcinadol** (**117**).[27]

In a similar vein, the pyridazine based antidepressant agent **minaprine** (**122**) also departs in structure from most other drugs in this class. Friedel–Crafts acylation of benzene with itaconic anhydride (**118**) leads to ketoacid **119**. Condensation with hydrazine leads to formation of the hydrazine and hydrazide bonds; the double bond shifts into the ring to give the fully unsaturated pyridazinone **120**; this compound is then converted to the chloride (**121**) in the usual way. Displacement of halogen by the amine on 3-(*N*-morpholino)ethylamine affords minaprine (**122**).[28]

The role of the pyridone ring as a pharmacophore in cardiotonic agents such as amrinone (**82**) has been noted earlier; very analogous activity is obtained with

C_6H_6, $AlCl_3$ → CO_2H H_2NNH_2 → N—NH

118 119 120

$POCl_3$

H_2N N O ←

122 121

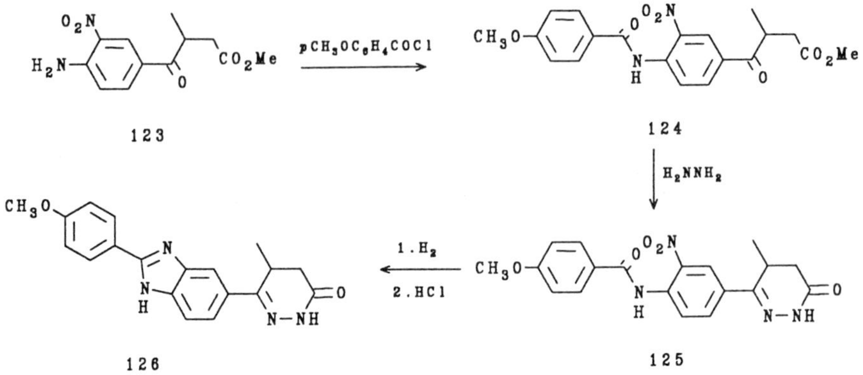

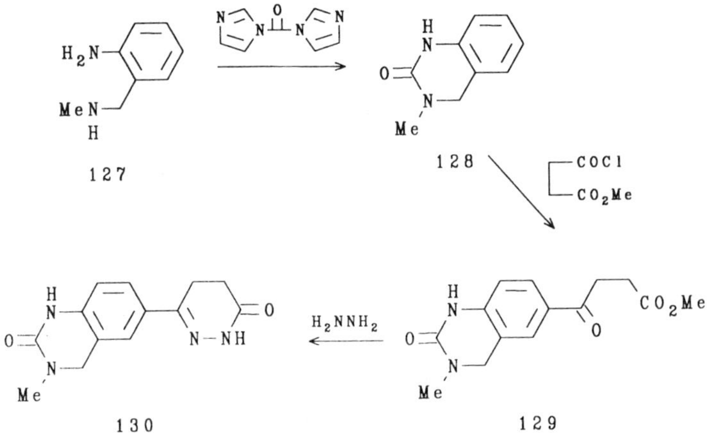

pyridazinone derivatives. The synthesis of the first of these agents, **pimobendan (126)**, starts with the acylation of the amino group in the nitroaniline **123** with anisoyl chloride to give the amide **124**. The ketoacid function is then converted to a pyridazinone (**125**) by treatment with hydrazine. Catalytic hydrogenation then converts the nitro group into an amine; cyclization of the resulting ortho amino amide by means of strong acid leads to formation of the corresponding benzimidazole. Thus, pimobendan (**126**)[29] is obtained.

The quinazolinone moiety **128** for the cardiotonic agent **prinodoxan (130)** is formed by reaction of the diamine **127** with carbonyl dimidazole. Friedel–Crafts acylation of the product with the half-acid chloride from methyl succinate gives the corresponding ketoester **129**. The pyridazinone **130** is then obtained by condensation of **129** with hydrazine.[30]

B. Pyrimidines

The era of modern antibiotics relies very heavily on metabolic differences between prokariotic and eukariotic species. As noted in Chapter 2, the selective toxicity of

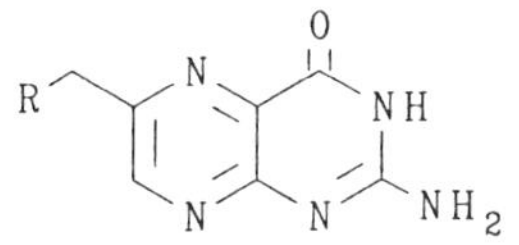

R = Glutamoyl PABA

131

sulfonamides, for example, is due largely to the fact that bacteria obtain folates by de novo synthesis, while mammals obtain these compounds from the diet. Folic acid (**131**) in fact needs further transformation to give the active compound; the pyrazine ring in **131** is thus converted by two sequential reductions to its tetrahydro derivative, folinic acid. A series of pyrimidine based antibiotics depend on the fact that these compounds show preferential binding and consequent antagonism to bacterial over mammalian folate reductase enzymes. The pyrimidine ring in these drugs presumably mimics the corresponding fused ring of folic and dihydrofolic acid.

Base-catalyzed condensation of 4-chlorophenylacetonitrile with ethyl propionate leads to the corresponding acylated derivative, which is shown as its enol tautomer **132**. Reaction of **132** with diazomethane, a reagent not likely to be used in commercial production, leads to the enol ether **133**. Construction of the pyrimidine ring then involves a fairly standard scheme that consists of condensation of a 1,3-dicarbonyl component, or its equivalent, with an amidine containing moiety. In the case at hand, reaction of **133** with guanidine can be visualized as proceeding to a transient interme-

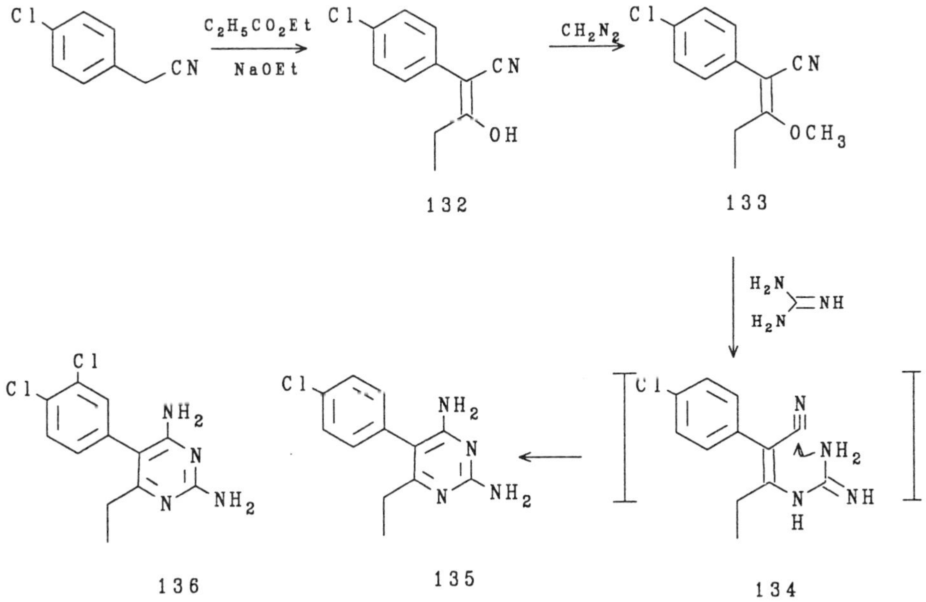

$C_2H_5CO_2Et$ / NaOEt → 132; CH_2N_2 → 133; $(H_2N)_2C{=}NH$ → [134] → 135

132 133 136 135 134

diate such as **134** by addition–elimination to the enol ether. Addition of the primary amine to the nitrile will lead to cyclization. Bond rearrangement to give the aromatic isomer leads to the antibacterial agent **pyrimethamine** (**135**)[31]. The somewhat selective toxicity to cancer cells of many antineoplastic drugs depend on the faster rate of growth of these cells. Antimetabolites have thus been intensively investigated for this application. The folate reductase antagonist **etoprine** (**136**), which is prepared by an analogous route, has been investigated clinically because of its appreciable activity against the mammalian enzymes.

Interposition of a methylene group between the two rings leads to agents that show somewhat higher selectivity for bacterial enzymes. The preparation of one of these agents, **trimethoprim** (**141**), starts by condensation of trimethoxybenzaldehyde (**137**) with 3-ethoxypropionitrile. The first step in the reaction with guanidine probably involves replacement of the allylic ethoxy group in **138** by a guanidine nitrogen to give an intermediate such as **139**. This compound then undergoes the usual cyclization reaction (**140**) and bond reorganization to give trimethoprim (**141**).[32] Compound **141** finds extensive use in fixed combinations with the sulfonamides achieving improved antibacterial efficacy by inhibiting sequential steps in the synthesis of folates.

A somewhat different route is used to prepare an analogue that bears additional oxygen. In this case, the sequence starts by base-catalyzed formylation of the hydrocinnamic acid derivative **142** with ethyl formate. Condensation of the product **143** with guanidine leads to a pyrimidone (**144**), with the cyclization involving ester–amide interchange between guanidine and the ester. Reaction of **144** with phosphorus oxychloride leads to formation of the chlorinated derivative **145**. Displacement of halogen by ammonia gives the antibacterial compound **tetroxoprim** (**146**).[33]

A compound that includes the aminopyrimidine ring and quaternary salt present in thiamine shows preferential inhibition of absorption of that cofactor by *coccidia* over uptake by vertebrates. The compound is thus used in poultry, where coccidiosis is an economically important disease. Condensation of ethoxymethylenemalononitrile (**147**) with an amidine leads to the aminopyrimidine **149**, probably via the intermediate addition–elimination product **148**. The nitrile group is then reduced to the methylamino derivative **150** by means of lithium aluminum hydride. Exhaustive methylation, for example, by reaction with formaldehyde and formic acid followed by methyl iodide leads to the quaternary methiodide **151**. The quaternary salt is then displaced by bromine, and the resulting allylic halide **152** is replaced by 2-picoline. Thus, **amprolium** (**153**)[34] is obtained.

Excessive levels of thyroxin, for example, in Grave's syndrome, can lead to a host of serious problems. These levels can be brought back to normal by the rather simple thiopyrimidine **propylthiouracil** (**156**), better known by its acronym **PTU**. This drug is available in a single step from reaction of the substituted acetoacetate **154** with thiourea.[35] Although the order of the two required steps is not clear, the sequence can be visualized by assuming the intermediacy of the ester amide interchange intermediate **155**.

One of the most potent known hypotensive agents incorporates an *N*-oxide function in addition to the amidine moiety usually associated with vasodilators. It was adventitiously noted in the course of clinical trials for the treatment of refractory hyperten-

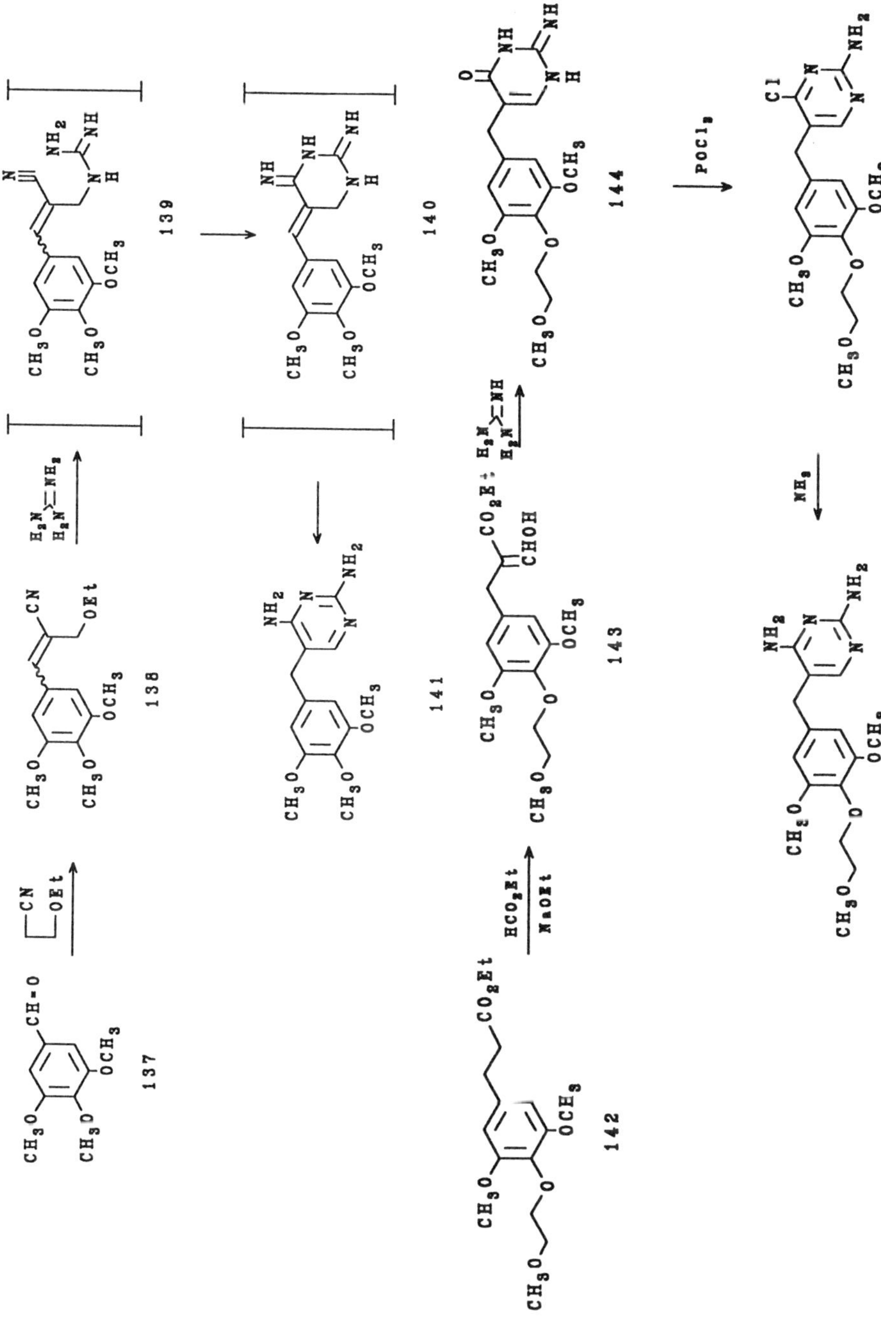
137
138
139
140
141
142
143
144
145
146
POCl3
NH3
HCO2Et
NaOEt

147 → 148 → 149 —LAH→ 150 —1. HCHO, HCO_2H; 2. MeI→ 151 —NaBr→ 152 → 153

154 + $H_2NC(S)NH_2$ → 155 → 156

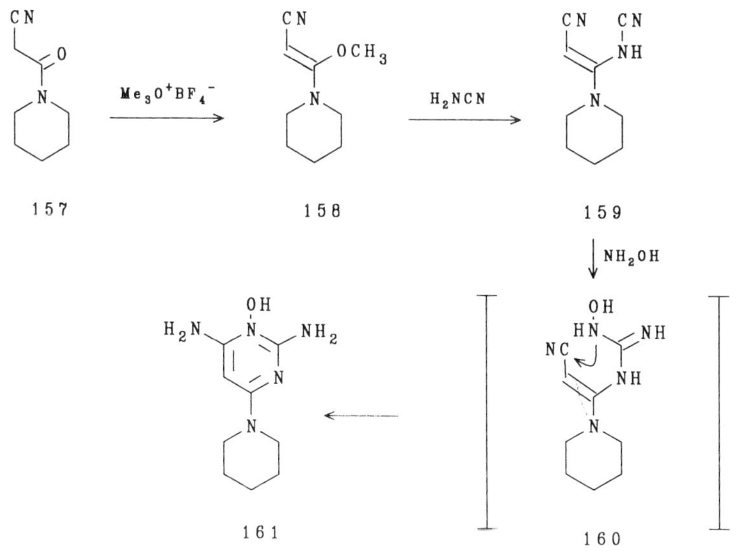

sion that use of the compound stimulated unusual new hair growth. This finding was confirmed by clinical observations after the drug was approved as an antihypertensive. Further studies showed that the hair growth stimulation could be obtained in appropriate cases when the drug was administered topically; an important consideration for an agent that exerts such profound effects on systemic administration. The compound **minoxidil** (**161**) is now available as Rogaine® (FDA rejected Regain as a trade name) as a nonprescription drug for treatment of male pattern baldness. An expeditious synthesis starts with the cyanomalonamide **157**; this compound is converted to its methyl enol ether **158**, for example, by reaction with trimethyloxonium fluoroborate. Reaction of this intermediate with cyanamid leads to the imine **159**. Condensation with hydroxylamine can occur at either cyano group; addition to cyanamid nitrile will lead to the formation of the hypothetical hydroxyguanidine **160**. Addition to the remaining nitrile leads to minoxidil (**161**) after bond rearrangement.[36]

C. Pyrimidine Nucleosides

Viruses differ from bacteria and fungi in a most fundamental way in that they are not able to reproduce independently. A virion basically consists of a chain of DNA, or RNA in the case of a retrovirus, packaged with a small group of specialized proteins. The virus actually replicates by taking over the infected cells' reproductive apparatus, in effect causing the cell to synthesize new virions. Chemotherapy of viral disease must thus rely on very subtle biochemical differences between normal and infected cells instead of the large divergences in biochemistry between microbial and host cells, which form the basis of antibiotics. The somewhat looser discriminatory power of the enzymes in viral cells for the nucleotides involved in replication has made it possible to identify a number of closely related false substrates; the resulting antagonists have provided the bulk of today's antiviral agents; additional compounds are discussed in Section 3 on triazines and on purine nucleosides in Chapter 12.

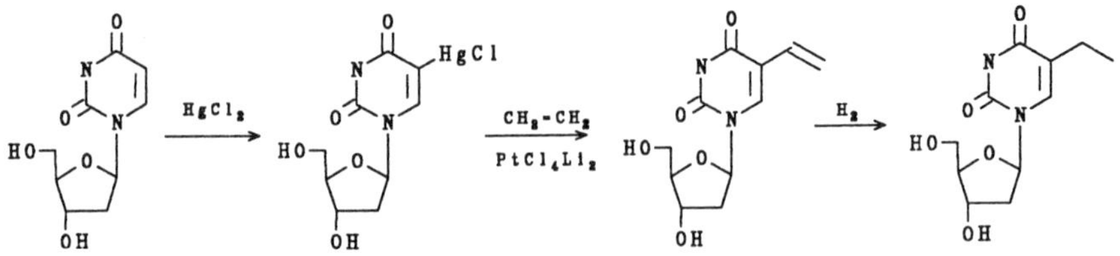

The nucleotide uridine (**162**) provides the starting material for one of these agents; treatment of **162** with mercuric chloride leads to mercuration of the pyrimidine ring to afford **163**. Reaction of this organometallic derivative with ethylene in the presence of a platinum salt leads to the alkylated product **164**. Catalytic hydrogenation of the double bond then affords the antiviral compound **edoxudine** (**165**).[37]

A small series of unusual virus, most notably the human immunodeficiency virus (HIV), which is the cause of AIDS, incorporate an RNA rather than a DNA chain. The first step in viral replication thus involves translation of the genome into DNA by an enzyme designated reverse transcriptase (transcriptase regulates the more usual translation of DNA to RNA, which precedes protein synthesis). This enzyme has proven a fruitful source for anti-HIV drugs, a number of which are nucleosides. The first step in the synthesis of the anti-HIV drug **zidovudine** (**168**), more familiarly known as AZT, involves reaction of thymidine (**166**), shown as the enol tautomer with the powerful dehydrating agent diethylaminosulfur tetrafluoride (DAST). The stereochemistry of the resulting cyclic ether **167** strongly suggests that the initial step involves transformation of the sugar hydroxyl to a good leaving group; this group is then displaced by the pyrimidine enol with consequent inversion. Treatment of the intermediate with sodium azide leads to ring opening and a second inversion step. Thus the reverse transcriptase inhibitor zidovudine (**168**)[38] is obtained.

Nucleosides that incorporate unusual substituents can also show cytotoxic activity in higher organisms by a mechanism analogous to the one that leads to antiviral activity. Such compounds have been investigated as antineoplastic agents in the expectation that they would be selectively cytotoxic in the quickly dividing less regulated cancer cells. Thus, inversion of the hydroxyl group to the 2′α-position of

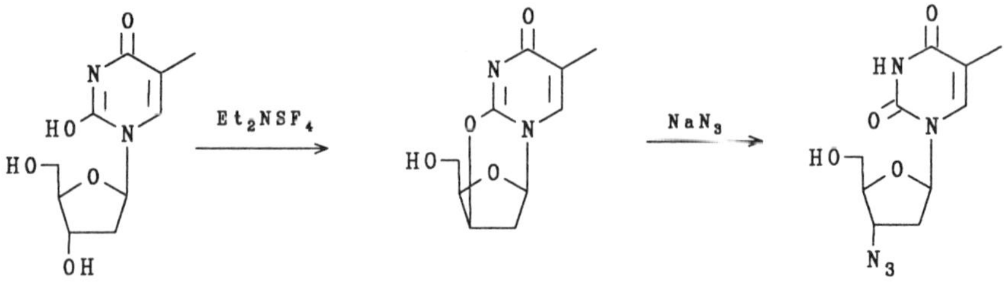

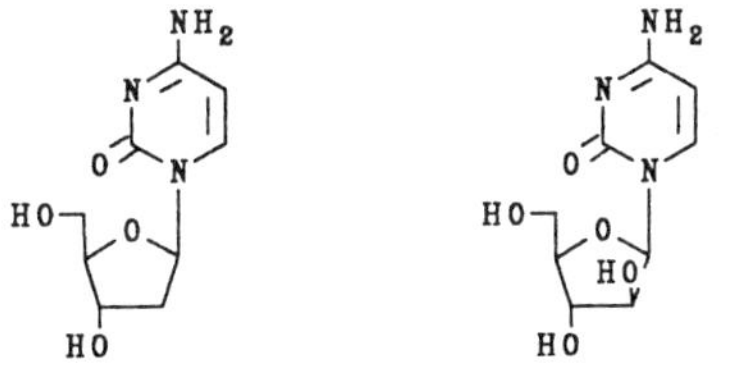

169 170

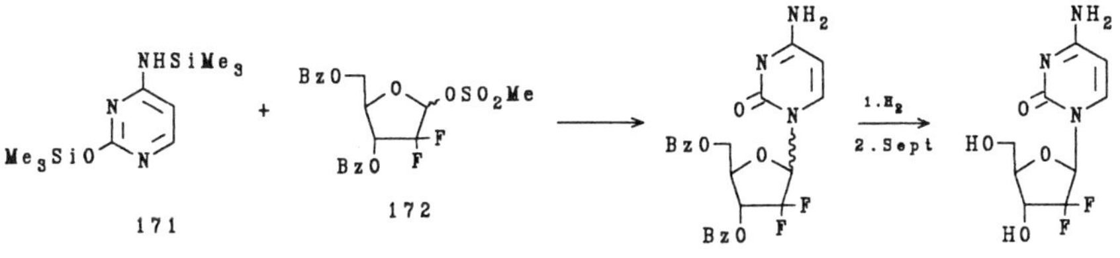

171 172 173 174

$Bz - C_6H_5CH_2$

cytidine leads to the anticancer drug cytidine arabinoside, **cytarabine** (**170**). Replacement of both protons at the 2-position by fluorine similarly leads to an antineoplastic agent. Several decades ago it was found that conversion of pyrimidine bases to their silyl ethers enhanced reactivity of ring nitrogen. Thus, exposure of the bis-silyl ether **171** from 4-aminopyrimidine-2-one to the anomeric mesylate from the protected fluorosugar **172** leads to formation of **173** as a 1:1 mixture of anomers. The benzyl protecting groups are then removed by catalytic hydrogenation; separation of the

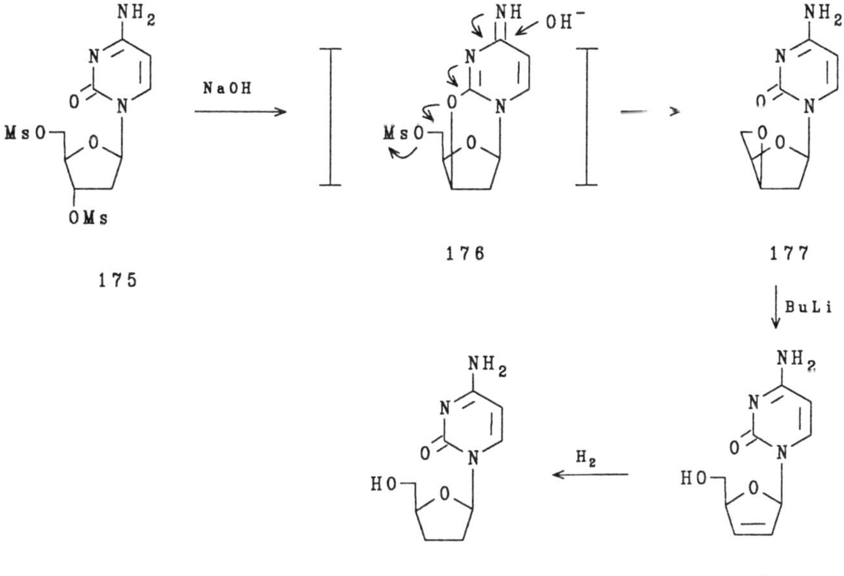

175 176 177 178 179

compound that corresponds to the anomer with the natural β configuration affords **gemcitabine** (**174**).[39]

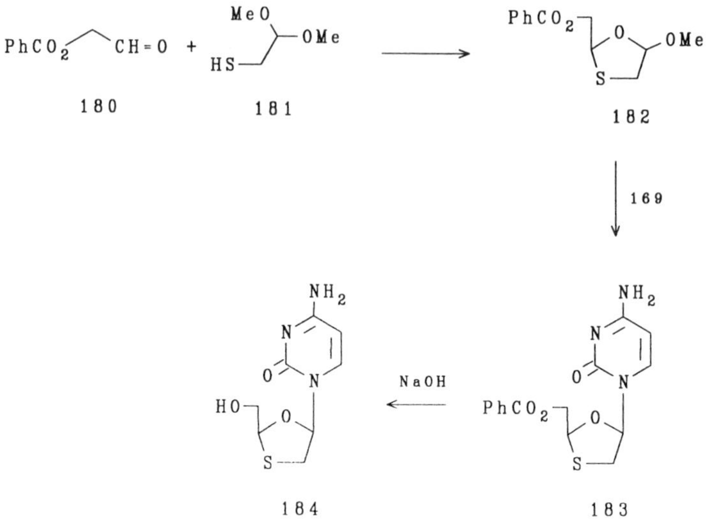

The 2′,3′-dideoxy derivative of cytidine, DDC, also shows clinical anti-HIV activity. The synthesis starts by conversion of 2′-deoxycytidine, **169** to its bis-methanesulfonate ester **175**. Reaction of the mesylate with sodium hydroxide leads to the fused oxetane ether **177**. This reaction consists formally of hydrolysis of the 5′ mesylate to an alcohol followed by backside displacement of the 4′-mesylate. Evidence suggests, however, that the first step involves formation of a bridged cyclic ether such as **176**; addition of hydroxide to the imine system would result in breaking the bridge bond; the resulting alkoxide would then displace the adjacent mesylate to form **177**. Reaction of **177** with strong base leads to ring opening of the oxetane and formation of the double bond (**178**). Catalytic hydrogenation completes the synthesis of the reverse transcriptase inhibitor **zalcitabine** (**179**).[40]

The isostere in which sulfur replaces the methylene group at the 4′-position in the furan ring also provides an anti-HIV reverse transcriptase inhibitor. The initial step consists in formation of the thioacetal **182** from glyoxal benzoate **180** and the methyl acetal of thioglyoxal **181**. Reaction of **181** with the silyl ether **169** predominantly leads to the desired anomer **183** of the glycosylated pyrimidine. The benzoate protecting group is then removed by treatment with an ion exchange resin to afford **lamivudine** (**184**).[41]

D. Pyrimidones

The great majority of antiulcer compounds fall neatly into the category of histamine H_2 antagonists or inhibitors of the sodium–potassium pump, which drives gastric acid secretion. A relatively simple pyrimidone, on the other hand, does not fit either category. This compound inhibits acid secretion in animal models and also interest-

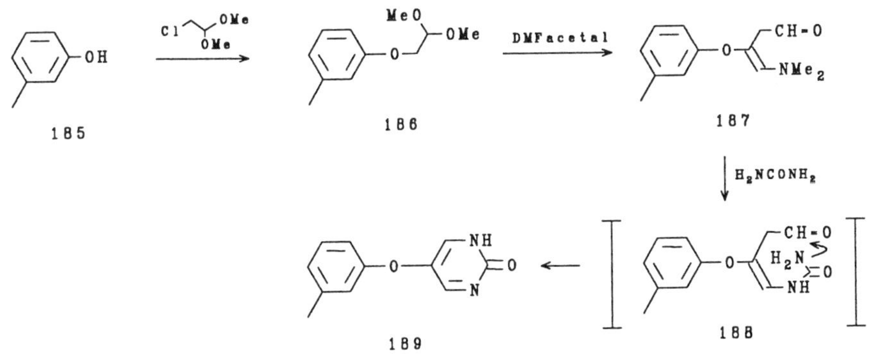

185 186 187 188 189

ingly acts as a bronchodilator in histamine challenged animals. The synthesis of this agent begins with the alkylation of *m*-cresol (**185**) with the dimethyl acetal from chloroacetaldehyde. Reaction of **186** with DMF acetal affords the formylation product **187** on workup. Treatment of **187** with urea can be visualized as proceeding to **188** by displacement of the dimethylamino group with urea nitrogen. Ring closure with the aldehyde carbonyl results in formation of the pyrimidone ring. Thus **tolimidone** (**189**)[47] is obtained.

The key to changing the activity of histamine related compounds from agonists to antagonist, as noted in Chapter 9, involved modifying the side chain amine to a thiourea-like function. The large amount of work devoted to H_2 blockers revealed that guanidine function embedded within a 2-aminopyrimidone would serve as a surrogate thiourea, as exemplified by oxmetidine, which is found in Chapter 8. Base-catalyzed condensation of picoline-5-aldehyde (**190**) with ethyl malonate gives the corresponding acrylic ester; the initial adduct decarboxylates in the course of the reaction. Catalytic hydrogenation then affords the propionic ester **191**. Treatment with ethyl formate in the presence of sodium ethoxide affords the formylated product **192**. The reaction of the resulting 1,3-dicarbonyl compound with nitroguanidine proceeds in a manner analogous to the one described above for urea to give the pyrimidone **193**. The nitro group transforms the amino group at the 2-position into a leaving group; thus

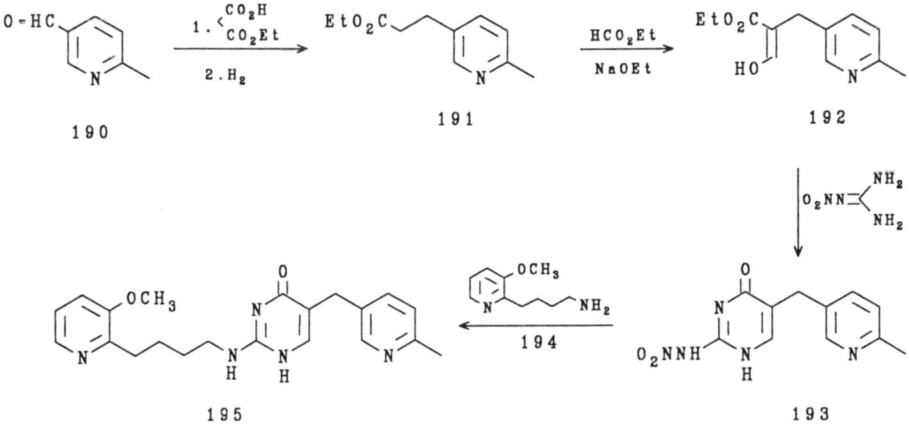

190 191 192 193 194 195

displacement by the primary amine in pyridylbutylamine (**194**) affords the alkylation product and thus **icotidine** (**195**).[43]

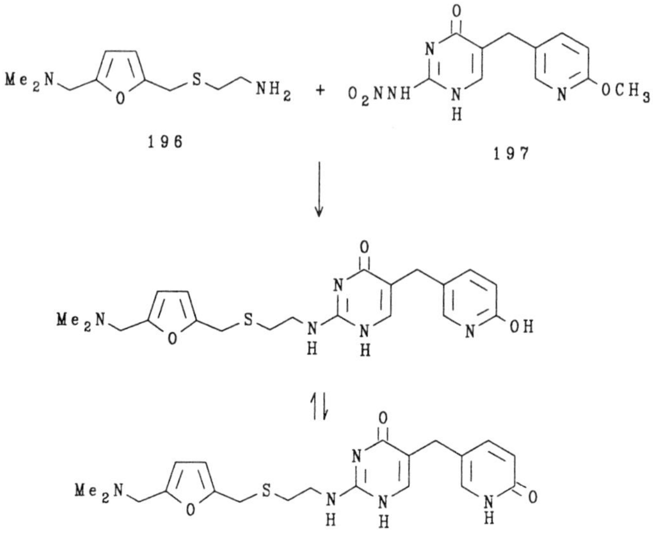

In a similar vein, displacement of the nitrated amino group in pyrimidone (**197**), prepared in the same way as **193**, by using the methoxyl pyridinealdehyde with the ranitidine nucleus **196** affords the H_2-blocker **donetidine** (**198**)[44] after O-demethylation.

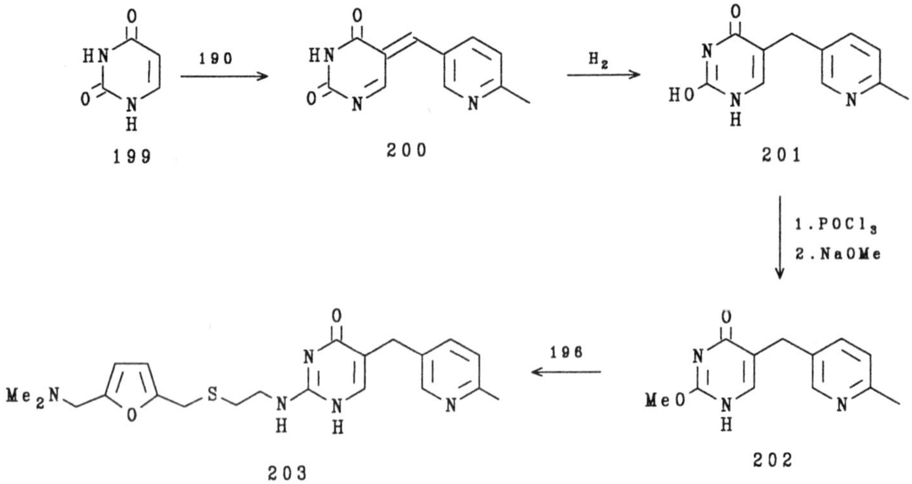

An alternate approach to the preparation of the side chain involves reaction of picoline-5-aldehyde with the pyrimidone **199**. Benzal-like condensation of the aldehyde with the enamide function in **199** results in formation of the bicyclic product **200**, which is shown as its enolate; the double bond is then removed by catalytic hydrogenation. The enol hydroxyl is then converted to the chloride in the usual way by treatment with phosphorus oxychloride; displacement of halogen by means of sodium methoxide gives the corresponding methyl ether **202**. Displacement of **202** by the terminal amino group in the ranitidine nucleus **196** gives the antiulcer agent **lupitidine** (**203**).[45]

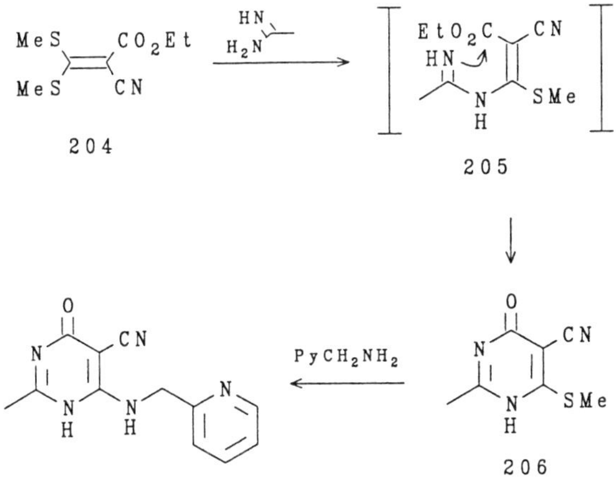

A pyridone obtained by formal incorporation of an additional ring nitrogen in the pyridone ring of the cardiotonic agent milrinone (**80**), along with some juggling of the side chain, exhibits the same activity. The reactive starting material **204** is available by base-catalyzed condensation of ethyl cyanoacetate with carbon disulfide followed by treatment of the intermediate with methyl iodide. Reaction of the product with acetamidine can be visualized as involving replacement of one the thiomethyl groups by amidine nitrogen. The initial step would thus yield the addition–elimination product **205**. Attack on the ester carbonyl by the second amidine nitrogen will lead to the pyrimidone intermediate **206**. Reaction of **206** with 2-methylaminopyridine then leads to replacement of the thiomethyl group by the more basic amine. This affords the cardiotonic agent **pelrinone** (**207**).[46]

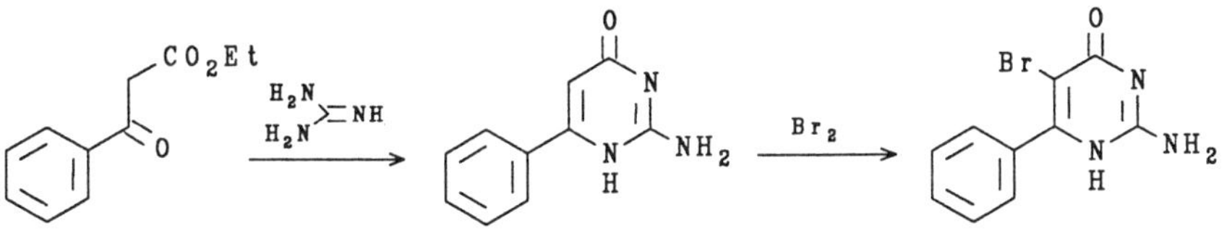

A relatively simple pyrimidone, **bropirimine** (**210**), has been extensively studied as an antitumor agent and immune modulator as a consequence of its unusual activity as an interferon inducer. Condensation of ethyl benzoylacetate (**208**) with guanidine leads to the pyrimidone **209** by the usual sequence. Treatment of **209** with bromine leads to bromination at the sole open position on the heterocyclic ring yielding bropirimine (**210**).[47]

E. Triketopyrimidines: The Barbiturates

The barbiturate sedative–hypnotic agents probably constitute the oldest class of medicinal agents whose synthesis was not prompted by some biologically active natural product. The first compound in this series, **barbital** (**213**) ($R^1 = R^2 = Et$), has been in continuous use since at least 1903. The persistent use of these compounds is somewhat puzzling in view of the now well-recognized shortcomings. The drugs possess significant abuse potential, which readily crosses over with alcohol abuse. The quality of sleep induced by these drugs is held to be poor and resistance to the hypnotic potential seems to develop relatively quickly in most individuals. All active barbiturates fit the general formula **213**, with obligatory substituents at both R groups. The general method for the preparation of these compounds involves base-catalyzed condensation of a malonic ester (**212**), or its cyanoacetate equivalent with urea. An alternative consists of reaction with guanidine to form the imino derivative **214** followed by acid hydrolysis. The same reaction with thiourea gives thiobarbiturates such as **215**. These last compounds are quickly metabolized into inactive species, making them ideal for short-term administration.

The straightforward step used to form the ring system means that the chemistry involved in the preparation of the scores of barbiturates that have been used clinically in fact devolves on the malonic ester synthese (**211** → **212**). It should be noted that little success has been achieved in changing the side-effect spectrum of these drugs. The main differences between the various agents involves their pharmacokinetic properties; these properties in turn are manifested as variances in bioavailablity by parenteral and oral routes as well as in time to onset and duration of action.

F. Pyrazines and Piperazines

Only a very few therapeutic agents are based on the pyrazine ring, in marked contrast to the large number of entities that contain a pyrimidine ring. The antitubercular

$R^1CH(CO_2Et)_2$ (211) $\xrightarrow[NaOEt]{R^2X}$ $R^1R^2C(CO_2Et)_2$ (212) $\xrightarrow[NaOEt]{H_2NCONH_2}$ 213

212 $\xrightarrow[NaOEt]{H_2NCSNH_2}$ 215

212 $\xrightarrow{(H_2N)_2CNH}$ 214 $\xrightarrow{H_3O^+}$ 213

211 212 213 215 214

216 → (CH-O / CH-O) → 217 → ($KMnO_4$) → 218 → (1. heat; 2. EtOH/H^+) → 219 → (NH_3) → 220

antibiotic **pyrazinamide** (**220**) probably acts by a similar mechanism as its pyridine parent **isoniazide**. The tonnage chemical, *o*-phenylenediamine (**216**) provides a convenient route to pyrazines. Thus, condensation of this diamine with glyoxal leads to quinoxaline (**217**). Treatment of the heterocycle with a strong oxidant such as permanganate leads to selective cleavage of the benzene ring and formation of the dicarboxylic acid **218**. Heating of **218** leads to loss of one of the carboxyl groups. The resulting carboxylic acid (**219**) is then converted to its ethyl ester with ethanolic hydrogen chloride; amide ester interchange with ammonia then affords pyrazinamide (**220**).[48]

One shortcoming of the thiazide diuretics is their tendency to cause excessive loss of potassium. The so-called potassium sparing diuretics, the most important of which is based on a pyrazine ring, are often used when loss of potassium is a potential problem. The dicarboxylic acid **218**, described above, provides the starting material for this drug as well. Treatment of the ester obtained from esterification of the acid with ammonia affords the corresponding bis-amide **221**. Exposure of **221** to a single equivalent of bromine in aqueous base leads to selective Hoffmann rearrangement of but one of the two amide groups and formation of the amino–amide **222**. Alcoholysis of **222** leads to conversion of the carboxyl group to its ester; reaction with sulfuryl chloride results in chlorination of the two open-ring positions to the dichloro compound **223**. Reaction of **223** with ammonia leads to displacement of the halogen para to the carboxyl group and formation of the corresponding diamine; ammonia concur-

218 → (1. EtOH/H^+; 2. NH_3) → 221 → (Br_2, NaOH) → 222 → (1. EtOH/H^+; 2. SO_2Cl_2) → 223 → (NH_3) → 224 → (H_2N–C(=NH)–NH_2) → 225

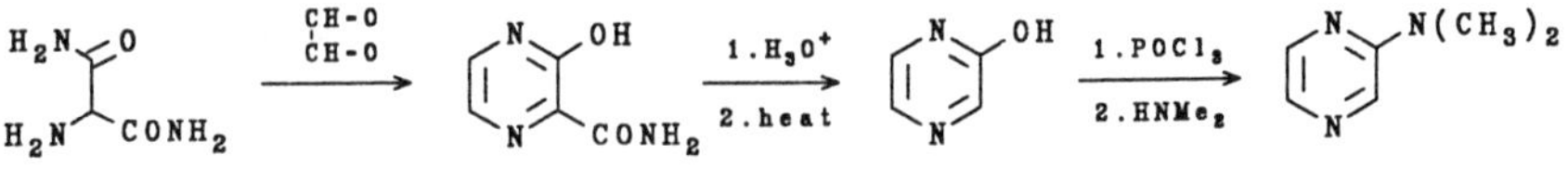

rently converts the ester back to the amide to afford **224**. Treatment of **224** with guanidine leads to formation of an acyl–guanidine function by an exchange reaction. Thus, **amiloride** (**225**)[49] is obtained.

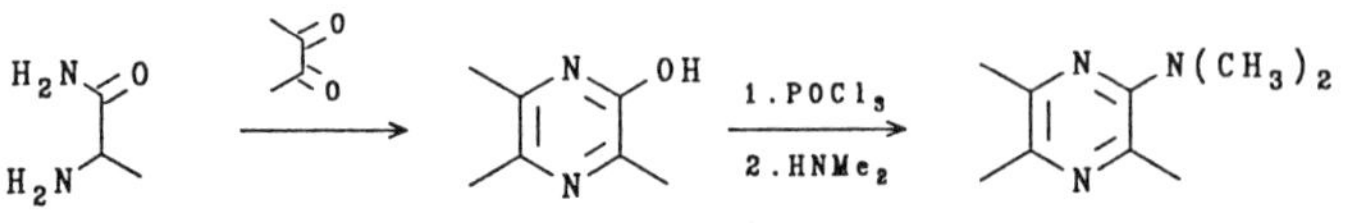

Amides that bear an amino group at the 2-position provide a more direct route to less highly substituted pyrazines. Thus, reaction of 2-aminomalonamide (**226**) with glyoxal leads in a single step to the pyrazine **227**. The superfluous carboxyl is removed in this case by first hydrolyzing the amide to its corresponding acid and then thermolyzing that intermediate to give **228**. The hydroxyl group is then converted to the chloride by means of phosphorus oxychloride. Displacement of chlorine with dimethylamine gives **ampyzine** (**229**).[50] This compound is described as a central CNS stimulant.

The synthesis of the trimethyl analogue of **229** represents the most direct approach to the pyrazine system. In this case, condensation of (*R,S*)-alaninamide (**230**) with diacetyl gives the pyrazinol **231** directly. The hydroxyl group is then converted to the corresponding dimethylamine by the same sequence as above.[51] The product, **triampyzine** (**232**), shows predominantly anticholinergic activity.

Piperazines, in contrast to pyrazines, abound in medicinal agents; this moiety quite often occurs in side chains serving as a surrogate tertiary amine or as an ethylenediamine. A number of these compound have already been covered in passing earlier chapters. The examples that follow have been chosen to illustrate the varied synthetic strategies that have been used to incorporate these structural fragments. In addition, the "spirone" anxyolytic agents are included in some detail since the common

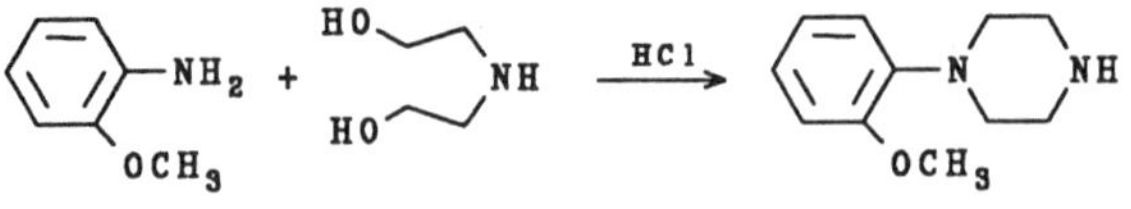

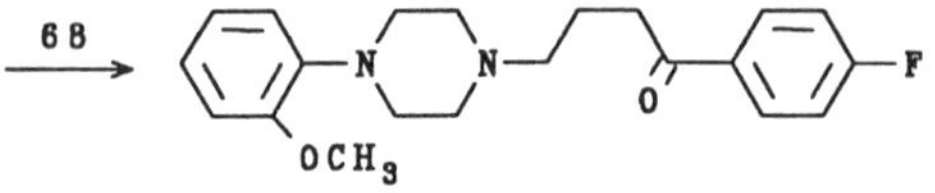

BzN NH $\xrightarrow{ClCH_2CO_2Et}$ BzN N—CO_2Et

237 238

$Bz{-}C_6H_5CH_2{-}$; $R{-}3,4,5{-}MeO_3C_6H_2CH{-}CH{-}$

$\xrightarrow[2.\ RCOCl]{1.\ H_2}$ CH_3O, CH_3O, CH_3O —CH=CH— O N N—CO_2Et

239

occurrence of the piperazines in this class suggests that the moiety may in this case form part of a pharmacophore.

The piperazine ring in the antipsychotic agent **fluanisone** (**236**) apparently serves the same role as the piperidine in the prototype butyrophenone **haloperidol** (**61**). Reaction of *o*-anisidine (**233**) with diethanolamine (**234**) and hydrogen chloride in all probability proceeds by alkylation of the aniline by the nitrogen mustard, which is formed in situ. Alkylation of **235** with the haloperidol intermediate **68** affords fluanisone (**236**).[52]

Selective reaction of but one of the nitrogens on piperazine can be assured by the use of protecting groups. Thus, alkylation of the benzyl derivative **237** with ethyl chloroacetate gives the alkylation product **238.** The protecting group is then removed by hydrogenation over palladium. Acylation of the thus obtained secondary amine with 3,4,5-cinnamoyl chloride affords **cinepazet** (**239**),[53] a compound that shows antianginal activity.

A somewhat more complex antianginal compound incorporates the xylylamide function of the local anesthetic-based antiarrhythmic agents as well as a moiety reminiscent of β-blockers. This synthesis in fact begins by ring opening of the oxirane in the familiar β-blocker intermediate **240** with piperazine to give the monoalkylation product **241**. (The fact that the resulting propanolamine side chain amine is tertiary make it questionable whether the compound has adrenergic activity.) Alkylation of the remaining free amine with ethyl chloroacetate then affords the intermediate **242**. Ester-amide interchange using 2,6-xylidine then gives **ranolazine** (**243**).[54]

The structures of the original venerable antihistamines often included an ethylenediamine side chain on the assumption that this would better mimic histamine itself. This side chain was modified to a piperazine in some of the more effective agents. Alkylation of the free amine in the monourethane **246** from piperazine (**245**) with benzhydryl chloride **244** gives the tertiary amine **247**; the protecting group is then removed by sequential saponification in base and treatment with mild acid. Alkylation of the intermediate **247** with 2-chloroethoxyethanol affords the potent H_1 antihistamine **hydroxyzine** (**248**).[55] One of the principal metabolites of this drug consists of the carboxylic acid from oxidation of the terminal alcohol. This product was found to be an orally effective antihistamine; the compound was further nonsedating since the polar acid group prevents its penetration into the CNS. This compound can be prepared

240

241

$ClCH_2CO_2Et$

242

243

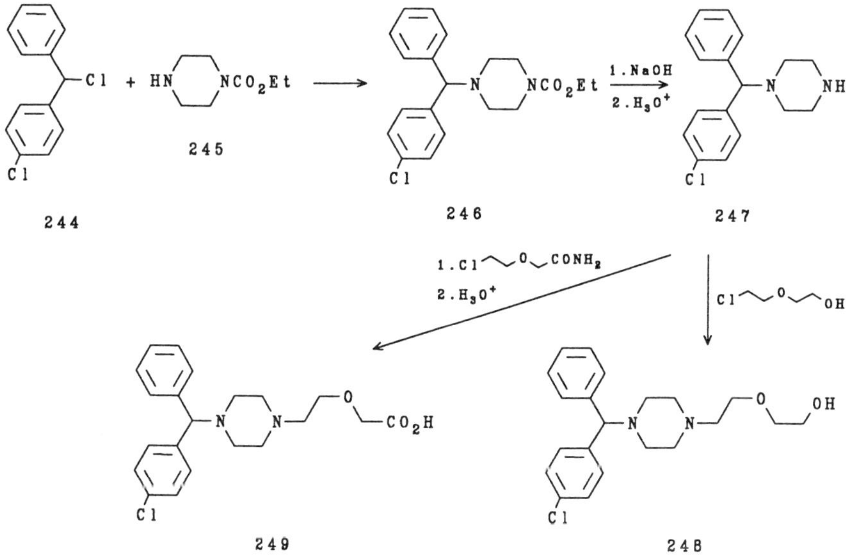

by alkylation of intermediate **247** by means of 2-chloroethoxyacetamide followed by hydrolysis of the amide group; this reaction gives the nonsedating antihistamine **cetirizine** (**249**).[56]

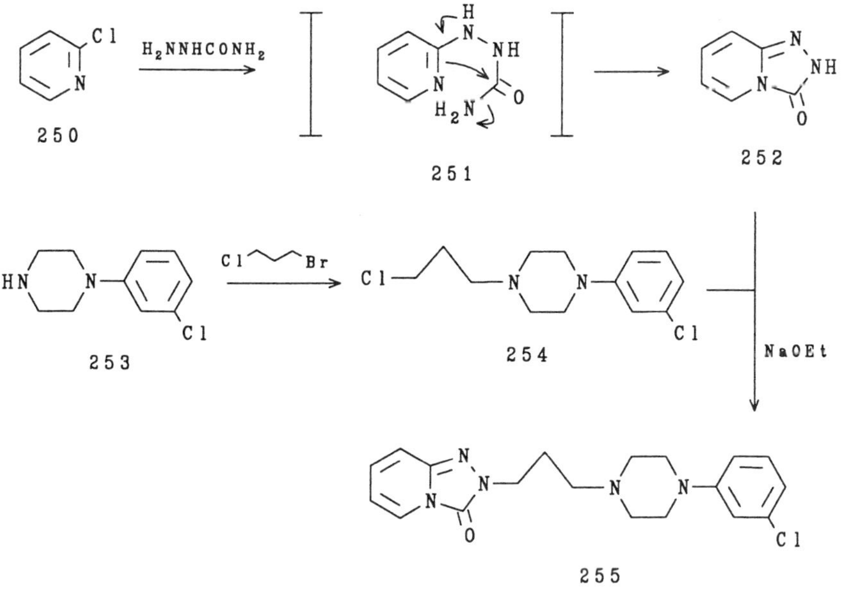

The classical tricyclic antidepressant drugs, which are discussed in Chapter 13, manifest a set of structure specific side effects that are thought to be due to their shared anticholinergic activity. Two of the first antidepressants agents that depart from the tricyclic motif, and do not manifest the side effects typical of the older drugs, are built around aryl piperazine moieties. Note that a three-carbon chain connects the piperazine fragment with the heterocyclic moiety in both these compounds. The seemingly complex bicyclic heterocyclic nucleus included in **trazodone** (**255**) is in fact constructed in a single step by reaction of 2-chloropyridine (**250**) with semicarbazone. The first step can be envisaged as formation of the displacement product **251**; displacement of amide amine by pyridine nitrogen will then afford **252**. The second half of the molecule is obtained by alkylation of *N*-(3-chlorophenyl)piperazine (**253**) with 3-bromo-1-chloropropane to afford **254**. Reaction of the anion from **252** with chloroalkylamine (**254**) leads to the alkylation product. There is thus obtained trazodone (**255**).[57]

Preparation of the closely related antidepressant **etoperidone** (**259**) illustrates a different approach to the piperazine ring. The first step in this synthesis involves alkylation of the anion from triazinone (**256**) with 3-bromo-1-chloropropane to afford **257**. Displacement of chlorine in this intermediate by the amino group in diethanolamine affords the corresponding bis-hydroxyethyl derivative; the hydroxyl groups are then converted to chlorides by treatment with thionyl chloride to give the nitrogen mustard **258**. Reaction of **258** with *m*-chloroaniline leads to piperazine ring formation. Thus, etoperidone (**259**)[58] is obtained.

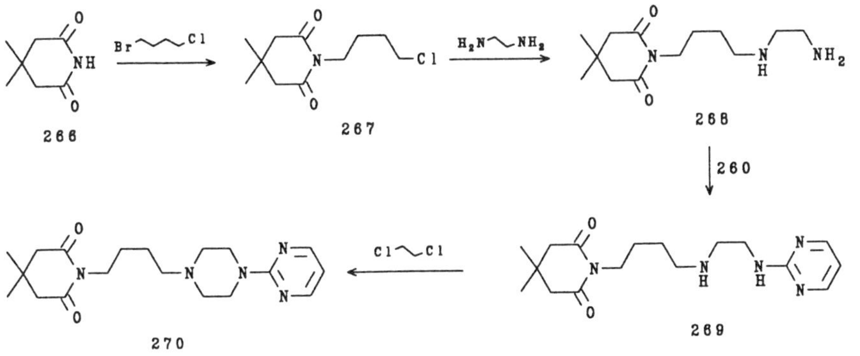

266 267 268 260 270 269

Piperazines also form an integral part of a series of relatively new anxiolytic drugs. These drugs act via a quite different mechanism than do the better known benzodiazepines, which are discussed in Chapter 12; they do not, for example, bind to benzodiazepine receptors and do not show typical side effects of the older tranquilizers. The preparation of the prototype **buspirone** (**265**) starts with displacement of chlorine in 2-chloropyrimidine (**260**) by piperazine to afford the monoalkylation product **261**. Alkylation of the remaining free amino group with 4-chlorobutyronitrile gives the nitrile **262**; the nitrile group is then reduced to the primary amine **263** with lithium aluminum hydride. The spiro cyclic glutaric anhydride **264** used in the next step can be obtained by an unusual aldol-like condensation of ethyl acetate with cyclopentanone; the second acetate fragment being incorporated by conjugate addition to the dehydration product of the initial aldol product; saponification followed by treatment with acetic anhydride completes the scheme. Reaction of primary amine **263** with the anhydride leads to formation of the corresponding glutarimide and thus buspirone (**265**).[59]

The strategy for preparation of a nonspiro cyclic analogue involves a string of alkylation reactions. Thus, treatment of 3,3-dimethyl glutarimide (**266**) with 4-bromo-1-chlorobutane gives the chlorobutyl intermediate **267**. Use of **267** to alkylate ethylenediamine gives the intermediate **268**; reaction with 2-chloropyrimidine leads to replacement of chlorine by the terminal primary amino group to give **269**. The central piperazine ring is then built by sequential alkylation of the two secondary amino groups

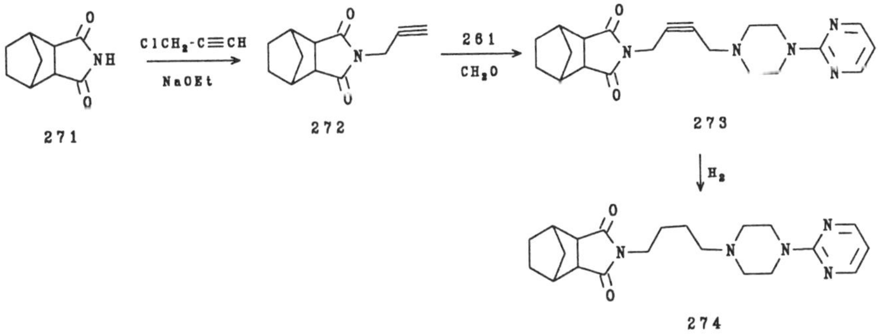

271 272 261 273 274

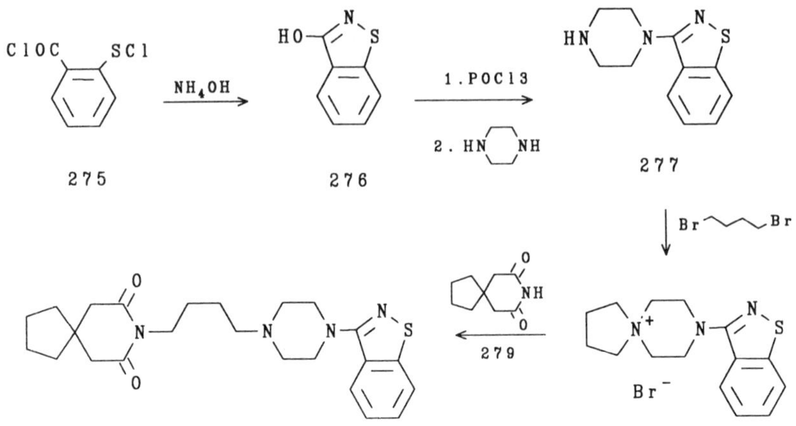

in **269** with ethylene dichloride. Thus, the anxiolytic agent **gepirone** (**270**) is obtained.[60]

The reduction product **271** from Diels–Alder addition of maleimide and cyclopentadiene provides the starting material for another analogue. Reaction of the anion from this compound with propargyl chloride gives the terminal acetylene **272**, a compound that now includes a reasonably acidic proton. The Mannich reaction provides the means for attaching the piperazine containing moiety as well as extending the connecting chain to the required four atoms. Thus, reaction of **272** with 2-pyrimidylpiperazine (**261**) and formaldehyde gives intermediate **273**. The acetylene group is then reduced by catalytic hydrogenation to afford **tandospirone** (**274**).[61]

Interestingly, a benzoisothiazole moiety can replace the terminal pyrimidine ring in this series. Reaction of thiosalicylic acid disulfide with thionyl chloride and then chlorine gives the dichloride **275**; this compound is cyclized with ammonium hydroxide to benzoisothiazolone (**276**), shown as its enol. The hydroxyl is then converted to chlorine in the usual way with phosphorus oxychloride; displacement of the newly introduced halogen by piperazine gives the monoalkylation product **277**. Alkylation with 1,4-dibromobutane then affords the spirocyclic quaternary salt **278** from double alkylation on the same amino group. Reaction of **278** with the anion from glutarimide **279** (from **264** and ammonia) leads to ring opening and consequently alkylation on imide nitrogen. Thus, **tiospirone** (**280**)[62] is obtained.

3. RINGS WITH THREE HETEROATOMS: THE TRIAZINES

The cytotoxic of nucleosides that contain "unnatural" substituents such as cytarabine (**170**) have, as noted previously, been used as antineoplastic agents. This activity may be enhanced by incorporation of an additional nitrogen atom in this compound, turning the pyrimidine ring into a 1,3,5-triazine. The first step in the synthesis of this agent consists of the standard glycosylation reaction of the triazine silyl ether **282** with the benzylated chloro arabinoside **281**. Catalytic reduction of the product **283** leads to removal of the benzyl protecting groups to give **284**. The triazine ring is, however, partly reduced as well during this reaction. The sensitivity of the final compound precludes direct oxidation to restore the double bond. Instead, the compound is converted to the silyl ether **285**; air oxidation followed by deprotection leads to the antitumor agent **fazarabine** (**286**).[63]

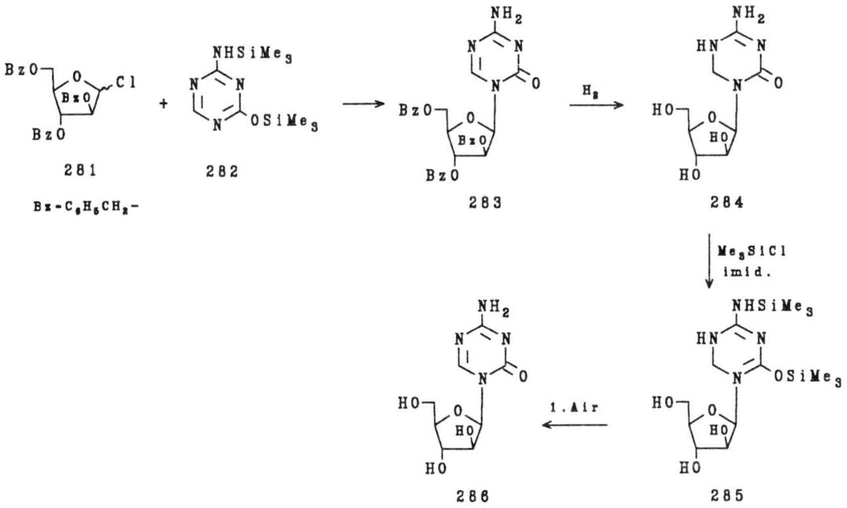

The route used to prepare the 3′-desoxy analogue relies on building the triazine ring onto the sugar. Thus, reaction of the benzoate protected chloro sugar **287** with silver

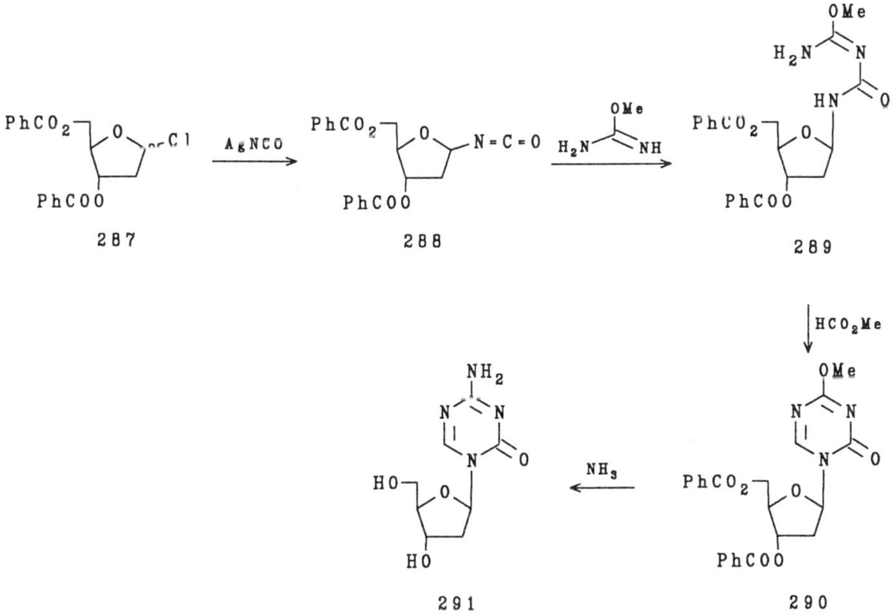

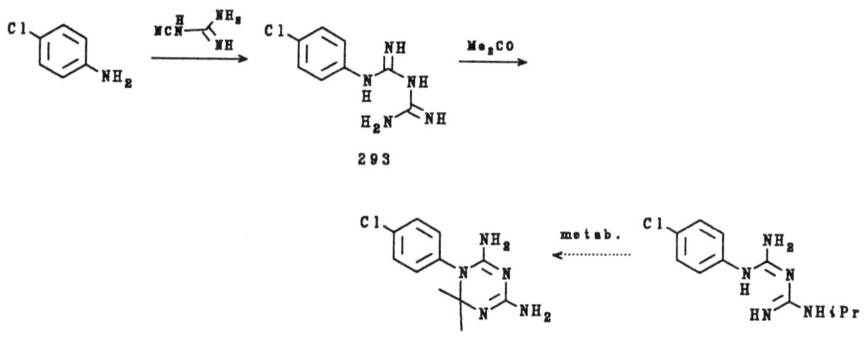

293

294

292

isocyanate affords the displacement product **288**, apparently mainly as the desired anomer. Condensation of **288** with the *O*-methylurea gives the product **289** from addition of the urea amino group of the cumulene carbonyl. Treatment of **289** with ethyl formate completes the elaboration of the triazine ring affording **290**. Ammonolysis of the product both replaces the ring methyl ether by an amino group and frees the hydroxyl groups by cleaving the benzoate esters. Thus, the antineoplastic agent **decitabine** (**291**)[64] is obtained.

A partly reduced 1,3,5-triazine, **cycloguanil** (**294**), shows similar folate reductase inhibiting activity to the amino pyrimidines exemplified by pyrimethamine (**135**); the compound **294** has been used as an antimalarial agent. The precursor to this drug was an open-chain biguanide, **proguanil** (**292**), discovered in the course of a screening operation; investigation of the metabolic fate of **292** revealed that the active agent was the oxidation product in which the methine carbon on the isopropyl group had cyclized to form a triazine. The route for preparing the triazine synthetically begins by

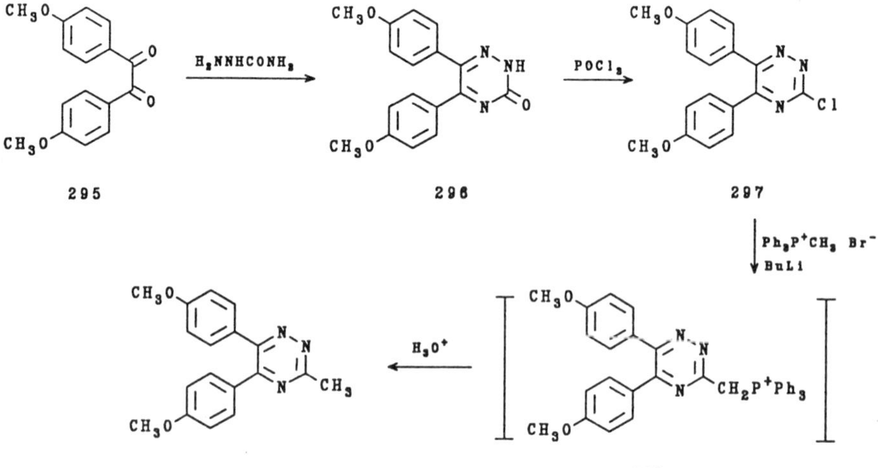

295

296

297

299

298

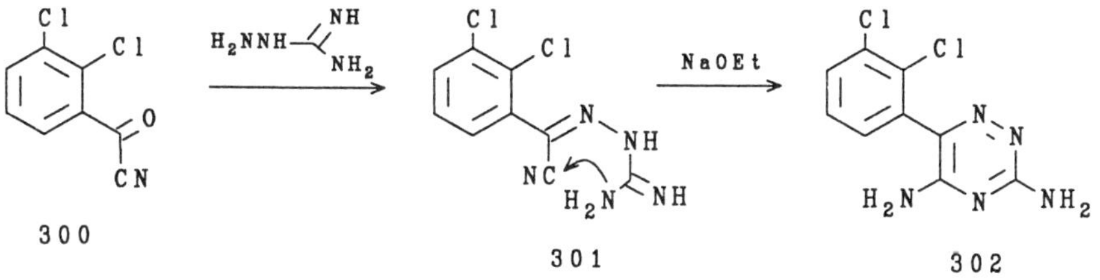

condensation of *p*-chloroaniline with dicyanamide to afford the biguanide **293**. Reaction of **293** with acetone results in formation of a cyclic aminal and thus cycloguanil (**294**).[65]

The cyclooxygenase inhibiting activity of 1,2-dianisyl heterocyclic compounds, for example, flumizole and itazagrel (see Chapter 8), is retained when the ring is expanded to a 1,2,4-triazine. The basic ring system (**296**) is obtained in a single step by condensation of the benzil **295** with semicarbazone. Reaction of this product with phosphorus oxychloride gives the corresponding chloro derivative **297**. This halogen is then replaced by a methyl group in a reaction characteristic of heterocyclic systems. Thus, treatment of **297** with the ylide from methyl triphenylphosphonium bromide gives the phosphonium salt **298**, produced by displacement of halogen by the anion on the methyl group. Hydrolysis of **298** leads to loss of phosphorus as triphenylphosphine oxide and consequent formation of **anitrazafen** (**299**).[66]

Preparation of the anticonvulsant agent **lamotrigine** (**302**) illustrates an alternate approach to 1,2,4-triazines. Condensation of acyl cyanide **300** with dicyanamide gives the imine **301** as the initial product. Treatment of **301** with base leads to addition of the guanidino anion to the nitrile and thus formation of the triazine ring.[67]

REFERENCES

1. Liberman, D.; Rist, N.; Grumbach, F.; Cals, S.; Moyeux, M.; Rouaix, A. *Bull. Soc. Chim. Fr.* **1958**, 687.
2. Hoffman, C.; Faure, A. *Bull. Soc. Chim. Fr.* **1966**, 2316.
3. Sherlock, M.H.; U. S. Patent **1974**; *Chem. Abstr.* **1975**, *82*, 16705.
4. Barth, W.E. German Offen. **1972**, 2204195; *Chem. Abstr.* **1972**, *77*, 151968.
5. Ash, A.B.; LaMontagne, M.P.; Markovac, A. U. S. Patent **1975**, 3886167; *Chem. Abstr.* **1975**, *83*, 97041.
6. Coker, A.B.; Findlay, J.W.A. Eur. Patent Appl. **1983**, 85959. *Chem. Abstr.* **1984**, *100*, 6345.
7. Bossert, F.; Vater, W. *Naturwissinschaften* **1971**, *58*, 578.
8. Berntsson, P.B.; Carlsson, A.I.; Gaarder, J.O.; Ljung, B.R. Swed. Patent **1985**, 442298. *Chem. Abstr.* **1987**, *106*, 4878.
9. Meyer, H.; Wehniger, E.; Bossert, F.; Scherling, D. *Arzneim.-Forsch.* **1983**, *33*, 106.
10. Sato, Y. U. S. Patent **1980**, 4338322.
11. Huang, F.C.; Lin, C.J.; Jones, H. Eur. Patent Appl. **1984**, 109049. *Chem. Abstr.* **1984**, *101*, 110747.

12. Semeraro, C.; Micheli, D.; Pieraccioli, D.; Gaviraghi, G.; Borthwick, D.A. German Offen. **1986**, 3529997; *Chem. Abstr.* **1986**, *105*, 97322.
13. Semeraro, C.; Micheli, D.; Pieraccioli, D.; Gaviraghi, G.; Borthwick, D.A. German Offen. **1987**, 3628215; *Chem. Abstr.* **1987**, *107*, 23240.
14. Janssen, P.A.J.; Van de Westeringhe, C.; Jageneau, A.H.M.; Demoen, J.A.; Hermans, B.F.K.; Van Daele, G.H.P.; Schellekens, K.H.L.; Van der Eycken, C.A.M.; Niemegeer, C.J.E. *J. Med. Chem.* **1959**, *1*, 281.
15. Yamamoto, H.; Okamoto, T.; Sasajima, K.; Nakao, M.; Maruyama, I.; Katayama, S. German Offen. **1971**, 2033909; *Chem. Abstr.* **1971**, *74*, 99879.
16. Janssen, P.A.J. Belg. Patent **1963**, 626307. *Chem. Abstr.* **1964**, *60*, 10690.
17. Janssen, P.A.J.; Soudijn, W.; Van Wijngaarden, I.; Dreese, A. *Arzneim.-Forsch.* **1968**, *18*, 282.
18. Janssen, P.A.J. U. S. Patent **1966**, 3238216. *Chem. Abstr.* **1966**, *65*, 8924.
19. Lesher, G.Y.; Opalka, C.J. U. S. Patent **1979**, 4107315; *Chem. Abstr.* **1979**, *90*, 103844.
20. Singh, B. U. S. Patent **1983**, 4413127.
21. Tagmann, E.; Sury, E.; Hoffmann, K. *Helv. Chim. Acta* **1952**, *35*, 1541.
22. Hoffmann, K.; Urech, E. U. S. Patent **1958**, 2848455. *Chem. Abstr.* **1959**, *53*, 7103.
23. Boss, A.I.; Clissold, D.W.; Mann, J.; Markson, A.J.; Thickitt, C.P. *Tetrahedron* **1989**, 6011.
24. Schnider, O.; Frick, H.; Lutz, A.H. *Experientia* **1954**, *10*, 135.
25. Liberman, D.; Rouaix, A. *Bull. Soc. Chim. Fr.* **1959**, 1793.
26. Lam, B.L. Eur. Patent Appl. **1982**, 47164; *Chem. Abstr.* **1982**, *96*, 217866.
27. Anonymous *Jpn. Kokai* **1987**, 62029575; *Chem. Abstr.*, **1987**, *106*, 176427.
28. Chou, T.S.; Heath, P.C.; Patterson, L.E.; Poteet, L.M.; Lakin, R.E.; Hunt, A.H. *Synthesis* **1992**, 565.
29. Nicolas, C.; Verny, M.; Maurizis, J.C.; Payard, M.; Faurie, M. *J. Labeled Cmpd. Radiopharm.* **1986**, *23*, 837.
30. Kuhla, D.E.; Campbell, H.F.; Studt, W.L; Faith, W.C.; Molino, B.F. *PCT Int. Appl.* **1987**, 8703201; *Chem. Abstr.* **1988**, *108*, 6037.
31. Russell, P.B.; Hitchings, G.H. *J. Am. Chem. Soc.* **1951**, *73*, 3763.
32. Stenbuck, P.; Hood, H.M. U. S. Patent **1962**, 3049544.
33. Liebenow, W.; Prikryl, J. Fr. Patent **1974**, 2227868; *Chem. Abstr.* **1975**, *82*, 156363.
34. Sarett, L.M. et al. *J. Am. Chem. Soc.* **1960**, *82*, 2974.
35. Andesron, G.W.; Halverstadt, I.H.; Miller, W.H.; Roblin, R.O. *J. Am. Chem. Soc.* **1945**, *67*, 2197.
36. McCall, J.M.; TenBrink, R.E.; Ursprung, J.J. *J. Org. Chem.* **1975**, *40*, 40.
37. Bergstrom, D.E.; Ruth, J.L. *J. Am. Chem. Soc.* **1976**, *98*, 1587.
38. Imazawa, M.; Eckstein, F. *J. Org. Chem.* **1978**, *43*, 3044.
39. Hertel, L.W.; Kroin, J.S.; Misner, J.W.; Tustin, J.M. *J. Org. Chem.* **1988**, *53*, 2406.
40. Horwitz, J.P.; Chua, J.; Noel, M.; Donatti, J.T. *J. Org. Chem.* **1967**, *32*, 817.
41. Storer, R.; Clemens, I.R.; Lamont, B.; Noble, S.A.; Williamson, C.; Beaulieu, B. *Nucleosides Nucleotides* **1992**, *12*, 225.

42. Lipinski, C.A.; Stam, J.G.: Pereira, J.N.; Ackerman, N.R.; Hess, H.J. *J. Med. Chem.* **1980**, *23*, 1026.
43. Brown, T.H.; Durant, G.J.; Ganellin, C.R. Can. Patent **1981**, 1106375; *Chem. Abstr.* **1982**, *96*, 35288.
44. Adger, B.M.; Lewis, N.J. Eur. Patent Appl. **1985**, 141560; *Chem. Abstr.* **1985**, *103*, 178273.
45. Lam, B.L.; Prigden, L.N. Eur. Patent Appl. **1982**, 58055; *Chem. Abstr.* **1983**, *98*, 16719.
46. Bagli, J.F. Eur. Patent Appl. **1985**, 130735; *Chem. Abstr.* **1985**, *103*, 6358.
47. Skulnik, H.I.; Weed, S.D.; Eidson, E.E.; Renis, H.E.; Wierenga, W.; Stringfellow, D.A. *J. Med. Chem.* **1985**, *28*, 1864.
48. Kushner, S.; Dallian, H.; Sanjuro, J.L.; Bach, F.L.; Safir, S.R.; Smith, V.K.; Williams, J.H. *J. Am. Chem. Soc.* **1952**, *74*, 3617.
49. Cragoe, E.J.; Woltersdorf, O.W.; Bicking, J.B.; Kwong, S.F.; Jones, J.H. *J. Med. Chem.* **1967**, *10*, 66.
50. Cheesman, G.W.H. *J. Chem. Soc.* **1960**, 242.
51. Meltzer, R.L.; Lutz, W.B. U. S. Patent **1967**, 3320126; *Chem. Abstr.* **1968**, *67*, 76300.
52. Janssen, P.A.J. U. S. Patent **1961**, 2997472.
53. Fauran, F.; Huguet, G.; Raynaud, G.; Pourias, B.; Turin, M. Br. Patent **1969**, 1168108; *Chem. Abstr.* **1970**, *72*, 12768.
54. Kluge, A.F.; Clark, R.D.; Strosberg, A.M.; Pascal, J.C.; Whiting, R.L. Eur. Patent Appl. **1984**, 126449; *Chem. Abstr.* **1985**, *102*, 166777.
55. Morren, H. U. S. Patent **1959**, 2899436. *Chem. Abstr.* **1960**, *54*, 3465.
56. Baltes, E.; De Lannoy, J.; Rodriguez, L. Eur. Patent Appl. **1982**, 58146; *Chem. Abstr.* **1982**, *98*, 34599.
57. Palazzo, G.; Silvestrini, B. U. S. Patent **1968**, 3381009; *Chem. Abstr.* **1968**, *69*, 52144.
58. Angelini, F. Austrian Patent **1977**, 336021; *Chem. Abstr.* **1977**, *87*, 85033.
59. Ferguson, H.C.; Wu, Y.H.; Rayburn, J.W.; Allen, L.E.; Kissel, J.W. *J. Med. Chem.* **1972**, *15*, 477.
60. Ormaza, V.A. Spanish Patent **1986**, 550086; *Chem. Abstr.* **1987**, *99*, 39628.
61. Yevich, J.P.; New, J.S.; Smith, D.W.; Lobeck, W.G.; Catt, J.D.; Minelli, M.S.; Eison, M.S.; Taylor, D.P.; Riblet, L.A.; Temple, D.L. *J. Med. Chem.* **1986**, *29*, 359.
62. Ishizumi, K.; Kojima, A.; Antoku, F. *Chem. Pharm. Bull.* **1991**, *39*, 2288.
63. Winkley, M.W.; Robins, R.K. *J. Org. Chem.* **1970**, *35*, 491.
64. Piml, J., Sorm, F. *Coll. Czech. Chem. Commun.* **1964**, *29*, 2576.
65. Modest, E.J. *J. Org. Chem.* **1956**, *21*, 1.
66. Lacefield, W.B.; Ho, P.P.K. Belg. Patent **1977**, 839469; *Chem. Abstr.* **1977**, *87*, 68431.
67. Baxter, M.G.; Elphick, A.R.; Miller, A.A.; Sawyer, D.A. Can. Patent **1982**, 1133938. *Chem. Abstr.* **1983**, *98*, 89397.

CHAPTER 10

FIVE-MEMBERED HETEROCYCLES FUSED TO A BENZENE RING

1. COMPOUNDS WITH ONE HETEROATOM

A. Benzofurans

The bicyclic array in the few therapeutic agents based on the benzofuran ring plays the role of a rigid support for functional groups. As was also the case with the monocyclic furans, this ring system does not seem to form a pharmacophore.

One classic approach to the construction of the benzofuran ring involves aldol condensation of an ether of salicylaldehyde containing a β-carbonyl group. Thus, treatment of the ether (**1**) from alkylation of salicylaldehyde with chloroacetone with sodium ethoxide gives the acyl benzofuran **2**. The carbonyl group is then removed by Wolff–Kishner reduction with hydrazine and potassium hydroxide to give **3**. Aluminum chloride catalyzed acylation of the product with anisoyl chloride proceeds to give ketone **4**; the methyl ether is then cleaved using pyridine hydrochloride to afford the phenol **5**. Treatment of **5** with a basic mixture of iodine and iodide ion leads to iodination of the ortho positions and formation of the coronary vasodilator **benziodarone** (**6**).[1]

The salicylaldehyde (**8**) required for the synthesis of a somewhat more complex benzofuran is obtained by acylation of the phenol **7** with hexamethylene tetramine (HMTA) in trifluoroacetic acid. It is likely that the reaction proceeds via an iminoformyl species formed by dissociation of HMTA in the strongly acidic medium. The initially formed imine is hydrolyzed to the observed aldehyde on workup. In this case, the benzofuran ring is formed, by reaction of **8** with ethyl bromomalonate in the presence of sodium ethoxide. The reaction can be visualized as first proceeding to give the displacement product **9**; addition of the anion from this compound to the adjacent aldehyde will give the intermediate **10**. The dicarboxylic acid formed when **10** is saponified undergoes sequential decarboxylation and dehydration to afford the carboxylated benzofuran. Thus, the platelet aggregation inhibitory **furagrelate** (**11**)[2] is obtained.

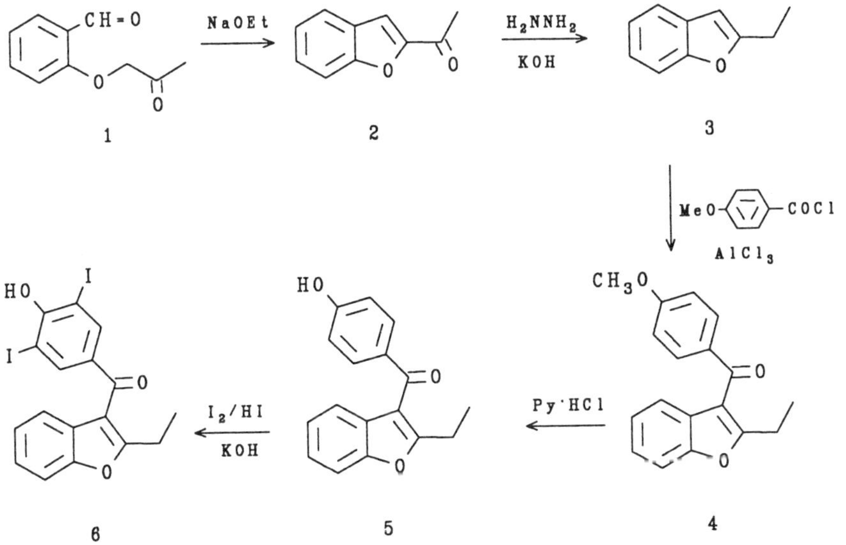

Substitution of a propionic acid fragment onto a benzofuran affords an NSAID "profen", which again illustrates the flexibility of the SAR in this series. In this case, the benzofuran ring is prepared by an acid catalyzed cyclodehydration reaction. Alkylation of *o*-bromophenol (**12**) with phenacyl bromide gives the corresponding ether **13**. Treatment of **13** with polyphosphoric acid leads to formation of the benzofuran **14**. Reaction of **14** with magnesium metal affords the corresponding Grignard reagent; addition of methyl pyruvate to the organometallic reagent leads to addition to the ketone and thus formation of the tertiary carbinol **15**. This compound is then dehydrated to the olefin by means of *p*-toluenesulfonic acid. The olefin is then reduced by means of catalytic hydrogenation. Saponification of the ester affords the NSAID **furaprofen** (**16**).[3]

The fermentation product **griseofulvin** (**22**) was, until the discovery of the "conazoles" (see Chapter 8), the mainstay for treatment of fungal infections, and is in fact still in use for this indication. This drug is in all probability produced commercially by large scale fermentation. A concise synthesis has, however, been developed; at least one drawback to its practical use is the fact that it produces racemic material while the natural product is chiral. Construction of fragment **19** intended for the spiro ring first involves 1,2-addition of the lithium reagent from methoxyacetylene **17** to crotonaldehyde **18**; oxidation of the newly formed allylic alcohol with manganese dioxide gives the Michael double acceptor **19**. Preparation of the dihydrobenzofuran fragment begins with acylation of the highly substituted phenol **20** with chloroacetyl chloride in the presence of aluminum chloride. Treatment of the intermediate chloroketone with

PhCOCH$_2$Br, NaOH; PPA; 1. Mg, 2. MeCOCO$_2$Me; 1. TSA, 2. H$_2$, 3. NaOH

12 13 14 15 16

CH$_3$OC≡CH + 18 (CH=O); 1. BuLi, 2. MnO$_2$; 17 18 19

20 (CH$_3$O, CH$_3$O, OH, Cl); 1. ClCH$_2$COCl, AlCl$_3$, 2. NaOAc; 21; t-BuOK; 22

sodium acetate leads to cyclization and formation of **21**. Reaction of **21** with potassium *tert*-butoxide affords the corresponding carbanion; this compound is thought to first add to the enone in **19**. The anion from reaction with a second equivalent of base then adds to the ynone function to form the spiro ring. The fact that the product from this reaction has the same relative stereochemistry as the natural product is attributed to the better overlap of the enolate with the triple bond in the transition state leading to this isomer. Thus, the product from this reaction is ± griseofulvin (**22**).[4]

B. Indoles

The relative abundance of indole-based therapeutic agents is attributable only in part to the fact that this nucleus forms part of a pharmacophore for selected CNS agents.

The indole moiety, however, likely simply serves as a rigid bicyclic support in the case of the majority of agents discussed below.

The presence of an acidic proton forms a recurring feature among the clinical successful NSAIDs. Antiinflammatory activity has, however, been observed in various test systems with compounds devoid of this structural feature. A number of heterocyclic compounds bearing 1,2-dianisyl substituents, examples of which have been discussed in Chapters 8 and 9, showed such activity. The first of this structural class was in fact the indole **indoxole** (**25**), a compound that bears an interesting structural relation to the naphthalene estrogens and their antagonists (Chapter 6). The key reaction for the preparation of this agent and several of those that follow consists of the Fischer indole synthesis, a reaction that has been recently reviewed.[5] In broad outline, the first step in this quite complex transform involves protonation of the starting hydrazone with migration of the double bond to form a species such as **A**. This compound then undergoes an electrocyclic rearrangement in which the carbon α to the hydrazone bonds to the ortho position of the aromatic ring with scission of the *N,N*-hydrazine bond to form the bis-imine **B**. The ring imine then adds to its side chain counterpart to give the fused indoline **C**. (The direction of this reaction is not crucial, since reverse addition should lead to the same overall product.) Loss of the elements of ammonia leads finally to an indole **D**. Reaction of desoxyanisoin (**23**) with phenylhydrazine goes in a straightforward manner to the hydrazone **24**. Treatment of **24** with acetic acid leads to formation of the indole ring and formation of **indoxole** (**25**).[6]

A simple indole derivative, serotonin (5-hydroxytryptamine; 5-hydroxy-3-β-aminoethylindole), is an important neurotransmitter in the brain; a variety of mental diseases and in particular depression have been traced to inappropriate levels of receptor sensitivity to this compound. The 5-hydroxytryptamine moiety thus has been used in the design of CNS drugs in an attempt to insure interaction with brain receptors. Condensation of dimethoxyphenylhydrazine (**26**) with the product (**27**) from alkylation of 4-(2-methoxyphenyl)piperazine and 3-bromopropyl methyl ketone, gives the hydrazone **28**. Treatment of **28** with hydrogen chloride leads to rearrangement to an

H N⁺H₂ N H → H N⁺H₂ NH → N H NH₂ → N H

A B C D

OCH₃ O OCH₃ —PhNHNH₂→ [N–N H OCH₃ OCH₃] —AcOH→ N H OCH₃ OCH₃

23 24 25

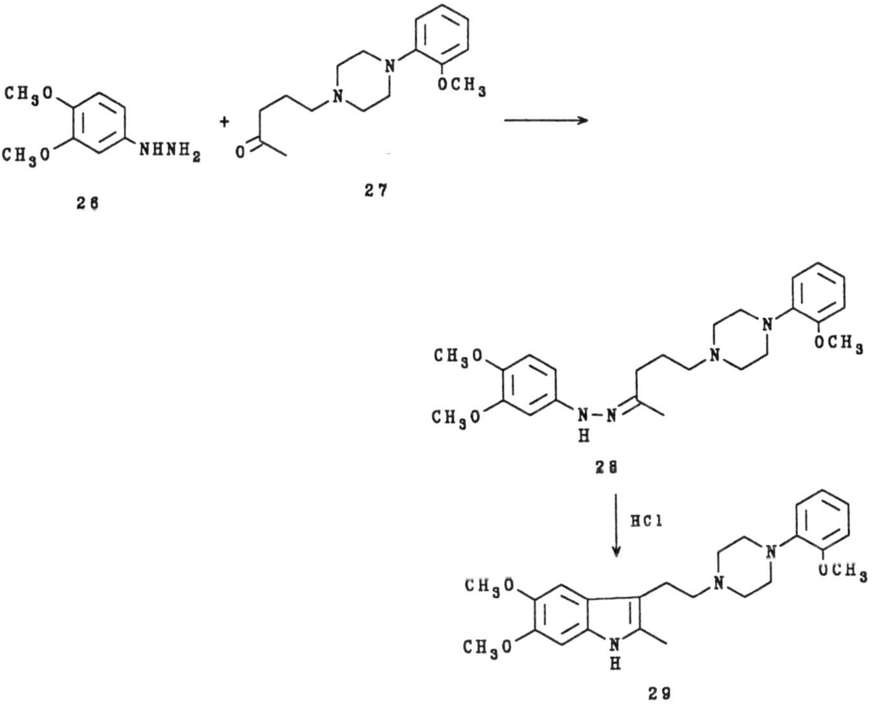

indole and formation of the antipsychotic agent **milipertine** (**29**),[7] a compound that shares with serotonin both the oxygen at the 5-position and an ethylamino side chain at the 3-position.

A quite recently introduced compound that is effective against migraine, **sumatriptan** (**34**), bears a closer resemblance to serotonin; the acidic phenol group in the natural product is replaced by a comparably acidic sulfonamide with an interposed methylene group and the terminal amine is dimethylated. The preparation starts by conversion of the aniline **30** to its diazonium salt with nitrous acid; reduction with stannic chloride then affords the corresponding arylhydrazine **31**. Condensation with 3-cyanopropionaldehyde gives the hydrazone **32**. Treatment of **32** with hydrogen chloride leads to rearrangement to the indole **33**. The nitrile is then reduced to the primary amine by catalytic hydrogenation. Reaction of the amine with excess formalin and sodium borohydride results in formation of the *N*,*N*-dimethylated derivative, yielding sumatriptan (**34**).[8]

The Fischer indole synthesis is quite tolerant of additional functional groups in the starting material. Thus reaction of 4-dimethylaminocyclohexanone (**35**) with phenylhydrazine in acetic acid leads directly to **cyclindole** (**36**),[9] a compound described as an antidepressant. A slightly different approach is used to prepare the fluorinated analogue. In this case, the tricylic indole **37** is obtained by reaction of 2,4-difluorophenylhydrazine with 4-hydroxycyclohexanone. The hydroxyl group in the product (**37**) is then converted to its tosylate by reaction with *p*-toluenesulfonyl

CH_3NHO_2S (30) NH_2 —1. HONO, 2. $SnCl_4$→ CH_3NHO_2S (31) $NHNH_2$ —O-HC‿CN→ CH_3NHO_2S (32) N–N=‿CN

32 —HCl→ 33 (CH_3NHO_2S, CN)

33 —1. H_2, 2. $CH_2O/NaBH_4$→ 34 (CH_3NHO_2S, $N(CH_3)_2$)

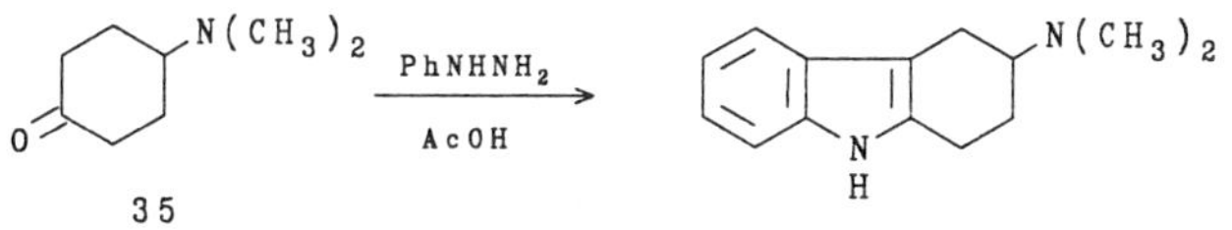

chloride. Displacement with dimethylamine then yields the antipsychotic agent **flucindole** (**38**).[10]

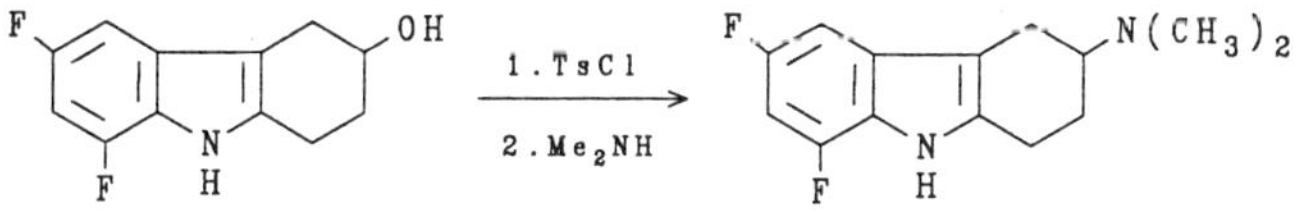

Additional substitution on nitrogen does not interfere with the Fischer indole reaction. It should, however, be noted that the direction of imine addition (**B** → **C**) must follow the pathway depicted above. In the simplest example, reaction of *N*-benzylphenylhydrazine (**39**) with *N*-methylpyrid-4-one (**40**) directly affords the H_1-antihistamine **mebhydrolin** (**41**).[11]

Indomethacin (**45**) constitutes one of the first and most potent NSAIDs discovered in the second part of this century. One of the many synthesis for this drug illustrates the versatility of the classic indole synthesis. Thus, condensation of the chlorobenzoyl arylhydrazide (**42**) with levulinic acid (**45**) leads to the hydrazone **44**. This reaction rearranges to indole **45** on treatment with strong acid.[12]

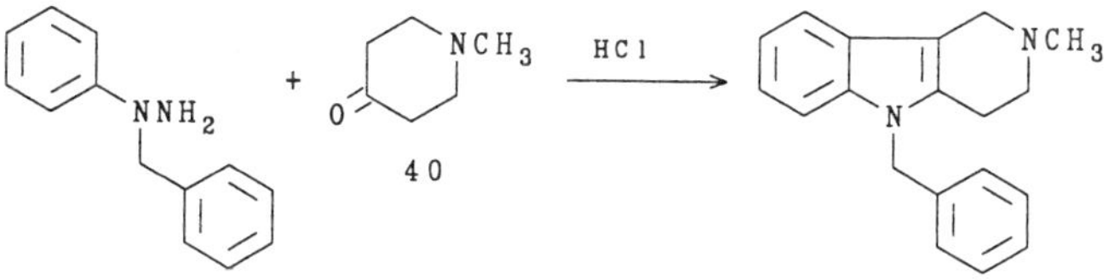

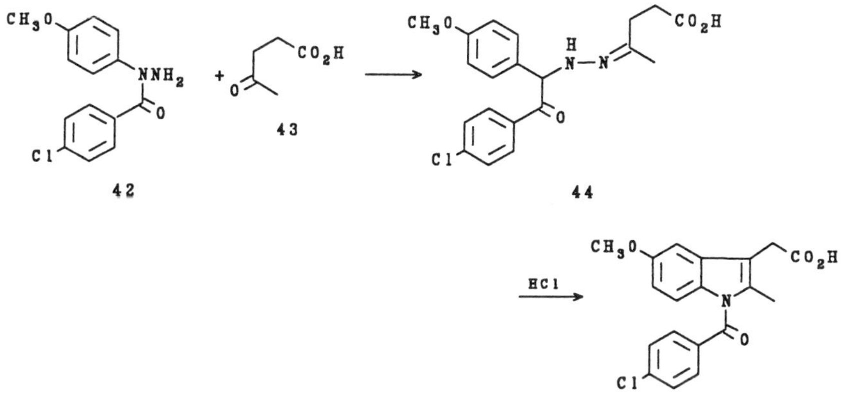

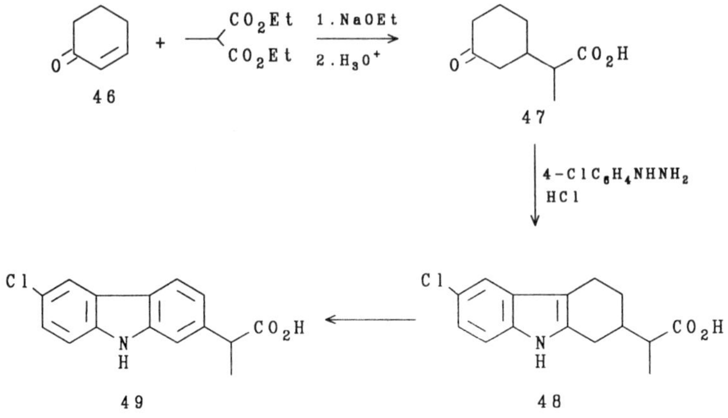

A fused tricyclic ring system based on an indole provides yet another NSAID. Michael addition of the anion from diethyl methylmalonate to cyclohexenone **46** followed by acid hydrolysis of the product gives the cyclohexanone **47**, which incorporates the characteristic "profen" 2-substituted carboxylic side chain. Sequential reaction with *p*-chlorophenylhydrazine and strong acid gives the fused indole derivative **48**. The carboxylic acid group is then esterified and the cyclohexane ring is aromatized by some unspecified means. Saponification of the ester gives the free acid and thus affords **carprofen** (**49**).[13]

The 3-position on indoles is quite reactive toward electrophiles as a consequence of its partial enamine-like character. The methoxyindole (**50**) thus readily undergoes Mannich reaction with formaldehyde and dimethylamine to afford the aminomethylated derivative **51**. Treatment of intermediate **51** with potassium cyanide leads to displacement of dimethylamine and formation of the nitrile **52**, possibly by an elimination–addition sequence involving a 3-exomethylene–indolenine intermediate. The protons on the methylene group adjacent to the nitrile are quite acidic and readily removed. Reaction of **52** with methyl carbonate in the presence of sodium methoxide gives the carbomethoxylated derivative **53**. Catalytic hydrogenation leads to reduction of the nitrile to a primary amine. Thus, the antihypertensive agent **indorenate** (**54**)[14] is obtained.

The discovery of the fungal metabolite **lovastatin** (**55**) has led to a new class of clinically effective cholesterol lowering drugs. This agent and its analogues block an enzyme HMG–CoA reductase, which is involved in the synthesis of mevalonate, an early precursor of cholesterol. Extensive work has demonstrated that the mevalonate-like lactone ring, or its ketoacid precursor, is essential for activity; considerable structural freedom exists as to the nature of the remainder of the molecule. The nonchiral synthesis of an indole based analogue starts with a different strategy for forming the indole ring. Thus, cyclodehydration of the alkylation product **56** from *N-iso*-propylaniline with phenacyl bromide with *p*-toluenesulfonic acid affords the indole **57**. The side chain is added to the reactive 2-position by what may be viewed as a vinylogous Villsmeyer reaction. Treatment of **57** with 3-dimethylaminoacrolein and phosphorus oxychloride give the substituted acrolein **58** on workup. The side chain

CH_3O 50 $\xrightarrow[CH_2O]{Me_2NH}$ CH_3O, $N(CH_3)_2$ 51 $\xrightarrow{KCN}$ CH_3O, CN 52 $\xrightarrow[NaOMe]{(MeO)_2C{=}O}$ CH_3O, CO_2CH_3, CN 53 $\xrightarrow{H_2}$ CH_3O, CO_2CH_3, NH_2 54

is then extended by addition of the more nucleophilic terminal enolate of the acetoacetate dianion obtained by sequential reaction of ethyl acetoacetate with sodium hydride and butyllithium. On reaction with **58**, this dianion affords the chain extended product **59**. The newly introduced carbonyl group is then reduced with a mixture of sodium borohydride and triethyl borane to give the diol in which the hydroxyl groups bear the desired relative configuration. Saponification of the ester group leads to the acid, isolated as its sodium salt, racemic **fluvastatin** (**60**).[15] The commercial drug, it should be noted, consists of a single enantiomer.

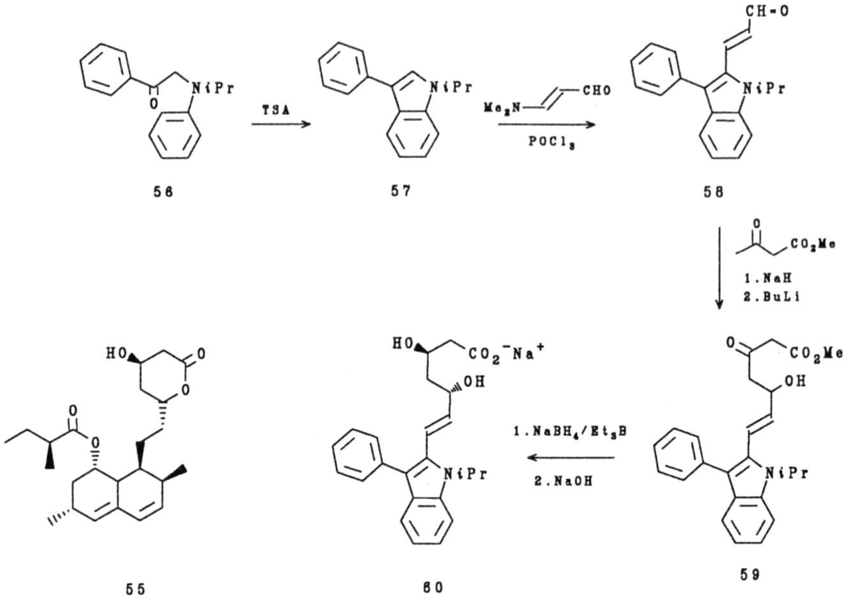

C. Indolines and Isoindolines

Partly reduced counterparts of the indole nucleus provide the basis for several agents with varied biological activities. A pair of closely related *N*-phenyl derivatives have both shown antidepressant activity in test systems. The apparent preference for the monomethyl amine suggests that these compounds act by the same mechanism as the classical tricyclic antidepressants, where the second amine is the more active species. The first step in the preparation of the common intermediate **63** to these compounds consists of acylation of diphenylamine (**61**) with chloropropionyl chloride. Cyclization of **62** under Friedel–Crafts conditions gives the desired indolinone **63**. Reaction of the carbanion obtained on treatment of **63** with 2-chloropropyldimethylamine then gives the alkylation product **64**. Note that, in spite of this extra step, the scheme is greatly simplified by starting with the very readily available tertiary amine. The superfluous methyl group is then removed by reaction of **64** with ethyl chloroformate in the current version of the Von Braun reaction. Thus, **amedalin** (**65**)[16] is obtained.

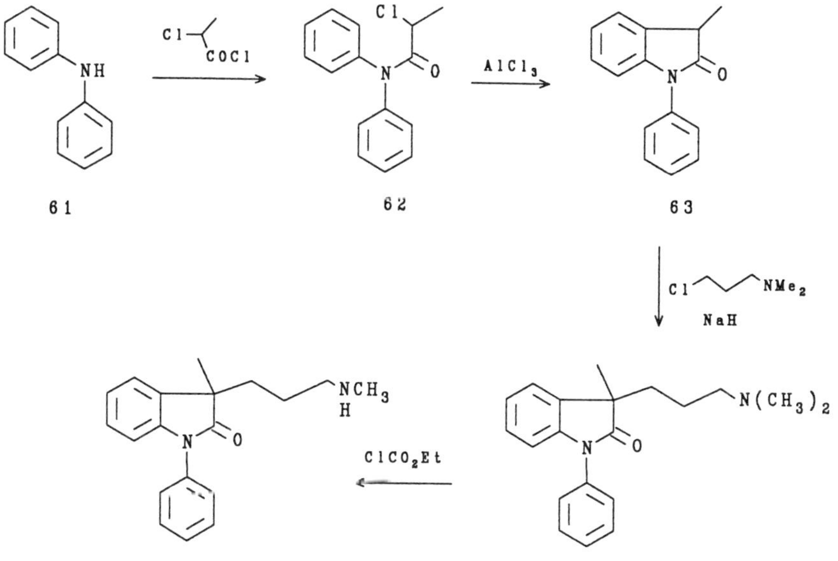

61 62 63

64 65

Reduction of the amide carbonyl group to the amine by any of several methods, for example, diborane, affords the antidepressant **daledalin** (**67**).[16]

The typical NSAIDs noted thus far rely on a carboxylic acid for the required acidic proton. This function can, however, also be supplied by a highly activated enolic system; the prototype **piroxicam** and its analogues are discussed in more detail in Chapter 11. Another highly enolizable compound based on the indolone nucleus also displays similar activity. The preparation begins with conversion of the indolone **67** to its carbamate **68** by successive conversion to the urethane by treatment with ethyl chloroformate and then ammononlysis of this intermediate. The extra electron-with-

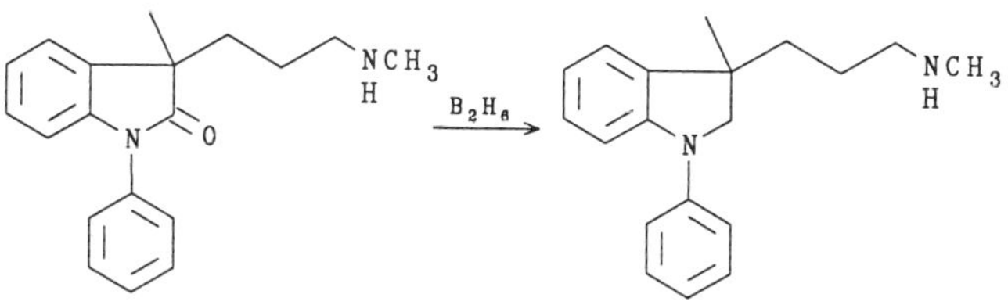

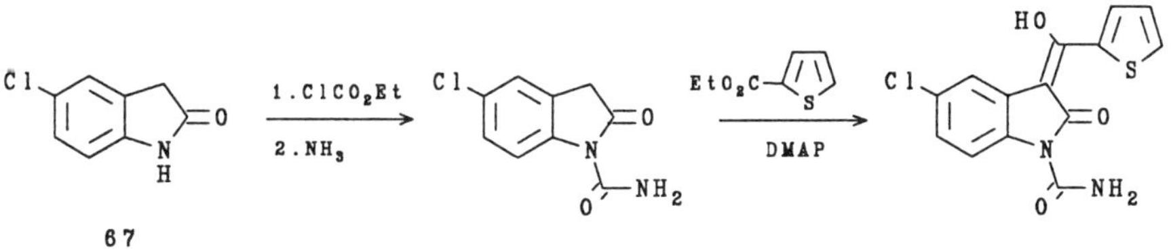

67 68 69

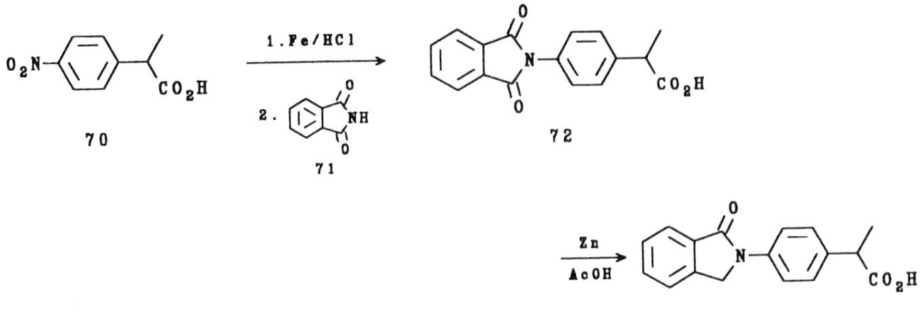

drawing power of the newly introduced function apparently increases acidity of the benzylic methylene group sufficiently so that it can be removed with 2,6-dimethylaminopyridine (DMAP). Reaction of **68** with 2-carbethoxythiophene in the presence of this catalyst gives the β-dicarbonyl condensation product **tenidap** (**69**),[17] shown as its enol tautomer.

The isoindolone ring system forms the nucleus for one of the more traditional "profen" NSAIDs. Reduction of the nitro group in arylpropionic acid **70** gives the corresponding aniline; this compound, on reaction with succinimide, (**71**) gives the product **72** in which the aniline has exchanged with ammonia. Interestingly, treatment of the new imide with zinc in acetic acid leads to reduction of but one of the carbonyl groups to afford the indolone **73** (**indoprofen**).[18]

Imide exchange provides the starting material for a sulfonamide diuretic agent whose structure incorporates many of the features of the agents noted in Chapter 2. Reaction of the succinimide **74** with cyclohexylamine provides the intermediate **75**. Tin in hydrochloric acid converts that to the lactum **76**. Nitration of the product (**76**) proceeds as would be predicted at the position meta to the carbonyl group; this compound is then reduced to the aniline **77** by means of stannic chloride. The amino group is then converted to the diazonium salt with nitrous acid; Sandmeyer reaction with sulfur dioxide in the presence of cuprous chloride affords the corresponding sulfonyl chloride. Now, this compound is treated with ammonia to give the diuretic **clorexolone** (**78**).[19]

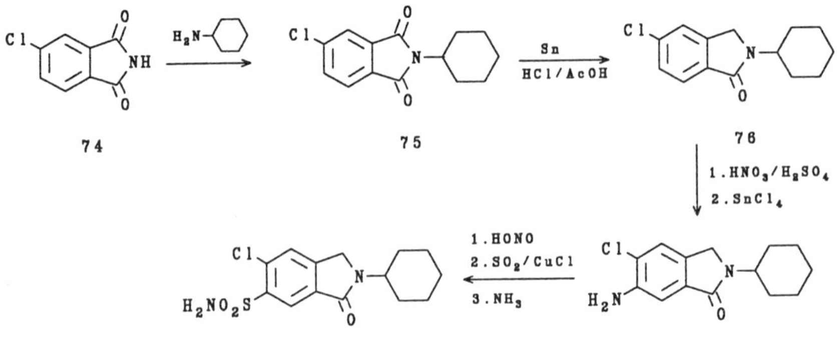

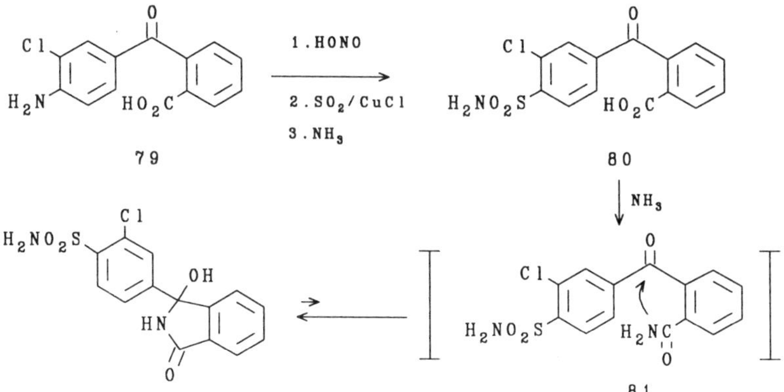

A related sulfonamide is classed as an isoindolone by virtue of typical benzoylbenzoic acid "pseudoacid" isomerism. The amino group in benzoylbenzoic acid **79** is converted to a sulfonamide by essentially the same sequence as used above to give **80**. The carboxylic acid is then converted to the amide **81** by sequential conversion to the acid chloride and reaction with ammonia. Now, the amide nitrogen adds to the benzoyl group to give isoindolone **82** in a reaction typical of ortho benzoyl benzoates.[20] This closed form of the widely used diuretic agent **chlorthalidone** (**82**) greatly predominates over the open isomer with which it can equilibrate.

2. COMPOUNDS WITH TWO HETEROATOMS

A. Indazoles

The readily prepared indazole ring system has found surprisingly little use as a nucleus for therapeutic agents. The preparation for the common intermediate **86** for two of these agents starts with the reaction of the *N*-benzyl derivative **83** from methyl anthranilate with nitrous acid to give the *N*-nitroso derivative **84**. Reduction by means of sodium thiosulfate leads to the transient hydrazine **85**, which undergoes spontaneous internal hydrazid formation to form **86**. The enolate from reaction of this amide with sodium methoxide gives the product from alkylation on oxygen. Thus, treatment of the enolate with 3-chloro-1-dimethylaminopropane gives **benzydamine 87**,[21] a compound that has been extensively investigated for its unusual antiinflammatory activity. Alkylation of this same anion with methyl chloroacetate affords the corresponding ester. This compound gives the more classical NSAID **bendazac** (**88**) after hydrolysis of the ester grouping.[22]

Partial reduction of the aromatic ring also leads to a compound that acts as a rather untypical analgesic antiinflammatory agent. The starting thioamido ketone **89** is available from acylation of cyclohexanone with methyl isothiocyanate. Reaction of **89**, shown as its enol tautomer with methylhydrazine, can be envisaged to involve initial formation of the addition–elimination product **90**. Reaction of the more basic nitrogen with the thiocarbonyl function then leads to formation of a pyrrazole ring. Thus, **tetrydamine (91)**[23] is obtained.

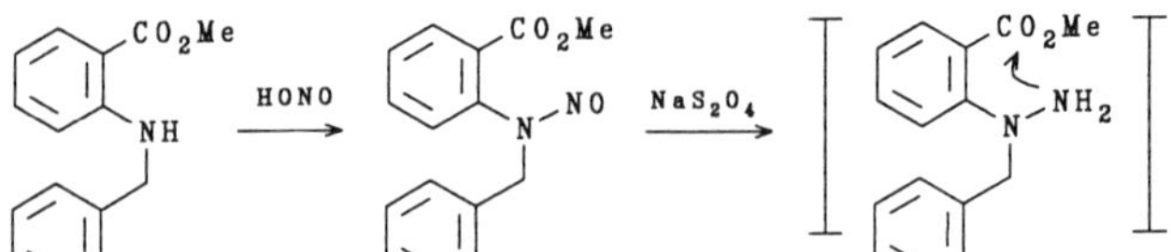

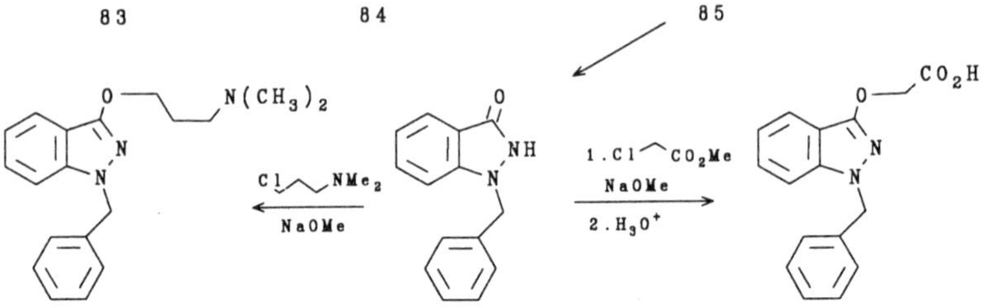

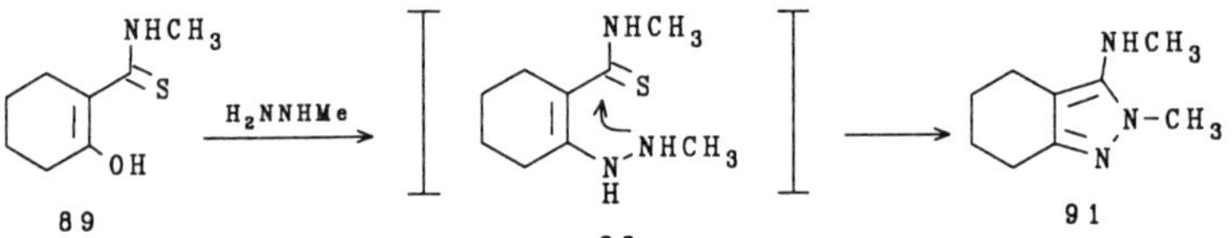

B. Benzimidazoles

A sizeable, if largely disparate, group of therapeutic agents are based on the benzimidazole nucleus; this heterocyclic system does, however, provide a unifying theme for the subset of anthelmintic compounds that include this moiety and are discussed at the end of this section.

The standard starting material for most benzimidazoles consists of *o*-phenylenediamine (**92**) or its derivatives. Reaction of **92** with chloroacetic acid can be rationalized by invoking initial formation of the chloromethyl amide **93**. Imide formation with the remaining free amino group closes the ring to afford the 2-chloromethyl imidazole **94**. Displacement of halogen with pyrrolidine affords the alkylation product **95**. The proton on the fused imidazole nitrogen is then removed by reaction with sodium hydride. Treatment of the resulting anion with α,4-dichlorotoluene gives the H_1 antihistaminic agent **clemizole (96)**.[24]

Aromatic nucleophilic displacement of the highly activated chlorine in 2,4-dinitrochlorobenzene (**97**) by means of *N,N*-diethylethylenediamine gives the corresponding aniline **98**. Reaction of **98** with ammonium sulfide interestingly leads to selective reduction of the nitro group adjacent to the newly introduced amino group to afford the diamine **99**; the selectivity may be due to localization of the reducing function by initial sulfide salt formation with the ortho amino group. Condensation of the diamine **99** with the iminoether **100** from 4-ethoxyphenylacetonitrile leads to benzimidazole

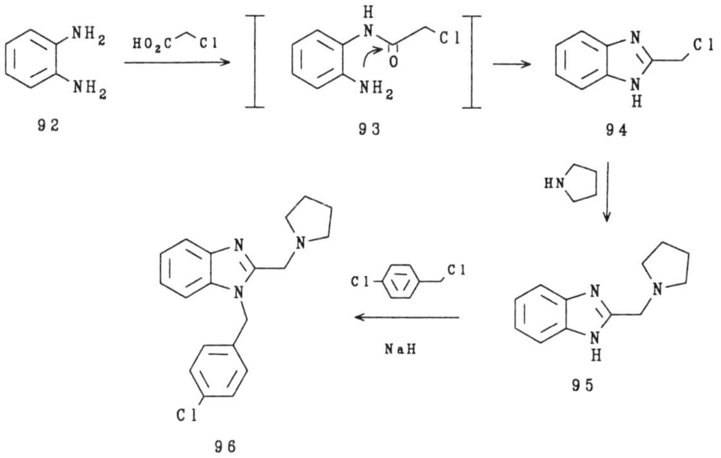

92 93 94 95 96

ring formation.[25] The product **etonitazene** (**101**) is a potent opioid analgesic. The structure of **101** at first sight differs markedly from the opioids discussed in Chapter 7; it will, however, provide a good fit when overlaid with these compounds, though this requires some departures from energy minimized conformations.

Most of the clinically effective antiviral agents consist of subtly modified nucleosides. The side effects of many of these drugs, thought to be due to the very intimate involvement of nucleosides in many biological regulatory processes, has led to a continuing search for nonnucleoside antiviral compounds. One of these compounds, **enviroxime** (**106**), is based on a benzimidazole nucleus. Reaction of benzoylphenylenediamine (**102**) with cyanogen bromide probably proceeds initially to give the *N*-cyano intermediate **103**. Attack on the nitrile group by the adjacent amine closes the ring with consequent formation of the 2-aminobenzimidazole **104**. Treatment of **104** with sodium hydride leads to selective formation of the anion on the ring nitrogen meta to the carbonyl group; one might have predicted ionization of the other, less electron

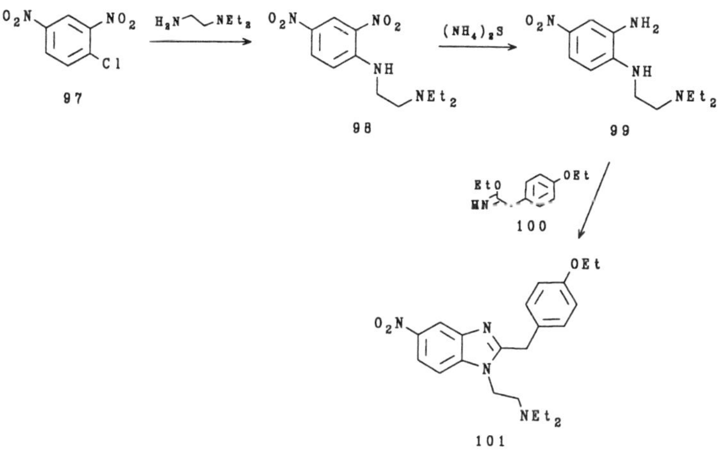

97 98 99 100 101

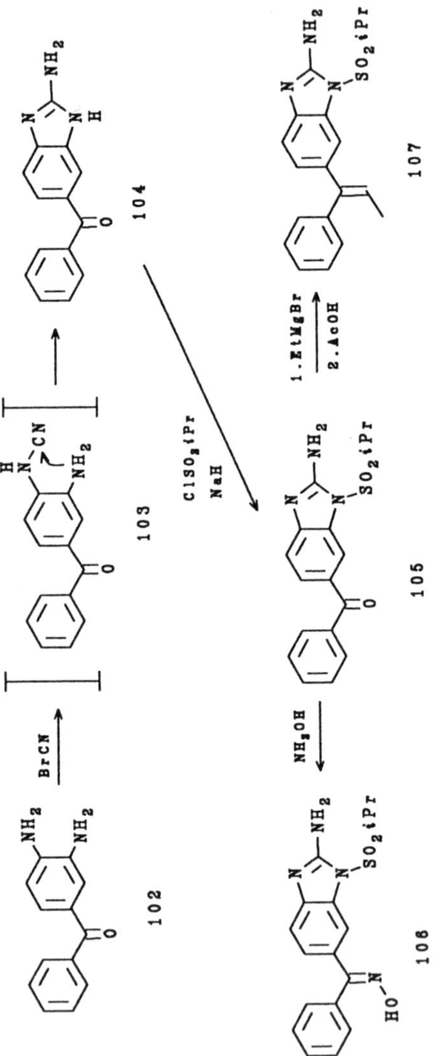

102
BrCN
103
104
ClSO2iPr
NaH
105
1. EtMgBr
2. AcOH
107
NH2OH
108

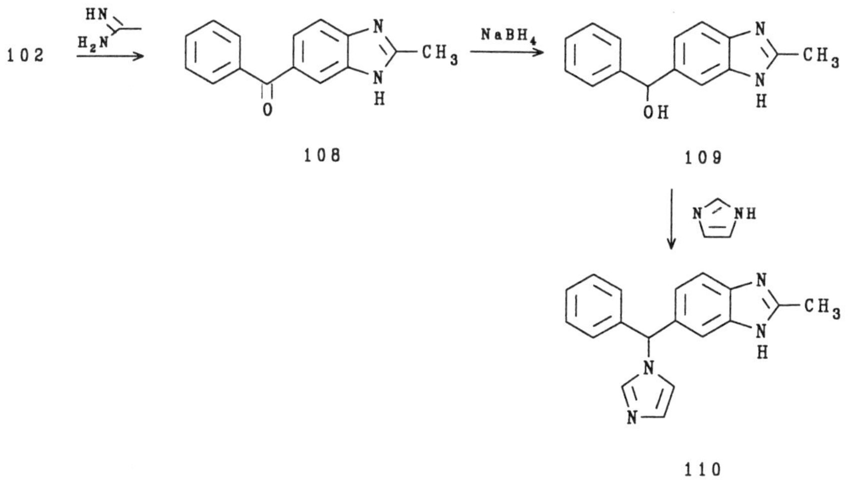

dense ring position. Addition of isopropylsulfonyl chloride leads to formation of the sulfonamide **105**. The ketone is then converted to its oxime by reaction with hydroxylamine hydrochloride and sodium acetate. The predominant product is the slightly less sterically encumbered (*E*) isomer, enviroxime (**106**).[26] Antiviral activity is retained when the oxime is replaced by the sterically similar ethylidene group. This compound is obtained by first condensing the ketone **105** with an organometallic such as ethylmagnesium bromide, followed by dehydration of the resulting alcohol with acetic acid. This compound affords predominantly the (*E*) isomer **enviradene** (**107**).[27]

Condensation of the same starting material (**102**) with acetamidine, affords the corresponding 2-methylbenzimidazole **108**. Reduction of the carbonyl group leads the benzhydryl alcohol (**109**). The benzhydryl hydroxyl readily undergoes displacement by nitrogen on imidazole to afford **irtemazole** (**110**),[28] a compound that can be useful in treating gout by promoting excretion of uric acid.

Most of the current crop of nonsedating H_1 antihistaminic drugs contain some structural feature that will prevent their crossing the blood–brain barrier; no such obvious feature is present in the widely used drug **astemizole** (**116**). The preparation of this structurally complex compound starts by addition of *o*-phenylenediamine (**92**) to the protected isothiocyanate **111** to afford the thiourea **112**. The requisite imidazole ring is then formed by treatment with base, a reaction that probably involves addition–elimination of the amine to the thiourea function in the enol form. The imidazole ring nitrogen is then converted to its anion; alkylation with 4-fluorobenzyl bromide gives the intermediate **114**. The protecting group on piperidine nitrogen is then removed by sequential saponification of the urethane followed by acidification with consequent decarboxylation of the amine carbonate. Alkylation of the newly revealed basic nitrogen with 4-methoxyphenethyl bromide leads to astemizole (**115**).[29]

The antiulcer histamine H_2 antagonists, discussed in Chapter 8, which act by blocking histamine stimulated gastric acid secretion, have proven so effective and safe that they are now approved for nonprescription use in the United States. The benzimidazoles discussed below, which achieve the same overall therapeutic effect, act more directly on stomach acid by inhibiting the sodium–potassium pump enzyme responsible for acid secretion. Reaction of methoxylated *o*-phenylenediamine **116** with potassium ethylxanthate, from sodium ethoxide and carbon disulfide, gives the benzimidazole-2-thiol **117**. In a convergent synthesis, the hydroxymethylpyridine **118** is converted to the corresponding chloride **119**. Reaction of **119** with thiol **117** leads to formation of the thioether. Controlled oxidation of sulfur with a single equivalent of peracid leads to the sulfoxide **omeprazole** (**120**).[30]

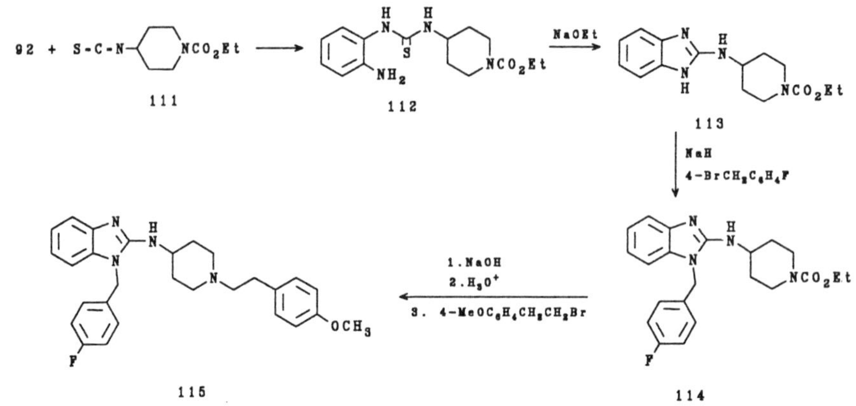

An analogue of **120** shifts one of the methoxyl groups from the benzimidazole to the pyridine ring. Preparation of the requisite intermediate illustrates some typical pyridine chemistry. Thus, aromatic nitration of the pyridine *N*-oxide (**121**) leads to the 4-nitro intermediate **122**. The presence of the N-oxide at the 4-position activates the nitro group toward nucleophilic displacement, a transform achieved by reaction with sodium methoxide that leads to **123**. Heating **123** with acetic anhydride leads to the Polonovski rearrangement and the formal shift of oxygen from the ring nitrogen to the methyl group at the 2-position. Saponification of the product finally leads to the key intermediate **124**. This compound is then converted to the chloride **125**. Reaction of **125** with benzimidazole-2-thiol followed by oxidation of the resulting thioether affords the antiulcer agent **pantoprazole** (**126**).[31]

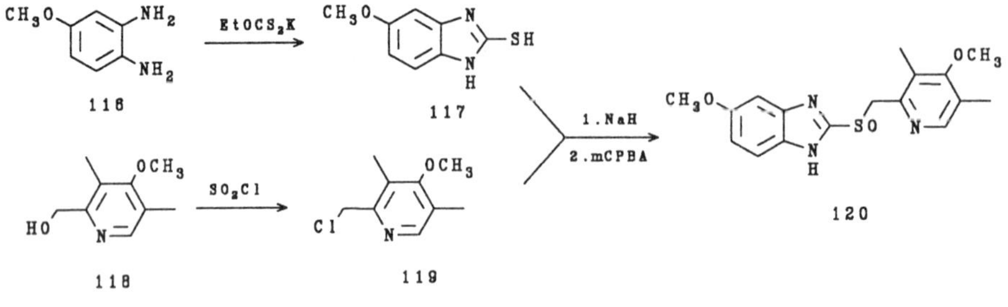

Helminths comprise a very large class of worm-like organisms, many of which parasitize humans. These organisms include nematodes, which are responsible for debilitating gastrointestinal infections. The benzimidazoles provided the first truly broad spectrum antinematodal agents. The efficacy of this class of drugs combined with the development of drug resistant strains led to the development of scores of analogues. The few anthelmintic benzimidazoles discussed below thus do not truly reflect the size of the research effort and have been selected on the basis of chemistry. Note that chemical considerations also caused a distortion of chronology since the first drug to be introduced (in 1961) was in fact **thiabendazole** (**146**). An appropriately substituted *o*-phenylenediamine constitutes the key intermediate to the preparation of all but that last benzimidazole.

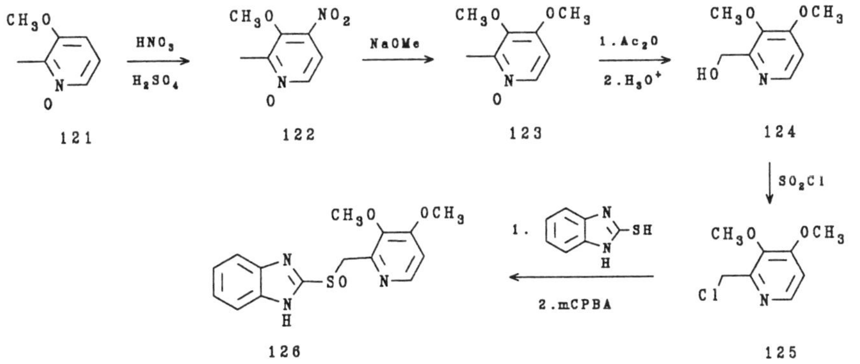

Alkylation of the phenolic group with propyl bromide in aminophenol **127** in the presence of potassium hydroxide gives the ether **128**. Nitration with the usual mixture of nitric and sulfuric acid proceeds at the position ortho to the amido group; saponification followed by reduction of the nitro group then gives the desired diamine **129**. Reaction of this intermediate with the S-methyl ether of thiourea can be visualized as proceeding first from the guanidine **130**, which is obtained by addition to the imine double bond followed by elimination of methyl mercaptan. Cyclization completes construction of the 2-aminobenzimidazole **131**. Acylation with methyl chloroformate results in formation of a urethane on the amino group in contrast to alkylation (**113** → **114**), which proceeds on ring nitrogen. Thus, **oxbendazole** (**132**)[32] is obtained.

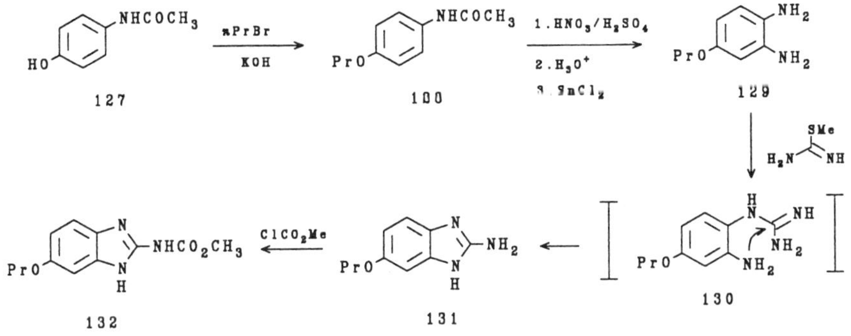

Nucleophilic aromatic substitution provides the key reaction to building the phenylenediamine in a somewhat more complex compound. The starting material is obtained by nitration of the fluorothiophenone **133**. Reaction of **134**, where the good leaving group fluorine is activated by electron-withdrawing groups at the both 2- and 4-positions with amide anion, affords the aniline **135**. The nitro group is then reduced by catalytic hydrogenation to give diamine **136**. Reaction of **136** with the S-methylthiourea, which already includes the requisite methylurethane function, leads directly to the antinematodal benzimidazole **nocodazole** (**137**).[33]

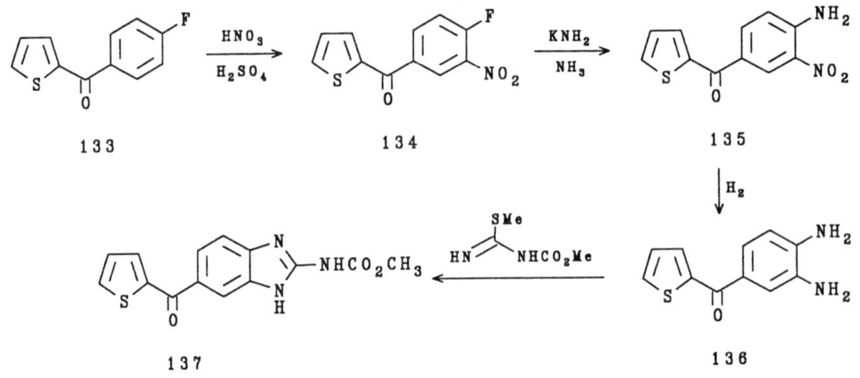

Nucleophilic aromatic displacement is invoked for incorporation of the side chain in yet another benzimidazole. Thus, treatment of 2,5-dinitroacetanilide (**138**) with the anion from mercaptocyclohexane leads to the unusual displacement of one of the nitro groups and formation of thioether **139**. This intermediate is then converted the diamine by sequential reduction and hydrolysis of the amide to give **140**. Condensation with the same thiourea derivative as above affords **dribendazole** (**141**).[34]

Preparation of the prototype drug departs from the phenylenediamine strategy used in all previous examples. Condensation of thiazolo nitrile (**142**) with aniline, which is catalyzed by aluminum chloride, affords the amidine addition product **143**. This compound is converted to its reactive *N*-chloro derivative **144** by reaction with sodium hypochlorite. Treatment of **144** with a base such as potassium hydroxide directly leads to the cyclization product and thus the benzimidazole **thiabendazole** (**146**).[35] This reaction can be rationalized by invoking abstraction of chloride to leave behind a nitrene species such as **145** as the first step; the reactive nitrogen would then readily insert in the CH bond at the ortho position.

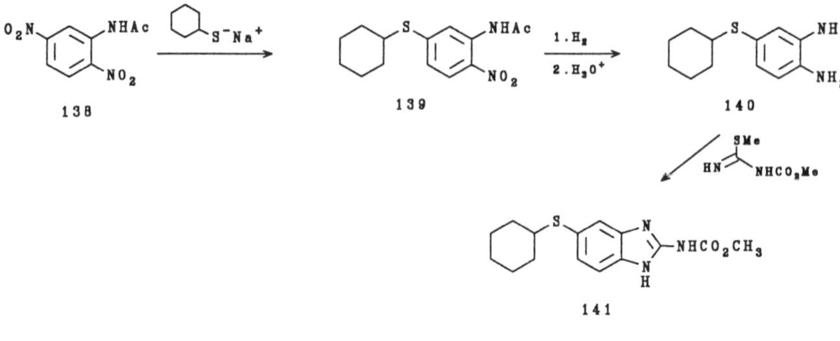

Further substitution of this compound led to a somewhat more potent antinematodal drug. Nitration of **146** under the usual conditions leads to the corresponding nitro derivative **147**. The nitro group is then reduced and the corresponding amine is acylated with isopropyl chloroformate. Thus, **cambendazole** (**148**)[36] is obtained.

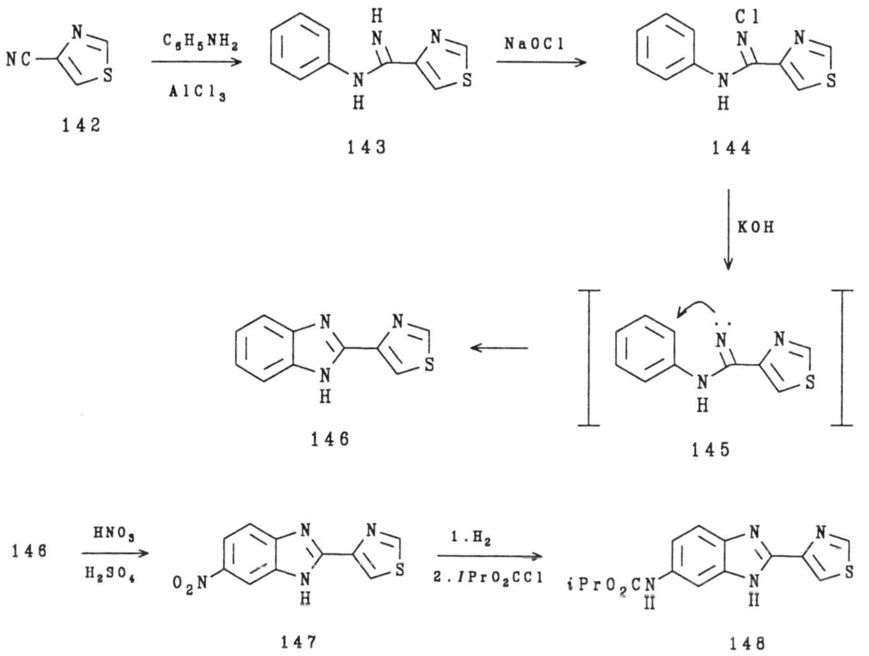

C. Benzothiazoles

Anthelmintic activity is retained when of one of the ring nitrogens in the 2-aminobenzimidazoles is replaced by sulfur. Treatment of the anilinothiol **149** with the thiocarbamoyl chloride from methyl urethane in the presence of triethylamine can be envisaged to yield the corresponding thioamide as the first intermediate; this compound is shown below as its enol tautomer **150**. Thiol exchange with the ring mercaptan will then close the ring to afford a benzothiazole. Thus, the anthelmintic compound **tioxidazole** (**151**)[37] is obtained.

Replacement of the urethane carbonyl function by an aromatic ring leads to a benzothiazole that is described as an immune function modulator. In an analogous approach to that used above, anilinothiol (**152**) is condensed with the thiocarbamoyl chloride **153** again in the presence of base. This reaction leads directly to **frentizole** (**154**).[38]

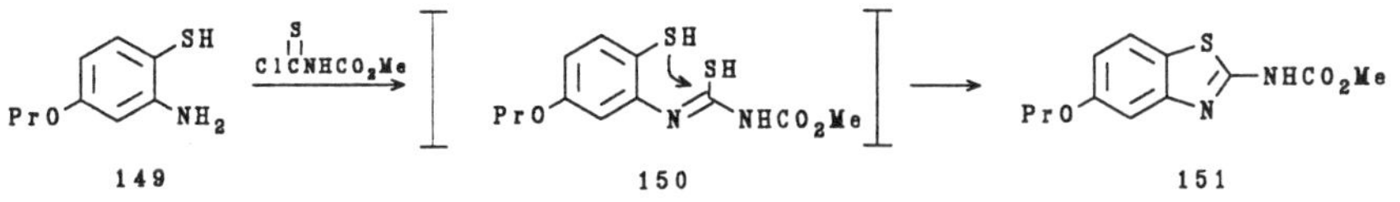

The presence of phosphorus is quite rare among therapeutic agents despite its widespread occurrence in the various organic compounds involved in biochemical

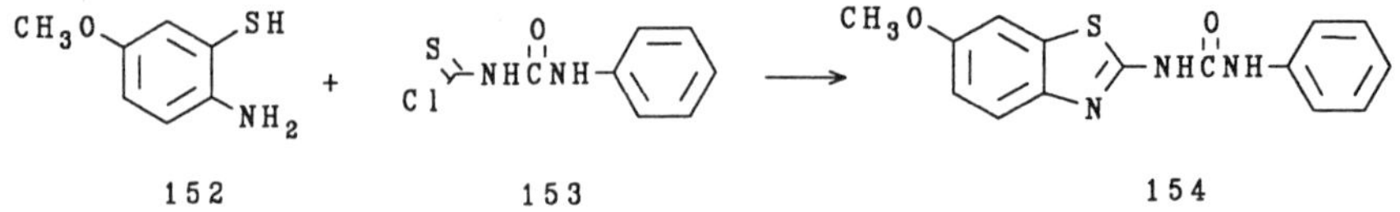

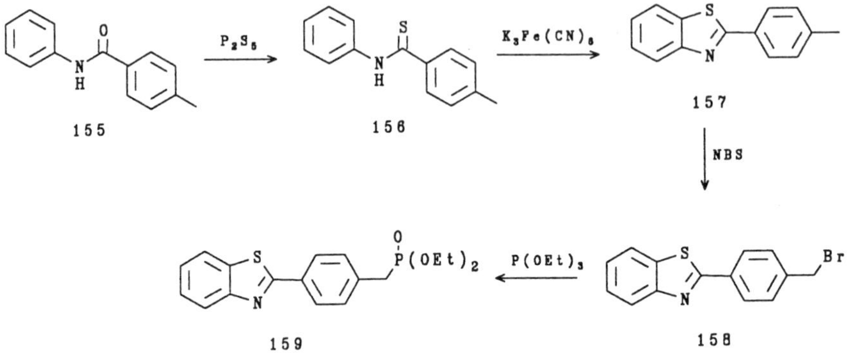

processes. In fact, this may be due to an undeserved reputation of toxicity for phosphorus containing compounds based on the nerve gases and the phosphorus based antitumor alkylating agent cyclophosphamide. The preparation of a phosphonate containing benzothiazole is notable for the different strategy used for preparing the heterocyclic system. Reaction of the anilide **155** with phosphorus pentasulfide leads to the corresponding thioamide **156**. Potassium ferricyanide is a well-known oxidizing agent that has been used in natural product synthesis to bring about radical induced phenol coupling reactions. The reaction of **156** with this reagent, which probably proceeds by way of its enol, affords an analogous coupling product, the benzothiazole **157**. The methyl group on the pendant benzene ring is then brominated with (NBS) to give **158**. Treatment of **158** with triethyl phosphite leads to formation of the phosphon-

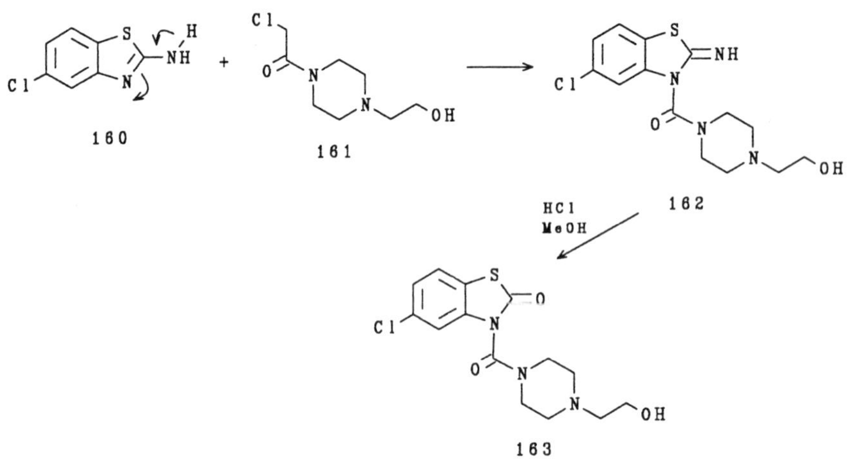

ate **159** by way of the Arbuzov reaction. This product, **fostedil (159)**,[39] is described as a calcium channel blocker that exhibits vasodilator activity.

Interestingly, alkylation of the 2-aminobenzothiazole occurs on the ring nitrogen with a simultaneous shift of the double bond to the exocyclic position in neutral protic solvents. Thus, treatment of chloroaminothiazole **160**, with the complex chloroacetamide (**161**) leads to the 2-imino product of ring alkylation **162**. Hydrolysis of **162** under acidic conditions gives the corresponding benzothiazol-2-one.[40] This product, **tiaramide** (**163**), is described as an antiasthmatic agent that also displays antiinflammatory activity.

REFERENCES

1. Buu Hoi, N.P.; Beaudet, C. U. S. Patent **1961**, 3041042.
2. Johnson, R.A.; Nidy, E.G.; Aiken, J.W.; Crittenden, N.J.; Gorman, R.R. *J. Med. Chem.* **1986**, *29*, 1461.
3. Bertola, M.A.; Quaz, W.J.; Robertson, B.W.; Marx, A.F.; Van der Laken, C.J.; Koger, H.S.; Phillips, G.T.; Watts, P.D. Eur. Patent Appl. **1987**, 233656; *Chem. Abstr.* **1987**, *108*, 20548.
4. Stork, G.; Tomasz, M. *J. Am. Chem. Soc.* **1964**, *86*, 471.
5. Hughes, D.L. *Org. Prep. Proced. Int.* **1993**, *25*, 607.
6. Szmuszkovicz, J.; Glenn, E.M.; Heinzelman, R.V.; Hester, J.B.; Youngdale, G.A. *J. Med. Chem.* **1966**, *9*, 527.
7. Laskowski, S.C. Fr. Patent **1968**, 1551082; *Chem. Abstr.* **1970**, *72*, 43733.
8. Oxford, A.W. German Offen. **1986**; *Chem. Abstr.* **1986**, *105*, 78831.
9. Mooradian, A. U. S. Patent **1976**, 3959309. *Chem. Abstr.* **1976**, *85*, 123759.
10. Mooradian, A. German Offen. **1973**, 2240211. *Chem. Abstr.* **1973**, *78*, 136069.
11. Horlein, U. *Chem. Ber.* **1954**, *87*, 463.
12. Lombardino, J.G. In *Nonsteroidal Antiinflamatory Drugs*, Lombardino, J.G., Ed.; Wiley-Interscience: New York, 1985, p. 267.
13. Berger, L.; Corraz, A.J. German Offen. **1974**, 2337040; *Chem. Abstr.* **1974**, *80*, 108366.
14. Scut, R.N.; Safdy, M.E.; Hong, E. German Offen. **1980**, 2921978; *Chem. Abstr.* **1980**, *92*, 128724.
15. Prous, J.; Castaner, J. *Drugs Future* **1991**, *16*, 804, and references cited therein; this source also describes enantiospecific synthesis for this drug.
16. Canas-Rodriguez, A.; Leeming, P.R. *J. Med. Chem.* **1972**, *15*, 762.
17. Schulte, R.G.; Ehrgott, F.J. Eur. Patent Appl. **1991**, 421749. *Chem. Abstr.* **1991**, *115*, 71392.
18. Nannini, G.; Giraldi, P.N.; Malgara, G.; Biasoli, G.; Spinella, F.; Logemann, W.; Dradi, E.; Zanni, G.; Buttinoni, A.; Tommassini, A. *Arzneim.-Forsch.* **1973**, *23*, 1090.
19. Cornish, E.V.; Lee, G.E.; Eragg, W.R. *Nature* (*London*) **1963**, *197*, 1296.
20. Graf, W.; Girod, E.; Schmid, E.; Stoll, W.G. *Helv. Chim. Acta* **1969**, *42*, 1085.
21. Lisciani, R.; Scorza Barcellona, P.; Silvestrini, B. *Eur. J. Pharmacol.* **1968**, *3*, 157.
22. Plazzo, G. U. S. Patent **1969**, 3470194. *Chem. Abstr.* **1970**, *72*, 110697.

23. Massaroli, G.; Del Corona, L.; Signorelli, G. *Boll. Chim. Farm.* **1969**, *108*, 706.
24. Jerchee, D.; Fischer, H.; Kracht, M. *Liebigs Annalen der Chemie* **1952**, *575*, 173.
25. Hunger, A.; Kebrle, J.; Rossi, A.; Hoffmann, K. *Experientia* **1957**, *13*, 400.
26. Wikel, J.H.; Paget, C.J.; DeLong, D.C.; Nelson, J.D.; Wu, C.Y.E.; Paschal, J.W.; Dinner, A.; Templeton, R.J.; Chaney, M.O.; Jones, N.D.; Chamberlin, J.W. *J. Med. Chem.* **1980**, *23*, 368.
27. Crowell, T.A. U.S. Patent **1984**, 4258198. *Chem. Abstr.* **1984**, *100*, 121077.
28. Rayemaekers, A.H.M.; Freyne, E.J.E.; Sanz, G.C. Eur. Patent Appl. **1988**, 260744; *Chem. Abstr.* **1988**, *109*, 73437.
29. Janssens, F.; Luyckx, M.; Stokbroekx, R.; Torremans, J. U. S. Patent **1981**, 4219559. *Chem. Abstr.* **1981**, *94*, 30579.
30. Lindberg, P.; Nordberg, P.; Alminger, T.; Braudstrom, A.; Wallmark, B. *J. Med. Chem.* **1986**, *29*, 1327.
31. Kohl, B.; Sturm, E.; Senn-Billfinger, J.; Simon, A.W.; Krueger, U.; Schaefer, H.; Rainer, G.; Figala, V.; Klemm, K. *J. Med. Chem.* **1992**, *35*, 1049.
32. Actor, P.P.; Pagano, J.F. Br. Patent **1968**, 1123317; *Chem. Abstr.* **1968**, *69*, 722.
33. Van Gelder, J.L.H.; Rayemaekers, A.H.M.; Roevens, L.F.C. German Offen. **1971**, 2029637; *Chem. Abstr.* **1971**, *74*, 100047.
34. Gyurik, R.J.; Kingsbury, W.D. U. S. Patent **1981**, 4258198. *Chem. Abstr.* **1981**, *95*, 7284.
35. Grenda, V.J.; Jones, R.E.; Gae, G.; Sletzinger, M. *J. Org. Chem.* **1959**, *30*, 259.
36. Hoff, D.R.; Fischer, M.H.; Bochis, R.J.; Lusi, A.; Waksmunski, J.R.; Egerton, J.R.; Yakstis, J.J.; Cuckler, A.C.; Campbell, W.C. *Experentia* **1970**, *26*, 550.
37. Paget, C.J.; Sands, J.L. German Offen. **1970**, 2003841; *Chem. Abstr.* **1970**, *73*, 87920.
38. Paget, C.J.; Sands, J.L. U. S. Patent **1978**, 4088768. *Chem. Abstr.* **1978**, *90*, 137814.
39. Yoshino, K.; Kohno, T.; Uno, T.; Morita, T.; Tsikamota, G. *J. Med. Chem.* **1986**, *29*, 820.
40. Masuoka, T. *Jpn. Kokai* **1980**, 55145673. *Chem. Abstr.* **1980**, *94*, 192376.

CHAPTER 11

SIX-MEMBERED HETEROCYCLES FUSED TO A BENZENE RING

1. COMPOUNDS WITH ONE HETEROATOM

A. Coumarins

Epidemiological investigations of outbreaks of cattle deaths due to hemorrhage in the 1930s led to the identification of spoiled sweet clover as the causative agent. The natural product isolation and structural studies that followed unambiguously pointed to the condensation product (**1**) of coumarin with formaldehyde as the cause for these deaths. This finding was confirmed when **1**, now known as **dicoumarol** (**1**) was discovered to be is a potent inhibitor of blood coagulation in mammalian species. It has subsequently been established that this agent and its congeners inhibit the synthesis by the liver of a series of peptide factors involved in the intricate blood clotting cascade; this finding accounts for their lack of efficacy in vitro.

Anticoagulants play an important role in postsurgical recovery by preventing wound induced thrombus formation. Dicoumarol (**1**) has now been largely superseded in clinical use by an analogue that contains a single coumarin moiety. The shorter biological half-life of this agent, **warfarin** (**5**), allows more accurate control of blood

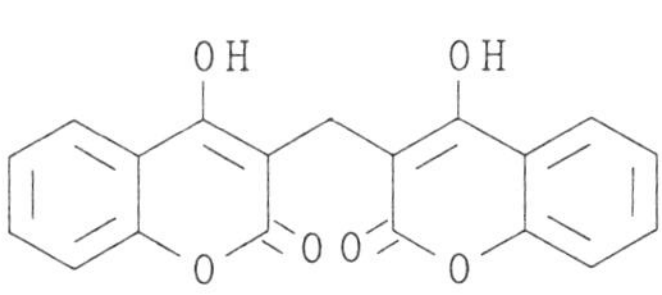

OH OH

O O O O

1

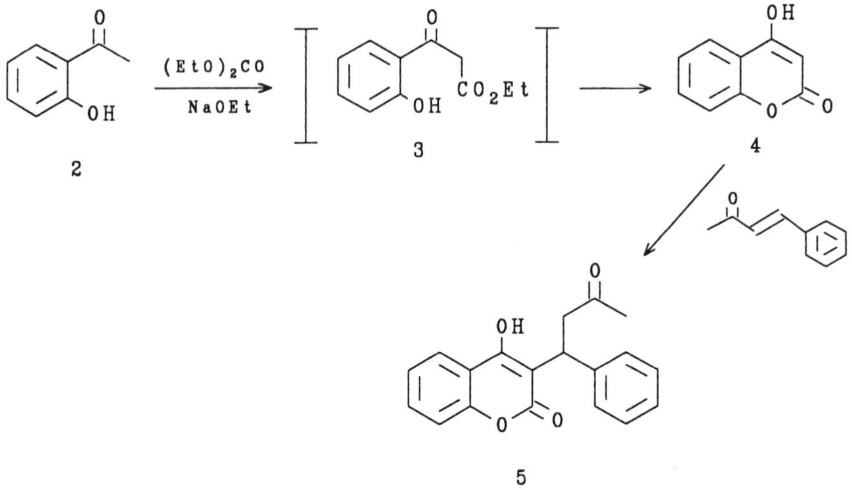

levels and thus a better chance to avoid the danger of hemorrhage due to elevated clotting times. Note that both compounds have been used extensively as rat toxins. The generic name warfarin alludes to the fact that the drug was developed at the Wisconsin Alumni Research Foundation (WARF), which was initially funded by patent royalties from the Steenbock process for increasing Vitamin D activity in milk by UV irradiation.

The succinct synthesis of this agent starts with condensation of *o*-hydroxyacetophenone (**2**) with ethyl carbonate to give the β-ketoester **3** as the presumed intermediate. Attack of the phenoxide on the ester grouping will lead to cyclization and formation of the coumarin **4**. Conjugate addition of the anion from that product to methyl styryl ketone gives the corresponding Michael adduct and thus warfarin (**5**).[1]

A somewhat more complex coumarin, which includes basic nitrogen, has been used as a coronary vasodilating agent. The key reaction in the synthesis of this compound involves Friedel–Crafts-like reaction of the alkylation product **6** from ethyl acetoacetate with 2-chlorotriethylamine and resorcinol to give initially an acylation product such as **7** (no stereochemistry implied). This compound cyclizes to the coumarin **8** under reaction conditions. The remaining free phenolic group is then alkylated with ethyl bromoacetate to give the corresponding ether and thus **chromonar** (**9**).[2]

Recognition of the photosensitizing effect of the naturally occurring furanocoumarin psoralin (desmethoxy **16**) led to trials on its utility for the treatment of skin diseases such as psoriasis. The partial effectiveness of this compound led to the preparation of synthetic analogues. The two commercially available drugs **methoxsalen** (**16**) and **trioxsalen** (**22**) are used in a procedure that goes by the acronym PUVA (psoralen and UV A irradiation) for treatment of psoriasis and other skin diseases.

The preparation of **16** begins with phosphorus oxychloride mediated acylation of pyrogallol (**10**) with chloroacetic acid to give the chloromethyl ketone **11**. Treatment of this intermediate with potassium carbonate leads to displacement of halogen by the phenoxide and formation of a benzofuranone ring; the carbonyl is then reduced

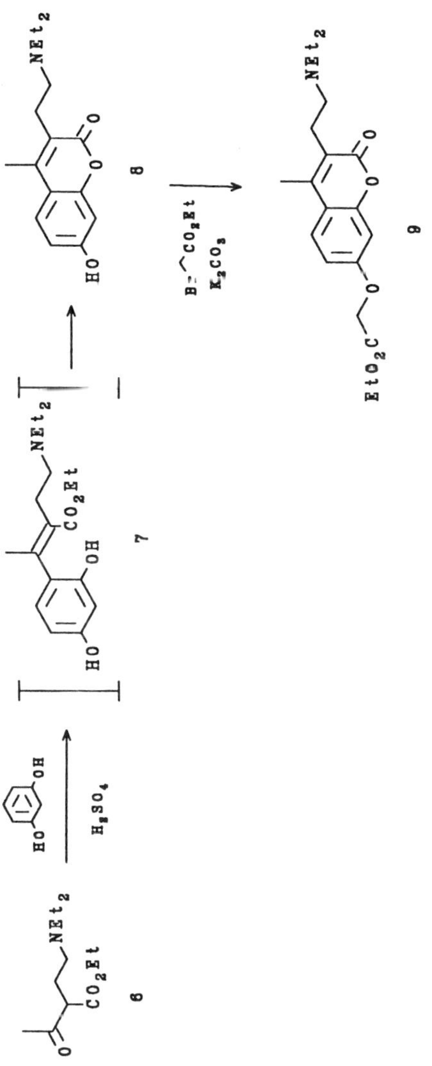

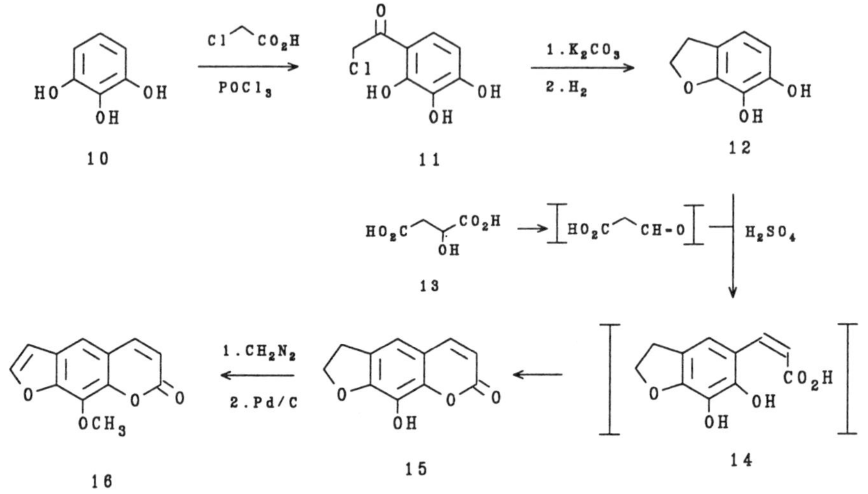

catalytically to a methylene group to afford the benzodihydrofuran **12**. This product is then allowed to react with malic acid (**13**) in concentrated sulfuric acid. The overall reaction can be viewed as involving initial decarboxylation of **13** to give formylacetic acid; the enol from the latter then attacks the highly activated aromatic ring to give the cinnamic acid derivative **14** (no stereochemistry implied). Esterification completes the formation of the coumarin and yields **15**. The remaining free phenolic group is then etherified with diazomethane and the resulting methyl ether is dehydrogenated over palladium on charcoal. Thus, methoxsalen (**16**)[3] is obtained.

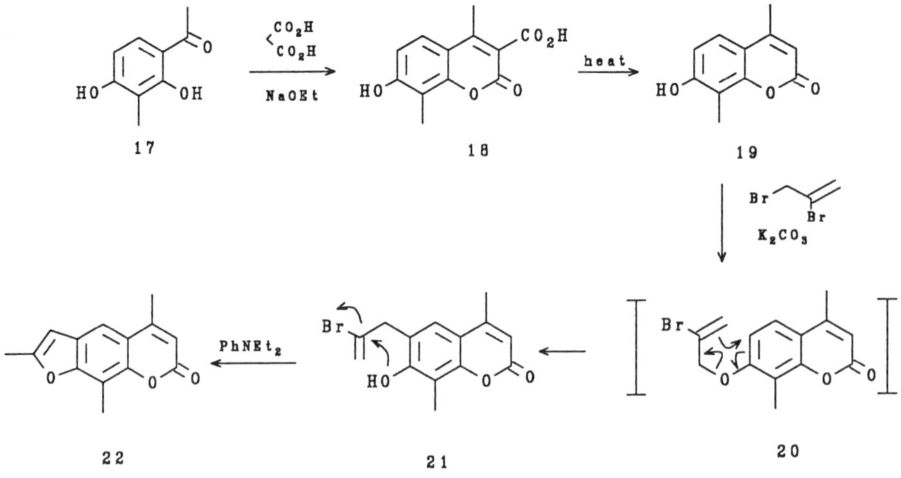

The strategy used to prepare the analogue trixosalen (**22**) differs in the order in which the heterocyclic rings are built. Condensation of malonic acid and the substituted hydroxyacetophenone **17** probably proceeds to give initially the malonylidene compound from Knoevnagel-like reaction with the carbonyl group. This compound cyclizes to the coumarin **18** under the reaction conditions. The superfluous acid is then caused to decarboxylate by heating to afford **19**.[4] Reaction of **19** with 2,3-dibromopropylene first affords the simple *O*-allyl ether **20**. Compound **20** then undergoes electrocyclic Claisen rearrangement to the *C*-allylated derivative **21**. Heating **21** in a high-boiling basic solvent such as diethylaniline results in displacement of bromine by alkoxide and formation of a furanylidene ring. The double bond then shifts into the ring to afford a furan and thus trioxsalen (**22**).[5]

B. Chromones

The chromone **cromolyn sodium** (**27**) was the forerunner of a class of antiallergic and antiasthmatic drugs that act at one of the earliest stages of the allergic reaction. Detailed experiments, actually conducted after the drug's clinical effectiveness had been confirmed, suggested that the compound inhibited the release of mediators of the allergic reaction from mast cells. This drug is in fact applied topically to the lung by insufflation as its sodium salt. The preparation starts much like that of β-blockers by reaction of the phenol **23** and epichlorohydrin with the important difference that the reaction is conducted with a controlled amount of the epoxide. The initially formed glycydil ether **24** thus reacts with a second phenoxide ion to afford the double ether **25** of glycerol. This product is then condensed with diethyl oxalate in the presence of base. The initially formed acylation product **26** then undergoes internal hydroxyl exchange to form a coumarone ring. The structure shown for the initial adduct is presented for illustrative reasons only; the sequence is just as likely to proceed by stepwise addition–cyclization of the two halves of **25**. Saponification of the product, without the usual neutralization step, affords cromolyn sodium (**27**).[6]

The Kostanecki reaction, which involves esterification of *o*-acylphenol followed by aldol cyclization of the resulting ketoesters, provides a convenient entry to chromones. As a matter of passing interest, the subclass of 2-phenyl derivatives obtained by use of benzoyl chloride are named flavones. Thus, reaction of dihydroxypropiophenone (**28**) with benzoyl chloride in the presence of sodium benzoate can be visualized as involving initial formation of the ester **29**. Internal condensation of this dicarbonyl

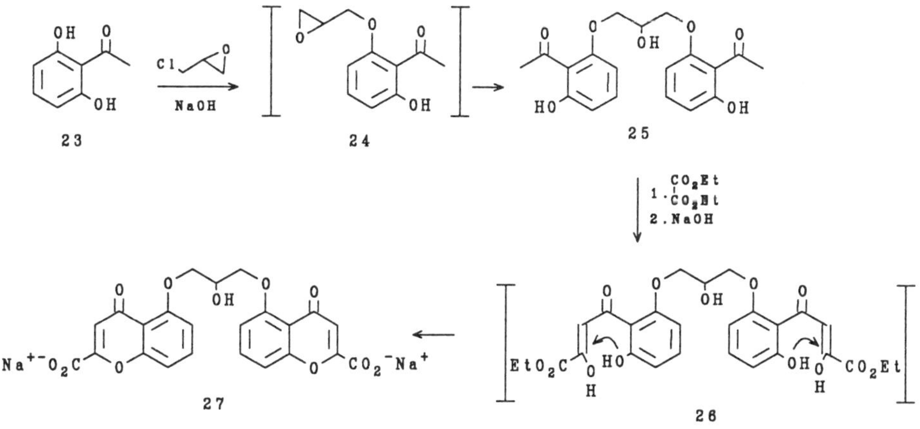

23 24 25 26 27

NaOH

1. CO_2Et–CO_2Et 2. NaOH

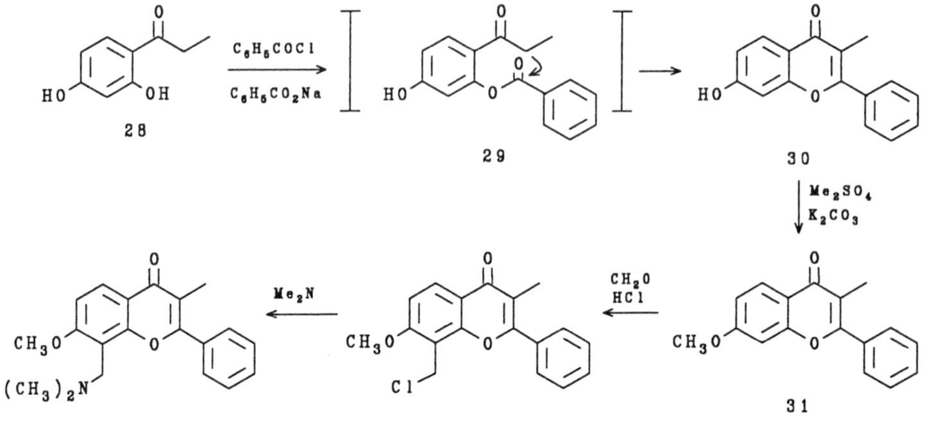

derivative, which can proceed in only one direction, provides a new ring and results in the flavone **30**. The remaining free phenolic group is then converted to its methyl ether **31** by means of dimethyl sulfate. Reaction with formaldehyde in the presence of dry hydrogen chloride serves to introduce a chloromethyl group (**32**), interestingly at the more hindered position. Displacement of chlorine by dimethylamine affords the respiratory stimulant **dimefline** (**33**).[7]

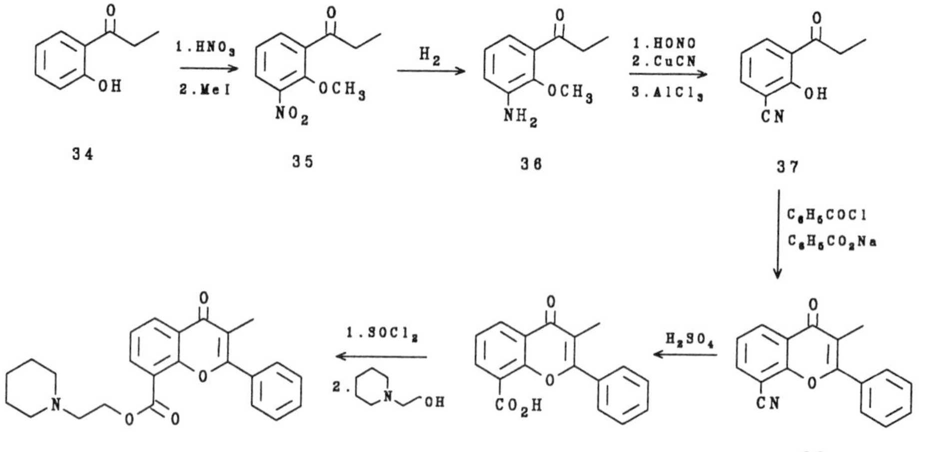

Nitration of hydroxypropiophenone (**34**) followed by conversion of the phenol to its methyl ether by means on methyl iodide provides the intermediate **35**; the nitro group is then reduced to the corresponding amine **36** by catalytic reduction. This group is then replaced by a nitrile group by successive conversion by means of nitrous acid to the diazonium salt followed by treatment with cuprous cyanide. Reaction with aluminum chloride removes the methyl ether to afford the ortho acylphenol **37**. This compound is converted to the chromone **38** as shown above by reaction with benzoyl chloride and sodium benzoate. The nitrile is hydrolyzed to the carboxylic acid **39** by means of sulfuric acid. The acid is then converted to its acid chloride by means of thionyl chloride and this compound is treated with 2-(*N*-piperidyl)ethanol. Thus, **flavoxate** (**40**),[8] a muscle relaxant whose name reflects its flavone nucleus is obtained.

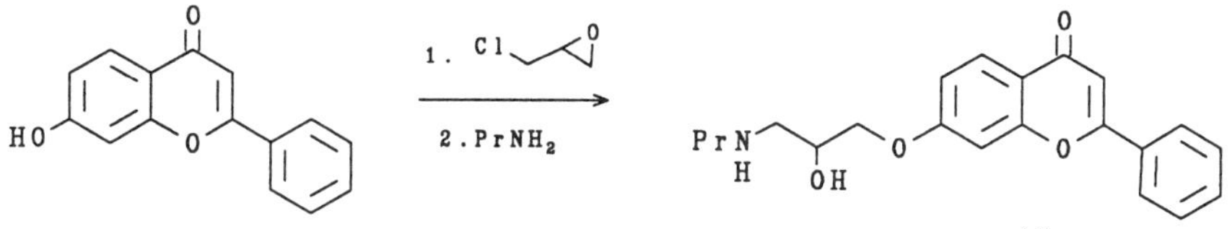

The extremely wide structural latitude for the aromatic nucleus for β-blockers has been noted repeatedly. Although the antihypertensive agent **flavodilol** (**42**) is seemingly one more contributor to this theme, this compound seems not to interact with β-adrenergic receptors. This agent is obtained by subjecting **41** to the standard glycidyl ether formation followed by reaction with propylamine.[9]

C. Benzopyrans

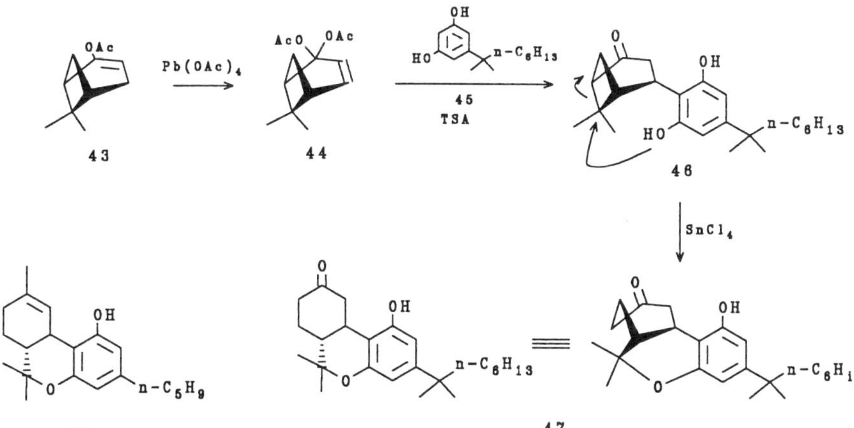

The well-known active principle of marijuana, tetrahydrocannabinol (THC) (**48**) exhibits a wide spectrum of activities in the CNS, the most notable of which is its euphoriant and hallucinogenic effects. The possible antiemetic activity of the drug has led to its limited approval for this use in special situations. A synthetic analogue has been investigated clinically for this indication. Like the classical syntheses of the natural product itself,[10] the route to the analogue also involves coupling of a terpene with a resorcinol derivative as a key step. The starting enol acetate **43** is obtained by treatment of nopinone, itself available by oxidation of the natural product β-pinene, with acetic anhydride. Oxidation with lead tetraacetate leads to the acetic acid acetal **44** of the dehydrogenated ketone. Reaction of **44** with the resorcinol derivative **45** in the presence of toluenesulfonic acid leads to alkylation of the aromatic ring and formation of the coupling product **46**. The oxonium ion formed by treatment of **46** with the strong Lewis acid stannic chloride attacks the quaternary center on the bicyclic fused cyclobutane ring from the backside; this reaction forms the new pyran ring and at the same time opens the strained four-membered ring. The trans stereochemistry of the new ring fusion in the product **nabilone** (**47**) follows from the stereochemistry of the starting material and the direction of attack.[11]

The leukotrienes, products of the lipoxygenase branch of the arachidonic acid cascade (see Chapter 1) are now recognized as important mediators of allergic responses. Many of the antagonists of these compounds consist of long-chain acids that bear a vague resemblance to the fatty chains of the leukotrienes. A tetrahydrobenzopyran ring provides the nucleus for one of these antagonists. The requisite ring system is formed by reaction of the dihydroxyacetophenone **49** with ethyl oxalate in the presence of strong base; formation of the chromone **50** follows an analogous course to the one outlined for **cromolyn** (**27**). Exhaustive hydrogenation reduces both the double bond and the carbonyl group to afford **51**. Acylation with acetyl chloride in the presence of aluminum chloride then affords the ketone **52**. The free phenol in the product is next alkylated with 5-acetoxy-1-bromopentane; the ester is then saponified and the resulting alcohol is converted to its mesylate by reaction with methanesulfonyl chloride to give **53**. Displacement of this good leaving group with the alkoxide from the complex resorcinol derivative **54** results in formation of the ether from the

HO OH 49 → (CO_2Et, CO_2Et / NaOEt) → HO O CO_2Et 50 → (H_2) → HO O CO_2Et 51 → (CH_3COCl / $AlCl_3$) → HO O CO_2Et 52 → (1. $AcO(CH_2)_5Br$ / 2. NaOH / 3. MsCl) → $CH_3SO_3(CH_2)_5$ O O CO_2H 53 → (OH OH 54) → O OH O O O CO_2H 55

nonchelated phenolic group at the ring position para to the carbonyl group. Thus, **ablukast** (**55**)[12] is obtained.

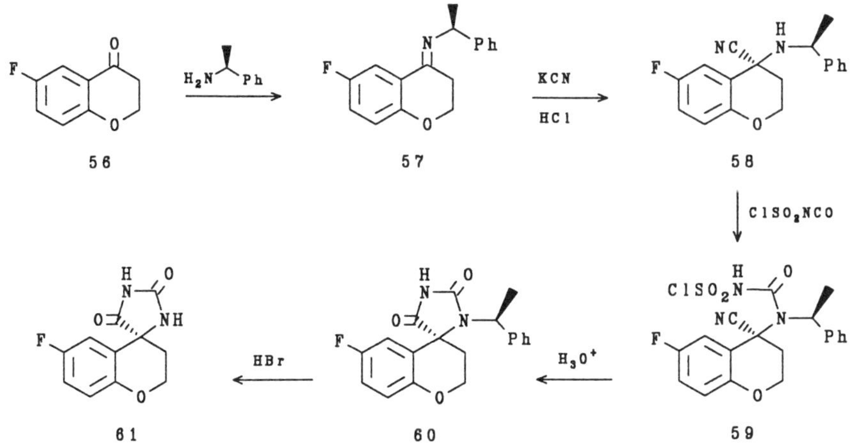

The cataracts that can appear even in those diabetics whose disease is under control has been attributed to accumulation of sorbitol in the eye. This accumulation results from reduction of glucose by elevated levels of the enzyme aldose reductase, which accompanies the disease. Inhibitors of this enzyme have been investigated as a means for controlling such cataracts. Known agents, as would be expected with enzyme inhibitors, tend to show marked differences in potency between optical isomers. The enantioselective synthesis of one of these compounds begins with the formation of an imine of dihydrochromone (**56**) with the (*S*) form of the chiral auxiliary α-methylbenzylamine; this compound will also in the end provide one of the required nitrogen atoms. Reaction of the imine (**57**) with hydrogen cyanide leads to formation of the α-aminonitrile **58**; the adjacent chirality leads to formation of this derivative as a single enantiomer. The remaining two carbons for the spirocyclic ring are then incorporated by addition of chlorosulfonylisocyanate to form the derivatized urea **59**. Reaction with hydrochloric acid leads to loss of the chlorosulfonyl group with consequent addition to the nitrile of the thus revealed terminal primary amide group. The first-formed imine hydrolyzes in the reaction mixture to afford the spirocyclic hydantoin **60**. Treatment with hydrogen bromide cleaves the benzylic bond causing loss of the phenethyl moiety, affording **sorbinil** (**61**) as a single enantiomer.[13]

A structurally unusual β-blocker that uses a second molecule of itself as the substituent on nitrogen is included here in spite of the ubiquity of this class of compounds. Exhaustive hydrogenation of the chromone **62** leads to reduction of both the double bond and the carbonyl group, as in the case of **50**. The carboxylic acid is then reduced to an aldehyde by means of diisobutylaluminum hydride (DIBAL-H). Reaction of **63** with the ylide from trimethylsulfonium iodide gives the oxirane **64** via the addition–displacement process discussed earlier (see Chapters 3 and 8). Treatment of an excess of the epoxide with benzylamine leads to addition of two equivalents of this compound with the basic nitrogen. The product is then debenzylated by catalytic reduction over palladium to afford **nebivolol (65)**.[14] The presence of four chiral centers in the product predicts the existence of eight chiral pairs.

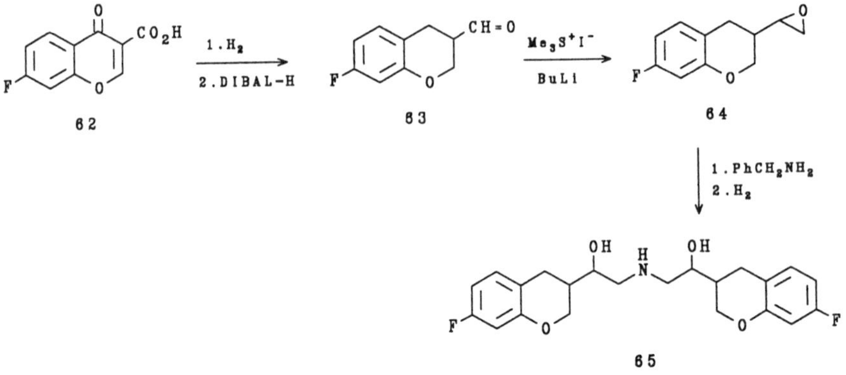

A benzoisopyran provides the nucleus for an appetite suppressing agent that incorporates in its structure a significant portion of the prototype anorexigenic agent, amphetamine. Carefully controlled reaction of the lactone **66** with DIBAL-H leads to the lactol **67**. This compound is then treated with nitromethane in the presence of strong base. The carbanion reacts first with the open form of the lactol to give the addition product **68**; this compound would be expected to dehydrate to the nitrovinyl derivative **69**. Cyclization to the benzoisopyran **70** may occur either by ether formation from the diol **68** or the conjugate addition of alkoxide to the derivative **69**. Catalytic reduction of the nitro group leads to the primary amine and thus to **fenisorex (71)**.[15]

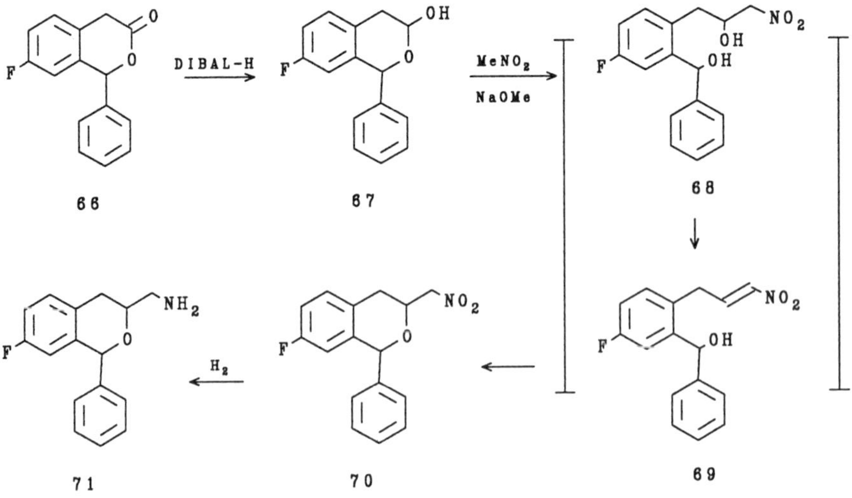

D. Quinolines

1. Antimalarial Compounds

Malaria constitutes one of the most widespread infectious diseases in humans; over 270 million new infections and 2 million deaths from the disease are estimated to occur worldwide each year. The causative protozoan plasmodia, comprised largely of the species *falciparum* and *vivax*, undergo a complex life history cycling between mosquitoes and vertebrates as hosts. This multiplicity of forms complicates approaches to chemotherapy. In spite of the fact that this is one of the earliest recorded human diseases, the role of the *anopheles* mosquito as the infecting agent was not recognized until 1898. Some progress was made in controlling the disease in the immediate post-World War II period, particularly in the United States through mosquito control using the now banned insecticide, DDT. The first and still widely used drug for treating this disease comes from the adventitious finding of the antimalarial activity of cinchona bark in the early seventeenth century. The active drug **quinine** (**77**), which forms the major part of the 25 alkaloids from the bark, was isolated in pure form by Pelletier in 1820 though the structure was not determined before the present century. In fact, much of the research on new synthetic antimalarial drugs was spurred by war needs. More recent work has been prompted by the emergence of *plasmodia*, which are resistant to currently available drugs. Note that the prevalence of malaria in tropical and subtropical regions results in the high frequency of disease in Third World countries. Their low per capita incomes means that few inhabitants, or for that matter governmental health departments, will be able to afford even the cheapest antimalarial drugs. To a considerable extent, this circumstance has also detracted from commercially guided research on antimalarial drugs.

The total synthesis of quinine (**77**) by Woodward and Doering,[16] which confirmed the structure of this drug, marked the first of an extensive series of elegant syntheses of natural products that led ultimately to the award of a Nobel prize. The key step in a more recent concise synthesis of this molecule is based on recognition of the nucleophilic character of ylides. Thus, reaction of the choroquinoline (**72**) with the phosphorane from methyltriphenylphosphonium iodide initially leads to displacement of chlorine by the negative charge on the methylene group; reaction of the newly formed quinolylmethyl phosphonium intermediate with excess phosphorane will lead to generation of the ylide **73**. Condensation of this reagent with the aldehyde in the piperine **74** leads to the olefin **75**. The acetyl protecting group is then removed by hydrolysis of the amide; conjugate addition of the basic free piperidine amino group to the reactive double bond conjugated to quinoline leads to formation of the quinuclidine ring to afford desoxyquinine (**76**).[17] Oxidation of **76** in the presence of base leads to a statistical mixture of quinine (**77**) and its epimer **quinidine**,[18] a compound that has been used extensively as an antiarrhythmic drug.

A sizeable group of antimalarial compounds is based on the quinoline nucleus; the high frequency with which this moiety occurs is probably due more to bias toward selection of candidates that include a fragment present in quinine than to systematic molecular dissection in the massive random screening programs sponsored by US Government agencies; two of the more important early synthetic antimalarial drugs,

CH_3O Cl N 72 $Ph_3P^+CH_3I^-$ BuLi [CH_3O PPh_3 N 73] O-HC N Ac 74

CH_3O HO H N 77 ← CH_3O H N 76 ← NaOH CH_3O N Ac 75

quinacrine and **methacrine**, are in fact based on an acridine nucleus; their discussion is deferred to Chapter 13. One of the classical entries to 4-substituted quinolines involves condensation of an aniline with a 1,3-dicarbonyl fragment. Reaction of diethyl 2-ketoglutarate, shown as its enol tautomer **79** with 3-chloroaniline (**78**), leads to the imine **80**. Heating **80** in a high-boiling solvent leads to displacement of an ethoxyl fragment from the ester with consequent cyclization; saponification of the product leads to the acid **81**. Heating **81** leads to loss of the carboxyl group; the enol group at the 4-position is then converted to the chloride **82** by the now familiar reaction with phosphorus oxychloride. Displacement of halogen by the primary amine in 2-amino-5-(diethylamino)pentane leads to the very widely used antimalarial compound **chloroquine** (**83**)[19]; the utility of this drug is threatened by the increasing development of resistant strains of plasmodia.

Inclusion of aromatic rings as part of the side chains results in quite potent agents, possibly because the rigid rings better define the position of the basic nitrogen. Reaction of *p*-hydroxyacetanilide (**84**) with formaldehyde and dimethylamine affords the corresponding Mannich product; hydrolysis of the acetamide leads to the aniline **85**. Treatment of **85** with dichloroquinoline (**82**) leads to displacement of chlorine on the heterocyclic ring and formation of **amidoquine** (**86**).[20]

Cl NH_2 78 + EtO_2C HO CO_2Et 79 → EtO_2C Cl N CO_2Et 80 1. heat 2. NaOH → OH Cl N CO_2H 81

1. heat 2. $POCl_3$ ↓

Cl Cl N 82 ← H_2N NEt_2 HN NEt_2 Cl N 83

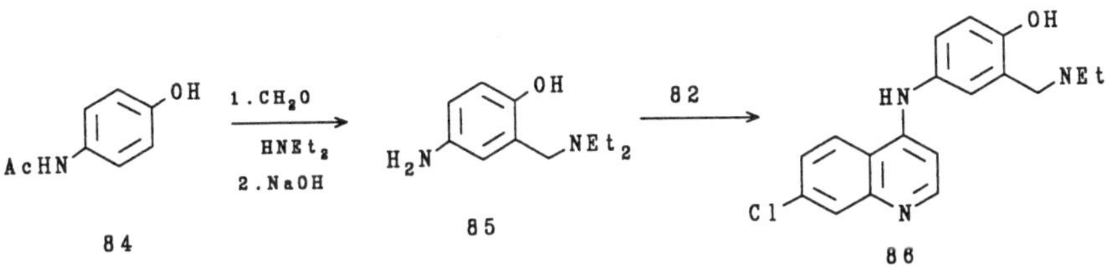

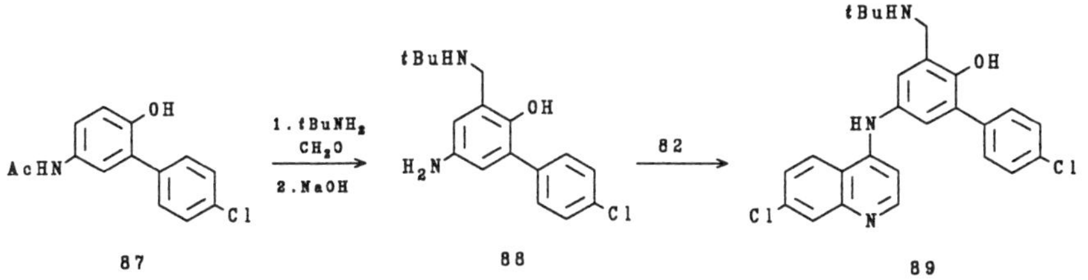

In a closely related example, Mannich reaction of the somewhat more complex phenol **87** with formaldehyde and *tert*-butylamine gives the aminomethylated product **88** after removal of the acetamide protecting group. Alkylation with the dichloroquinoline (**82**) in this case also proceeds on aniline nitrogen. The selectivity over the more basic secondary side nitrogen can probably be ascribed to steric hindrance about the latter. Thus, **tebuquine** (**89**)[21] is obtained.

The starting quinoline (**91**), for analogues that include a methyl group at the 3-position, is prepared by a modification of the scheme that consists of using ethyl 3-methyl-2-ketoglutarate (**90**) as the starting material. Alkylation of **91** with 2-amino-5-(diethylamino)pentane yields **sontoquine** (**92**).[22]

In fact, the first synthetic quinoline antimalarial agents carried the basic side chains at the 8- rather than the 4-position as found in quinine, lending further credence to their origin from random screening instead of molecular dissection programs. The starting quinoline **95** is in this case obtained by reaction of the substituted aniline **93** with glycerol in sulfuric acid and nitrobenzene. It is known that the first step consists of oxidation and dehydration of glycerol to acrolein. Studies on the mechanism of the quinoline synthesis from unsaturated carbonyl compounds and anilines (Skraup reaction) suggest that the first intermediate consists of a Michael adduct-like adduct such as **94**. This compound cyclizes to a dihydroquinoline under strongly acidic conditions; the presence of oxidants leads to aromatization of the heterocyclic ring to give quinoline **95**. The nitro group is then reduced to an amine by any of several means such as hydrogenation to give **96**. Reductive alkylation of this primary amino group with 1-diethylaminopentan-4-one gives the antimalarial drug **pamaquine** (**97**).[23]

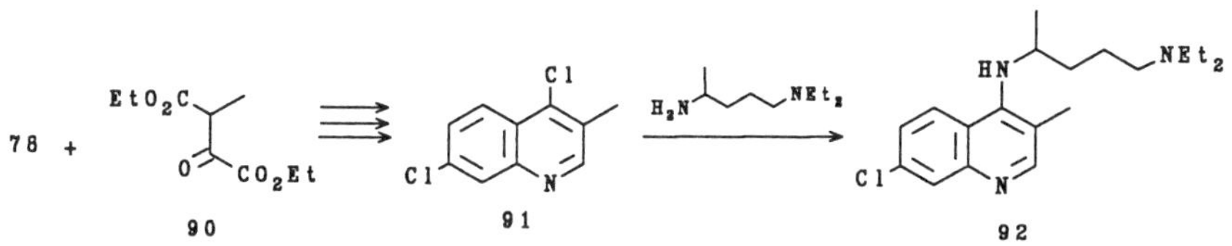

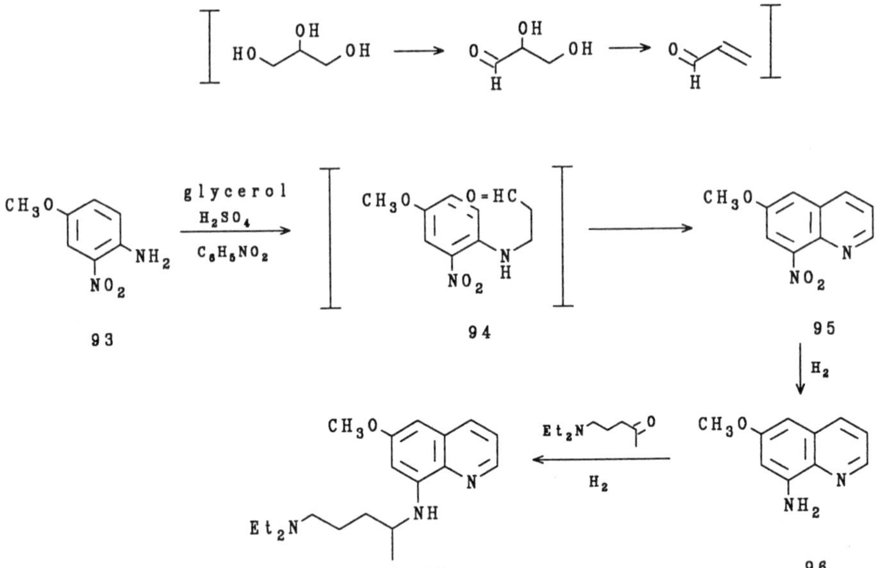

One of the most recently introduced antimalarial drugs, the product of a U.S. Army sponsored program at Walter Reid, represents a return to the structure of **quinine** in that the basic nitrogen is connected to the quinoline at a hydroxymethyl group at the 4-position on the quinoline ring. The unexpectedly long half-life of this drug, **mefloquine** (**104**), in serum, 385 ± 150 hr (16 days!),[24] permits a dosage regime that consists of a single oral tablet taken at weekly intervals. A very concise synthesis starts with the displacement of bromine in 5-bromohexene (**98**) by potassium phthalimide to give the protected amine **99**. Palladium mediated coupling of this olefin with the substituted 4-bromquinoline **100** affords the vinylation product **101**. The double bond is then oxidized to the corresponding epoxide **102** with peracid. The phthalimide protecting group is removed in the usual way by treatment with hydrazine to afford the transient primary amine **103**. This function then opens the oxirane with consequent formation of a piperidine ring; opening via the alternate epoxide bond is disfavored since it would lead to a seven-membered ring. Thus, the antimalarial drug mefloquine (**104**)[25] is obtained.

2. Other Quinolines

The quinoline nucleus also appears in many of the compounds that have activity against other parasitic protozoa and amoeba. The 7,8-dialkoxy-4-hydroxyquinoline-carboxylates, for example, comprise an important class of drugs that are toxic to coccidia, a protozoan that can devastate commercial poultry flocks. The structures of these compounds interestingly foreshadow the wide spectrum quinolone "acin" antibiotics discussed below. Compounds within the cocciodostatic series vary mainly in the nature of the substituents on the oxygen atoms on the carbocyclic ring. The key step in the syntheses of these agents is the formation of the quinoline ring. A typical example involves the addition to diethyl ethoxymethylenemalonate (EMME) of the

98 + KN(phthalimide) → 99 (NPhth)

99 + 100 $\xrightarrow{(oMeC_6H_4)_3P,\ Pd(OAc)_2}$ 101

101 $\xrightarrow{mCPBA}$ 102

102 $\xrightarrow{N_2H_4}$ [103] → 104

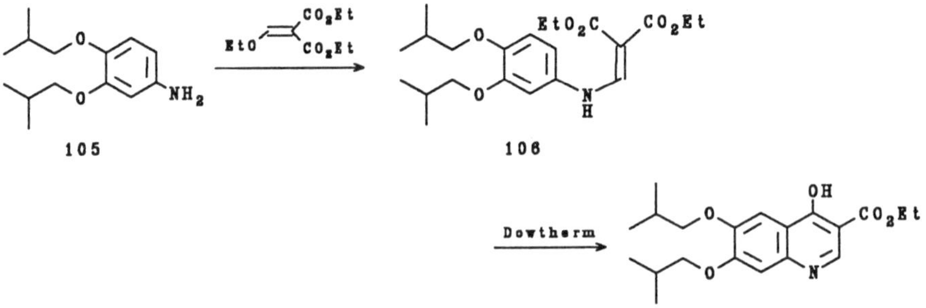

bis-*iso*-butylalkoxyaniline (**105**), which is obtained by reduction of the nitration product from the catechol ether; elimination of ethanol affords the adduct **106**. Heating **106** in a high-boiling ether such as Dowtherm leads to cyclization by displacement of ethoxide and formation of a transient quinolone; this spontaneously enolizes to the hydroxyquinoline **buquinolate** (**107**).[26]

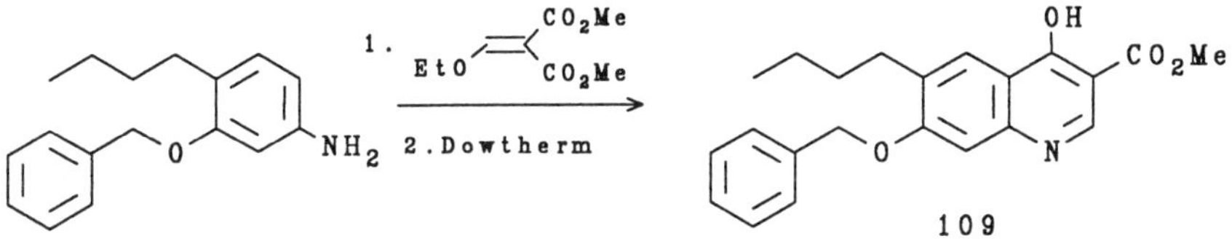

The reaction follows the same course when one of the alkoxy groups in the starting material is replaced by alkyl. Thus, condensation of aniline **108** with EMME followed by heating of the product in Dowtherm leads to the poultry coccidiostat **nequinate** (**109**).[27]

The antiprotozoal spectrum is apparently changed by replacing the alkoxy substituent by nitrogen; the resulting analogue manifests mainly antiplasmodial rather than coccidiostatic activity. That the nitration of the aniline **110** perhaps surprisingly proceeds at the position meta to the amino group can be rationalized by the fact that the electron density will be higher para to the alkyl group than para to the protonated aniline. Catalytic reduction of the newly introduced nitro group leads to the diamine **111**. This compound affords **amquinate** (**112**), when subjected to the quinoline-forming reaction sequence.[28]

Antimalarial activity also predominates in a quinoline that bears a diaminoalkyl side chain at a rather different position from the agents noted in Section 11.D.2. Thus, Mannich condensation of the hydroxyquinoline **113**, with formaldehyde and *N,N*-diethylpropylenediamine, affords **clamoxyquin** (**114**).[29]

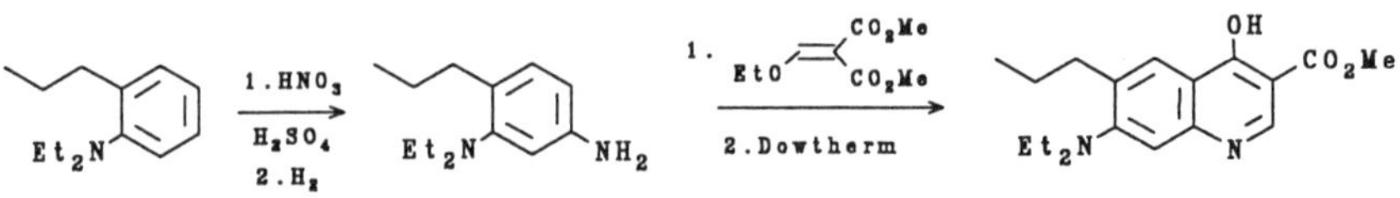

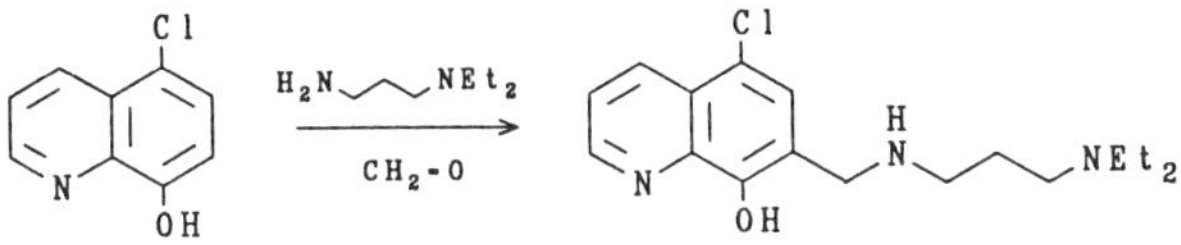

The quinoline nucleus sometimes serves as a surrogate for aromatic moieties; this finding is illustrated by two NSAIDs that are closely related to the anthranilic acid "fenac" compounds discussed in Chapter 2. The glycerol esters in these examples are probably included to avoid irritation of gastric mucosa by exposure to the carboxyl groups; serum esterases would quickly cleave these groups. The first of these compounds, **glafenine** (**116**), is prepared by displacement of chlorine in the chloroquine intermediate **82** by the amino nitrogen in glyceryl anthranilate (**115**).[30]

Preparation of the analogue that contains a trifluoromethyl group first involves subjecting *o*-trifluoromethyl aniline (**117**) to the same set of transformations used to prepare **82**; the compound is thus condensed with EMME, cyclized, decarboxylated, and finally converted to the 4-chloroquinoline **118** by reaction with phosphorus oxychloride. Displacement of chlorine with methyl anthranilate then affords the intermediate **119**. Ester interchange of **119** with glycerol leads to the glyceryl ester. Thus, the NSAID **flocatfenine** (**120**)[31] is obtained.

Similar considerations apply to leukotriene antagonists; quinoline rings provide a role comparable to the benzopyran in the antagonist ablukast (**55**), which was discussed earlier in this chapter. Reaction of the product **121** from side chain chlorination of 2-methylquinoline with the phenoxide from 3-nitrophenol leads to the corresponding ether; the nitro group is then reduced, for example, by catalytic hydrogenation to give the aniline **122**. Acylation of the amino group with trifluromethylsulfonyl chloride (triflyl chloride) gives the sulfonamide derivative **ritolukast** (**123**);[32] the acidity of the sulfonamide proton in this compound is comparable to that of carboxylic acids.

Preparation of a somewhat more complex leukotriene antagonist begins by aldol condensation of the methyl carbanion from the quinoline **124** with *m*-phthalaldehyde to give the stilbene-like derivative **125**; dimer formation is presumably inhibited by use of excess aldehyde. Reaction of **125** with *N*,*N*-dimethyl-3-mercaptopropionamide in the presence of hexamethylsilazane affords the silyl ether **126** of the hemimercaptal.

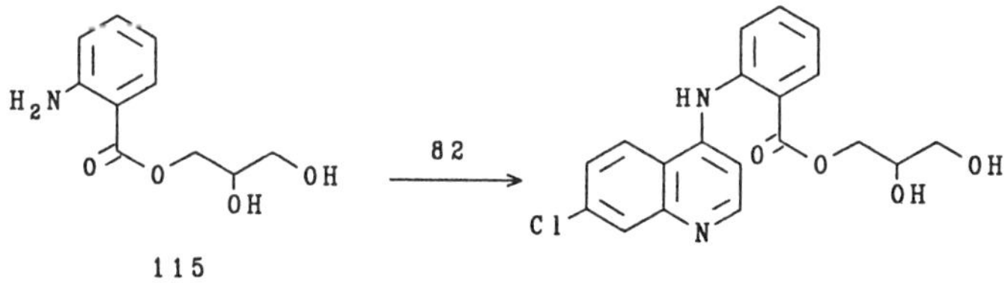

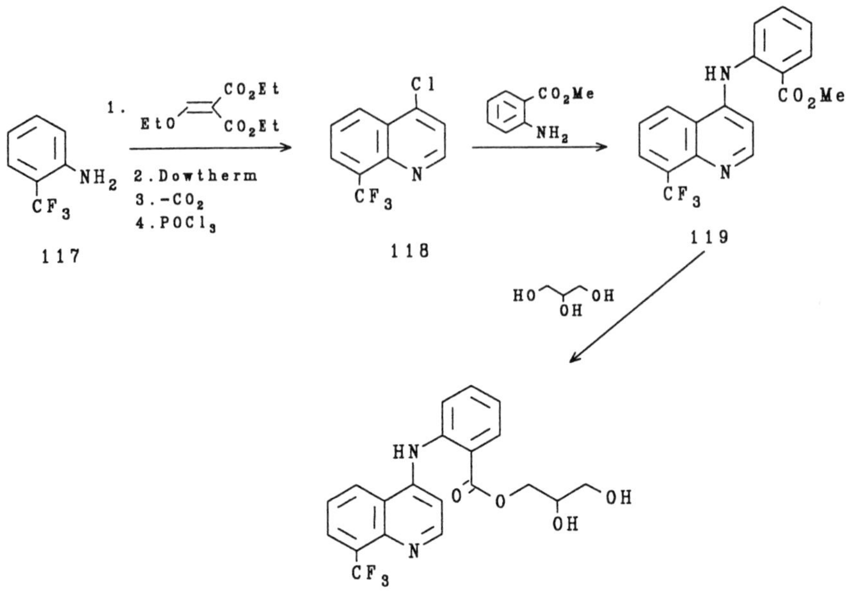

Treatment of **126** with ethyl 3-mercaptopropionate leads to replacement of the silyl ether by sulfur and formation of the corresponding thioacetal. Saponification of the ester group leads to the carboxylic acid and thus to **verlukast** (**127**).[33]

Schistosomes are yet another of the helminths that parasitize humans. Infection is most often caused by entry through breaks in the skin during immersion in infected water. The very debilitating disease, schistosomiasis, is quite prevalent in the Third World and is often associated with irrigation schemes. The structure of one of the more effective antischistosomal drugs includes a partly reduced quinoline. The synthesis starts by chlorination of the methyl group on the heterocyclic ring in dimethylquinoline (**128**); displacement of chlorine with isopropylamine gives the intermediate **129**. High-pressure catalytic hydrogenation leads to selective reduction of the heterocyclic ring and formation of the tetrahydro derivative **130**; nitration under standard conditions is directed by the same considerations that apply to the aniline **110** above. This product shows reasonable antischistosomal activity in its own right and was probably at one time considered for clinical development. Pharmacokinetic experiments revealed that the metabolite from hydroxylation of the methyl group had superior activity to its parent. Microbiological experiments revealed that the mold *Aspergillus sclerotiorum* effects the same oxidation. Thus, fermentation of the methyl compound **131** with this organism results in introduction of the hydroxyl group and formation of **oxamniquine** (**132**).[34]

A roundabout route is used to prepare tetrahydroquinolines with reduced carbocyclic rings since direct reduction, as noted above, adds hydrogen to the heterocyclic ring. In fact, the key reaction consists of a variant of the Hantzsch pyridine synthesis. Condensation of the imine **133** from dihydroresorcinol with ethoxymethylenepropionaldehyde can be envisaged as proceeding through the addition–elimination product

121 → (1. HO–C$_6$H$_4$–NO$_2$; 2. H$_2$) → 122 → (ClO_2SCF_3) → 123

124 → (OHC–C$_6$H$_4$–CHO, LDA) → 125

125 → (HS⁀CONMe$_2$, $(Me_3Si)_2N$) → 126

126 → (1. HS⁀CO_2Et; 2. NaOH) → 127

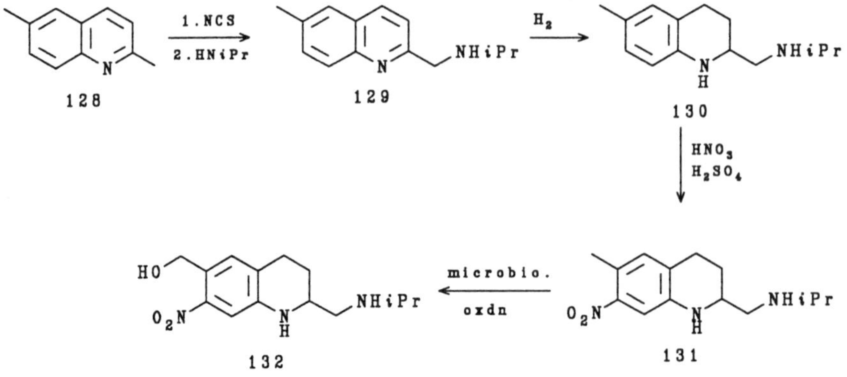

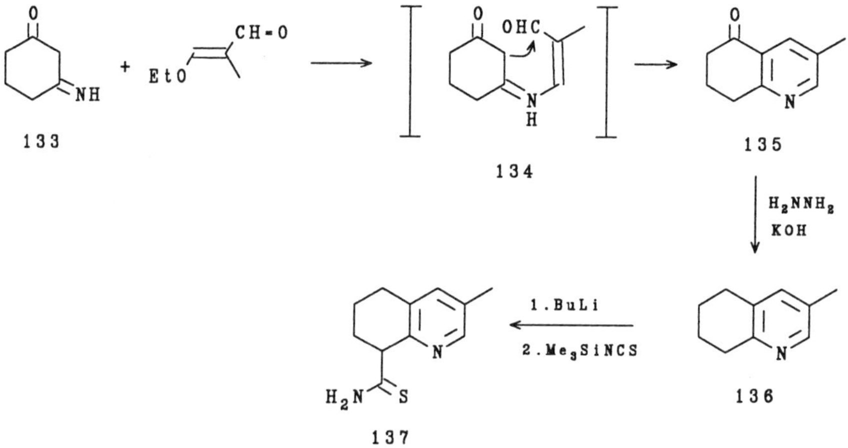

134. Aldol condensation will then result in ring closure and after bond reorganization, formation of the fused pyridine **135**. The carbonyl group in this product is then removed by Wolff–Kishner reaction with hydrazine and potassium hydroxide to give **136**. Reaction of **136** with butyllithium proceeds preferentially at the 8-position, where the metalated center can be stabilized by the adjacent ring nitrogen. Addition of the carbanion to trimethylsilylisothiocyanate leads to formation of a thioamide function. Thus, **tiquinamide** (**137**)[35] is also obtained. The antiulcer activity displayed by this compound was attributable largely to its anticholinergic activity. Drugs that act by this mechanism have been abandoned with the advent of the very specific histamine H_2 blockers and gastric antisecretory drugs.

3. Antibacterial Quinolones

The antibiotics that have led to the major advances in the treatment of infectious disease all depend on their selective toxicity to microorganisms by attack on biochemical processes unique to these organisms. Examples that have already been discussed include inhibition of PABA by sulfonamides and folates by DHFR inhibitors; the antagonism of bacterial cell wall synthesis by β-lactams is detailed in Chapter 14.

Recent work has shown that the quinolone antibiotics act directly on DNA-mediated microbial replication. Rearrangement of sections of supercoiled DNA so as to permit transcription is a necessary first step in this process. A set of enzymes known as topoisomerases control the topology of this process in both bacteria and higher organisms. The enzyme in essence maintains the integrity of the DNA chain even in the presence of temporary cuts required for uncoiling. It has been found that the quinolones specifically interact with a subclass of topoisomerases known as gyrases, that are crucial for bacterial replication.

The lead compound for this series, **nalidixic acid** (**140**), which is actually a naphthyridine rather than a quinolone, dates back to over three decades; the compound found its niche as a rather effective drug for treatment of urinary tract infection. The rediscovery of this general structural class in the early 1980s led to the development of a series of potent broad spectrum antibiotics; close to 20 of these compounds have been granted U. S. nonproprietary names.

The chemistry used to prepare the antischistosomal hydroxy-quinolines provided the initial entry to this series. Thus, addition–elimination of aminopicoline (**138**) to EMME gives the corresponding enamino ester; thermal cyclization of this compound leads to the hydroxyquinoline **139**. Reaction of the ambident anion from this compound leads to alkylation via the keto tautomer and thus formation of the *N*-alkylated derivative. Saponification of the ester then gives nalidixic acid (**140**).[36] Incidentally, it has been shown that the presence of the strong Michael acceptor function in this series plays little role in the mechanism of action in these compounds.

The same scheme affords a true quinolone when applied to an aniline. Condensation of piperonylamine (**141**) with EMME followed by cyclization of this intermediate gives the quinolinol (**142**). Alkylation as shown above followed by saponification affords the antibiotic **oxolinic acid** (**143**).[37]

The presence of fluorine at the 6-position and a basic function at the 7-position are hallmarks of the more potent recent quinolones. The starting material **146** for one of these agents is prepared by application of the same scheme as shown above to the substituted aniline **144**. Nucleophilic aromatic displacement with *N*-methylpiperazine

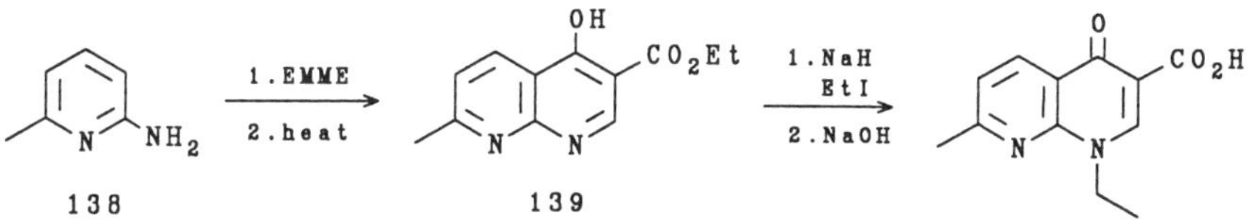

138 —(1. EMME; 2. heat)→ 139 (OH, CO_2Et) —(1. NaH, EtI; 2. NaOH)→ 140 (O, CO_2H)

141 (NH_2) —(1. EMME; 2. heat)→ 142 (OH, CO_2Et) —(1. NaH, EtI; 2. NaOH)→ 143 (O, CO_2H)

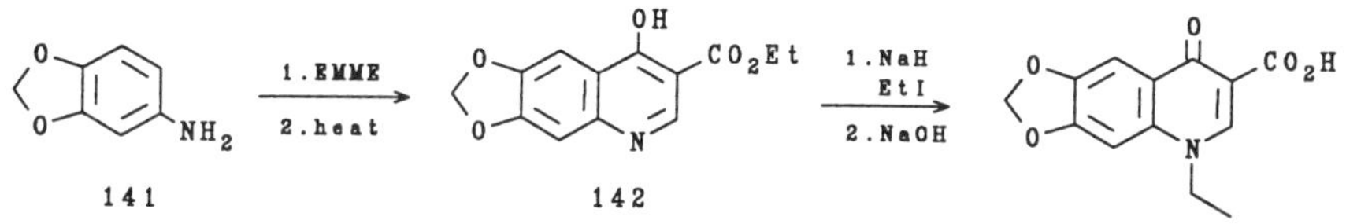

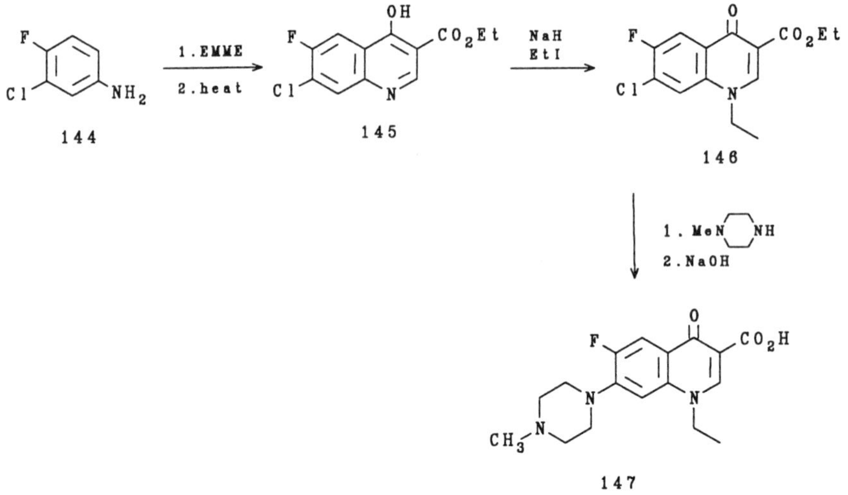

proceeds at the 7-position due to activation by the carbonyl group para to the chlorine. Saponification of the displacement product leads to **pefloxacin** (**147**).[38]

Very much the same strategy is used to synthesize a quinolone that contains an additional fused ring. In this case, note that the use of a secondary amine leads directly to a quinolone, doing away with the need for the alkylation step. The tricylic system is obtained directly by condensation of the tetrahydroquinoline **148** with EMME followed by thermal cyclization. Saponification of this intermediate then affords **ibafloxacin** (**149**).[39]

The hydroxyquinoline **145** also provides the starting material for a quinolone that incorporates a hydrazine function. Reaction of **145** with 2,4-dinitrophenyl *O*-hydroxylamine ether in the presence of potassium carbonate leads to scission of the weak N–O hydroxylamine bond by the transient anion from the quinolone; the excellent leaving character of 2,4-dinitrophenoxide adds driving force for the overall reaction resulting in alkylation on nitrogen to form the hydrazine **150**. The primary amine is then converted to the formamide by reaction with the mixed acetic-formic anhydride. Alkylation of this intermediate with methyl iodide followed by removal of the formamide affords the monomethylated derivative **151**. Chlorine at the 7-position is then displaced by *N*-methylpiperazine and the product is saponified. Thus, **amifloxacin** (**152**)[40] is obtained.

One alternate approach to construction of the quinolone nucleus relies on forming the heterocyclic ring by an internal aromatic nucleophilic displacement reaction. The

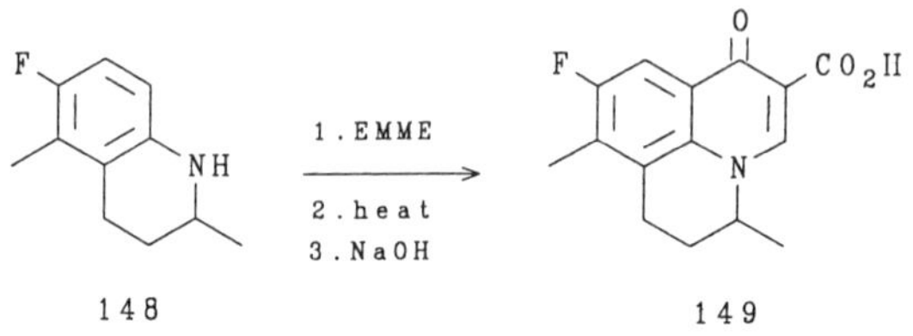

145 + O_2N-C6H3(NO2)-ONH_2 → (K_2CO_3) 150 → (1. Ac_2O/HCO_2H; 2. CH_3I) 151 → (1. MeN(piperazine)NH; 2. NaOH) 152

requisite ketoester **154** can be obtained by base catalyzed acylation of the methyl group on the acetophenone **153** with methyl carbonate; an alternate procedure for this transform, used to prepare **159**, involves ethyl magnesium carbonate as the reagent. Condensation of the product **154** with methyl orthoformate in the presence of acetic anhydride then gives the methoxymethylene derivative **155**. Addition–elimination with *p*-fluoroaniline replaces the methoxyl group by the aromatic amine to give **156**. The anion on nitrogen obtained on treatment of **156** with base displaces nitrogen on the adjacent ring to form the pyridone ring, yielding quinolone **157**. Displacement of the remaining chlorine with *N*-methylpiperazine followed by saponification affords **difloxacin** (**158**).[41]

This approach seems to offer considerable flexibility in that the piperazine moiety may be incorporated at an early stage. Treatment of the tetrafluorobenzoyl acetate (**159**) with *N*-methylpiperazine results in displacement by nitrogen of the more activated fluorine at the 4-position to form **160**. Condensation of **160** with DMF acetal leads to formation of the enamide derivative **161**.

153 → ($(MeO)_2CO$, MeONa) 154 → ($HC(OMe)_3$, Ac_2O) 155 → (p-fluoroaniline) 156 → (NaH) 157 → (1. MeN(piperazine)NH; 2. NaOH) 158

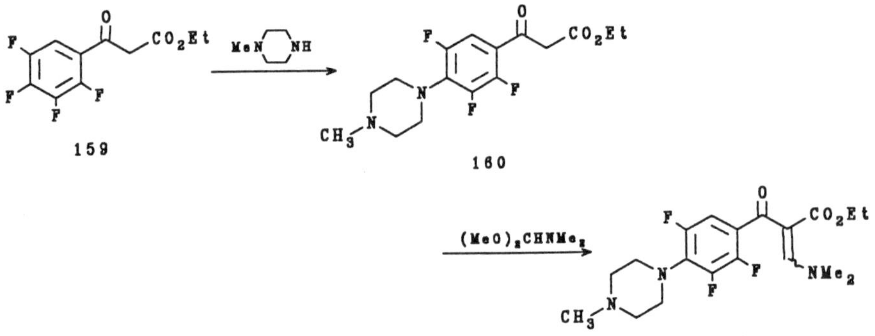

Reaction of **161** with 2-methylethanolamine (alaninol) leads to amine exchange and formation of the intermediate **162**. This compound cyclizes to the quinolone **163** on reaction with potassium fluoride in DMF, which often acts as a strong base. Treatment of **163** with sodium hydride leads to the corresponding alkoxide; the anion displaces the adjacent fluorine in spite of its lack of activation to form the oxygen-containing ring. The same intermediate can be obtained directly from **162** by reaction with potassium fluoride under more strenuous conditions. Saponification of the ester group completes the synthesis of **ofloxacin** (**164**).[42]

A variation on this theme involves construction of the enamide intermediate by direct acylation on carbon of an enamine as in the synthesis of the *N*-ethyl analogue (**169**) of the widely used quinolone **ciprofloxacin**. Condensation of the enamine **166** from ethyl formylacetate with the benzoyl chloride **165** in the presence of triethylamine gives the acylation product **167**. Reaction of **167** with strong base leads the resulting nitrogen anion to displace chlorine on the benzene ring and thus form the quinolone **168**. Incorporation of the *N*-methylpiperazine function followed by saponification of the product leads to **enrofloxacin** (**169**).[43]

Both methods for forming the heterocyclic ring in quinolones involved cyclization into the carbocyclic ring. A closely related quinolone, which displays cardiovascular

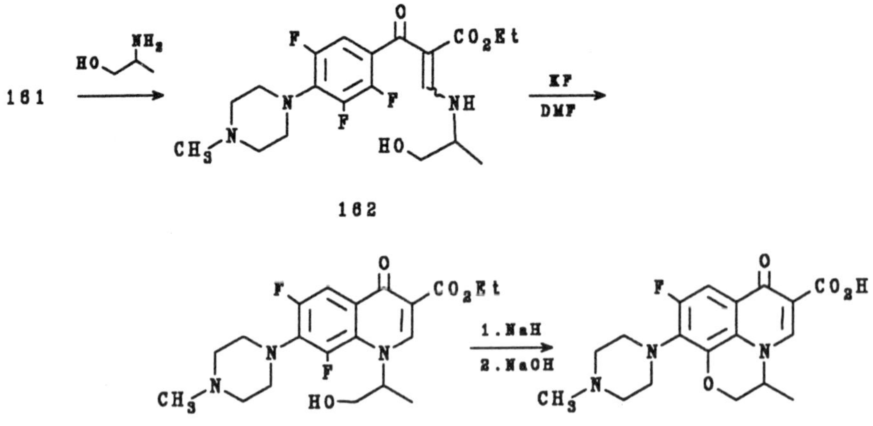

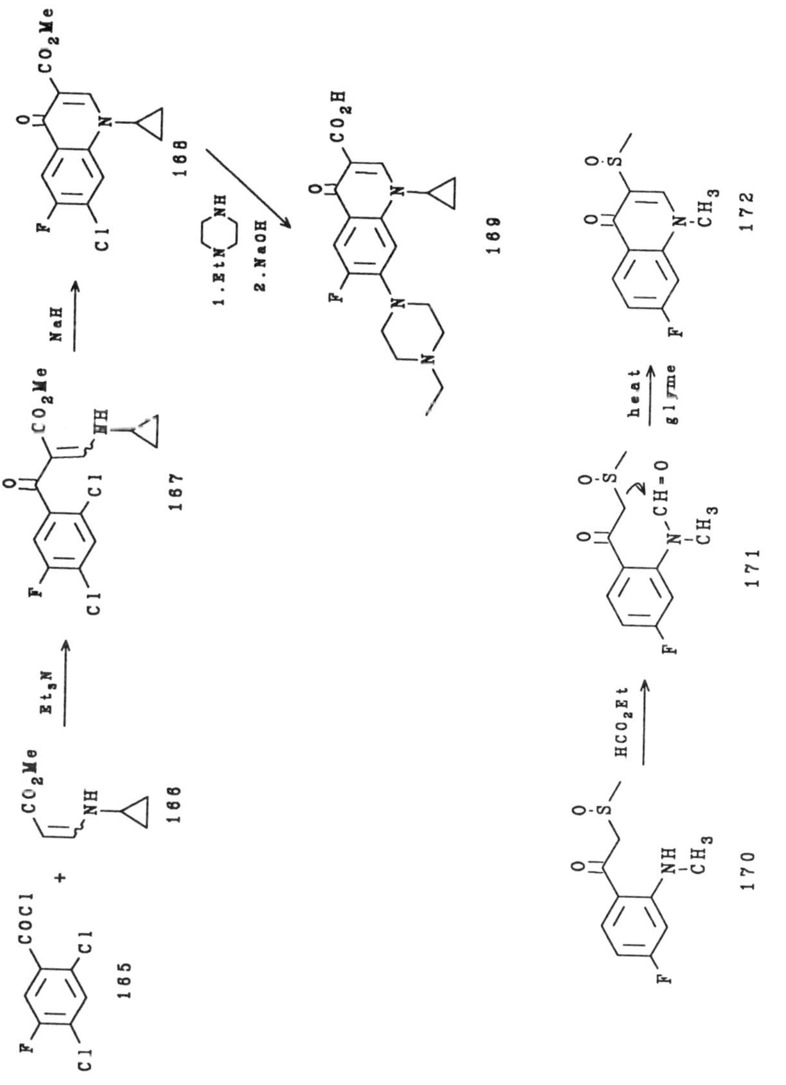
165
166
Et3N
167
NaH
168
1. EtN NH
2. NaOH
169
170
HCO2Et
171
heat
glyme
172

rather than antibiotic activity, is constructed by a condensation that closes the bond at the 2,3-position in the heterocyclic ring. The starting material **171** is obtained by reaction of the aminoacetophenone derivative **170** with ethyl formate. Heating this product in ethyleneglycol methyl ether leads to an aldol-like cyclization and formation of a quinolone ring. The product **flosequinan** (**172**),[44] displays vasodilator and cardiotonic activities.

E. Isoquinoline and Its Derivatives

The ubiquity of the isoquinoline structure among natural products might lead to the expectation of a correspondingly large number of therapeutic agents that incorporate this nucleus. The relatively small number of isoquinolines that shows useful biological activity is thus somewhat of a surprise. Note that the examples given below are included more for illustration of the chemistry of this heterocyclic system than for their therapeutic importance since most have long since passed out of use.

Papaverine (**176**) is the best known drug that incorporates the isoquinoline nucleus. This natural product, that accompanies the opioids in various *papaver* species, exhibits muscle relaxing properties in various isolated muscle strip preparations; on the strength of this observation and some related data, the drug has been used as a spasmolytic agent and as a vasodilator to improve cerebral blood circulation. The synthesis of this drug, which is discussed below, was first described by Pictet and Finkelstein[45] and illustrates what is now known as the Pictet–Spengler reaction. The first step consists of reaction of homoveratrylamine (**173**) with homoveratroyl chloride to give the amide **174**. Treatment of **174** with strong Lewis acid, for example, phosphorus oxychloride, leads to cyclodehydration and formation of the dihydroisoquinoline **175**. Dehydrogenation of **175** with palladium affords papaverine (**176**).

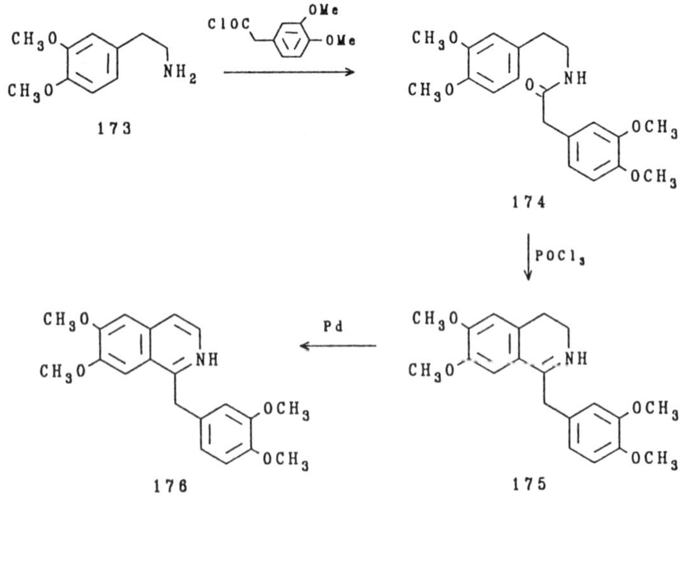

CH_3O CH_3O NH_2 173 $ClOC$ OMe OMe → CH_3O CH_3O O NH OCH_3 OCH_3 174

↓ $POCl_3$

CH_3O CH_3O NH OCH_3 OCH_3 176 ← Pd CH_3O CH_3O NH OCH_3 OCH_3 175

ClOC 177 Cl → (Ph⁀NH$_2$) → 178 (NH, O, O, Cl) → (P_2O_5) → 179 (N, O, Cl)

173 → ($\langle^{CO_2Et}_{COCl}$) → CH_3O, CH_3O, NH, O, CO_2Et 180 → (1. $POCl_3$ 2. H_2) → CH_3O, CH_3O, NH, CO_2Et 181 → (CO_2Et, CO_2Et, CH_2O) → CH_3O, CH_3O, N, EtO_2C, EtO_2C, CO_2Et 182

→ (1. H_3O^+ 2. $-CO_2$ 3. $EtOH_2^+$) → CH_3O, CH_3O, N, EtO_2C, CO_2Et 183 → (NaOEt) → CH_3O, CH_3O, N, O, CO_2Et 184 → (1. H_3O^+ 2. $-CO_2$) → CH_3O, CH_3O, N, O 185

A dihydroisoquinoline was at one time investigated in some detail as an antiviral compound. Acylation of β-phenethylamine with 2-(4-chlorophenoxy)acetyl chloride (**177**) gives the corresponding amide **178**. Cyclodehydration of **178**, this time using phosphorus pentoxide, gives **famotine** (**179**).[46]

The Pictet–Spengler reaction provides the key intermediate in the preparation of a fused tricyclic compound **tetrabenazine** (**185**), which has been used as an antipsychotic agent and has more recently been investigated as a possible antagonist for some of the deleterious side effects of dopamine antagonists. Reaction of the arylethylamine **173** with the half-acid chloride from methyl malonate gives the expected amide **180**. This compound is then subjected to the cyclodehydration reaction and the product is reduced to the tetrahydroisoquinoline **181** by catalytic hydrogenation. Treatment of **181** with diethyl isopropylmalonate and formaldehyde leads to a rather unusual Mannich reaction that results in formation of the homologation–alkylation product **182**. Hydrolysis of the ester followed by decarboxylation of the diacid and then reesterification gives the diester **183**. Base-catalyzed Dieckmann cyclization leads to formation of the carbethoxycyclohexanone ring. The remaining carbethoxyl group in the product (**184**) is then removed by repeating the hydrolysis and decarboxylation sequence. Thus, tetrabenazine (**185**)[47] is finally obtained.

A concise synthesis of a tetrahydroisoquinoline is designed to take into account the fact that the benzaldehyde **186** is more readily available than the required homologue **189**. Reaction of **186** with ethyl chloroacetate and sodium hydride initially leads to simple addition of the anion to the carbonyl group. The second step in this Darzens reaction consists in internal displacement by the resulting alkoxide of the adjacent chlorine to afford the glycidic ester **187**; Treatment of **187** with base results in formation of the sodium salt **188**. This salt is then allowed to react with 3,4-dihydroxy-β-phenethylamine (dopamine) under Pictet–Spengler conditions. The strong acid present in the medium brings about as the first step the decarboxylative rearrangement of the free acid from salt **188** to the arylacetaldehyde **189**. This aldehyde then adds to the amino group in dopamine to form a carbinolamine. This species or the resulting Schiff base **190** then attacks the highly activated aromatic ring. This reaction, which amounts to a cyclodehydration, results in the formation of **trimethoquinol** (**191**).[48] This compound is one of a small group of compounds devoid of an aminoalcohol moiety that act as β-adrenergic agonists; consequently, it has been investigated as a bronchodilator.

Pictet–Spengler reaction on 3,4-dimethoxy-β-phenethylacetamide provides the starting material **192** for a tetrahydroisoquinoline. The reaction in this case is in fact used as a means for placing an acetyl group ortho to the alkyl side chain since a different strategy is used to form the final heterocyclic ring. Reaction of the imine **192** with acetic anhydride leads to acylation on nitrogen and a shift of the double bond to the exocyclic position to give **193**. Hydrolysis of this anhydro form of a carbinolamine *N*-acetate leads to ring opening and formation of the acetophenone **194**. Base-catalyzed aldol condensation with *p*-chlorobenzaldehyde gives the benzal derivative **195**. Treatment of **195** with strong acid first leads to hydrolysis of the amide function; the newly freed primary amine then condenses with the carbonyl group to form an imine and thus the dihydroisoquinoline **196**. Catalytic hydrogenation is then used to reduce both the side chain double bond and the imine function; the resulting secondary amine is then methylated by any one of several methods, such as reaction with formalin and formic acid. Thus **methopholine** (**197**),[49] a weak opioid analgesic agent, is obtained.

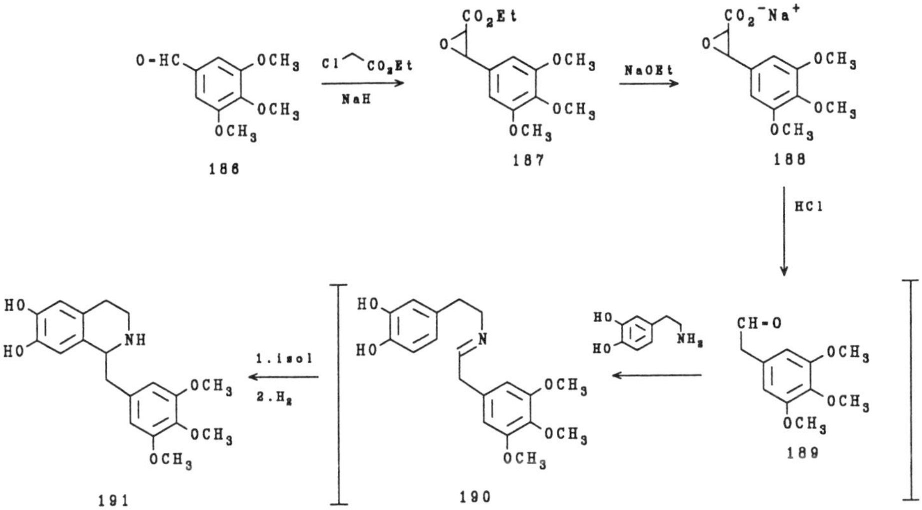

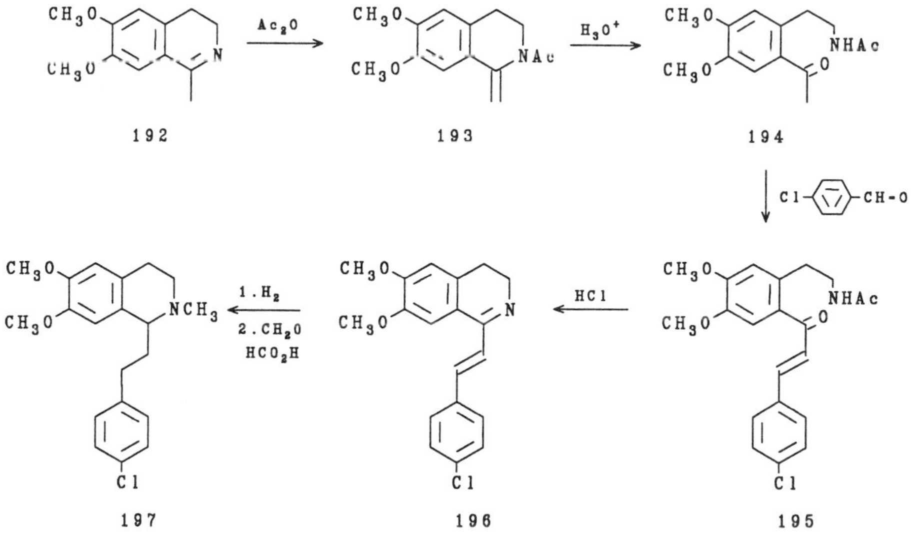

A commercial NSAID and some of its analogues, which depend on an enolized heterocyclic rings for an acidic proton, are considered in more detail later in this chapter. The preparation of a simplified example starts by conversion of homophthalic acid **198** with ammonia to its imide **199**. Reaction of **199** with *p*-chlorophenylisocyanate in the presence of triethylamine results in addition of the anion to the cumulene to give the amide.[50] This product, **tesicam**, is shown as both extreme tautomers (**200** and **201**) in what is probably a very mobile equilibrium. The multiple enol forms that can be visualized probably contribute to the acidity of this compound.

2. COMPOUNDS WITH TWO HETEROATOMS

A. Benzodioxans

The majority of the small number of biologically active compounds based on a fused ring system, that includes two oxygen atoms, consist of 1,4-benzodioxans. Interestingly, all but one of the compounds discussed below interact with the adrenergic system. The simplest, **piperoxan (206)**,[51] was in fact one of the first of the α-blockers to be studied in any detail. The synthesis of **206** illustrates the standard entries into this heterocyclic system. Condensation of catechol (**202**) with epichlorohydrin in the presence of aqueous base can be visualized as proceeding initially to the epoxide **203**, which is formed by either direct displacement of chlorine or ring opening of the oxirane followed by displacement of chlorine by the resulting alkoxide to form a new epoxide. Attack on the side chain epoxide by the phenoxide anion in **203** will lead to the key intermediate **204** after neutralization. Reaction of **204** with thionyl chloride leads to the corresponding chloro derivative **205**. Displacement of the newly introduced halogen with piperidine affords piperoxan (**206**). Guanidine derivatives were among the first drugs used as antihypertensive agents. The use of these agents was accompanied by a host of side effects attributed to the fact that these agents acted as sympathetic blockers. Preparation of a guanidyl benzodioxane antihypertensive agent starts with a small variation of the above scheme; the alcohol **204** is first converted to its mesylate; displacement with guanidine affords **guanoxan** (**207**).[52]

A more fundamental variation on this scheme leads to a benzodioxan containing an oxirane side chain that can be used for subsequent elaboration. The use of salicylaldehyde as a starting material instead of catechol simplifies the initial alkylation step. Thus, reaction of this compound with 1,4-dichloro-2-butene leads to the ether **208** of unspecified configuration. Reaction of the product (**208**) with peracid leads to

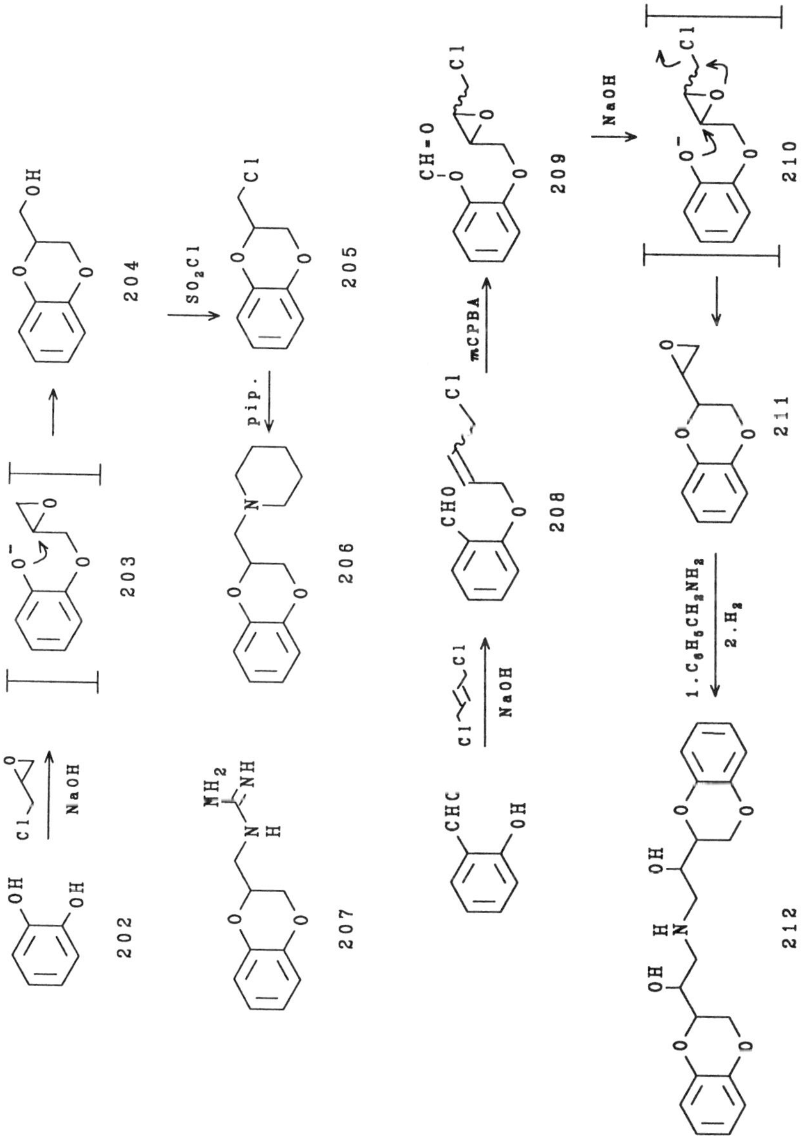
202
NaOH
203
204
SO2Cl
205
pip.
206
207
NaOH
208
mCPBA
209
NaOH
210
211
1. C6H5CH2NH2
2. H2
212

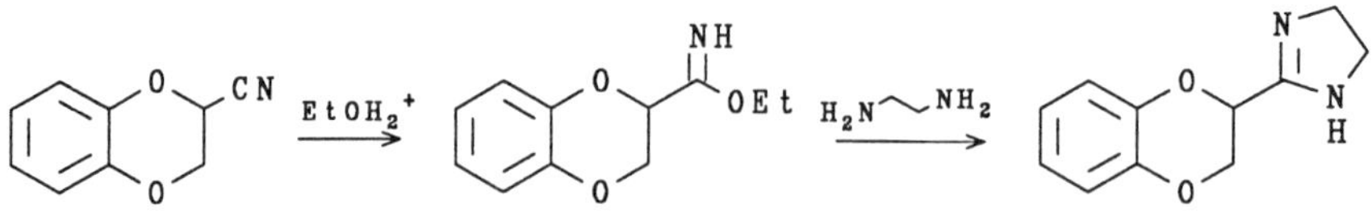

both epoxidation of the olefin and Bayer–Villiger scission of the aldehyde to give the formate **209**. Saponification of the ester with mild base initially affords the alkoxide **210**. This compound undergoes a chain of epoxide opening and closing reactions to give the product **211**. Note that compound **211** is an oxygenated analogue of the intermediate **64**, which is used for the synthesis of nebivolol (**65**) described earlier in this chapter. Proceeding in analogous fashion, reaction of **211** with benzylamine followed by reductive removal of the benzyl protecting group, affords the β-blocker **bendacalol** (**212**).[53]

A much simpler compound includes both a benzodioxan nucleus and the imidazoline function associated with α-adrenergic agonists such as clonidine. As in the standard approach for preparing imidazolines, treatment of the nitrile **213** with alcoholic hydrogen chloride leads to the iminoether **214**. Reaction of this intermediate **214** with ethylenediamine then affords **idazoxan** (**215**),[54] a compound that interacts with α-adrenergic receptors.

Attachment of a base-bearing side chain to the carbocyclic ring of a benzodioxan gives another compound that acts as an α-adrenergic blocker. Mannich reaction of the methyl ketone in **216**, obtainable by acetylation of benzodioxan proper with phenylpyrrolidine and formaldehyde, leads directly to **proroxan** (**218**).[55]

The benzodioxan ring also serves as the aromatic moiety for one of the ubiquitous analogues of the "spirone" anxiolytic agents discussed in Chapter 9. The requisite intermediate **219** is obtained by reducing the cyano group in nitrile **213** with lithium aluminum hydride. Alkylation with the "spirone" side chain chloride **220** affords **binospirone** (**221**).[56]

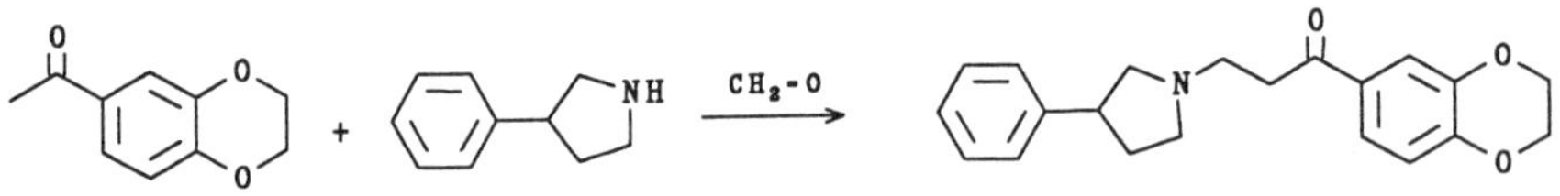

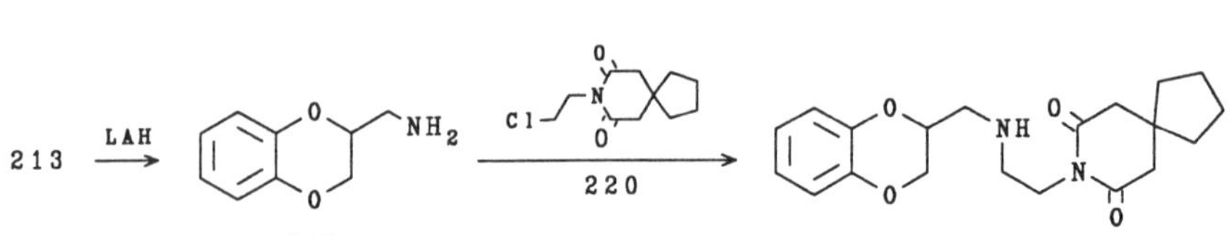

B. Cinnolines and Phthalazines

Internal diazonium coupling provides the key reaction to the preparation of two cinnolines, that have been investigated as potential therapeutic agents. Treatment of the 2-aminoacetophenone **222** with nitrous acid leads to the corresponding diazonium salt **223**, which is depicted as its enol tautomer. Attack by the diazonium group on the electron-rich enol leads to coupling and formation of the 4-hydroxycinnoline derivative **224**. Reaction of **224** with bromine leads to aromatic halogenation and formation of bromide **225**. Nucleophilic aromatic displacement of halogen by means of cyanide gives the corresponding nitrile; alkylation of the anion from treatment of the intermediate with strong base takes place, as in the case of the analogous quinolols, on nitrogen to form the *N*-ethyl cinnolone **226**. Hydrolysis of the nitrile to the carboxylic acid affords **cinoxacin** (**227**),[57] an antibiotic that can be viewed as an aza analogue of the more familiar quinolones.

Pyrrazolones formed by the reaction of alkylmalonates with 1,2-diarylhydrazines provide the heterocyclic nucleus for NSAIDs related to phenylbutazone (see Chapter 8). The same function is provided in a somewhat more complex NSAID, in which the arylhydrazine function is embedded in a partly reduced cinnoline. The preparation of this agent starts with the conversion of aminobenzophenone (**228**) to its corresponding methyl carbinol with methylmagnesium bromide followed by acid catalyzed dehydration to the methylene derivative **229**. The diazonium salt **230**, which is obtained on reaction of the product with nitrous acid, undergoes internal coupling when treated with ammonium hydroxide to give the cinnoline **231**. Catalytic reduction probably proceeds by initial 1,4-addition of hydrogen; the product then spontaneously rearranges to the hydrazine derivative **232**. Condensation of **232** with diethyl amylmalonate leads to a fused pyrrazolone. Thus, **cintazone** (**233**)[58] is obtained.

Dihydralazine (**236**) is the least important of the two phthalazine antihypertensive agents; its preparation is, however, recorded first because of its simplicity. Thus, reaction of phthalhydrazide (**234**) with phosphorus oxychloride leads to the now very familiar conversion of the amide functions to enol chlorides (**235**). Displacement of

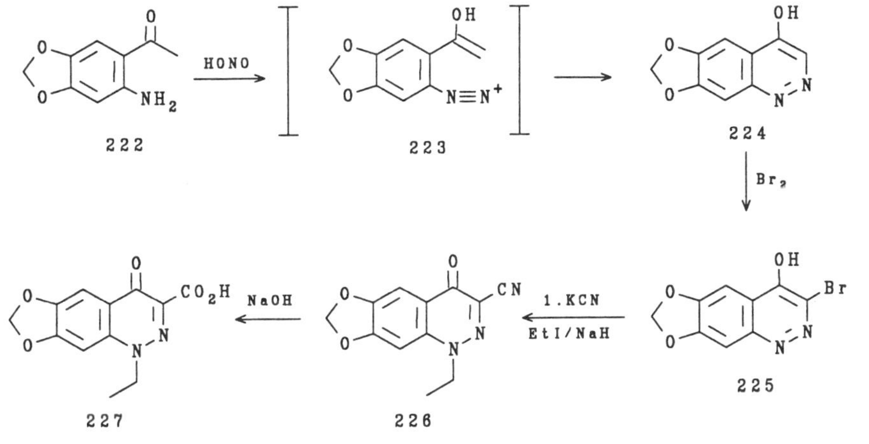

HONO

222 223 224

Br_2

1. KCN EtI/NaH

NaOH

227 226 225

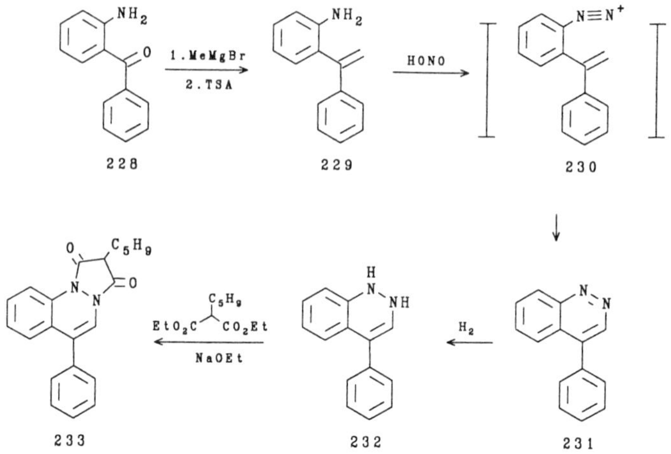

halogen with hydrazine leads directly to the antihypertensive agent dihydralazine (**236**).[59]

The preparation of a monosubstituted phthalazine hinges on the differing oxidation states of the starting carbonyl groups. Thus, reaction of acid-aldehyde **237** with hydrazine gives intermediate **238**, which contains a hydrazone as well as a hydrazide group. Exposure to phosphorus oxychloride converts **238** to an enol chloride to give the monofunctional derivative **239**. Hydrazinolysis of **239** leads to **hydralazine** (**240**).[59] This antihypertensive agent as well as its disubstituted analogue have been shown to be direct acting vasodilator by their relaxing effect on vascular smooth muscle; these drugs were once one of the mainstays for treating elevated blood pressure.

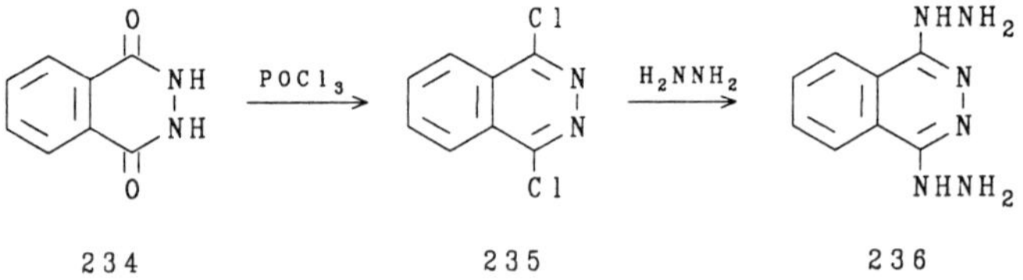

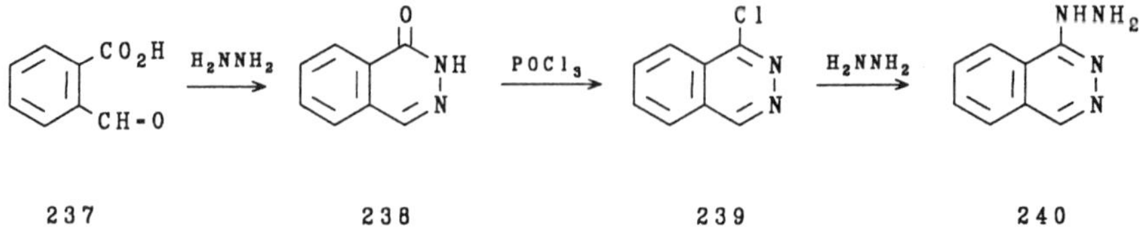

Replacement of the hydrazine function by a substituted 4-piperidinol leads to a compound that has been investigated as a cardiotonic agent as it increases the contractile force of cardiac smooth muscle. The preparation is quite analogous to those above; the key step involves displacement of chlorine from the substituted chloro-phthalazine (**241**) with the ethyl urethane from 4-piperidinol. Thus, **carbazeran** (**242**)[60] is obtained.

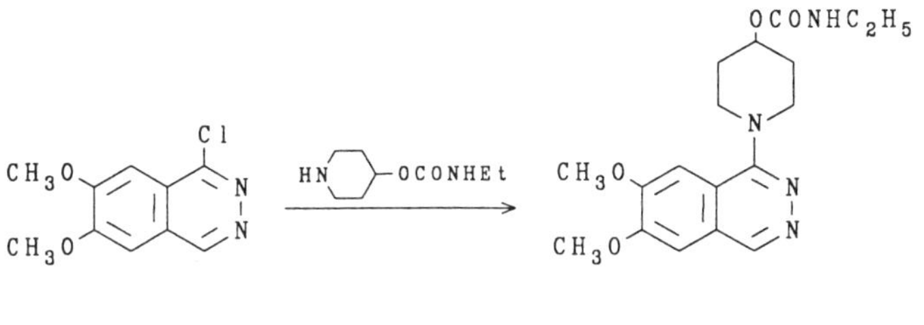

Several therapeutic agents are based on the phthalazinone nuclei in which the hydrazide carbonyl group persists in unmodified form. Reaction of ketoacid **243** with hydrazine leads to the phthalazinone **244**. Alkylation of the hydrazide nitrogen with 2-(chloroethyl)-*N*-methylpyrrolidine (**245**) surprisingly leads to incorporation of a seven-membered azepine ring rather than the expected ethylpyrroldine. This finding can be explained by keeping in mind that it is likely that the starting alkylating agents **245**, like many other 2-chloroethyl nitrogen mustards, exists, to some extent, in equilibrium with the cyclized quaternary form **246**. Attack at the bridgehead opposite the ammonium ion with simultaneous ring opening will lead to the observed product; this reaction affords the H_1 antihistamine **azelastine** (**247**).[61] The cyclic form of the *N*-chloroethyl amines, the so-called mustards, that correspond to **246**, consists of aziridinium salts; the product from reaction of this species is in this case identical with the one from direct displacement of halogen from the open form.

Reaction of phthalic anhydride (**248**) with the ylide from ethyl triphenylphosphoniumacetate leads to the condensation product **249**, which in effect consists of a cyclic enol anhydride. Treatment of **249** with hydrazine leads to the hydrazone–hydrazide **250**. Alkylation of the anion from removal of the hydrazide proton with the substituted benzyl bromide **251** followed by saponification affords the aldose reductase inhibitor **ponalrestat** (**252**).[62]

The more complex phthalic anhydride **253** is the starting point in the synthesis of a phthalazinone based platelet aggregation inhibitor. Reaction of **253** with malonic acid in pyridine involves as the first step ring opening of the anhydride by malonate anion and formation of the acylation product. The resulting product spontaneously loses carbon dioxide to afford the transient intermediate **254**; this species also decarboxylates to give the product of methylation of the more electrophilic of the phthalic

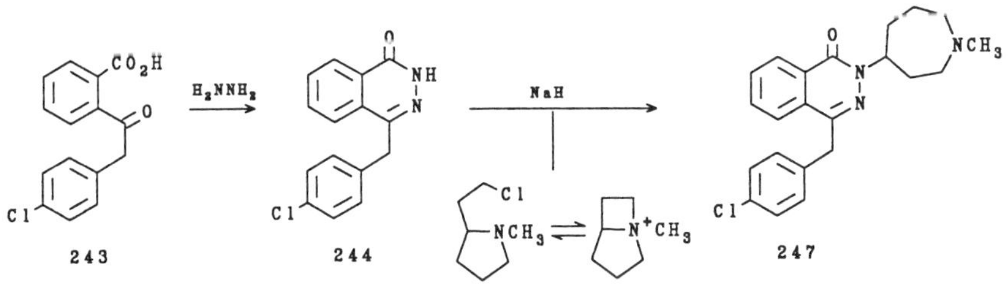

$Ph_3P^+CH_2CO_2Et$ H_2NNH_2 1. NaH 2. 251 3. NaOH

248 249 250 251 252

acid carbonyl groups. This intermediate is actually isolated in its pseudoacid form (**256**). The methyl group is then subjected to carefully controlled oxidation with potassium permanganate to give the ketoacid **257**. Reaction with hydrazine proceeds in a straightforward manner to give the phthalazinone **258**. Carboxylic esters are not, in the normal course of events, reduced by sodium borohydride. The presence of an adjacent imine nitrogen apparently changes the resistance to that reagent. Thus treatment of **258** with sodium borohydride leads to selective reduction of the ester on the heterocyclic ring to an alcohol while leaving that on the benzene ring unaffected. Thus, **oxagrelate** (**259**)[63] is obtained.

C. Quinazolines

A number of compounds based on the 6,7-dimethoxyquinazoline nucleus have shown effect on cardiovascular function or act as bronchodilators. The simplest and earliest of these agents, **quazodine** (**262**), has been investigated as a vasodilator, a cardiotonic agent, and a bronchodilator. It became clear from later work that this compound probably acts as a phosphodiesterase (PDE) inhibitor; the aminoquinazoline antihypertensive agents that were developed later, on the other hand, act as α-adrenergic blockers. Methods for forming the quinazoline ring system closely parallel those used for the monocyclic pyrimidines. The preparation of this first compound thus involves as the initial step formation of formamide **261** by reaction of the anilinoketone **260** with a mixture of formic acid and acetic anhydride. Treatment of the amide with ammonia then leads to the quinazoline and thus quazodine (**262**).[64]

Incorporation of a piperazine function on the heterocyclic ring leads to a compound in which bronchodilator activity predominates. Treatment of the amino-amide **263**

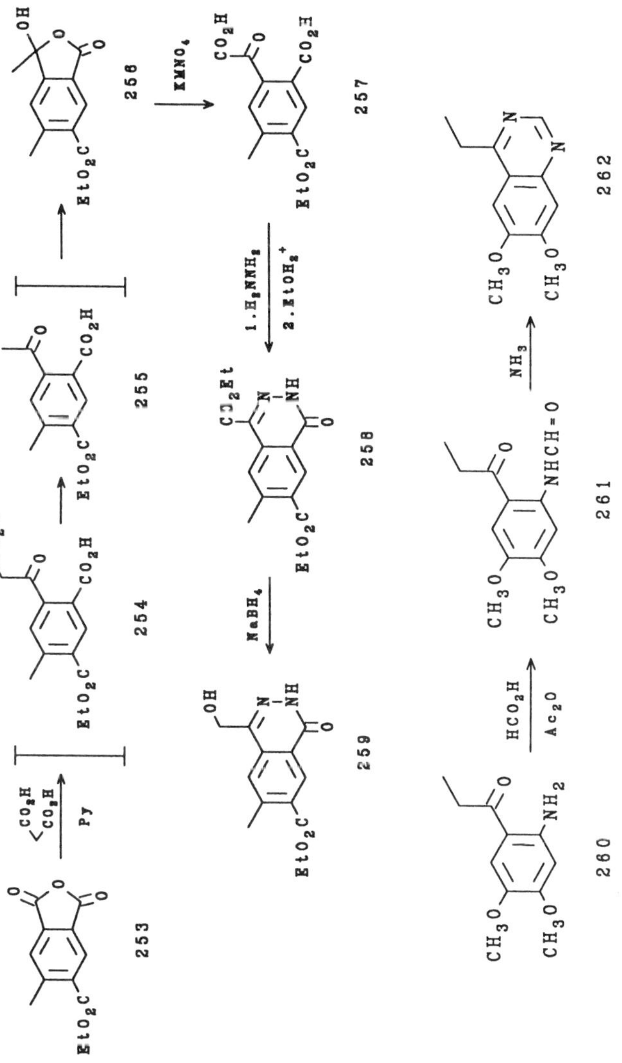
253
254
255
256
257
258
259
260
261
262
Py
KMnO4
1. H2NNH2
2. EtOH2+
NaBH4
HCO2H
Ac2O
NH3
EtO2C
CO2H
CO2Et
OH
NH2
NHCH=O
CH3O

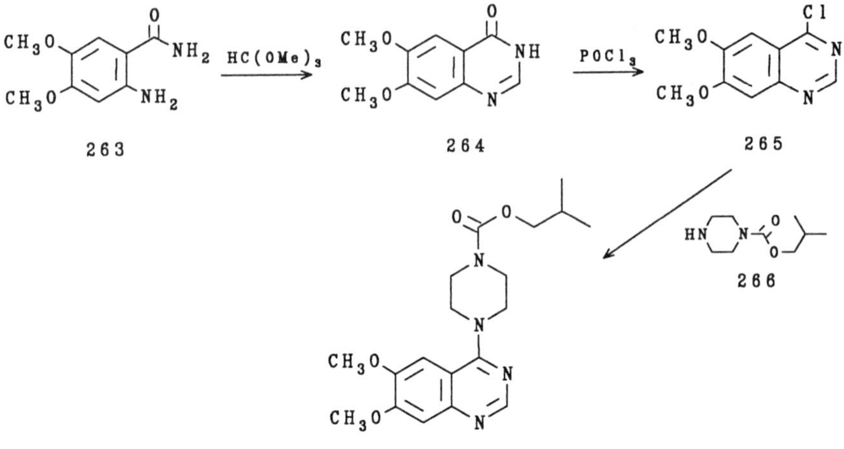

with trimethylorthoformate provides the additional carbon atom for formation of the quinazolone ring in **264**. Reaction with phosphorus oxychloride in effect converts the ring to its aromatic form **265** by locking in the former amide as an enol chloride. Displacement of the halogen with the isobutyryl urethane **266** from piperazine affords **piquizil** (**267**).[65]

It sometimes happens that novel drugs lead to the discovery of new pharmacological principles; remember that the science of pharmacology did have its origin in the study of the biological effects of drugs. Up to the advent of **prazosin** (**272**) it had been accepted that the side effects observed with α-adrenergic blockers, such as increase in heart rate and water retention, were due to reflex compensatory responses. The detailed examination of the mechanism of action of this new drug revealed that it too lowered blood pressure by blocking vasoconstriction due to α-adrenergic stimulation. This work also led to the finding that the drug interacted with a subset of receptors which did not cause the classic compensatory effects. Note that many additional subclasses of α- and, for that matter, β-adrenergic receptors, have been identified since. Condensation of amino-amide **263** with urea supplies the additional carbon atom in the form of a carbonyl group to afford the quinazolodione **268**; the ubiquitous phosphorus oxychloride reaction then converts **268** to the dichloride **269**. The halogen on the carbon next to the aromatic ring is apparently significantly more reactive than the one at the 2-position; treatment of this intermediate with ammonia at room temperature results in selective displacement of this halogen to afford **270**. Reaction of **270** with piperazine under more strenuous condition leads to replacement of the remaining chlorine and formation of **271**. Acylation of **271** with furoyl chloride affords the antihypertensive drug **prazosin** (**272**).[66]

Long-term clinical use of these presynaptic α-blockers revealed that they relieved the symptoms of prostatic hypertrophy. Several drugs in this class are now used specifically for this indication. The synthesis for the starting material **273** for one of these compounds is quite analogous to the one used for the dimethoxy quinazoline

263 $\xrightarrow{(NH_2)_2CO}$ 268 $\xrightarrow{POCl_3}$ 269 $\xrightarrow{NH_3}$ 270 → 271 → 272

273 $\xrightarrow{274}$ 275

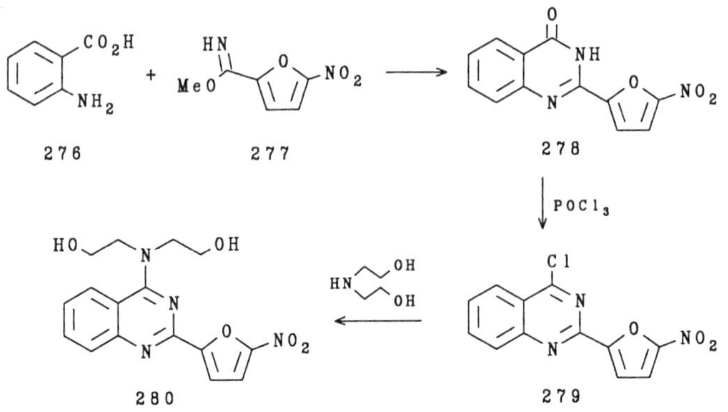

276 277 278 $POCl_3$ 279 280

(**270**). In this case, the remaining chlorine is displaced with the complete prefabricated side chain, the piperazine urethane **274**. Thus, **trimazosin** (**275**)[67] is obtained.

Condensation of anthranilic acid (**276**) with an iminoether represents another method of preparing quinazolones. The reaction with the iminoether **277** from 2-cyano-5-nitrofuran can be visualized as proceeding through formation of the amidine from addition–elimination of anthranilic acid; cyclization then affords the observed product **278**. This compound is then converted to the chloride **279** in the usual way. Displacement of chlorine with diethanolamine leads to formation of **nifurquinazol** (**280**),[68] one of the antibacterial nitrofurans (see also Chapter 8).

The use and design of inhibitors of folate synthesis for antitumor agents was discussed in Chapter 8 in connection with relatively simple monocyclic folate mimics. Bicyclic compounds that more closely resemble the pteridine present in the endogenous compounds would be expected to have at least as good activity as the simpler analogues. The quinazoline **trimetrexate** (**288**), that in addition contains an aromatic side chain that simulates the PABA moiety of folic acid, is under study as a cancer chemotherapy drug. The concise method for the preparation of the starting quinazoline involves aromatic nucleophilic substitution as a key step. Thus, reaction of the nitrile **281** with guanidine can be envisaged as initially proceeding the arylguanidine **282**, that results from displacement of chlorine activated by electron-withdrawing nitro and cyano groups. Addition of one of the guanidine amines to the nitrile will then lead to

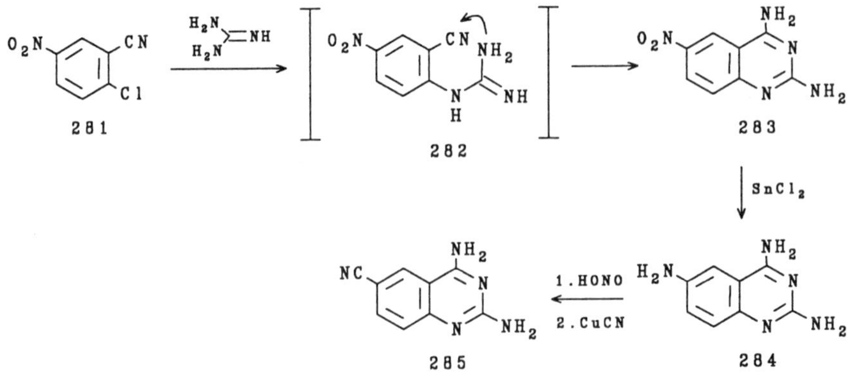

281 282 283 $SnCl_2$ 284 1. HONO 2. CuCN 285

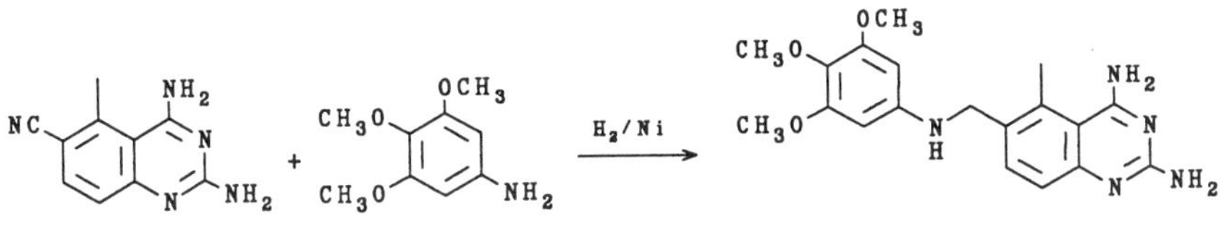

286 287 288

the diaminoquinazoline **283**; the nitro group is then reduced to the amine **284** by any of several methods, for example, stannous chloride in hydrochloric acid. It should be noted that the two amino groups on the quinazoline ring are virtually nonbasic; treatment of **284** with nitrous acid thus proceeds selectively on the newly introduced anilino group. Reaction of this diazonium salt with cuprous cyanide results in formation of the corresponding nitrile **285**.[69] The same reaction sequence starting with the benzonitrile containing a 5-methyl group will give **286**.

The final step in the sequence consists of hydrogenation of the nitrile **286** in the presence of trimethoxyaniline (**287**) over Raney nickel. A number of possibilities can account for this exchange of bases; the aniline could, for example, add to an intermediate imine from the nitrile followed by loss of ammonia from the resulting aminal. This reduction–exchange reaction affords trimetrexate (**288**).[70]

D. Quinazolones

One of the first quinazolone based drugs, the sedative hypnotic **methaqualone** (**291**), gained considerable notoriety as a drug of abuse under the alias "ludes," named after the original trade name Quaalude. This compound is prepared in a straightforward fashion by fusion of anthranilamide (**289**) with *o*-toluidine[71]; this reaction can be envisaged as proceeding via the diamide **290**.

Condensation of an aminoketone with urea leads to formation of a quinazolone by incorporation of carbonyl carbon and one amino group. This reaction with **292** can be rationalized by assuming initial formation of the urea exchange intermediate **293**. Cyclization will then give **fluproquazone** (**294**),[72] a non-opioid analgesic that shows NSAID-like activity in the absence of the typical acidic function.

A quinazolone moiety also provides the nucleus for a highly simplified leukotriene antagonist (cf. this compound with verlukast (**127**), discussed earlier in this chapter). Condensation of the anthranilate ester **295** with formamide leads to formation of the quinazolone **296**. Reaction of the salt from reaction of **296** with strong base and ethyl 3-bromoacrylate leads to vinylation on nitrogen by what is probably an addition–

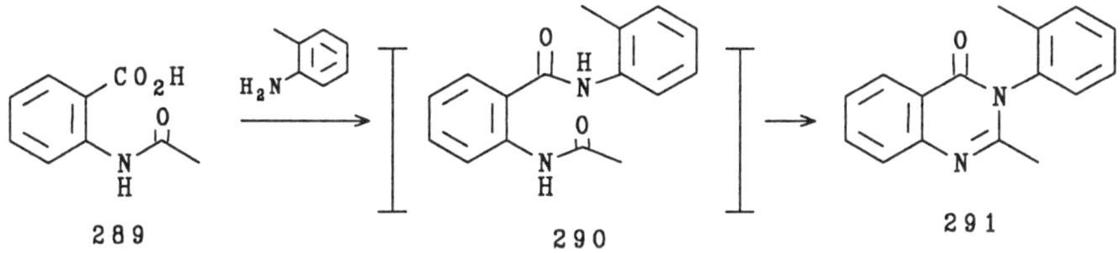

289 290 291

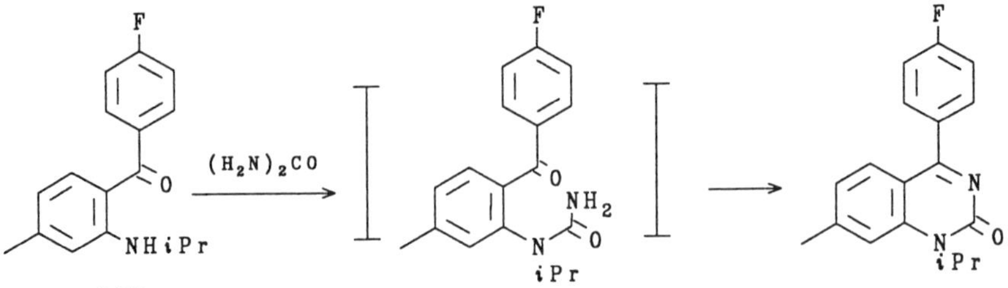

elimination sequence; the product is largely the (*E*) isomer. Saponification then affords **tiacrilast** (**297**).[73]

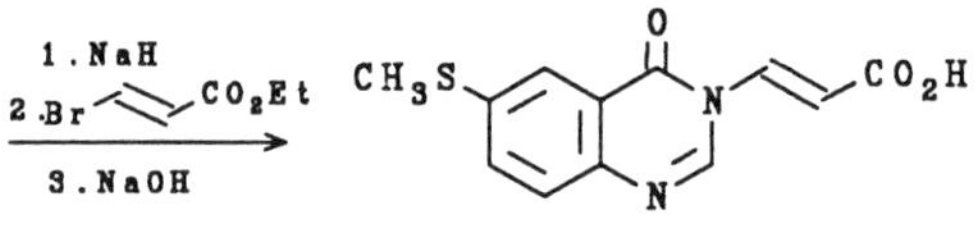

Aminoacetophenones can also act as starting materials for quinazolones. Conversion of the nitrobenzoic acid **298** to an acetophenone involves a sequence reminiscent of the one used for **oxagrelate** (**259**). Thus, reaction of the acid chloride from **298** with the anion from diethyl malonate affords the tricarbonyl intermediate **299**. This compound undergoes double decarboxylation on hydrolysis to give the methyl ketone **300**. The nitro group is then reduced by hydrogenation and the resulting aniline is converted to its urethane **301**. Fusion with ammonium acetate then forms the quinazolone ring, which affords **bemarinone** (**302**),[74] a cardiotonic agent with a somewhat atypical structure.

The discovery that cyclization of one of the sulfonamide groups and the adjacent amine in classic diuretics led to an increase in potency is discussed in Section 3. A similar increase in potency is also observed in the anthranilic acid-based diuretics. Thus fusion of anthranilic acid **303**, which may show diuretic activity in its own right, with propionamide leads to the quinazolone **304**. Reduction of **304** then gives diuretic **quinethazone** (**305**).[75]

Reaction of anthranilamide (**306**) with benzaldehyde, in a reaction clearly modeled on the sulfonamides, can be envisaged as first involving formation of the carbinolamine **307**; displacement of hydroxyl by amino leads to the aminal and thus the diuretic agent **fenquizone** (**308**).[76] The same product would of course result from initial formation of the carbinolamine with amide nitrogen.

The use of activated anthranilic acid derivatives facilitates the preparation of the amides in those cases where the amines are either unreactive or difficult to obtain.

298 → 1. SO_2Cl, 2. $CH_2(CO_2Me)_2$, NaOMe → 299 → NaOH → 300 → 1. H_2, 2. $ClCO_2Et$ → 301 → NH_4OAc → 302

303 → H_2NOC (propyl) → 304 → 305

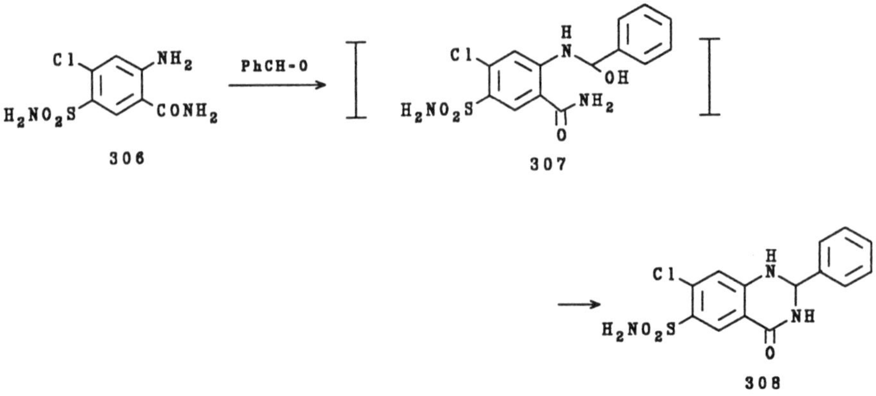

Thus, reaction of **303** with phosgene gives the reactive isatoic anhydride **309**. Condensation of this compound with *o*-toluidine leads to the acylation product **310**, which is formed with simultaneous loss of carbon dioxide. This compound is then converted to the quinazolone **311** by heating with acetic anhydride. Reaction with sodium borohydride in the presence of aluminum chloride selectively reduces the double bond to yield the diuretic agent **metolazone** (**312**).[77]

Isatoic acid itself (**313**), when allowed to react with the primary amine **314** gives the corresponding amide **315**, again with the expulsion of carbon dioxide. Treatment of **315** with phosgene results in the formation of a quinazolodione ring. Thus, the serotonin blocking compound **cloperidone** (**316**)[78] is obtained that is described as a sedative.

Condensation of ethyl anthranilates with isothiocyanates provides entry to a closely related compound in which the carbonyl at the 2-position is replaced by a thione. The sequence begins with the somewhat unusual alkylation of pyrrolidine nitrogen in **317** with 2-bromoethylamine. Reaction of the primary amine in the product **318** with thiophosgene leads to the isothiocyanate derivative **319**. Reaction of this reactive intermediate with methyl anthranilate leads initially to the transient addition product

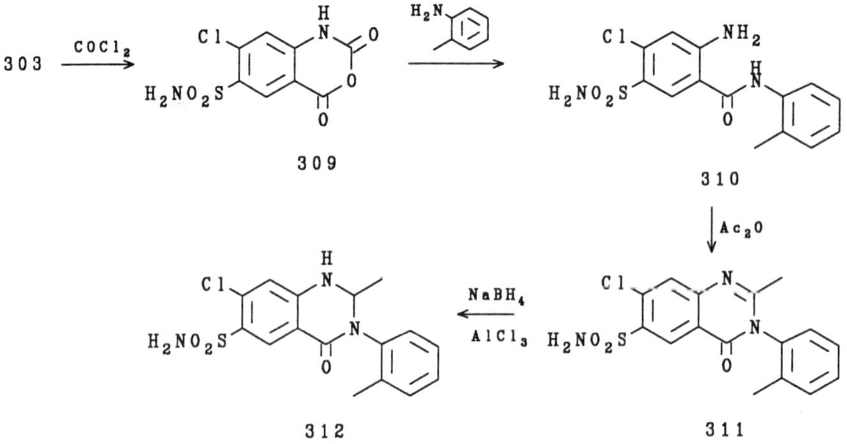

322 323 324 325 326 327

320. This compound then undergoes the customary internal ester exchange reaction to form the heterocyclic ring. Thus, the serotonin antagonist **altanserin** (**321**)[79] is obtained.

The cyclooxygenase inhibiting activity manifested by NSAIDs, as noted earlier, would seem to depend on the presence in the molecule of a group capable of supplying an acidic proton. The great majority of these drugs, as exemplified by the "profen" aliphatic acids and the "fenamate" benzoic acid derivative, incorporate a carboxylic acid. The acidity of appropriately substituted dicarbonyl groups, as noted above in connection with tesicam (**200**), can approach that of the more common carboxylates. An important series of NSAIDs incorporate a β-ketoamide function built into a 1,2-benzothiazine oxide ring. The synthesis of this heterocyclic nucleus begins with the base-catalyzed alkylation of saccharin (**322**) with ethyl chloroacetate. Treatment of this product with sodium methoxide leads to formation of a carbanion on the newly introduced side chain. This anion (**324**) then undergoes a ring expansion reaction, part of which involves an internal Claisen reaction with simultaneous bond migration. The product **325** now contains the desired β-dicarbonyl function. Alkylation of **325** with methyl iodide and sodium hydroxide interestingly occurs on nitrogen to give **326**, due to the greater acidity of the sulfonamide. Heating of **326** with 2-aminopyridine leads to exchange and formation of the amide. Thus, the NSAID drug **piroxicam** (**327**)[80] is obtained. Two closely related analogues are obtained by varying the heterocyclic amine used in this last step; 2-aminothiazole thus leads to **sudoxicam**, while 3-amino-5-methylisoxazole affords **isoxicam**.

3. COMPOUNDS WITH THREE HETEROATOMS

Continued research on the sulfonamide diuretic agents whose genesis from sulfanilamide is discussed in Chapter 2 revealed that some increase in potency was obtained by incorporating an amino group adjacent to one of the sulfonamide groups. This compound, which bears the trivial name of chloroaminophenamide, is prepared by reaction of *m*-chloroaniline (**328**) with chlorosulfonic acid under forcing conditions; ammonolysis of the thus obtained bis-sulfonyl chloride leads to the bis-sulfonamide

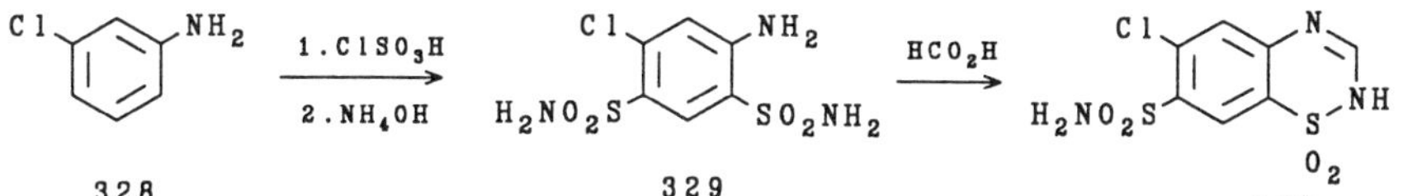

329, that shows somewhat improved potency over the desamino analogue. A marked increase in potency is, however, observed when the adjacent amino and sulfonamide groups are cyclized. This new functional array is analogous to the anthranilamides and shows similar chemistry. Thus, reaction of **329** with formic acid leads to a benzothiadiazine[81]; this drug, **chlorothiazide** (**330**), was the first orally effective potent diuretic that could be used without upsetting the patient's acid–base balance.

Omission of the open-chain sulfonamide function interestingly suppresses diuretic activity and results in a compound that acts as a direct vasodilator. The required starting material is prepared by first displacing chlorine at the 2-position in 2,4-dichloronitrobenzene by means of benzyl mercaptide to afford the thioether **332**. Oxidation of sulfur by means of aqueous chlorine results as well in debenzylation; the overall reaction finally yields the sulfonyl chloride **333**. Treatment of **333** with ammonia gives the corresponding sulfonamide **334**; catalytic hydrogenation then converts the nitro group to an aniline. Reaction of this amino sulfonamide with acetic anhydride closes the ring to afford the desired benzothiadiazine.[82] The product **diazoxide** (**336**) is a very potent antihypertensive agent whose properties seem to be very similar to those of minoxidil (Chapter 9); there are scattered reports that this compound also stimulates hair growth.

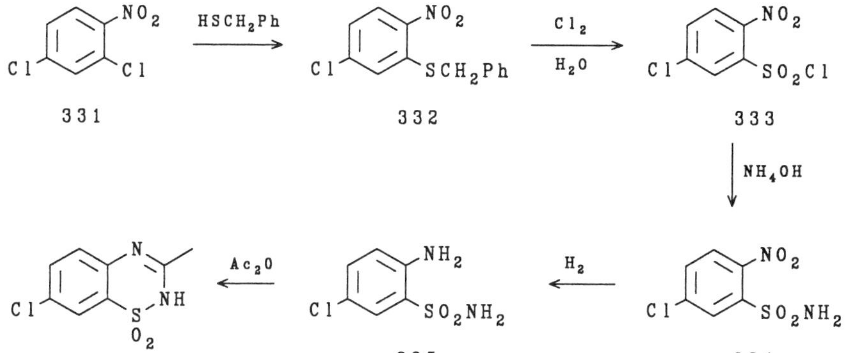

Reducing the double bond in the heterocyclic ring of chlorothiazide leads to a quite unexpected major increase in potency. While methods may exist for reducing the double bond in the heterocyclic ring, the compound, **hydrochlorothiazide** (**337**) is in fact readily available from reaction of the aminoanthranilamide (**329**) with formaldehyde.[83] The availability of this extremely well tolerated potent diuretic agent led clinicians to investigate the possibility that elevated blood pressure could be relieved by decreasing blood volume. This drug, now better known by its initials **HCTZ**, was in fact found to be effective in lowering blood pressure in close to one-half of all hypertensive patients. The mechanism by which this and other thiazide diuretics lower blood pressure is now known to be more complex than simply decreasing volume.

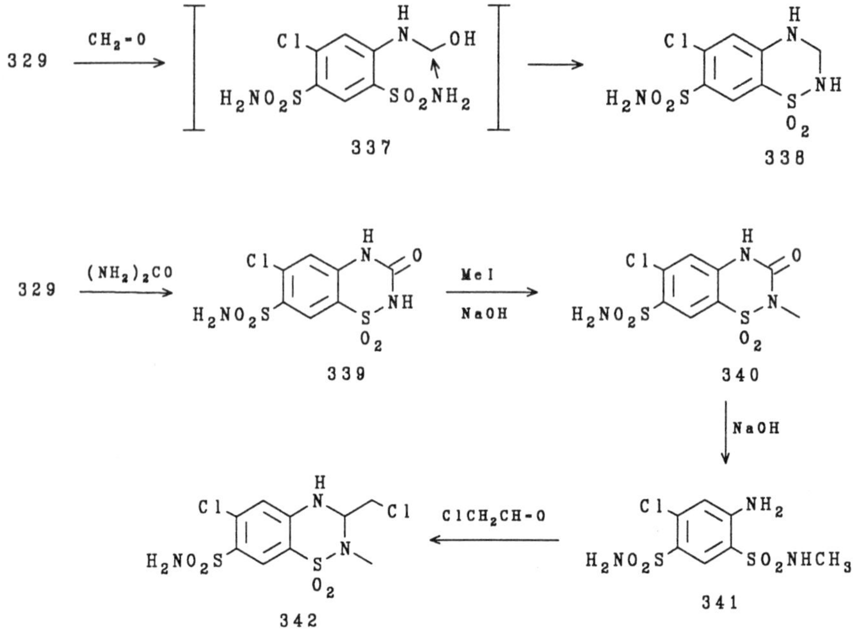

The straightforward nature of the last step in the previous scheme combined with the uncomplicated chemistry used to prepare both the starting aminosulfonamides led to the synthesis of probably hundreds of analogues; scores of compounds were prepared with modified side chains by replacing formaldehyde with various aldehydes. A dozen or more hydrothiazides are now available to the clinician. Some of the functionality on the ring can also be modified; a temporary ring is, for example, used to insure monoalkylation on the sulfonamide nitrogen. Thus, reaction of the familiar starting material **329** with urea leads to formation of **339**, that is essentially a cyclic urea. The sulfonamide nitrogen is alkylated by means of base and methyl iodide to give **340**. Base hydrolysis of the product restores the aminosulfonamide **341**. Condensation of **341** with chloroacetaldehyde gives the diuretic drug **methyclothiazide** (**342**).[84]

REFERENCES

1. Kawa, M.I.; Stahmann, M.A.; Link, K.P. *J. Am. Chem. Soc.* **1944**, *66*, 902.
2. Nitz, R.E.; Potzch, E. *Arzneim. Forsch.* **1963**, *13*, 243.
3. Langercrants, C. *Acta Chem. Scan.* **1956**, *10*, 647.
4. Kaufman, K.D. *J. Org. Chem.* **1961**, *26*, 117.
5. Kaufman, K.D.; Hewitt, L.E. *J. Org. Chem.* **1980**, *45*, 738.
6. Fitzmaurice, C.; Lee, T.B. U. S. Patent **1968**, 3419578.

7. da Re, P.; Verlicchi, L.; Setnikar, I. *Arzneim. Forsch.* **1960**, *10*, 800.
8. da Re, P.; Verlicchi, L.; Setnikar, I. *J. Med. Chem.* **1960**, *2*, 263.
9. Wu, E.S.C. Eur. Patent Appl. **1983**, 81621. *Chem. Abstr.* **1983**, *99*, 139772.
10. See, for example, Petrzilka, T.; Sikemeyer, C. *Helv. Chim. Acta* **1967**, *50*, 2111.
11. Archer, R.A.; Blanchard, W.B.; Day, W.A.; Johnson, D.W.; Lavagnino, E.R.; Ryan, C.W.; Baldwin, J.E. *J. Org. Chem.* **1977**, *42*, 2277.
12. Manchand, S.P.; Micheli, R.A.; Saposnik, S.J. *Tetrahedron* **1992**, *48*, 9391.
13. Sarges, R.; Howard, H.R.; Kelbaugh, P.R. *J. Org. Chem.* **1982**, *47*, 4081.
14. Van Lommen, G.R.; De Bruyn, M.L.F.; Schroven, M.F.J. Eur. Patent Appl. **1985**, 145067; *Chem. Abstr.* **1985**, *104*, 5773. *Drugs Future* **1989**, *14*, 957.
15. Klohs, M.W.; Petracek, F.J.; Bolger, J.W.; Sugisaka, N. So. African Patent **1973**, 72000748; *Chem. Abstr.* **1973**, *79*, 126310.
16. Woodward, R.B.; Doering, W.V. *J. Am. Chem. Soc.* **1945**, *67*, 860.
17. Taylor, E.C.; Martin, S.F. *J. Am. Chem. Soc.* **1972**, *94*, 6218.
18. Gutzwiller, J.; Uskokovic, M. *J. Am. Chem. Soc.* **1970**, *92*, 204.
19. Surrey, A.R.; Hammer, H.F. *J. Am. Chem. Soc.* **1946**, *68*, 113.
20. Burckhalter, J.H.; Tendrick, F.H.; Jones, E.M.; Holcomb, W.F.; Rawling, A.L. *J. Am. Chem. Soc.* **1948**, *70*, 1363.
21. Werbel, L.M.; Cook, P.D.; Elslager, E.F.; Hung, J.H.; Johnson, J.L.; Kesten, S.J.; McNamra, D.J.; Ortwein, D. F.; Worth, D.F. *J. Med. Chem.* **1986**, *29*, 924.
22. Steck, E.A.; Hallock, L.L.; Holland, A.J. *J. Am. Chem. Soc.* **1946**, *68*, 380.
23. Elderfield, R.C. *J. Am. Chem. Soc.* **1948**, *70*, 40.
24. Juma, F.D.; Ogeto, J.O. *Eur. J. Drug Metab. Pharmacokin.* **1989**, *14*, 15.
25. Adam-Molina, S. Eur. Patent Appl. **1984**, 103259; *Chem. Abstr.* **1984**, *88*, 22664.
26. Watson, J. Belg. Patent **1965**, 659237; *Chem. Abstr.* **1966**, *64*, 2071.
27. Bowie, R.A.; Cairns, J.P.; Jones, W.G.M.; Hayes, A.; Ryley, J.F. *Nature (London)* **1967**, 1349.
28. Clark, R.L.; Patchett, A.A.; Rogers, E.F. U. S. Patent **1968**, 3377352; *Chem. Abstr.* **1969**, *71*, 38821.
29. Burckhalter, J.H.; Brinigar, W.S.; Thompson, P.E. *J. Org. Chem.* **1961**, *26*, 4070.
30. Allais, A.; Meier, J. U. S. Patent **1966**, 3232944. *Chem. Abstr.* **1966**, *64*, 4410.
31. Allais, A. *Chim. Ther.* **1973**, *8*, 154.
32. Musser, J.H.; Kreft, A.F.; Bender, R.H.; Kubrak, D.M.; Carlson, R.P.; Chang, J.; Hand, J.M. *J. Med. Chem.* **1989**, *32*, 1176.
33. McNamara, J.M.; Leazer, J.L.; Bhupathy, M.; Amato J.S.; Reamer, R.A.; Reider, J.; Grabowski, F.J. *J. Org. Chem.* **1989**, *54*, 3718.
34. Richards, H.C. So. African Patent **1968**, 6803636; *Chem. Abstr.* **1969**, *71*, 30369.
35. Curran, A.C.W.; Shepherd, R.G. *J. Chem. Soc. Perkin Trans. 1.* **1976**, 983.
36. Lesher, G.Y.; Froelich, E.J.; Gruett, M.D.; Bailey, J.H.; Brundage, R.P. *J. Med. Chem.* **1962**, *5*, 1063.
37. Kaminsky, D.; Meltzer, R.I. *J. Med. Chem.* **1968**, *11*, 160.
38. Koga, H.; Murayama, S.; Suzue, S.; Irikura, T. *H. Med. Chem.* **1980**, *23*, 1538.
39. Stern, I.M. Eur. Patent Appl. **1984**; *Chem. Abstr.* **1984**, *101*, 110764.

40. Wentland, M.P.; Bailey, D.M.; Cornett, J.B.; Dobson, R.A.; Powles, R.G.; Wagner, R.B. *J. Med. Chem.* **1984**, *27*, 1103.
41. Chu, D.T.W.; Fernandes, P.B.; Claiborne, A.K.; Pihuleac, E.; Nordeen, C.W.; Maleczke, R.E.; Pernet, A.G. *J. Med. Chem.* **1985**, *28*, 1558.
42. Egawa, H.; Miyamoto, T.; Matsumoto, J. *Chem. Pharm. Bull.* **1986**, *34*, 4098.
43. Grohe, K.; Heitzer, H. *Liebigs Ann. Chim.* **1987**, 29.
44. Maclean, L.; Roberts, D.L.; Barron, K.; Nichol, K.J.; Harrison, A.E. Eur. Patent Appl. **1989**, 317149; *Chem. Abstr.* **1989**, *111*, 232593.
45. Pictet, A.; Finkelstein, M. *Chem. Ber.* **1909**, *42*, 1979.
46. Anonymous Neth. Patent Appl. **1966**, 6516328; *Chem. Abstr.* **1966**, *65*, 15351.
47. Brossi, A.; Lindlar, H.; Walter, M.; Schnider, O. *Helv. Chim. Acta* **1958**, *41*, 119.
48. Yamato, E.; Hirakura, M.; Sugasawa, S. *Tetrahedron (Supl. 8)* **1966**, *8*, 129.
49. Brossi, A.; Besendorf, H.; Pellmont, B.; Walter, M.; Schnider *Helv. Chim. Acta* **1960**, *43*, 1459.
50. Kadin, B.B. So. African Patent **1969**, 6803456; *Chem. Abstr.* **1969**, *70*, 115025.
51. Fourneau, E. U. S. Patent **1936**, 2056046.
52. Monro, A.M. *Chem. Ind.* **1964**, 1806.
53. Huebner, C.F.; Gschwend, H.W. U. S. Patent **1983**, 4380653; *Chem. Abstr.* **1983**, *99*, 53771.
54. Chapleo, C.B.; Davis, J.A.; Myers, P.L.; Redhead, M.J.; Stillings, M.R.; Welbourn, A.P.; Hampson, F.C.; Sugden, K. *J. Heterocycl. Chem.* **1984**, *21*, 77.
55. Dobrina, V.A.; Ioffe, D.V.; Kuznetsov, S.G.; Chigarev, A.G. *Khim. Pharm. Zh.* **1974**, *8*, 14. *Chem. Abstr.* **1974**, *81*, 91445.
56. Hilbert, M.; Gittos, M.W. Eur. Patent Appl. **1986**, 4380653; *Chem. Abstr.* **1986**, *105*, 24193.
57. White, W.A. German Offen. **1975**, 2065719; *Chem. Abstr.* **1975**, *83*, 58860.
58. Schatz, F.; Wagner-Jauregg, T. *Helv. Chim. Acta* **1968**, *51*, 1919.
59. Druey, J.; Ringler, B.H. *Helv. Chim. Acta* **1951**, *34*, 195.
60. Campbell, S.F.; Danilewicz, J.C.; Evans, A.G.; Ham, A.L. Br. Patent **1979**, 2006136. *Chem. Abstr.* **1979**, *91*, 193331.
61. Tasaka, K.; Akagi, M. *Arzneim.-Forsch.* **1979**, *29*, 488.
62. Brittain, D.R.; Wood, R. Eur. Patent Appl. **1979**, 2895. *Chem. Abstr.* **1980**, *92*, 76533.
63. Ishikawa, M.; Eguchi, Y. *Chem. Pharm. Bull.* **1980**, *28*, 2770.
64. Minielli, J.L.; Scarborough, H.C. Fr. Patent **1965**, M3207. *Chem. Abstr.* **1965**, *63*, 13287.
65. Cronin, T.H.; Hess, H.J. So. African Patent **1968**, 6706512. *Chem. Abstr.* **1969**, *70*, 68419.
66. Althuis, T.H.; Hess, H.J. *J. Med. Chem.* **1977**, *20*, 146.
67. Hess, H.J. German Offen. **1971**, 2120495; *Chem. Abstr.* **1972**, *76*, 127012.
68. Burch, H.A. *J. Med. Chem.* **1966**, *9*, 408.
69. Davoll, J.; Johnson, A.M. *J. Chem. Soc. C* **1970**, 997.
70. Elslager, E.F.; Johnson, J.L.; Werbel, L.M. *J. Med. Chem.* **1983**, *26*, 1753.
71. Jackman, G.B.; Petrow, V.; Stephenson, O. *J. Pharm. Pharmacol.* **1960**, *28*, 344.
72. Mattner, P.G.; Salmond, W.G.; Denzer, M. Fr. Patent **1973**, 2174828.
73. Aguire Ormaza, V. Spanish Patent **1986**, 549881. *Chem. Abstr.* **1987**, *107*, 236726.

74. Press, J.B.; Bandurco, V.T.; Wong, E.M.; Hajos, G.Z.; Konojia, R.M; Mallory, R.A.; Deegan, E.G.; McNally, J.J.; Roberts, J.R.; Coffer, M.L.; Gradan, D.W.; Lloyd, J.R. *J. Heterocycl. Chem.* **1986**, *23*, 1821.

75. Cohen, E.; Karberg, B.; Vaughan, J.R. *J. Am. Chem. Soc.* **1957**, *82*, 2731.

76. Cantarelli, G. *Il Framaco, Ed. Sci.* **1970**, *25*, 761.

77. Shetty, B.V.; Campanella, L.A.; Thomas, T.L.; Fedorchuk, M.; Davidson, T.A.; Michelson, L.; Volz, H.; Zimmerman, S.E.; Belair, E.J.; Truant, A.P. *J. Med. Chem.* **1970**, *13*, 886.

78. Hayao, S.; Havera, H.J.; Strycker, W.G; Leipzig, T.J.; Kulp, R.A.; Hrtzler, H.E. *J. Med. Chem.* **1965**, *8*, 807.

79. Aguiree Ormaza, V. Spanish Patent **1986**, 548996. *Chem. Abstr.* **1986**, *106*, 84630.

80. Lombardino, J.G.; Wiseman, E.H. *J. Med. Chem.* **1972**, *15*, 849.

81. Novello, F.C.; Sprague, J.M. *J. Am. Chem. Soc.* **1957**, *79*, 2028.

82. Rubin, A.A.; Roth, F.E.; Winburg, M.W.; Topliss, J.G.; Sherlock, M.H.; Sperber, N.; Black, J. *Science* **1961**, *133*, 2067.

83. DeStevens, G.; Werner, L.H.; Halamandaris, A.; Ricca, S. *Experientia* **1958**, *14*, 463.

84. Close, W.J.; Swett, L.R.; Brady, L.E.; Short, J.H.; Vernstein, M. *J. Am. Chem. Soc.* **1960**, *82*, 1132.

CHAPTER 12

SEVEN-MEMBERED HETEROCYCLIC RINGS FUSED TO BENZENE

The large number of entries in Chapter 11 underlines the abundance of biologically active benzo-fused six-membered heterocycles that are actually used, or have been investigated as therapeutic agents. Increasing the size of the fused ring by just one carbon leads to a remarkable drop in the number of candidates. The benzodiazepines that form the bulk of the material in this chapter, however, represent a distinct exception to that generalization. The remarkable commercial success of this class of anxiolytic agents has led to the synthesis of literally thousands of analogues. The very small selection of compounds discussed below represents a very small sample of even those that have been assigned nonproprietary names.

1. COMPOUNDS WITH ONE HETEROATOM

Dopamine, the free catechol corresponding to **1**, plays an important role as a neurotransmitter particularly in the CNS. The synthesis of a dopamine related sedative agent begins with the condensation of homoveratramine (**1**) with styrene oxide (**2**) to afford the carbinol **3**. Treatment of **3** with strong acid leads to attack on the electron-rich aromatic ring by the resulting carbocation. Thus, the benzazocine **4** is obtained. The secondary amine is then methylated by reaction with formaldehyde and formic acid to yield **trepipam** (**5**).[1]

Dopamine proper has long been used as an inotropic agent in acute treatment of congestive heart failure. Both this compound and a number of its analogues have a positive action on contractility as a consequence of their adrenergic agonist activity. Experimental data, which suggest differential biological effects for different dopamine

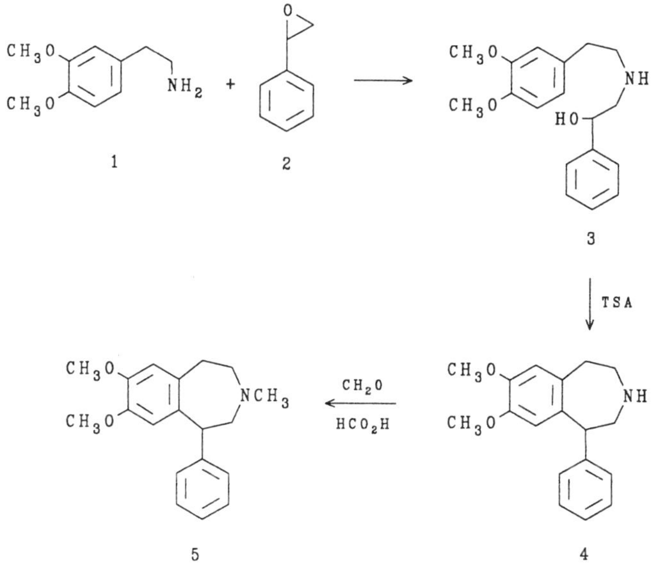

conformers, have led to extensive investigations of rigid analogues of this drug. Cardiovascular activity interestingly predominates in a close analogue of **5**, which more nearly resembles dopamine by including free phenols and a secondary amine. A condensation and cyclodehydration sequence similar to the one shown above starting

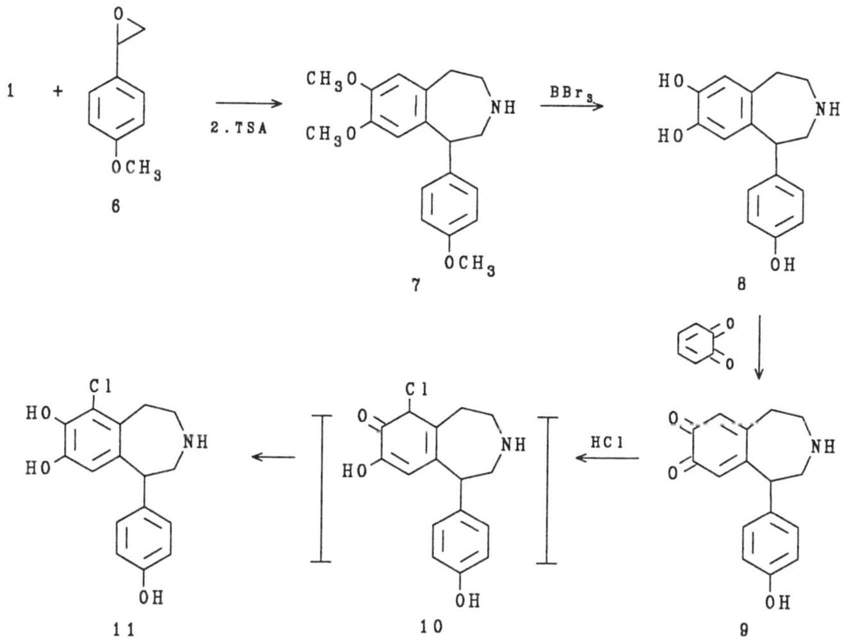

with 4-methoxystyrene oxide (**6**) leads to the benzazocine **7**. Treatment of **7** with at least four equivalents of boron tribromide, one for each Lewis base present, leads to cleavage of the methyl ethers to free phenols to give the product **8**. The catechol function is then oxidized selectively to the corresponding *o*-quinone **9** by a redox exchange with *o*-quinone itself. Addition of hydrogen chloride to the quinone can be envisaged as proceeding through a 1,4-addition intermediate such as **10**. Enolization with consequent aromatization then affords the dopamine agonist **fenoldopam** (**11**).[2]

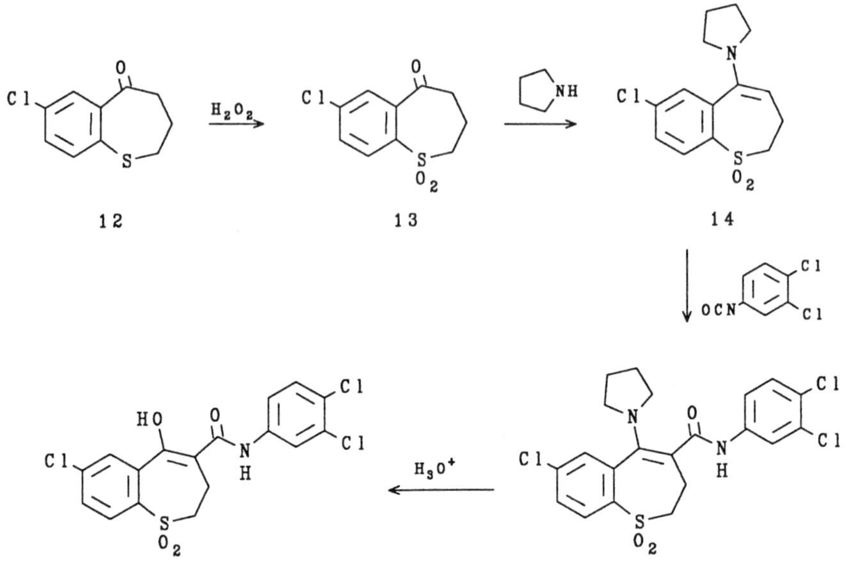

A benzazocine that includes the same β-ketoamide array as piroxicam (Chapter 11) retains NSAID activity. Oxidation of benzothiapinone (**12**), obtainable by cyclization of 4-(4-chlorophenylthio)-butyric acid, with hydrogen peroxide gives the corresponding sulfone **13**. This compound is then converted to its enamine **14** by reaction with pyrrolidine. Condensation of the intermediate with 3,4-dichorophenylisocyanate leads to the amide **15**. Hydrolysis of **15** with aqueous acid cleaves the enamine function to give the ketoamide and thus, **enolicam** (**16**).[3]

2. COMPOUNDS WITH TWO HETEROATOMS

A. Benzodiazepine Anxiolytic Agents

The discovery of the benzodiazepines provides one of the most apt illustrations of the role of serendipity in finding new classes of drugs. First, the case at hand involved the unexpected course for two successive chemical reactions; the availability of a new mouse behavioral screen just in time to test the newly synthesized compound completed the unexpected discovery sequence. The benzodiazepines, as noted above,

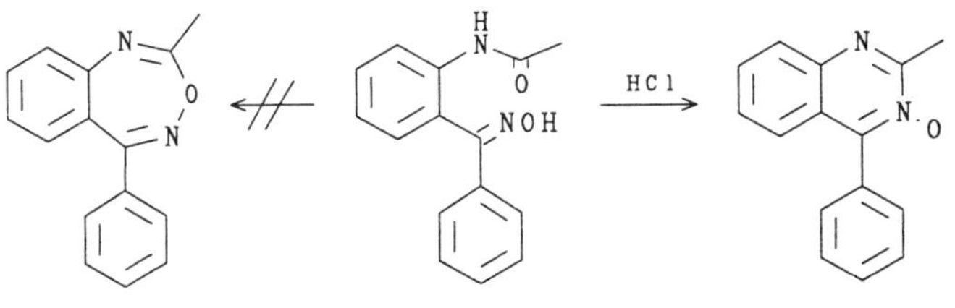

18 17 19

quickly gained an enormous market as antianxiety agents, as hypnotics, and, to a lesser extent, as muscle relaxing agents. The mood altering properties of these compounds almost inevitably resulted in their becoming drugs of abuse. Consequently, at least in the United States, they are controlled substances. Advances in radioreceptor assays led to the identification, in the late 1970s, of receptors that bind benzodiazepines. This receptor, which is actually part of a complex involved in γ-aminobutyric acid (GABA) regulation, has been subsequently shown to bind structurally unrelated antianxiety agents, for example, zolpidem (see Chapter 15).

The first starting material for benzodiazepines was prepared inadvertently in a synthesis aimed at the benzodiazoxepine **18**. The oxime acetamide **17**, from 2-aminobenzophenone, was thus treated with hydrogen chloride in the expectation that the new heterocycle would form by elimination of water between the oxime and the enol form of the amide. In fact, the product turned out to be the quinazoline *N*-oxide **19**, the product from addition of the nucleophilic oxime nitrogen to the amide carbonyl group.

The original preparation of an arylbenzodiazepine relied on a closely related quinazoline; a large number of alternate routes have, however, now been developed.[4] Reaction of the oxime from the aminobenzophenone **20** with chloroacetyl chloride gives the chloroacyl amide **21**; this compound is converted to the corresponding quinazoline *N*-oxide **22** on treatment with hydrogen chloride. Attempted displacement

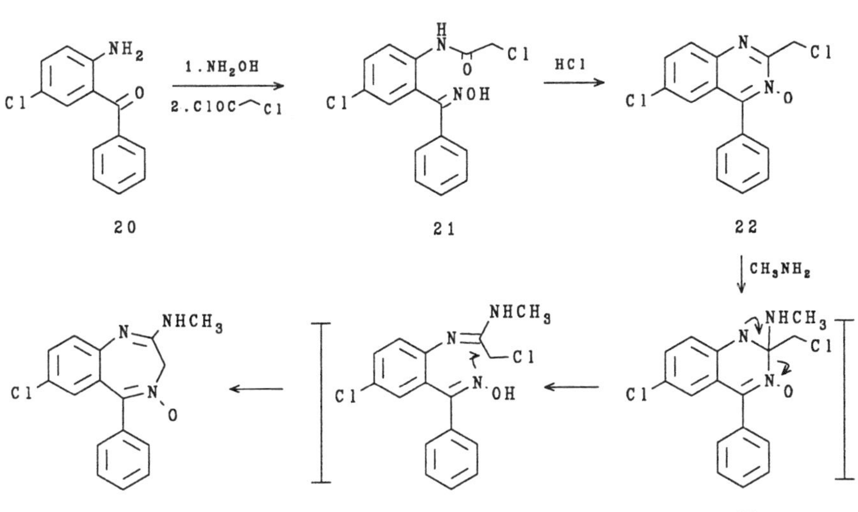

20 21 22

25 24 23

26 → ClOC⌒Cl → 27 → NH₃ or HMTA → 28

of the halogen in **22** with methylamine, probably intended to provide a simple analogue for screening, gives instead a benzodiazepine. The reaction can be visualized as initially involving formation of the adduct **23**; return of the electrons in the ring nitrogen will then lead to ring opening and formation on the amidino-oxime **24**. This species then undergoes internal displacement of chlorine by the basic oxime nitrogen with consequent formation of the seven-membered ring. This reaction affords **chlordiazepoxide** (**25**),[5] better known as Librium®, the first of a long series of benzodiazepines.

Full activity is retained in the face of considerable structural simplification; the amidine function in **25** can thus be replaced by a simple amide. One of the more straightforward approaches first involves acetylation of the aminobenzophenone **26** with chloroacetyl chloride to give the amide **27**. Heating **27** with ammonia or its latent equivalent, hexamethylenetetramine (HMTA), can be envisaged to involve initial displacement of chlorine to give a glycineamide. Cyclization by imine formation then affords **diazepam** (**28**),[6] more familiarly known as Vallium®. Support for this sequence comes from the observation that a modest yield of **28** can be obtained on heating **26** with glycine ethyl ester in pyridine.[7]

One of the more important routes for metabolism of benzodiazepines involves the introduction of a hydroxyl group at the 3-position in the heterocyclic ring. The fact that this product shows full activity opens the possibility that it may be the active species. A variant on the quinazoline *N*-oxide rearrangement provides the starting material **29** for one route to 3-hydroxylated compounds. Thus, reaction of **22** with sodium hydroxide can be visualized as first involving addition of hydroxide to give a carbinolamine equivalent of **23**; ring opening will give a chloroacetamide that will cyclize to the observed product **29**. Reaction of **29** with acetic anhydride leads to the formation of acetate **30** via the typical *N*-oxide Polonovski rearrangement. Saponification with mild base then gives **oxazepam** (**31**).[8]

22 → NaOH → 29 → Ac₂O → 30 → NH₄OH → 31

A somewhat different strategy is used for preparing a hydroxylated derivative from a dichloro analogue relies on the use of a carbethoxy group at the 3-position for activation. The first step consists of condensation of diethyl 2-aminomalonate with the aminobenzophenone **32**. The reaction probably follows a course similar to the one using ethyl glycinate to form an amide such as **33** as the first intermediate. Imine formation will then lead to the carbethoxy substituted benzodiazepine **34**. Bromination proceeds at the 3-position, which readily enolizes due to the proximity of two carbonyl groups. Methanolysis of the somewhat labile intermediate replaces the bromine by a methyl ether to give the intermediate **35**. Saponification of the ester group is followed by decarboxylation of the β-amidoacid, which removes the extraneous carbon atom at the 3-position to give **36**. The methyl ether is then cleaved to an alcohol by means of boron tribromide to afford **lorazepam** (**37**).[9]

The SAR in the benzodiazepine series is sufficiently flexible to tolerate a simple tertiary amine at the 1-position. An interesting scheme for preparing such compounds involves the use of aziridine for the required two-carbon fragment and a cyclodehydration reaction to form the ring. The concise sequence starts with the reaction of the anion from *p*-chloro-*N*-methylaniline (**38**) with the benzoic acid amide of aziridine. Opening of the reactive three-membered ring leads to the amide **39**, which now contains the required atoms. Treatment of this amide with phosphorus oxychloride leads to cyclodehydration and formation of the diazepine ring. Thus, the antianxiety agent **medazepam** (**40**)[10] is obtained.

An alternate and equally brief approach to the same compound begins with the reaction of the aziridino-benzophenone **41** with methyl iodide. The outcome of this reaction can be rationalized by assuming initial formation of the quaternary salt **42**.

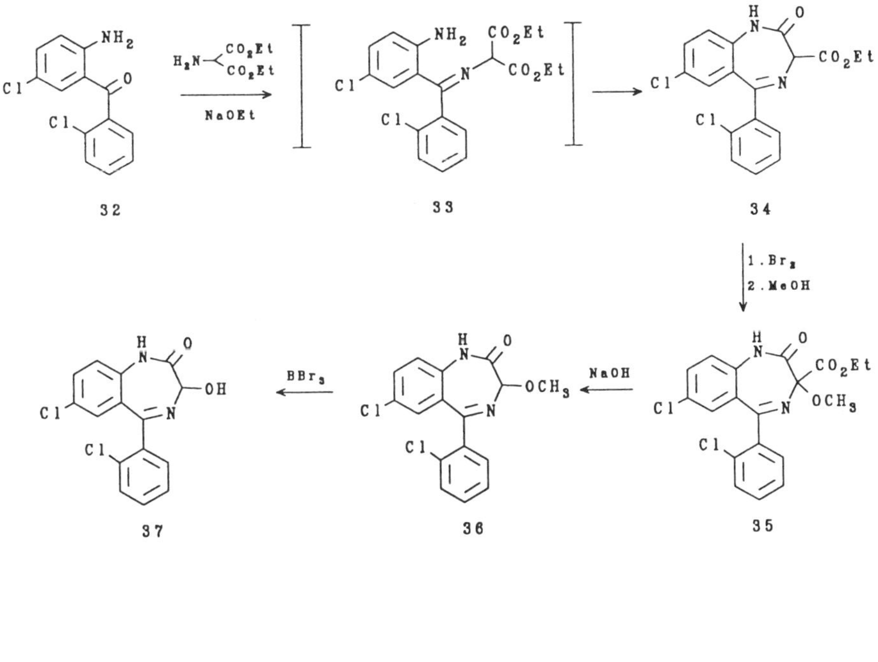

$H_2N{-}CH(CO_2Et)_2$ / NaOEt; 1. Br_2 2. MeOH; NaOH; BBr_3

32 33 34 35 36 37

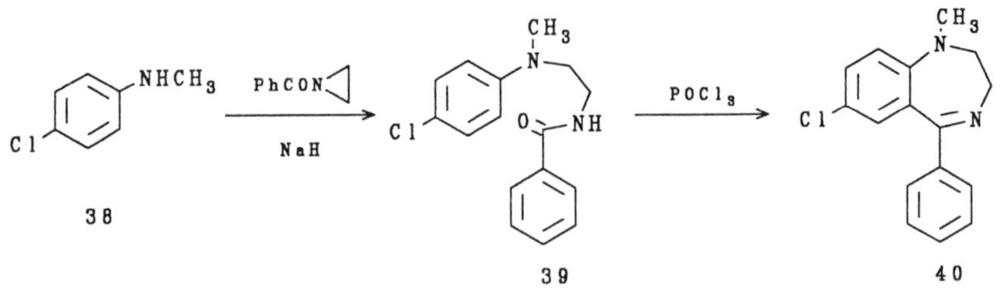

Attack on the strained ring by the iodide counterion will open the ring to afford the *N*-iodoethyl derivative **43**. This compound then affords medazepam (**40**) on reaction with HMTA.[11]

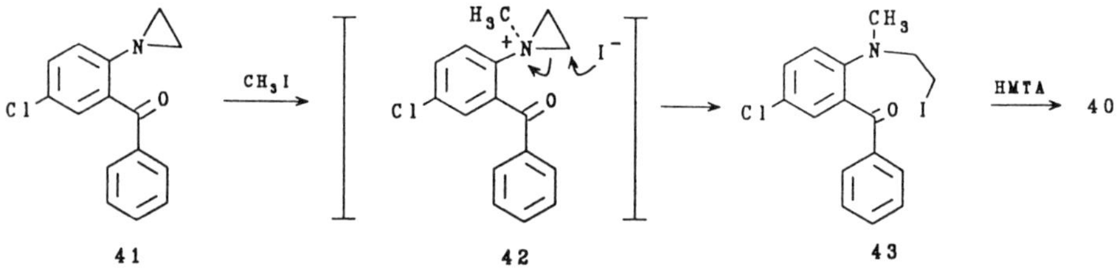

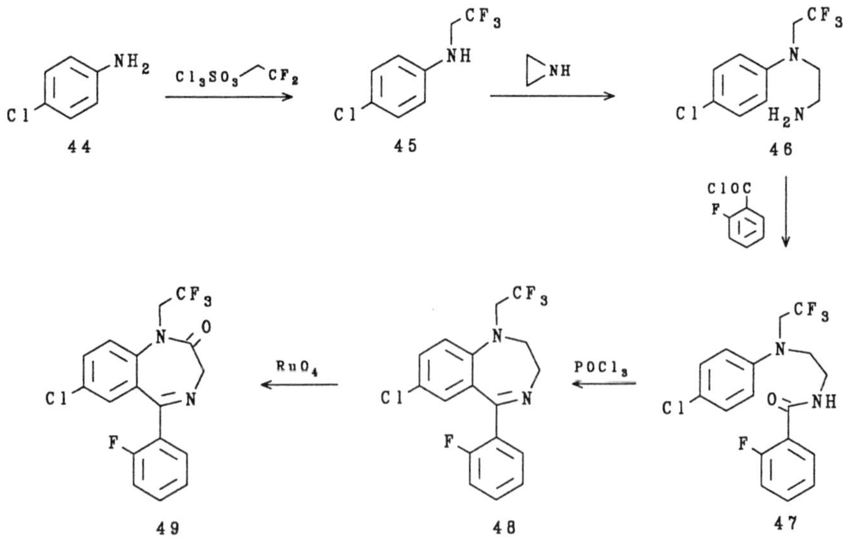

Alkylation of *p*-chloroaniline (**44**) with 2,2,2-trifluoroethyl trichloromethylsulfonate affords the corresponding trifluroethylated derivative **45**. Reaction of this anion from that with aziridine proper leads to formation of the diamine **46**. Acylation of **46** with *o*-fluorobenzoyl chloride proceeds to give the amide **47**. Cyclodehydration of **47** with phosphorus oxychloride gives the benzodiazepine **fletazepam** (**48**). Interestingly, ruthenium tetroxide preferentially oxidizes the 2-position rather than the 3-position adjacent to the imine function. This reaction results in the formation of (**49**).[12]

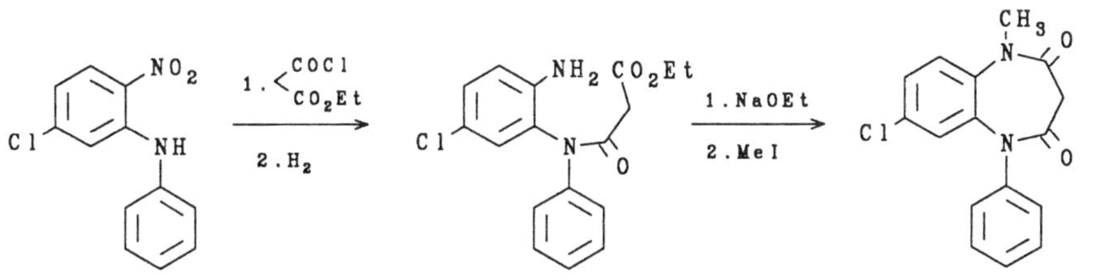

The compound in which nitrogen and carbon at the 4- and 5-positions are transposed (with formal reduction of the imine) displays much the same activity as the prototypes. Acylation of the nitrodiphenylamine **50** with ethyl malonyl chloride gives the corresponding amidoester; the nitro group is then reduced to the amine by catalytic hydrogenation to give the intermediate **51**. Reaction of **51** with strong base closes the benzodiazepine ring. Methylation of the secondary aniline by means of methyl iodide completes the synthesis of **clobazam** (**52**).[13]

Conversion of an amide to a thioamide enhances the reactivity of this function since it favors the enol form and provides a better leaving group for addition–elimination reactions (mercaptide vs. hydroxide). Thioamides obtained by treatment of diazepinones such as **53** or **58** with phosphorus pentasulfide provide starting materials for further modification of the benzodiazepine nucleus. [More recently developed reagents such as Lawesson's reagent or bis(tricyclohexyltin) sulfide provide a more convenient method for this transformation.] Thus, reaction of the thioamide **54** with *O*-allylhydroxylamine leads directly to the amidine, probably via an addition–elimination sequence of the thioenol tautomer of **54**. Thus, the antianxiety agent **uldazapam** (**55**)[14] is obtained.

In addition, the thioamide function provides a means for building additional heterocyclic rings onto the basic benzodiazepine nucleus. Many of the products show markedly enhanced potency compared to the simpler compounds discussed so far. This finding has not surprisingly led to the synthesis of hundreds of analogues in the search for agents with unique properties. Reaction of **54** with acetylhydrazide can be visualized as proceeding through the initial formation of amidine **56**; cylization with simultaneous bond reorganization will then afford the triazolobenzodiazepine.[15] This product, **triazolam** (**57**), is an extremely potent compound; it has been used largely as a hypnotic agent because of its apparent short half-life. This drug, under its trade name of Halcion® has gained considerable notoriety due to reports of bizarre (and unlikely) side effects including allegation that it induced criminal behavior.

The fused ring can also be built in stepwise fashion. Reaction of the thioamide **59** with hydrazine gives the versatile intermediate amidine **60**. Condensation of **60** with methylorthoformate leads to the formation of an unsubstituted fused triazole ring; this affords **estazolam** (**61**).[16] The same reaction using methyl orthoacetate gives the antianxiety agent **alprazolam** (**62**).[17]

53 → P_2S_5 → 54 → H_2NO → 55

54 → H_2NNHAc → [56] → 57

The presence of a basic amino group affords a compound that also shows some degree of antidepressant activity. Acylation of the hydrazine **60** with chloroacetyl chloride proceeds on the more basic nitrogen and gives the hydrazide **63**. Heating **63** in acetic acid closes the triazole ring to give the chloromethylated product **64**. Displacement of chlorine by means of dimethylamine then affords **adinazolam** (**65**).[18]

The thioamide function also provides the entry for the construction of a fused imidazole ring although the sequence is somewhat more complex due to the need to form a new carbon–carbon bond. Reaction of the thioamide **66** with methylamine

58 → P_2S_5 → 59 → H_2NNH_2 → 60 (NHNH$_2$)

60 → $HC(OMe)_3$ → 61

60 → $CH_3C(OMe)_3$ → 62

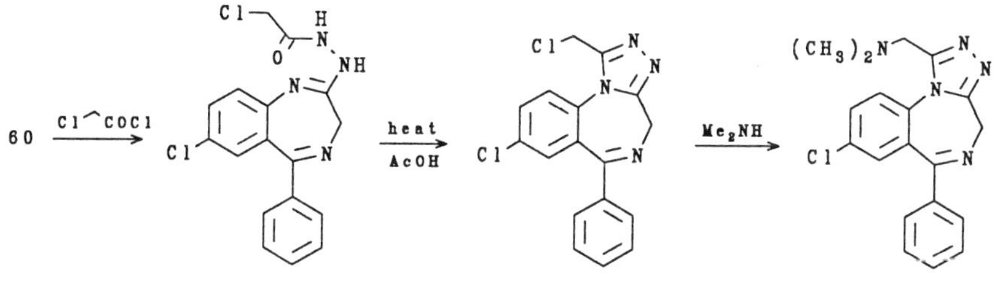

60 → (ClCH2COCl) 63 → (heat, AcOH) 64 → (Me2NH) 65

proceeds to give the corresponding amidine; this compound is transformed into a good leaving group by conversion to the *N*-nitroso derivative **67** by treatment with nitrous acid. Condensation of this intermediate with the carbanion from nitromethane leads to displacement of the *N*-nitroso group at the 2-position and formation of the methylnitro derivative; the double bond shifts into conjugation with the nitro group to afford **68**. Treatment with Raney nickel then reduces the nitrovinyl function all the way to an aminomethyl group. Reaction of the resulting diamine **69** with methyl orthoacetate then leads to formation of the fused imidazoline ring (**70**). Dehydrogenation of this ring with manganese dioxide converts it to an imidazole to give **midazolam** (**71**).[19]

An analogue of these fused benzodiazepines, where the benzene ring at the 5-position is omitted, shows benzodiazepine antagonist activity in both in vitro binding assays and in selected in vivo models. The benzodiazepinedione nucleus is obtained from the condensation of the fluorinated isatoic anhydride **72** with *N*-methylglycine (sarcosine). The first step probably involves acylation of the amino acid nitrogen by the activated anhydride carbonyl group. Loss of carbon dioxide from the resulting carbamic acid will lead to the amide **73**. This compound then cyclizes to the benzodiazepinedione **74**. Reaction of **74** with ethyl isocyanoacetate then leads to addition

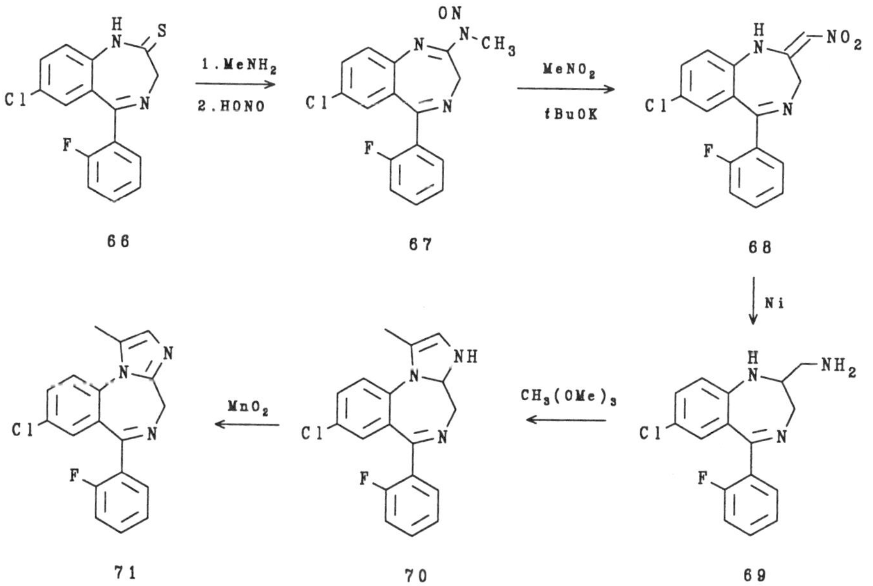

66 → (1. MeNH2, 2. HONO) 67 → (MeNO2, tBuOK) 68 → (Ni) 69 → (CH3(OMe)3) 70 → (MnO2) 71

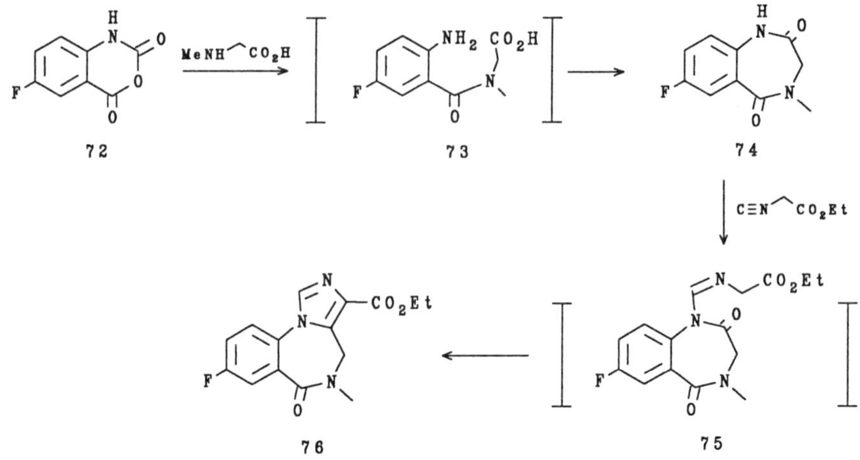

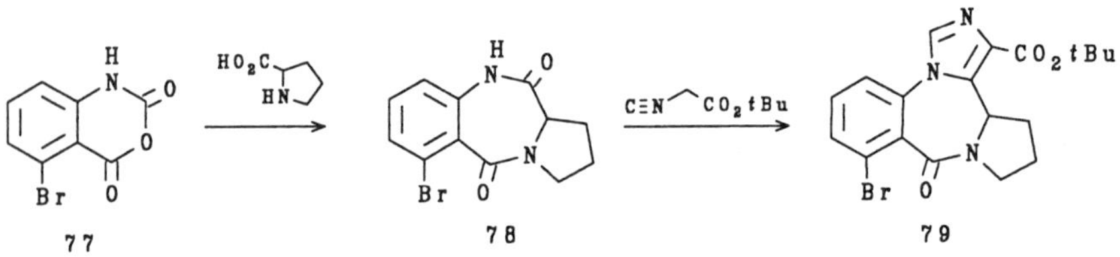

of the only free amide nitrogen to the isocyanide function to afford an intermediate such as the amidine **75**. The doubly activated acetate methylene group then condenses with the ring carbonyl group to form an imidazole, affording **flumazenil** (**76**).[20]

Fusion of an additional heterocyclic ring onto the above benzodiazepine restores at least partial agonistic activity. The benzodiazepine nucleus is built exactly as above, starting from **77** and using L-proline instead of sarcosine to afford **78**. The fused imidazole ring is again added by condensation with an isocyanoacetate, in this case the *tert*-butyl ester. This compound then affords **bretazenil** (**79**)[21] a compound that displays some antianxiety activity.

B. Other Seven-Membered Heterocycles Fused to a Benzene Ring

A very sizeable proportion of biological mediators and virtually all enzymes consist of peptides. The use of such compounds as therapeutic agents is severely constrained by biopharmaceutical considerations. Very few peptides can be administered orally because the compounds are in the first instance often destroyed in the GI tract; those that do survive are usually not absorbed because of their high molecular weight and/or polarity. Careful studies of the three-dimensional structure of peptides made possible by the increased availability of X-ray crystallographic structures and computer molecular modeling programs has made it possible to map the substructures on peptides that trigger responses by receptor and/or substrate interaction. This information can be used in appropriate cases to design smaller molecules that fit and interact with these

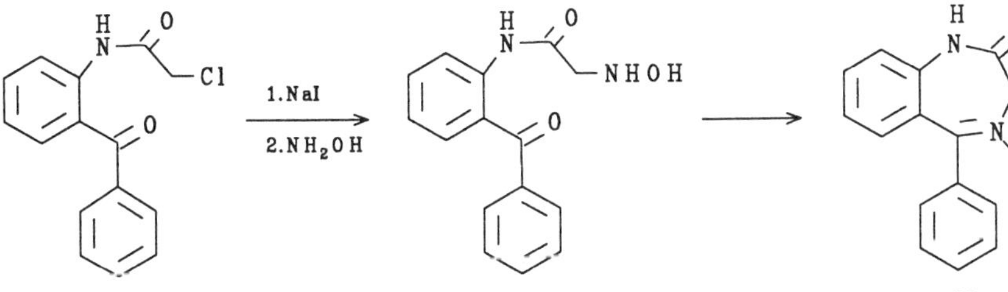

1.NaI
2.NH_2OH

NHOH

80 81 82

same sites. A benzodiazepine provides the backbone for one of the first successful small peptide mimics; the compound **devapezide** (**85**) acts as an antagonist to cholecystokinin, an enzyme with numerous actions including gastric stimulation.

The route to this agent uses a modification of the quinazoline approach for the construction of the required benzodiazepine *N*-oxide. This *N*-oxide function is included as a handle for eventual introduction of an amino group at the 3-position. Reaction of the product **80** from acylation of aminobenzophenone with chloroacetyl chloride with sodium iodide gives the corresponding iodide. Treatment with hydroxylamine leads to displacement of iodine and formation of the hydroxylamine **81**. The basic nitrogen then attacks the carbonyl group in a reaction reminiscent of the formation of the pyrimidine ring in minoxidil (Chapter 9). This reaction results in formation of the Schiff base, *N*-oxide **82**.

Reaction of **82** with acetic anhydride leads to the Polonovski rearrangement. Saponification of the initially formed acetate gives the 3-hydroxylated derivative **83**. The newly introduced function is then converted to the chloride by means of thionyl chloride; displacement with ammonia then gives **84**. Acylation with the acid chloride from indole-2-carboxylic acid leads to the corresponding amide. Reaction of this product with strong base leads to formation of an anion on the lactam nitrogen; this affords **devapezide** (**85**) on treatment with methyl iodide.[22]

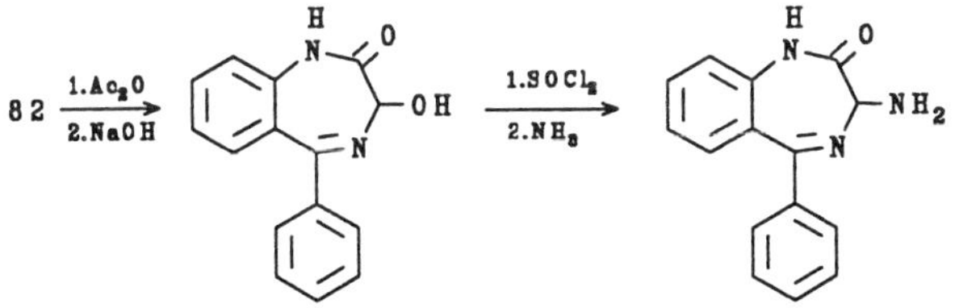

82 1.Ac_2O 2.NaOH

1.$SOCl_2$ 2.NH_3

83 84

ClOC

2.MeI/NaH

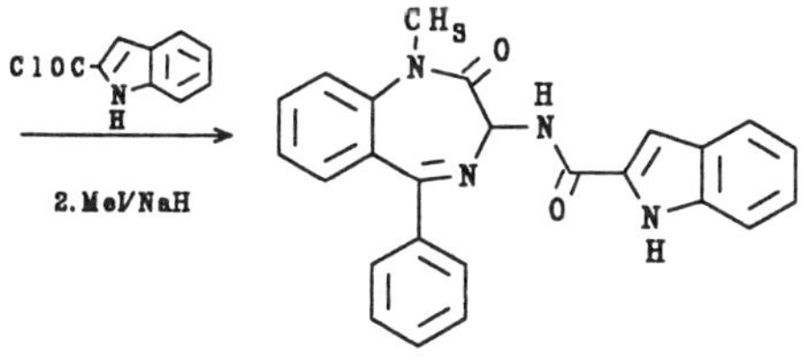

85

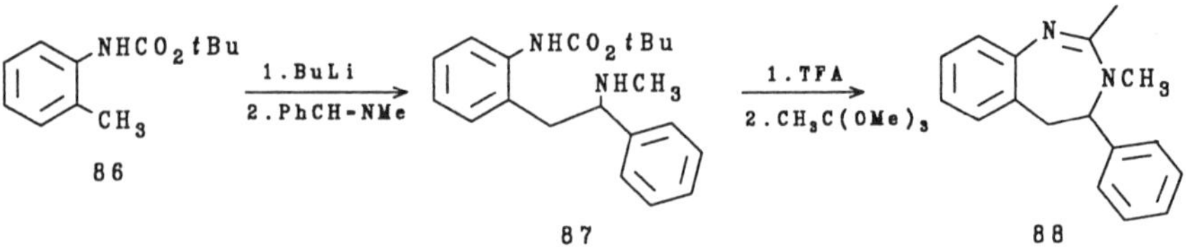

Fused tricyclic compounds have a venerable history as antidepressant drugs. A number of carbocyclic representatives are considered in Chapter 3, while those that consist of dibenzo heterocyclic compounds are noted in Chapter 13. Antidepressant activity is retained in a benzodiazepine in which the third ring is in an unfused form. The short synthesis starts with the addition of the carbanion from the toluidine amide **86** to the Schiff base from benzaldehyde to give the 1,2-diphenyl ethane derivative **87**. The protecting group is then removed by means of trifluoroacetic acid. Reaction of the resulting diamine with methyl orthoacetate adds the required extra carbon atom and leads to formation of the diazepine **dazepinil** (**88**).[23]

A triazole provides the third ring in an antidepressant agent that also includes the pendant piperazine ring often found in this class of CNS agents. Reaction of the iminoether **89** from *o*-nitrophenylacetonitrile with acethydrazide leads to formation of the triazole ring (**90**). The nitro group is then reduced to the amine **91** by catalytic hydrogenation. Treatment of **91** with carbonyl diimidzole (CDI) bridges the aniline and triazole functions to afford the urea **92**. This compound is converted to the enol chloride **93** by the standard phosphorus oxychloride reaction. Displacement of halogen by means on *N*-methylpiperazine leads to formation of **batelapine** (**94**).[24]

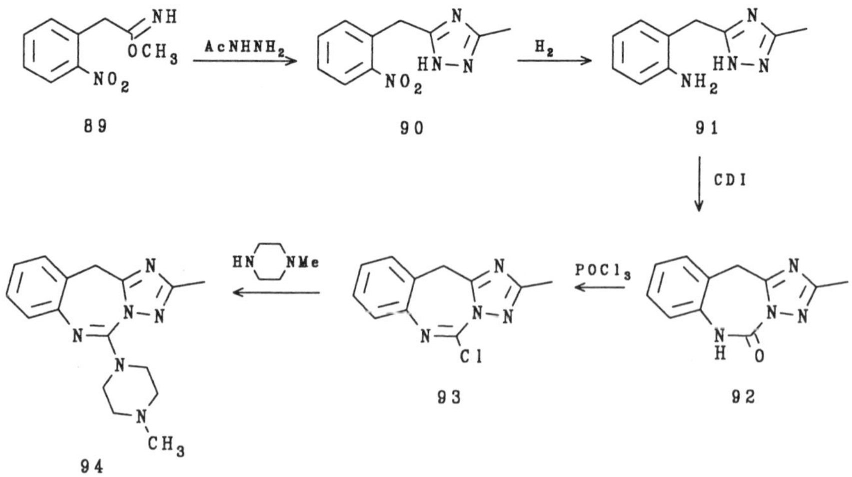

The synthesis of a compound in which the piperazine is incorporated in the form of an additional fused ring starts with the condensation of nitrobenzylamine **95** with 2,5-dimethoxytetrahydrofuran; this latent 1,5-dialdehyde reacts with the primary amino group to form the benzylpyrrole. The nitro group is then reduced catalytically to give the aniline **96**. Acid catalyzed reaction of **96** with the hemicetal from ethyl glyoxal leads to the tricyclic nucleus **97**; this reaction can be visualized as involving initial addition of aniline to the aldehyde to form a protonated imonium ion that then attacks the pyrrole to form the central ring. Acylation of the intermediate with chloroacetyl chloride then gives chloroamide **98**. Treatment of **98** with methylamine leads to the amide **99** by initial displacement of chlorine followed by ester amide interchange. Reaction of **99** with lithium aluminum hydride then leads to reduction of both amide groups and thus formation of the antidepressant agent **aptazepine** (**100**).[25]

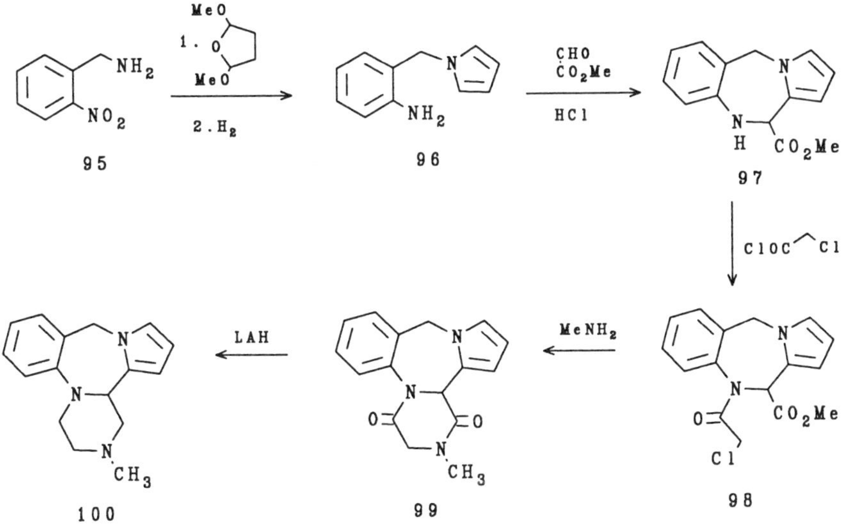

Yet a further variation on this theme consists in replacement of the bridging methylene group by sulfur to give a thiadiazepine as the central ring. The starting thiophene ether **102** is obtained by nucleophilic aromatic displacement of fluorine in

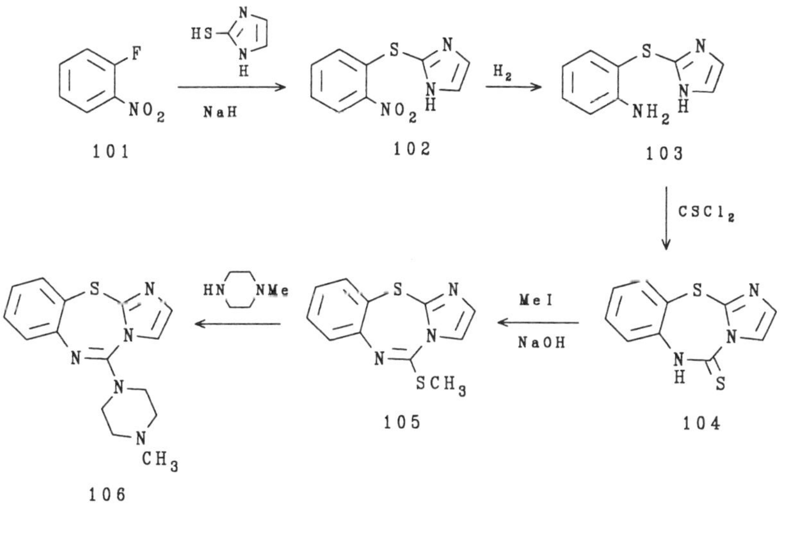

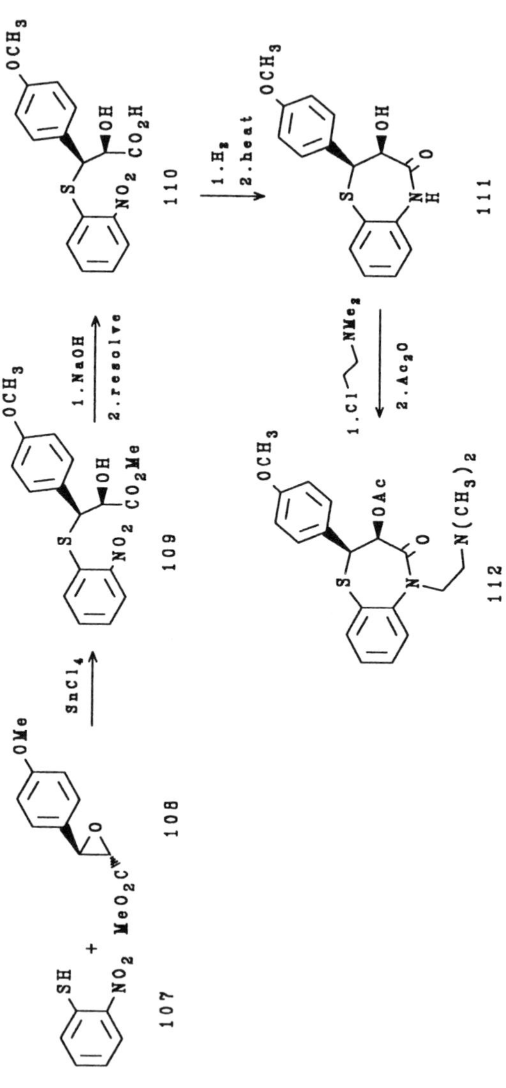
SH
NO2
+
MeO2C
O
OMe
107
108
SnCl4
S
NO2
OH
CO2Me
OCH3
109
1. NaOH
2. resolve
S
NO2
OH
CO2H
OCH3
110
1. H2
2. heat
S
N
H
O
OH
OCH3
111
1. Cl NMe2
2. Ac2O
S
N
O
OAc
OCH3
N(CH3)2
112

the nitrobenzene **101** by the anion from imidazole-2-thiol. The nitro group is then reduced to the aniline **103**. The amine groups are bridged much as above although in this instance it is by the use of thiophosgene to give the thiadiazepine **104**. Alkylation of the thioamide with methyl iodide locks that function into its enol form **105** and at the same time converts sulfur to the potential good leaving group, methyl mercaptan. Reaction of **105** with *N*-methylpiperazine thus results in formation of the displacement product **pentiapine** (**106**).[26]

A benzoazathiazepine provides the nucleus for a structurally unusual calcium channel blocker that is one of the growing class of drugs provided as the pure biologically active enantiomer. The key, and very carefully studied, reaction to the preparation of this compound consists of the opening of the racemic trans glycidic ester **108** with nitrothiophenol **107**. The reaction proceeds to give the threo hydroxyester **109**; tin salt catalysis insures the unusual highly specific cis opening of the oxirane ring. The product is then saponified to its acid and this is resolved as its cinchonine salt. The sequence proceeds with the isomer whose absolute configuration corresponds to **110**. The nitro group is then converted to the amine by catalytic hydrogenation; heating the product closes the amide ring to give **111**. The requisite side chain is then added by alkylation of **111** with 2-chloroethyldimethylamine. Acetylation of the free hydroxyl group with acetic anhydride completes the synthesis of **diltiazem** (**112**).[27]

REFERENCES

1. Kaiser, C.; Dandridge, P.A.; Garvey, E.; Hahn, R.A.; Sarau, H.M.; Setler, P.E.; Bass, L.S.: Clardy, J. *J. Med. Chem.* **1982**, *25*, 697.
2. Weinstock, J.; Ladd, D.L.; Wilson, J.W.; Brush, C.K.; Yin, N.C.F.; Gallagher, G.; McCarthy, M.E.; Silvestri, J.; Saran, H.M.; Flaim, K.M.; Ackerman, D.M.; Settler, P.E.; Tobia, A.J.; Hahn, R.A. *J. Med. Chem.* **1986**, *29*, 2315.
3. Rosen, M.H. U. S. Patent **1980**, 4185109. *Chem. Abstr.* **1980**, *93*, 46451.
4. For a comprehensive review of benzodiazepine chemistry, see Hester, J.B. In *Antianxiety Agents*, Berger, J.G., Ed., Wiley: New York, 1986, p. 51.
5. Sternbach, L.H.; Reeder, E. *J. Org. Chem.* **1961**, *26*, 4936.
6. Blazevic, N.; Kajfez, F. *J. Hetrocycl. Chem.* **1970**, 1173.
7. Sternbach, L.H.; Fryer, R.I.; Metlesics, W.; Reeder, E.; Sach, G.; Saucy, G.; Stempel, A. *J. Org. Chem.* **1972**, *37*, 3788.
8. Bell, S.C.; Childress, S.J. *J. Org. Chem.* **1962**, *27*, 1691.
9. Bell, S.C.; Childress, S.J. *J. Org. Chem.* **1968**, *33*, 216.
10. Wunsch, K.H.; Dettmann, H.; Schonberg, S. *Chem. Ber.* **1969**, *102*, 3891.
11. Mihalic, M.; Sunjic, V.; Kajfez, F.; Zinic, M. *J. Heterocycl. Chem.* **1977**, *14*, 941.
12. Steinman, M.; Topliss, J.G.; Alekel, R.; Wong, Y.S.; York, E.E. *J. Med. Chem.* **1973**, *16*,1354.
13. Weber, K.H.; Bauer, A.; Hauptmann, K.H. *Liebigs Ann.* **1972**, *756*, 128.
14. Hester, J.B.; Rudzik, A.D. *J. Med. Chem.* **1974**, *17*, 293.

15. Hester, J.B.; Chidester, C.G.; Duchamp, D.J.; MacKellar, F.A.; Slomp, G. *Tetrahedron Lett.* **1971**, 3665.
16. Meguro, K.; Kuwada, Y. *Chem. Pharm. Bull.* **1973**, *21*, 2375.
17. Meguro, K.; Tawada, H.; Miyano, H.; Sato, Y.; Kuwada, Y. *Chem. Pharm. Bull.* **1973**, *21*, 2382.
18. Hester, J.B.; Rudzik, A.D.; Von Voigtlander, P.F. *J. Med. Chem.* **1980**, *23*, 392.
19. Walser, A.; Benjamin, L.E.; Flynn, T., Mason, C.; Schwartz, R.; Fryer, R.I. *J. Org. Chem.* **1978**, *43*, 936.
20. Hunkeler, W.; Kyburz, E. Eur. Patent Appl. **1982**, 59390; *Chem. Abstr.* **1983**, *98*, 53949.
21. Hunkeler, W.; Kyburz, E. Eur. Patent Appl. **1982**, 59391; *Chem. Abstr.* **1983**, *98*, 53950.
22. Evans, B.E.; Rittle, K.E.; DiPardo, R.M.; Freidinger, R.M.; Whitter, W.L.; People, W.T.; Lendell, G.F.; Veber, D.F.; Anderson, P.S.; Chang, R.S.L.; Lotti, V.J.; Cerino, D.J.; Chen, T.B.; Kling, P.J.; Kunkel, K.A.; Springer, J.P.; Hirschfield, J. *J. Med. Chem.* **1988**, *31*, 2235.
23. Lee, G.E.; Lee, T.B.K. Eur. Patent Appl. **1982**, 66773. *Chem. Abstr.* **1983**, *98,* 197754.
24. Vlattas, I. Eur. Patent Appl. **1984**, 129509. *Chem. Abstr.* **1985**, *102,* 149293.
25. Boyer, S.K.; Fitchett, G.; Wasley, J.W.F.; Zaunius, G. *J. Hetrocycl. Chem.* **1984**, *21*, 833.
26. Della Vecchia, L.; Dellureficio, J.; Kisis, B.; Vlattas, I. *J. Hetrocycl. Chem.* **1983**, *20*, 1287.
27. Inoue, H.; Nagao, T. In *Chronicles of Drug Discovery*, Lednicer, D., Ed.; ACS Books: Washington, DC, 1993, Vol. 3, p. 207.

CHAPTER 13

HETEROCYCLES FUSED TO TWO BENZENE RINGS

1. COMPOUNDS WITH ONE HETEROATOM

A. Derivatives of Dibenzopyran and Dibenzoxepin

Anticholinergic agents constituted one of the very few drugs available for the treatment of gastric ulcers prior to the advent of the highly specific H_2 antihistamine gastric acid secretion inhibitors, or the more recent compounds that interfere with the action of cellular sodium–potassium pumps. These anticholinergic agents were often converted to characteristically poorly absorbed quaternary salts in order to confine the compounds to the desired site of action in the GI tract. The starting phenoxybenzoic acid (**1**) for one of these agents can be obtained by nucleophilic aromatic displacement of halogen from 2-chlorobenzoic acid. Friedel–Crafts cyclization, using, for example, concentrated sulfuric acid, leads to the dibenzopyranone **2**. Treatment of this compound with sodium in ethanol initially leads to a benzhydrol; this very labile hydroxyl is then reduced to give the parent dihydrodibenzopyran **3**. Reaction of the anion from **3** with carbon dioxide gives the acid **4** on acidification. The carboxylate salt from this intermediate is then alkylated with 2-chlorotriethylamine to afford the corresponding ester; reaction of this product with bromomethane leads to the quaternary methobromide and thus **methantheline bromide** (**5**).[1]

The dibenzopyranone ring system may be viewed as chromone with an additional fused benzene ring and thus generally related to the antiasthmatic mediator release inhibitor cromolyn (see Chapter 11). Two dibenzopyranones have in fact been investigated for this indication in the clinic. Friedel–Crafts cyclization of the substituted cresyloxybenzoic acid (**6**) in sulfuric acid leads to the dibenzopyranone **7**. The methyl group is then oxidized to a carboxylic acid by means of chromic acid. The acid is then converted to its sodium salt **xanoxate sodium** (**8**).[2]

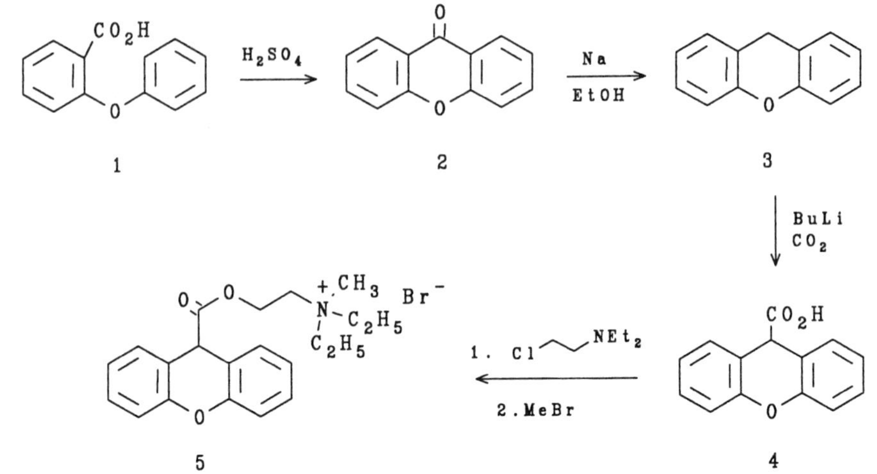

The preparation of the analogue in which a sulfoxide group replaces the isopropyl ether involves a more complex scheme. The starting bromobenzoate (**10**) in this case includes the pre-formed carboxyl group; replacement of halogen in this compound by the alkoxide from hydroquinone monomethyl ether **9** leads to the diphenyl ether **11**. The dibenzopyranone **12** is obtained by Friedel–Crafts cyclization of the saponification product from **11**. The methyl ether in the intermediate is cleaved with hydrogen bromide and the carboxylic acid is reconverted to an ethyl ester to afford **13**. The phenol is then converted to the corresponding thiophenol by an interesting oxygen–sulfur interchange reaction sequence.[3] The first step consists of conversion of the phenol to the thiocarbamate **14** by reaction with *N*,*N*-diethylthiocarbamoyl chloride. On heating, this intermediate undergoes oxygen-to-sulfur migration to give the isomeric thiocarbamate **15**, with consequent conversion of the phenol to a thiophenol.

Treatment of the rearrangement product with aqueous base leads to saponification of both the thiocarbamate and ester groups to afford the free thiophenol **16**. This compound is converted to the corresponding methyl ether **17** by means of methyl

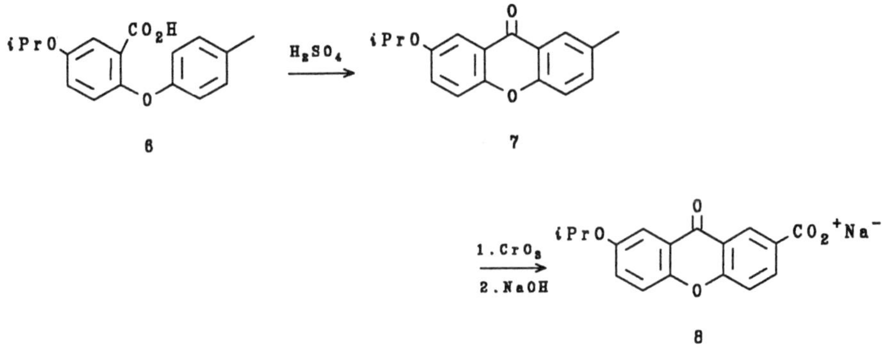

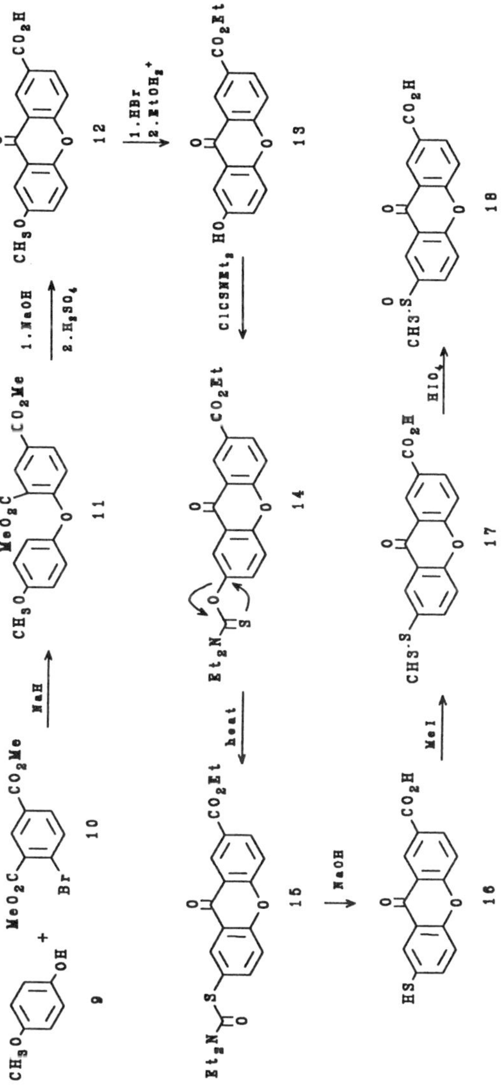

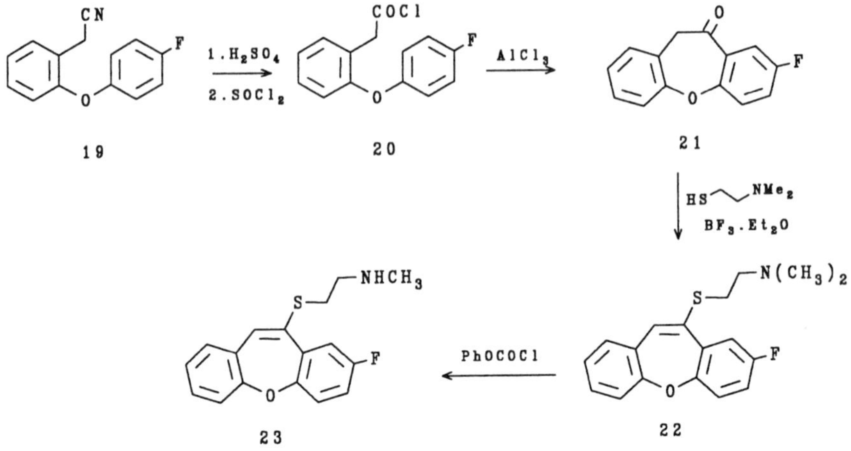

iodide. Oxidation of **17** with periodic acid then affords the mediator release inhibitor **tixanox** (**18**).[4]

Tricyclic antidepressant drugs often show nonspecific analgesic activity that is not unrelated mechanistically to the one displayed by either opioids or NSAIDs. The dibenzoxepin **fluradoline** (**23**), which is structurally related to the antidepressants discussed later in this chapter, was investigated as an analgesic rather than for its CNS activity. Hydrolysis of the cyano group in the starting nitrile **19** with sulfuric acid followed by treatment of the resulting acid with thionyl chloride affords the acid chloride **20**. This compound cyclizes to the dibenzoxepinone **21** on treatment with aluminum chloride. The carbonyl group in **21** readily converts to its enol form due to stabilization from conjugation of the resulting double bond with the adjacent aromatic rings. Thus, reaction of **21** with *N,N*-dimethylthioethanolamine leads to incorporation of the basic side chain in **22** via a thioenol ether linkage. One of the methyl groups on nitrogen is then removed by reaction with phenyl chloroformate in the modern version of the von Braun reaction to afford fluradoline (**23**).[5]

The nucleus for the prototype tricyclic antidepressant **imipramine** (**59**), which is discussed in Section 1.B, consists of a dibenzazepine system. This activity is retained when the nucleus is replaced by a dibenzoxepin; the propylamino side chain in this compound is attached to a trigonal sp^2 carbon rather than trigonal nitrogen. Nucleophilic displacement provides the key to entry to the required isomeric dibenzoxepinone **26**. Thus, replacement of halogen in the benzyl bromide **24** by phenoxide leads to the corresponding phenyl ether **25**. Friedel–Crafts cyclization in hot polyphosphoric acid leads to the dibenzoxepinone **26**. Condensation of **26** with the Grignard reagent from 3-chloro-1-(*N,N*-dimethylamino)propane initially leads to the tertiary carbinol. This compound dehydrates spontaneously to give the olefin that consists primarily of the trans isomer.[5] The stereoselectivity may result from the fact that the tricyclic ring system in **26** is actually puckered; the isomer from preferential addition to the ketone from the face of the molecule containing oxygen can form a hydrogen-bonded bicyclic intermediate that will favor formation of the trans isomer on dehydration. The final drug **doxepin** (**27**) consists of a 5:1 mixture of trans:cis isomers as the hydrochloride.

24 → (1. PhONa; 2. NaOH) → 25 → (PPA) → 26 → (Cl~~~NMe$_2$, Mg) → 27

24 → (NaO-C$_6$H$_4$-CH$_2$CO$_2$Na; 2. NaOH) → 28 → (PPA) → 29

Displacement of bromine from **24** with the phenoxide from *m*-hydroxyphenylacetic acid gives the phenyl ether **28**. Ring closure of **28** with polyphosphoric acid gives the dibenzooxepinone. **Isoxepac** (**29**), not unexpectedly, displays NSAID activity.[6]

B. Dibenzo Heterocycles with One-Ring Nitrogen Atom

The nausea and emesis caused by administration of many cancer chemotherapy agents could not until quite recently be controlled by standard antiemetic agents. The discovery of the serotonin antagonists such as **ondansetron** (**32**) has provided a means for controlling this treatment limiting side effect. The starting material (**30**) for this drug can be obtained by alkylation with the methyl iodide of the partly reduced carbazole from the Fischer indole condensation of dihydroresorcinol and phenylhydrazine. Mannich reaction of the product with formaldehyde and dimethylamine gives the aminomethylated derivative **31**. Treatment of **31** with 2-methylimidazole leads to replacement of dimethylamine by the heterocycle, either by direct S_N2 displacement or elimination to the exomethylene derivative followed by conjugate addition of the imidazole. Thus, ondansetron (**32**)[7] is obtained.

30 → (HNMe$_2$, CH$_2$O) → 31 → (2-methylimidazole) → 32

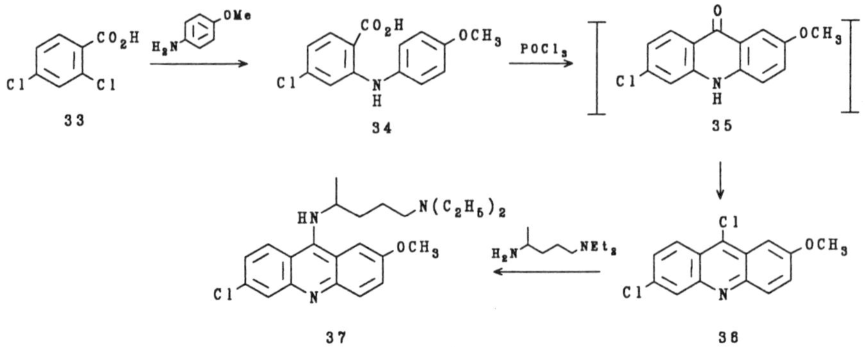

The great majority of antimalarial agents, as noted in Chapter 11, consist of substituted quinolines. It is an interesting circumstance that the drug that has been most widely used as a prophylactic and curative antimalarial is in fact an acridine. This agent, **quinacrine** (**37**), better known to Americans as Atabrine, became available just before World War II and was widely used in tropical areas by armies on both sides. The first step in the synthesis of this agent consists of nucleophilic aromatic displacement of chlorine from 2,4-dichlorobenzoic acid (**33**) to give the diphenylamine **34**. Treatment of **34** with phosphorus oxychloride proceeds initially to the acridone **35**. Excess reagent then converts what is in effect a vinylogous amide to its enol chloride **36** going to the aromatic heterocycle in the process. Displacement of chlorine with 4-amino-5-(diethylamino)pentane gives quinacrine (**37**).[8]

The fact that the ketone in acridones undergoes normal Grignard reactions indicates that it retains considerable carbonyl character. Thus, condensation of the acridone **38** with the organomagnesium derivative from 3-(dimethylamino)chloropropane gives the corresponding tertiary carbinol on workup; this readily dehydrates to the olefin **39** on heating; note that **39** can readily tautomerize to the aromatic acridine. Catalytic hydrogenation then gives the dihydroacridine **clomacran** (**40**);[9] this compound shows the same type of antipsychotic activity as the better known phenothiazines discussed below.

Reaction of diphenylamine (**41**) with acetone in strong acid leads to the formation of the acridan addition product **42**. The reaction can be envisaged as involving initial formation of a hydrated carbocation such as $Me_2C(OH)_2H^+$; this species attacks one of the aniline rings to give the corresponding tertiary carbinol; this compound undergoes normal cyclodehydration to give the observed product. Alkylation with 3-(dimethylamino)chloropropane gives **dimetacrine** (**43**);[10] this compound, which may be viewed as a ring contracted analogue of imipramine (**59**), also shows an antidepressant activity.

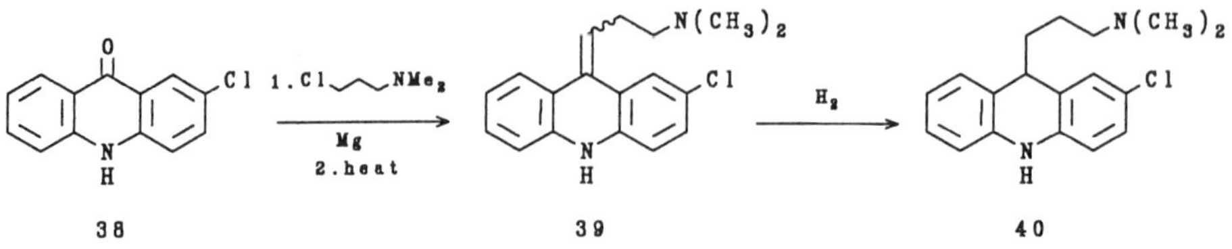

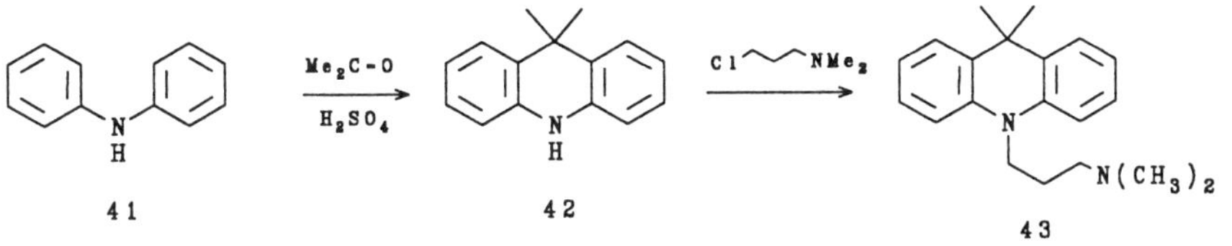

The devastating mental deterioration that characterizes Alzheimer's disease has been attributed to a process which results eventually in mishandling of the neurotransmitter acetylcholine. Inhibitors of acetylcholinesterase, the enzyme that catabolizes this substance, should help restore deficient levels of acetylcholine. Several cholinesterase inhibitors, which consist of partly reduced acridines, have shown some activity in treating Alzheimer's disease. The initial step in the synthesis of the first of these compounds consists of the sodium amide catalyzed condensation of isatin (**44**) with cyclohexanone. The reaction can be visualized by assuming the first step to involve attack of amide on isatin to give an amido-amide such as **45** (note that no attempt has been made to account for charges). This compound can then react with cyclohexanone to lead to the enamine **46**. Normal enamine chemistry will then give the observed product, acridine carboxamide (**47**). This intermediate is then subjected to Hofmann degradation; treatment with bromine in the presence of sodium hydroxide leads to **tacrine** (**48**).[11]

Inclusion of a hydroxyl group increases the hydrophilicity of the compound and also presumably improves the oral absorption over the very hydrophobic parent drug. In a conceptually related synthesis, *o*-aminobenzonitrile (**49**) is condensed with dihydroresorcinol to give the isolable enamine **50**. Treatment of **50** with the Lewis acid cuprous chloride leads to addition of the enamine to the nitrile and formation of an addition product such as **51**. This intermediate then tautomerizes to give the fully conjugated dihydroacridine **52**. The carbonyl group is then reduced with lithium aluminum hydride to afford **velnacrine** (**53**).[12]

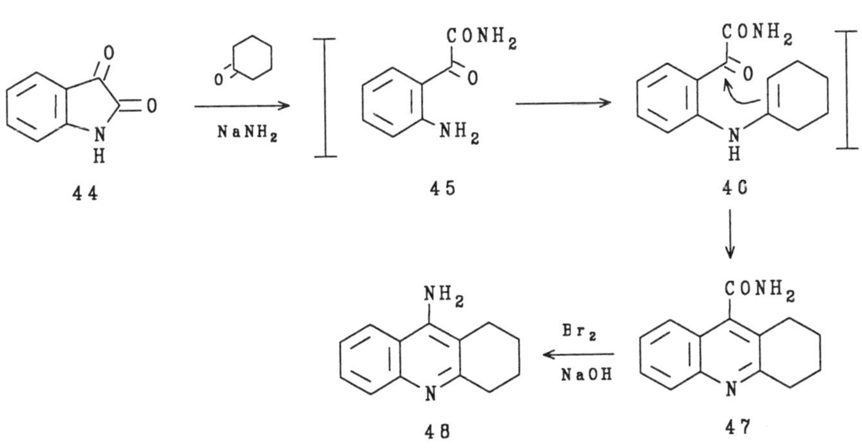

The prototype for the large class of tricyclic antidepressant compounds **imipramine** (**59**) can be viewed as a bioisostere of a phenothiazine in which an ethylene bridge replaces the central ring sulfur in the latter. Anecdotal accounts suggest that the actual activity of this drug was uncovered serendipitously while it was in trial as an antipsychotic agent. One of the early syntheses for this dibenzazepine starts with the self-alkylation of *o*-nitrobenzyl chloride. The first step consists in displacement of halogen from **54** by an anion from reaction of a second equivalent to form the transient intermediate **55**. Elimination of hydrogen chloride then leads to the stilbene **56** as a mixture of isomers. Catalytic hydrogenation then reduces both the nitro groups and the double bond to afford diamine **57**. Pyrolysis of **57** leads to formation of the dibenzazepine ring (**58**) by an unusual nucleophilic aromatic displacement reaction. The anion from treatment of **58** with sodium amide is then treated with 3-chloro-1-dimethylaminopropane to afford imipramine (**59**).[13] The metabolite **desipramine (60)** has the same activity as the parent; this observation is repeated for the majority of antidepressant compounds, which makes it difficult to identify the proximate active species. The monomethyl derivatives are obtained by the von Braun demethylation reaction or one of its modern equivalents.

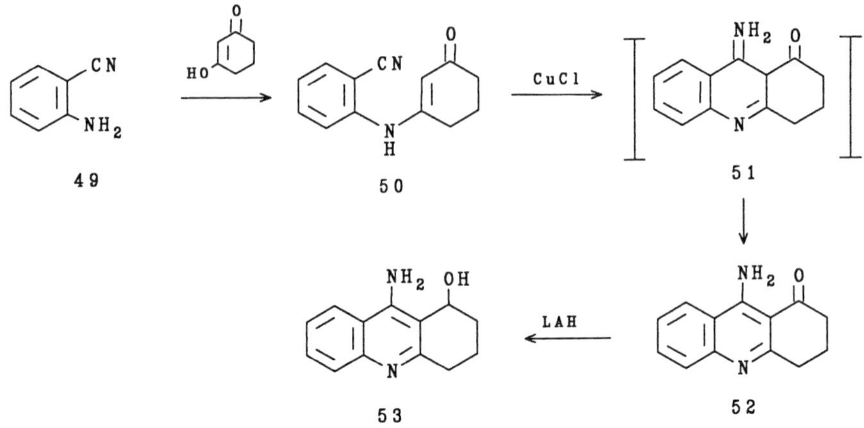

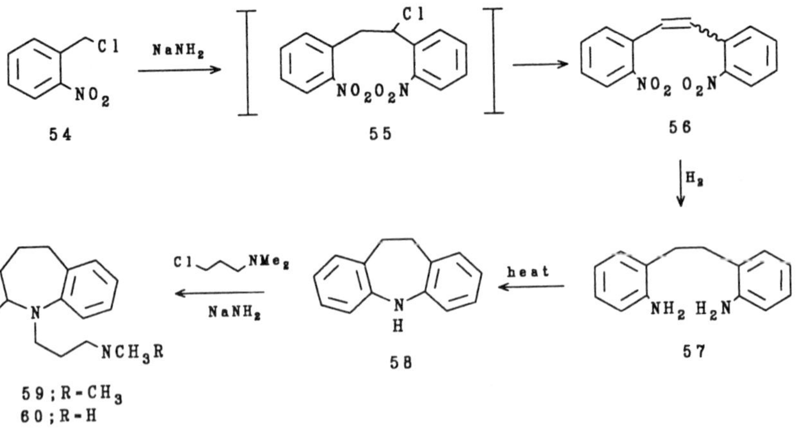

59; R-CH3
60; R-H

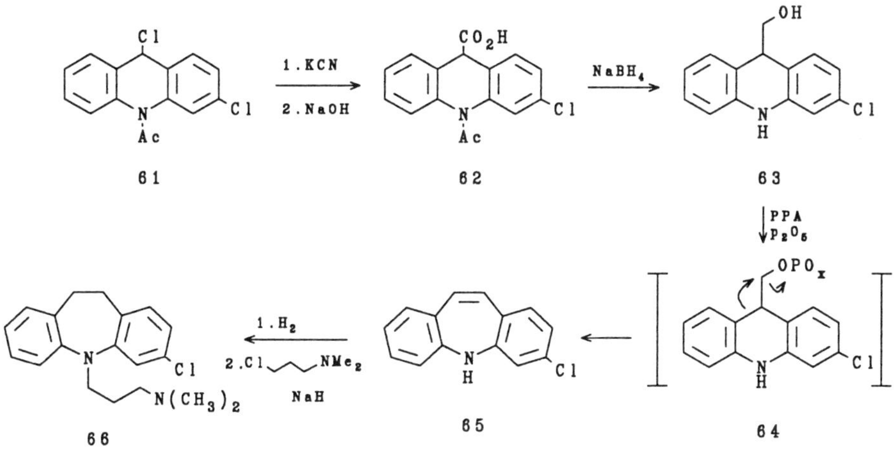

Ring expansion of more readily accessible acridines provides an alternate method for preparation of dibenzazepines. The starting halide **61** can be obtained by chlorination of the corresponding *N*-acetyl dihydroacridine. Displacement of chlorine by cyanide gives the nitrile; this compound is then hydrolyzed to the acid **62** with base; the amine is deacylated under the reaction conditions. The carboxyl group is then reduced to the carbinol by means of sodium borohydride to give the intermediate **63**. Reaction of **63** with polyphosphoric acid and phosphorus pentoxide probably leads to formation of phosphate esters such as **64**; these species then rearrange in a concerted manner or via a discrete carbocation to afford the ring enlarged product **65**. The double bond is then reduced by catalytic hydrogenation. Alkylation on nitrogen as shown above leads to the antidepressant **chlorimipramine** (**66**).[14]

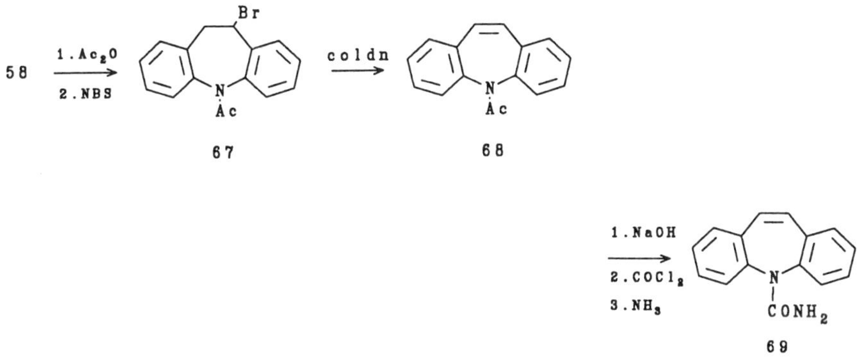

Reaction of dibenzazepine (**58**) with acetic anhydride gives the corresponding protected intermediate. Treatment of that with NBS leads to bromination at the benzylic position to give **67**. This compound is then dehydrohalogenated by means of collidine to afford the dehydro derivative **68**. The protecting group is then removed by basic hydrolysis. The resulting amine is converted to the carbamoyl chloride with phosgene; exposure of this reactive intermediate with ammonia gives the corresponding urea.[15] This product, **carbamazepine** (**69**), is a widely used anticonvulsant agent.

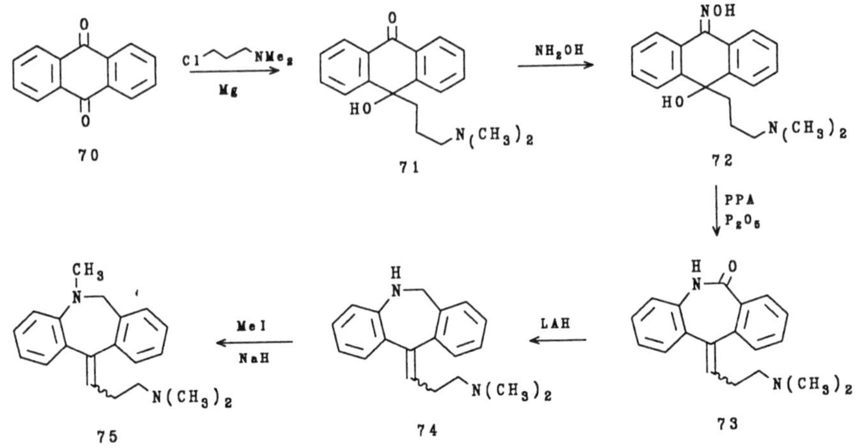

Many of the tricyclic antidepressants show some measure of undesired anticholinergic activity, which causes the side effects characteristic of this class of compounds. Anticholinergic activity predominates in a dibenzazepine in which the ring nitrogen is moved to the two atom bridge. Condensation of anthraquinone (**70**) with the Grignard reagent from 3-chloro-1-dimethylaminopropane, in the cold, leads to the monoadduct **71**; steric hindrance about the remaining carbonyl group may account for this selectivity. The ketone is converted to its oxime (**72**) by means of hydroxylamine. Treatment of **72** with a mixture of phosphorus pentoxide and polyphosphoric acid results in formation of the ring enlarged lactam by a classic Beckmann rearrangement; the labile tertiary benzhydryl alcohol dehydrates under these reaction conditions to give **73** as a mixture of isomers. The lactam is then reduced to the secondary amine **74** by means of lithium aluminum hydride. Base-catalyzed alkylation with methyl iodide then affords **elantrine** (**75**).[16]

C. Dibenzo Heterocycles with One Sulfur Atom

The thioxanthone ring system forms the nucleus for two of the more efficacious orally active antischitosomal drugs. One synthesis for these compounds starts with the chlorosulfonation of *p*-chlorotoluene (**76**) to give the sulfonyl chloride **77**. Treatment of **77** with zinc in aqueous sulfuric acid leads to reduction of the sulfonyl chloride to form the thiophenol **78**. The key to formation of this and related thioxanthones and thioxanthenes consists of the Ullman reaction, which involves copper catalyzed displacement of aromatic halogen. Condensation of *o*-chlorobenzoic acid with **79** under Ullman conditions thus affords the bis-aryl thioether **80**. This intermediate undergoes Friedel–Crafts cyclization to the thioxanthone **81** in sulfuric acid. The basic side chain is then incorporated by nucleophilic aromatic displacement of chlorine with *N*,*N*-diethylamino-1,3-diaminopropane. Thus, **lucanthone** (**82**)[17] is obtained. Interestingly, the metabolic hydroxylation product from this drug is more potent and better absorbed than its parent. This compound, **hycanthone** (**83**), can be obtained by fermentation of **82** with *Aspergillus scelorotium.*[18]

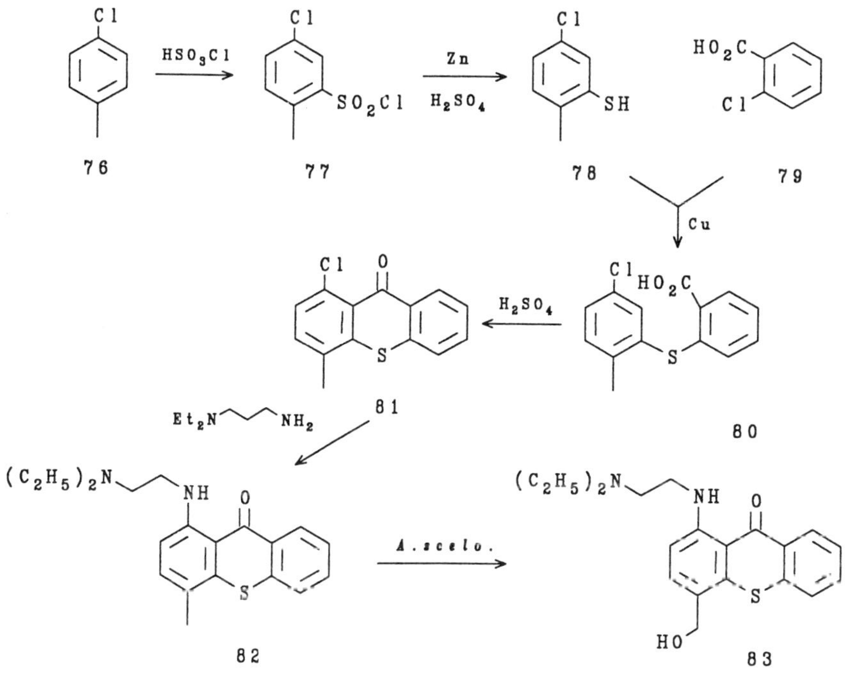

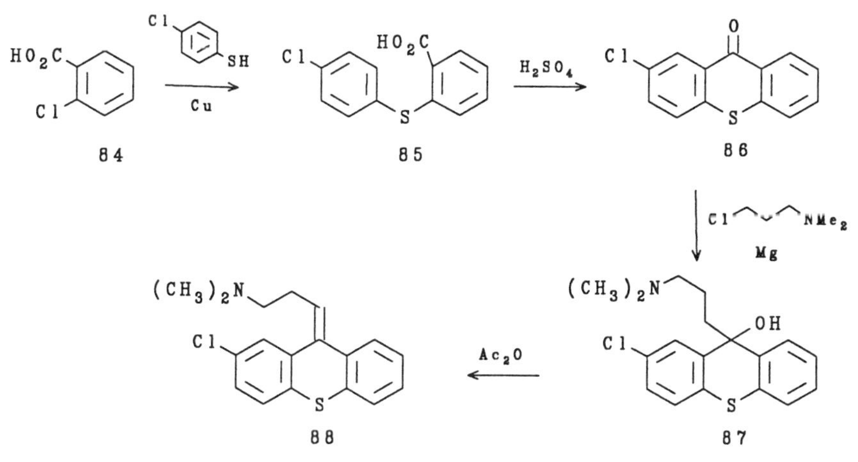

Replacement of the nitrogen atom that bears the side chain in the phenothiazine antipsychotic agents by trigonal carbon affords compounds that show quite comparable activity, at least one of which, **thiothixene** (**96**), is widely used in the clinic. Condensation of *p*-chlorothiophenol with chlorobenzoic acid **84** gives the thioether **85**. This compound is then cyclized to the thioxanthone **86** by means of sulfuric acid. Addition of the Grignard reagent from 1-chloro-3-dimethylaminopropane to the carbonyl group serves to add the required basic side chain. Dehydration of **87** by means of acetic anhydride gives the corresponding olefin as an isomeric mixture. The more potent (*Z*) isomer is then separated to afford **chlorprothixene (88)**.[19]

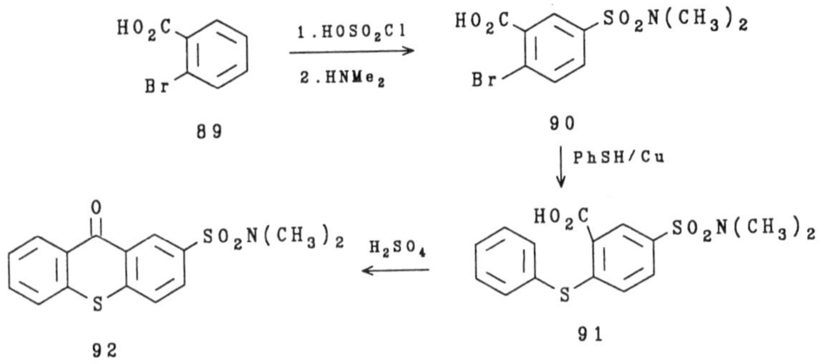

Reaction of the bromobenzoic acid **89** with chlorosulfonic acid leads to the corresponding sulfonyl chloride; this compound affords the sulfonamide **90** on treatment with dimethylamine. This product is then converted to a thioxanthone as shown above. Thus, coupling with thiophenol gives the thioether **91**; compound **91** gives the tricyclic intermediate **92** on exposure to sulfuric acid.

A somewhat different scheme is used to introduce the side chain in this case. The first step consists in reduction of the carbonyl group to methylene by catalytic hydrogenation over palladium to give **93**. The anion obtained by treatment of **93** with butyllithium is then acylated with methyl acetate to afford the methyl ketone **94**, which contains two of the required side chain carbon atoms. The additional carbon atom and the basic function are incorporated by means of a Mannich condensation. Thus, reaction of **94** with *N*-methylpiperazine and formaldehyde leads to the aminoketone **95**. The carbonyl group is then reduced with sodium borohydride and the resulting alcohol is dehydrated by reaction with phosphorus oxychloride in pyridine. In this case, the (*Z*) isomer is again responsible for most of the activity. This compound is isolated from the resulting mixture of olefins to afford **thiothixene (96)**.[20]

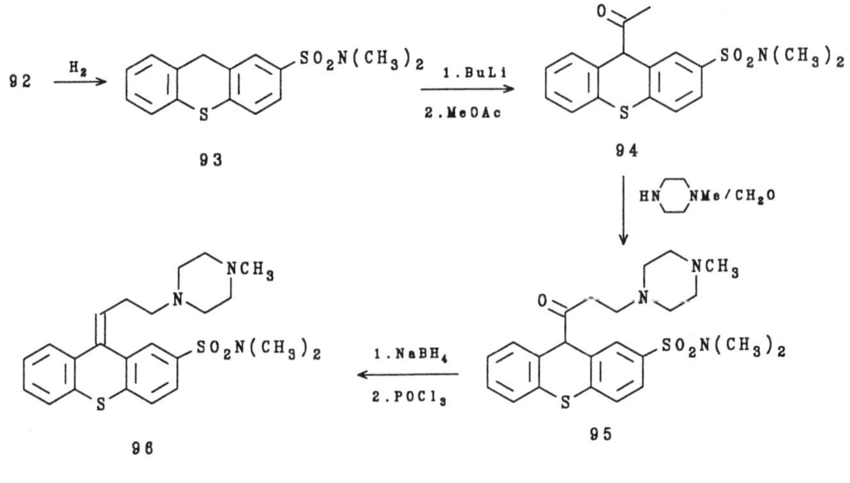

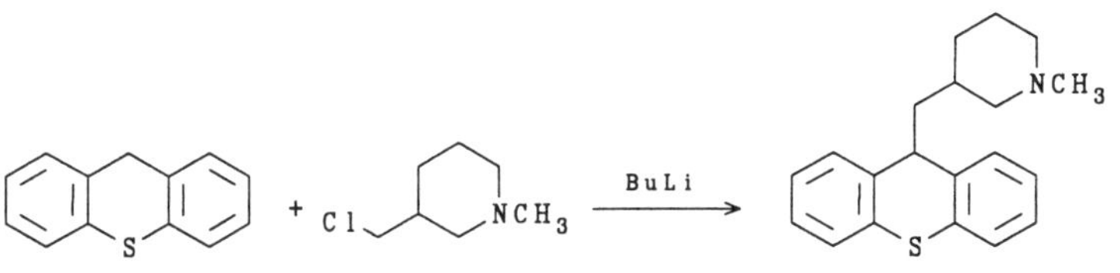

97 + 98 —BuLi→ 99

The anion from the parent thioxanthene provides the starting material for a simplified analogue that shows muscle relaxant rather than CNS activity. Alkylation of the anion obtained by treating **97** with butyllithium with substituted piperidine **98** affords **methixene** (**99**) in a single step.[21]

Antipsychotic activity is retained when heterocyclic ring is enlarged by one carbon. The synthesis of this agent is quite similar to that used for its oxygen analogue **doxepin**, **27**, though with reversed functionality. The sequence in the present case starts with alkylation of thiosalicylic acid **100** with benzyl chloride to give the thioether **101**. The product is then cyclized by means of polyphosphoric acid to give the ketone **102**. Condensation with the familiar Grignard reagent serves to introduce the side chain. Dehydration affords **dothicpin**, **104** as a mixture of isomers.[22]

The equivalence of sulfur and oxygen in this ring system carries over to NSAID's as well. Preparation of the sulfur analogue of isoxepac (**29**) starts with the alkylation of thiophenol **105** with benzyl chloride; cyclization of the intermediate thioether then affords the homothioxanthone **106**. The carboxyl side chain is then extended by means of the Arndt-Eistert homologation reaction. The acid is thus first converted to its acid chloride by means of thionyl chloride. Reaction with excess diazomethane leads to the diazoketone **107**. Treatment of that intermediate with silver benzoate and triethylamine leads the ketone to rearrange to an acetic acid. Thus, **tiopinac** (**108**)[23] is obtained.

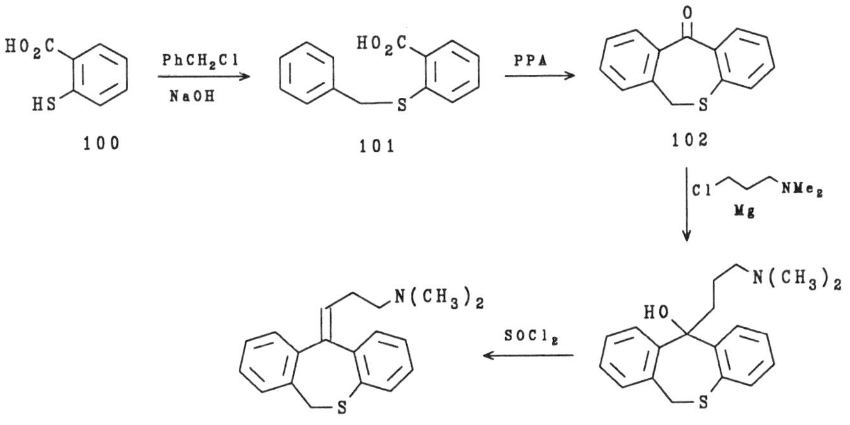

100 —$PhCH_2Cl$, NaOH→ 101 —PPA→ 102 —$Cl(CH_2)_3NMe_2$, Mg→ 103 —$SOCl_2$→ 104

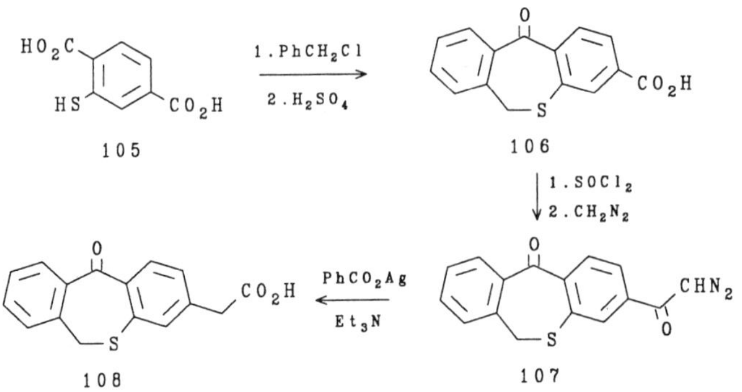

2. COMPOUNDS WITH TWO HETEROATOMS

A. Phenothiazines

The various topics discussed in this book have to this point been arranged as far as possible on the basis of chemical structure; compounds that contain one oxygen and one nitrogen atom have, for example, generally preceded those that contain two nitrogen atoms. The fact that virtually all entities that follow show CNS activity in combination with the circumstance that phenothiazines comprise a large part of this section, requires a departure from this standard approach. The serendipitous discovery of the antipsychotic activity of the phenothiazines in the early 1950s virtually opened the modern era of medicinal chemistry. These compounds will thus be discussed at the outset to preserve historical perspective.

One of the earliest classes of synthetic medicinal agents consists of the antihistamines derived from benzhydrol; the ethylamine moiety present in many of these compounds was included in the belief that it mimicked the same structural fragment in histamine. This observation may have led to substitution of this side chain on phenothiazine (**110**). In fact, the product **diethazine** shows respectable activity as an antihistamine (H_1 histamine antagonist in today's parlance). Note, however, that both this compound and some related ethylamino substituted phenothiazines also exhibit effects on the CNS manifested as sedative and antiemetic activity. The starting phenothiazine can be prepared by the so-called sulfuration reaction, which consists of treatment of diphenylamine (**109**) with sulfur and iodine. This transformation may take place via initial formation of the ortho iodo aniline derivatives, which then undergo displacement by sulfur. Alkylation of the nitrogen anion from reaction of **110** with a strong base such as sodium amide with 2-chlorotriethylamine affords **diethazine** (**111**).[24]

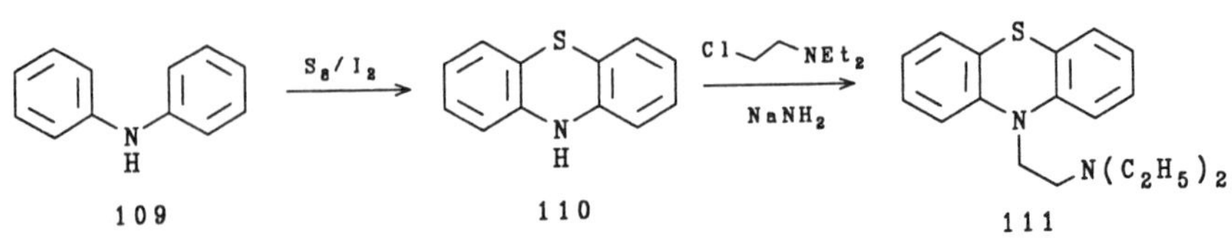

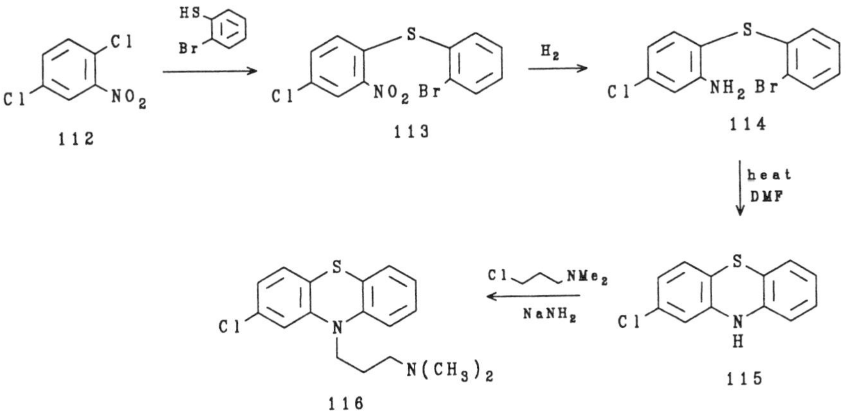

Inclusion of an extra methylene group in the side chain led to compounds with diminished antihistaminic activity; these compounds, however, showed increased CNS activity over that which had been observed with the ethylamino derivatives. This activity was qualitatively different from that of the antihistamines. The dearth of animal models for psychoses available at that time means that inspired insight led to the first clinical trials of these compounds as antipsychotic agents. The efficacy of **chlorpromazine** (**116**) and its many analogues led to a revolution in the treatment of mental disease. It is now recognized that this class of drugs acts as dopamine antagonists by binding receptors for this neurotransmitter in the same manner as haloperidol and related compounds discussed in Chapter 9, which were in fact developed after the phenothiazines.

An alternate and more controlled approach to the synthesis of phenothiazines involves sequential aromatic nucleophilic displacement reactions; this route avoids the formation of isomeric products sometimes observed with the sulfuration reaction when using substituted aryl rings. The first step in this sequence consists of displacement of the activated chlorine in the nitrobenzene **112** by the salt from *o*-bromothiophenol to give the thioether **113**. The nitro group is then reduced to form the aniline **114**. Heating **114** in a solvent such as DMF leads to displacement of bromine by amino nitrogen and formation of the chlorophenothiazine (**115**). Alkylation of the anion from this intermediate with 3-chloro-1-dimethylaminopropane affords chlorpromazine (**116**).[25]

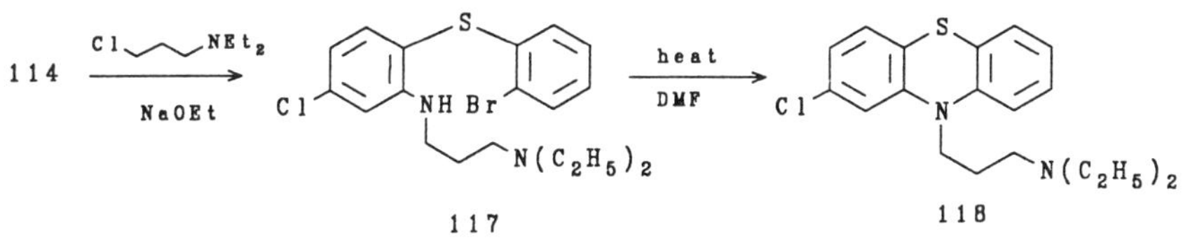

This general scheme is sufficiently flexible to permit the interchange of the order of some of the steps. Thus alkylation of aniline thioether **114** with 3-chloro-1-diethylaminopropane leads to the intermediate **117**. Ring closure, as shown above, by nucleophilic aromatic displacement, leads to the antipsychotic drug **chlorproethazine** (**118**).[26]

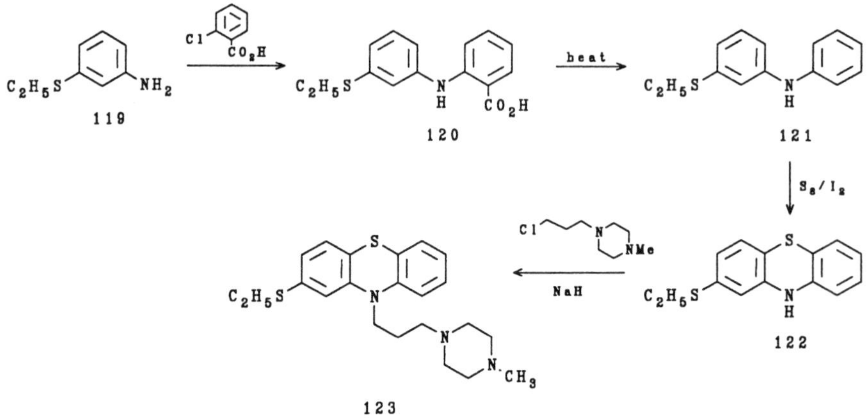

Preparation of a diarylamine required for the synthesis of a phenothiazine via the sulfuration reaction requires the use of an activated chlorobenzene. Displacement of chlorine from 2-chlorobenzoic acid by the nitrogen in the substituted aniline **119** gives the diphenylamine **120**. This anthranilic acid derivative loses carbon dioxide on heating to give the intermediate **121**. This compound gives predominantly the phenothiazine **122** on treatment with sulfur and iodine, which is probably aided by the presence of the electron-donating thioether at the para position. Alkylation of nitrogen, again via its anion, with 1-methyl-4-(3-chloropropyl)piperazine affords **thiethylperazine** (**123**).[27]

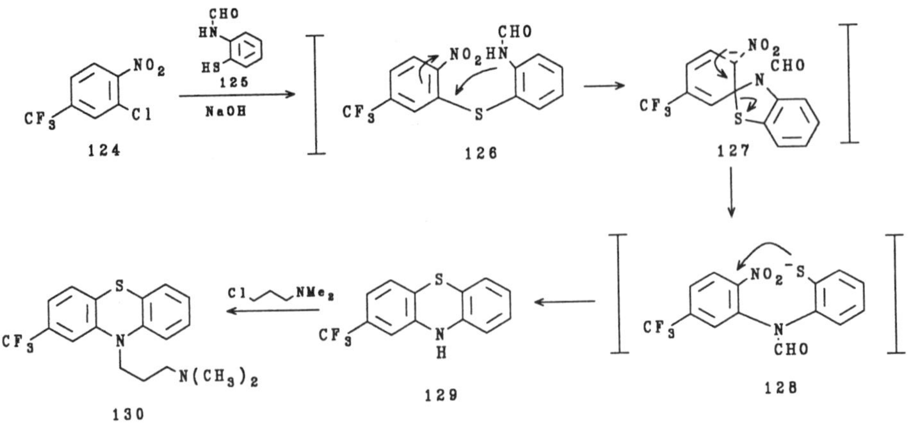

The Smiles rearrangement, which leads to transposition of sulfur and nitrogen, is used in the synthesis of 3-trifluoromethylphenothiazine (**129**). The overall reaction involves treatment of the nitrobenzene **124** with the thiophenol **125** in the presence of strong aqueous base to give phenothiazine (**129**). The first step in this complex sequence can be visualized as displacement of the highly activated chlorine in **124** by sulfur to give thioether **126**. In the presence of base, the anion from the formamide function in this product adds to the aromatic ring bearing the nitro group, to give the charged spiro cyclic intermediate **127**. The negative charge on the carbon bearing the nitro group then returns with opening of the spiran by scission of the carbon–sulfur bond to give thiophenoxide **128**. This anion then displaces the nitro group to form the desired phenothiazine (**129**); the formamide cleaves under reaction conditions. Alkylation of the anion from **129** with 3-chloropropyldimethylamine then affords **triflupromazine** (**130**).[28]

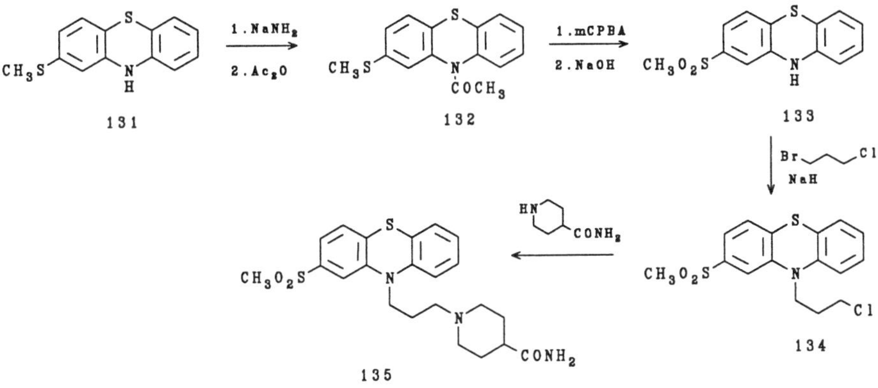

The reduced basicity of the phenothiazine nitrogen requires that even acylation proceed via the anion. The amide **132** from the methyl thioether **131** can be prepared, for example, by sequential reaction with sodium amide and acetic anhydride. Oxidation of **132** with peracid proceed preferentially on the more electron-rich alkyl thioether to give the sulfone; this reaction affords the phenothiazine **133** on hydrolysis of the amide. Complex side chains are most conveniently incorporated in a stepwise fashion. The first step in this sequence involves reaction of **133** as its anion with 1-bromo-3-chloropropane to give **134**. The use of this halide to alkylate piperidine-4-carboxamide affords the antipsychotic agent **metopimazine** (**135**).[29]

B. Dibenzodiazepines, Dibenzoxazepines, and a Dibenzothiazepine

Seven-membered rings that contain two heterocyclic atoms have also provided a significant number of CNS agents as foreshadowed in Chapter 12. The Ullman reaction, to give the diarylamine or diarylether starting materials, constitutes a key reaction in this series as well. As an example, copper catalyzed coupling of methyl *N*-methylanthranilate (**136**) with nitrobromobenzene gives the arylaniline **137**. The ester is then saponified and the nitro group is reduced to the corresponding amine **138**. This product cyclizes to the lactam **139** on heating. Strong base preferentially removes the proton on the lactam nitrogen to form an anion. Alkylation with 2-chloroethyldimethylamine then affords **dibenzepin** (**140**), a compound that shows antidepressant activity.[30]

The fact that antipsychotic agents control rather than cure schizophrenia means that most patients will be treated for their lifetime with dopamine antagonist drugs. Such long-term use is unfortunately often accompanied by the emergence of severe side effects that can be traced to the same antagonism of dopamine that makes the drugs effective in the first place. The undesired effects are it is thought a reflection of the lack of selectivity of the antidopamine action. The series of dibenzo heterocycles substituted with a pendant piperazine ring discussed below comprise a set of significantly better tolerated antipsychotic agents. This finding may be due to enhanced selectivity for a subset of dopamine receptors as well as antagonist action at serotonin receptors.

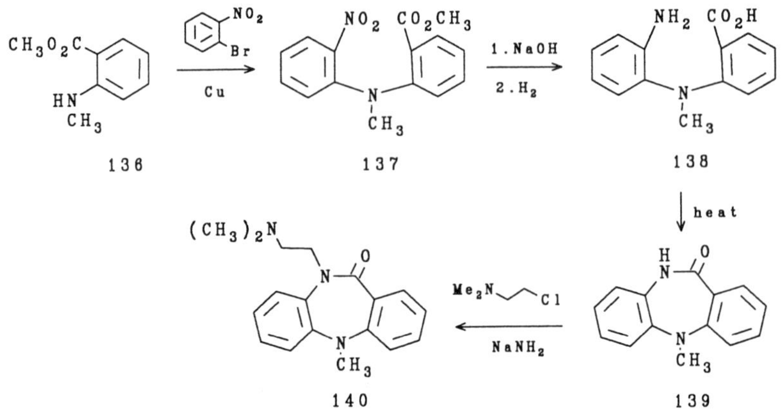

The first step in one synthesis of the antipsychotic drug **clozapine** (**145**) involves Ullman coupling of anthranilic acid (**141**) with 2,4-dichloronitrobenzene to give the substituted anthranilate **142**. The carboxyl group is then converted to the *N*-methylpiperazinamide **143** via a suitably activated intermediate, for example, the imidazolide obtained by reaction with CDI. The nitro group is then reduced to amine

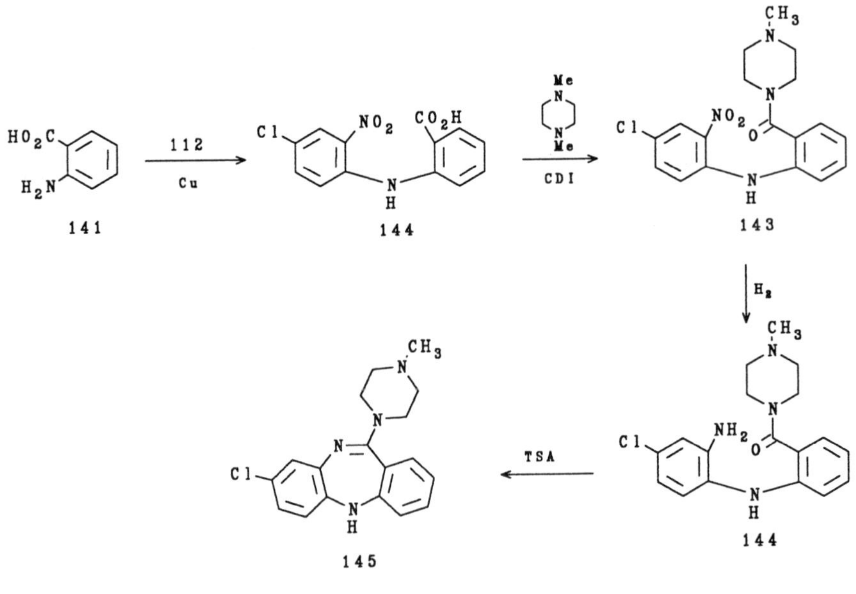

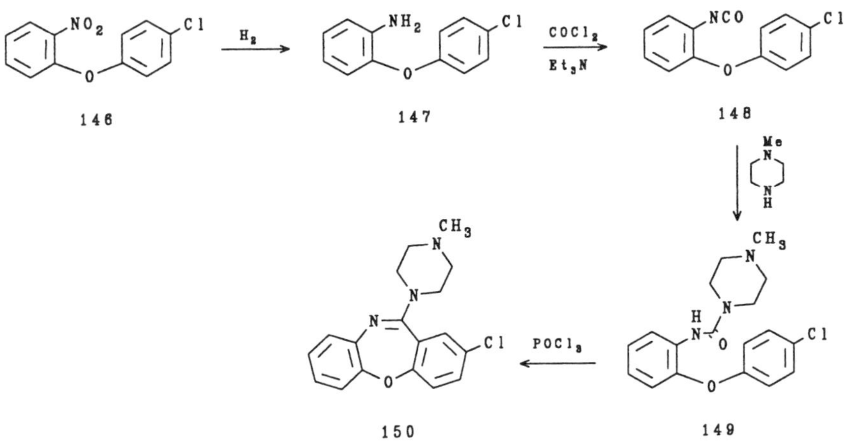

144 by means of catalytic hydrogenation. Intramolecular Schiff base formation catalyzed by toluenesulfonic acid then completes the synthesis of clozapine (**145**).[31]

A considerable degree of freedom prevails as to the nature of the one atom bridge since activity is retained when this is oxygen, sulfur, or even carbon. Displacement of chlorine in *o*-chloronitrobenzene by *p*-chlorophenoxide leads to the starting material **146**. Reduction of the nitro group leads to the aniline **147**. Reaction of **147** with phosgene in the presence of triethylamine gives the corresponding isocyanate **148**. Addition of *N*-methylpiperazine to the isocyanate function leads to a urea and thus the product **149**. Treatment of **149** with phosphorus oxychloride leads to a cyclodehydration reaction possibly via the imino chloride. Thus, the antipsychotic compound **loxapine** (**150**)[32] is obtained.

The preparation of the *N*-desmethyl analogue **amoxapine** (**155**) illustrates an approach in which the oxygen ether linkage is formed last. Reaction of the imidazolide **151** from 2,4-dichlorobenzoic acid and CDI with *o*-aminophenol gives the benzamide **152**. This compound is then converted to its imino chloride (**153**) with the ubiquitous phosphorus oxychloride. Treatment of **152** with piperazine leads to the amidine **154**

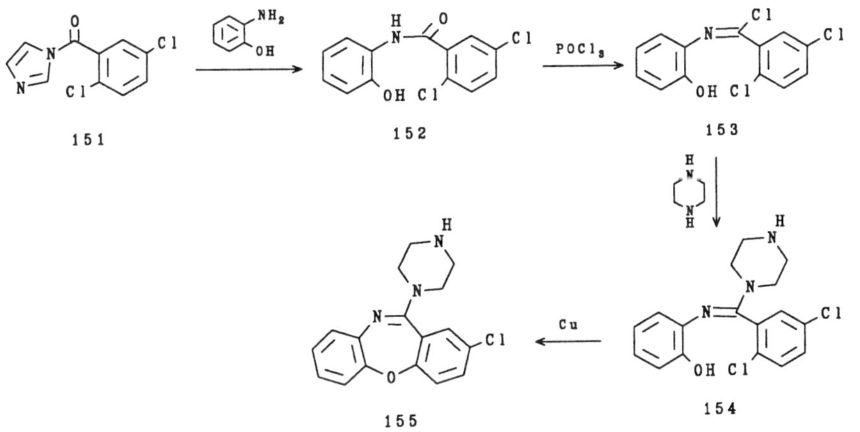

probably by an addition–elimination sequence. Copper catalyzed displacement of chlorine by phenoxide closes the ring. Thus, amoxapine (**155**)[33] is obtained.

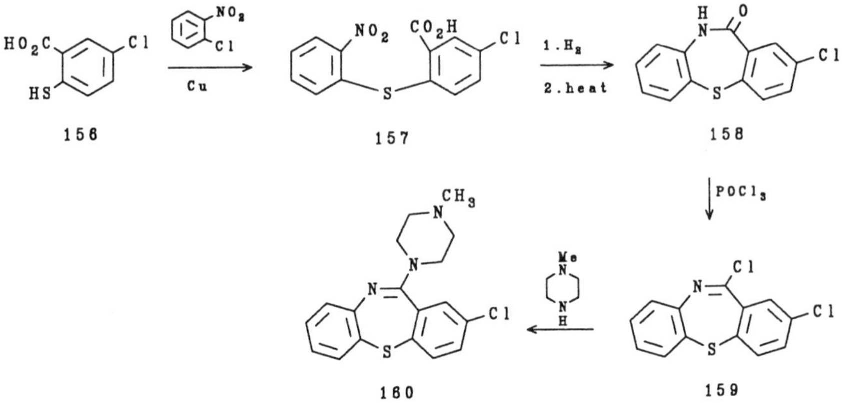

Yet another approach to these compounds consists of substituting the piperazine ring onto the preformed heterocyclic moiety. Ullman condensation of substituted thiosalicyclic acid **156** with *o*-chloronitrobenzene results in displacement of chlorine by thiophenoxide and formation of the thioether **157**. The nitro group in **157** is then reduced to an aniline; the resulting amino acid is then cyclized thermally to the lactam **158**. Treatment of **158** with phosphorus oxychloride gives the imino chloride **159**. Reaction with *N*-methylpiperazine leads to replacement of chlorine by nitrogen and formation of **clothiapine** (**160**).[32]

C. Azadibenzodiazepines

Benzene rings in drugs often serve simply as flat relatively electron-rich moieties. Many examples have been noted thus far where such rings can be replaced by heterocycles that have some degree of aromatic character. One of the benzene rings in dibenzepin (**140**) can thus be replaced by pyridine. In a one-pot reaction, condensation of the 2-chloronicotinic acid (**161**) with *o*-phenylenediamine leads to the lactam **162**. The order of the two steps, aromatic displacement and amide formation, has not been elucidated. Simple alkylation of the anion from the product with 3-chloro-2-(*N*,*N*-dimethylamino)propane affords the antidepressant agent **propizepine** (**163**).[34]

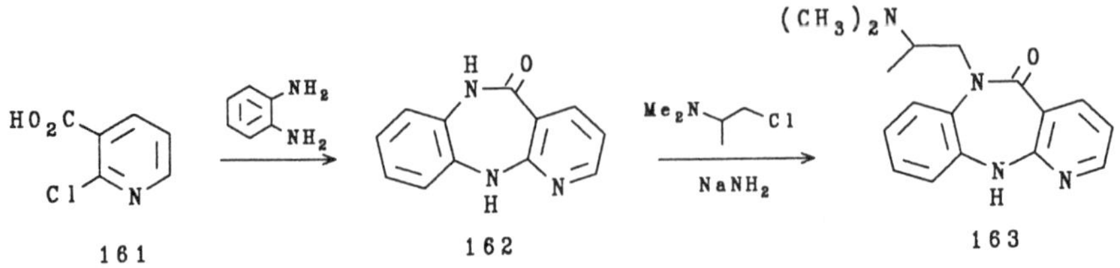

The antidepressant agent **tampramine** (**168**) can be viewed as a distant imipramine analogue that contains an extra benzene ring and two additional nitrogen atoms. The preparation of **168** starts by Ullman coupling of the chloropyridine **164** with aminobenzophenone (**165**) more familiarly used for benzodiazepine syntheses. Reduction of the nitro group in the product **166** leads to a diamine that readily cyclizes to form the benzodiazepine **167**. Alkylation of the anion from treatment of **167** with sodium hydride and 3-chloro-1-dimethylaminopropane affords tampramine (**168**).[35]

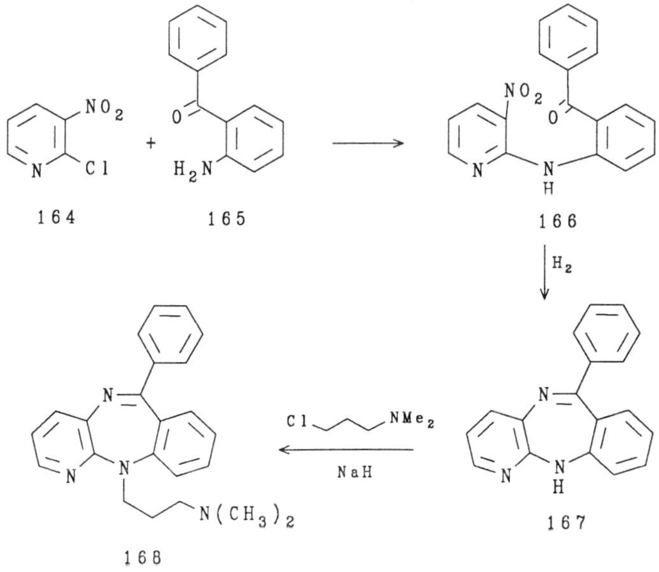

REFERENCES

1. Cusic, J.W.; Robinson, R.A. *J. Org. Chem.* **1951**, *16*, 1921.
2. Bays, D.E. German Offen. **1971**, 2058295. *Chem. Abstr.* **1971**, *75*, 98447.
3. Newman, M.S.; Karnes, H.A. *J. Org. Chem.* **1966**, *31*, 3980.
4. Pfister, J.R.; Harrison, I.T.; Fried, J.H. U. S. Patent 3949084. *Chem. Abstr.* **1976**, *85*, 21108.
5. Bloom, B.M.; Tretter, J.R. Belguim Patent **1964**, 641498. *Chem. Abstr.* **1966**, *64*, 719.
6. Ong, H.H.; Profitt, J.A.; Anderson, V.B.; Spaulding, T.C.; Wilker, J.C.; Geyer, H.M.; Kruse, H. *J. Med. Chem.* **1980**, *23*, 494.
7. Coates, I.H.; Bell, J.A.; Humber, D.C.; Ewan, G.B. German Offen. **1985**, 3502508. *Chem. Abstr.* **1985**, *104*, 19589.
8. Mietzsch, F.; Mauss, H. U. S. Patent **1938**, 2113357. *Chem. Abstr.* **1938**, *32*, 4287.
9. Zirkle, C.L. U. S. Patent **1964**, 3131190. *Chem. Abstr.* **1964**, *61*, 4326.
10. Holm, T. Br. Patent **1963**, 933875. *Chem. Abstr.* **1964**, *60*, 510.
11. Bielavsky, J. *Collect. Czech. Chem. Commun.* **1977**, *42*, 2802.
12. Schutzke, G.M.; Pierrat, F.A.; Cornfeldt, M.L.; Szewzack, M.R.; Huger, F.P.; Bores, G.M.; Hartounian, V.; Davis, K.L. *J. Med. Chem.* **1988**, *31*, 1278.
13. Schindler, W., Hafliger, F. *Helv. Chim. Acta* **1954**, *37*, 472.

14. Craig, P.N.; Lester, B.M.; Suggiomo, A.J.; Kaiser, C.; Zirkle, C.M. *J. Org. Chem.* **1961**, *26*, 135.
15. Schindler, W. U. S. Patent **1960**, 2948718.
16. Anonymous Belgium Patent **1965**, 652938; *Chem. Abstr.* **1966**, *64*, 19575.
17. Archer, S.; Suter, C.M. *J. Am. Chem. Soc.* **1952**, *74*, 4296.
18. Rosi, D.; Peruzotti, G.; Dennis, E.W.; Berberian, D.A.; Freele, H.; Archer, S. *J. Med. Chem.* **1967**, *10*, 867.
19. Sprague, J.M.; Engelhardt, E.L. U. S. Patent **1960**, 2951082.
20. Bloom, B.M.; Muren, J.F. Belgium Patent **1964**, 647066. *Chem. Abstr.* **1965**, *63*, 11512.
21. Schmutz, J. U. S. Patent **1959**, 2905590.
22. Zirkle, C.L. U. S. Patent **1971**, 3609167. *Chem. Abstr.* **1971**, *75*, 151694.
23. Ackrell, J.; Antonio, Y.; Fidenico, F.; Landros, R.; Leon, A.; Muchowski, J.M.; Maddox, M.L.; Nelson, P.H.; Rooks, W.H. *J. Med. Chem.* **1978**, *21*, 1035.
24. Charpentier, P. *C.R.* **1947**, *225*, 306.
25. Charpentier, Gailliot, P.; Jacob, R.; Gaudechon, J.; Buisson, P. *Compt. Rend.* **1952**, *235*, 59.
26. Buisson, P.; Gailliot, P. U. S. Patent **1956**, 2769002.
27. Bourquin, J.P.; Schwarb, G.; Gamboni, G.; Rischer, R.; Ruesch, L.; Uldimann, S.G.; Theuss, U.; Schenke, E.; Renz, J. *Helv. Chim. Acta* **1958**, *41*, 1072.
28. Yale, H.L.; Sowinsky, F.; Bernstein, J. *J. Am. Chem. Soc.* **1957**, *79*, 4375.
29. Jacob, R.M.; Robert, J.G. German Offen. **1959**, 1092476.
30. Hunziker, F.; Lauener, H.; Schmutz, J. *Arzneim.-Forsch.***1963**, *13*, 324.
31. Hunziker, F.; Fischer, E.; Schmutz, J. *Helv. Chim. Acta* **1967**, *50*, 1588.
32. Schmutz, J.; Kunzle, F.; Hunziker, F.; Gauch, R. *Helv. Chim. Acta* **1967**, *50*, 245.
33. Howell, C.F.; Hardy, R.A.; Quinones N.Q. Fr. Patent **1968**, 1508536. *Chem. Abstr.* **1969**, *70*, 57923.
34. Hoffmann, C.; Faure, A. *Bull. Soc. Chim. Fr.* **1966**, 2316.
35. Lo, Y.S.; Taylor, C.R. So. African Patent **1983**, 8209154. *Chem. Abstr.* **1984**, *100*, 121121.

CHAPTER 14

BETA-LACTAM ANTIBIOTICS

The first indication that molds produce substances that inhibit bacterial growth predates the discovery of the antibacterial drug prontosil by a good one-half of a decade. The compound responsible for this activity, **penicillin (1)** was, however, not actually isolated until the late 1930s and was not developed for clinical use until the early 1940s. The poor stability and short biological half-life of this drug initially led to the development of a series of salts intended to remedy these problems. The development of methods for producing the bare penicillin nucleus subsequently led to the synthesis of a host of analogues by manipulation of the amide side chain intended to approach this problem as well as the fact that the antibacterial spectrum was restricted to Gram-positive organisms. An intense worldwide search for other antibiotic producing molds, mostly in soil samples, led to the discovery of the beta-lactam, cephalosporin C, which is active against Gram-negative bacteria; the development of methods for obtaining the bare cepham nucleus led to clinically effective analogues. The vast number of penicillin and cephalosporin analogues that differ only in the amide substituent will be touched on only briefly, since most of the chemistry lies in methods for preparing the side chains.

The extremely low toxicity of these drugs to mammalian species follows directly from the mechanism by which beta-lactams as a class inhibit bacterial growth. This circumstance devolves on the fact that bacteria, like plants, depend on a cell wall for their structural integrity; a cell membrane that is fundamentally different fulfills this function in higher species. A crucial part of such cell walls consist of highly cross-linked peptidoglycans in which peptide chains that include D-alanine fragments provide most of the cross-links. In brief, the beta-lactams act as false substrates in the construction of the cross-linking peptides. The very reactive beta-lactam function leads to irreversible blockade and in effect inhibits the formation of cell walls required for

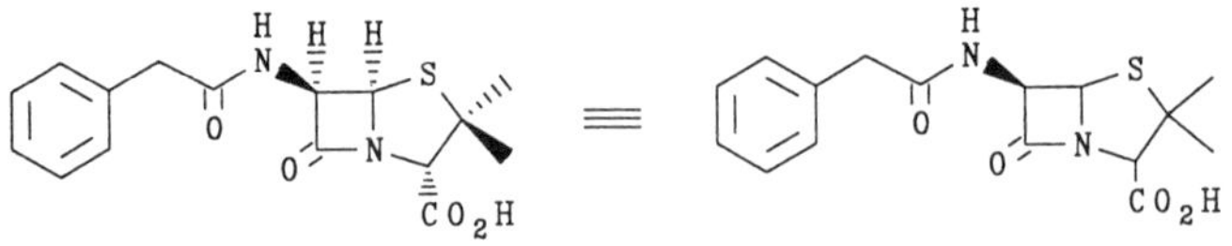

1

bacterial replication. The fact that the stereochemistry of the penicillins mimics that of D-amino acids further reduces toxicity, since the biochemistry of eukariotic species is based virtually exclusively on L-amino acids. Note that a small proportion of the population do manifest allergic reactions to this class of drugs.

1. PENICILLINS

Penicillin is but one of a series of closely related compounds isolated from fermentation broths of *Penicillium notatum*. This compound, also known as penicillin G or benzyl penicillin, is quite unstable and quickly eliminated from the body. Initial approaches to solving these problems, as noted above, consisted of preparing salts of the compound with amines that would form tight ion pairs, which in effect provided controlled release of the active drug. Research on fermentation conditions aimed at optimizing fermentation yields succeeded to the point where **1** and **penicillin V** (**85**), in which the phenylacetyl group is replaced by phenoxyacetyl, are now considered commodity chemicals. Another result of this research was the identification of fermentation conditions that favored the formation of the deacylated primary amine 6-aminopenicillanic acid (6-APA) (**5**), a compound that provided the key to semisynthetic compounds with superior pharmaceutical properties to the natural material. An elegant procedure for removal of the amide side chain proved competitive with 6-APA from fermentation. This method, which is equally applicable to penicillin V, starts by conversion of the acid to the corresponding silyl ester **2**. Treatment of **2** with phosphorus pentachloride in the presence of base leads to the imino chloride **3**; selectivity over the cyclic lactam is due to the fact that the latter cannot enolize. Solvolysis of this

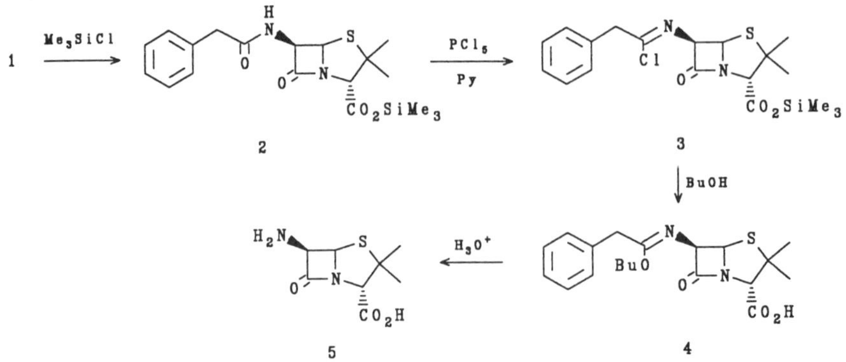

1 $\xrightarrow{Me_3SiCl}$ 2 $\xrightarrow[Py]{PCl_5}$ 3 $\xrightarrow{BuOH}$ 4 $\xrightarrow{H_3O^+}$ 5

product in butanol leads to the iminoether **4**; the silyl ether is converted to the acid under reaction conditions. Hydrolysis of **4** with aqueous acid cleaves the iminoether function to afford the free primary amine and thus 6-APA (**5**).[1]

The same functionality that is responsible for the activity of this class of antibiotics, the fused azetidone ring, makes for chemical reactivity and thus instability. In addition, many bacteria possess specialized enzymes, the beta-lactamases, which deactivate the drugs by cleaving this bond. Several semisynthetic analogues include quite bulky groups so as to provide steric hindrance about the beta-lactam function to decrease its lability. The chemical reactivity of the lactam function also means that special precautions must be used in the acylation reactions used to prepare analogues although the reactions themselves involve standard activated derivatives of the side chains such as acid chlorides or mixed anhydrides. (In the interest of uniformity, analogues are shown as carboxylic acids; it should be noted, however, that many beta-lactams are provided commercially as their potassium salts.) Acylation of 6-APA (**5**) with the acid chloride from 2,6-dimethoxybenzoic acid, for example, leads to **methicillin** (**6**)[2]; reaction with an activated derivative from 2-methyl-5-(2-chlorophenyl)isoxazole carboxylic acid gives **cloxacillin** (**7**).[3] These drugs show decreased sensitivity to beta-lactamase.

Acylation of 6-APA with an amino acid leads to a compound that bears some resemblance to a dipeptide. The coupling product with D-phenylglycine shows an enhanced antibacterial spectrum, possibly as a result of a better fit to the cell wall cross-linking enzyme. This compound also shows improved oral absorption. One synthesis starts with the acylation of 6-APA with the acid chloride **8** from D-2-azidophenylacetic acid. Catalytic reduction of the product **9** affords **ampicillin** (**10**).[4] The corresponding product from acylation with the 4-hydroxyphenylacetic acid is **amoxycillin**, a widely used drug that is reasonably well absorbed on oral administration.

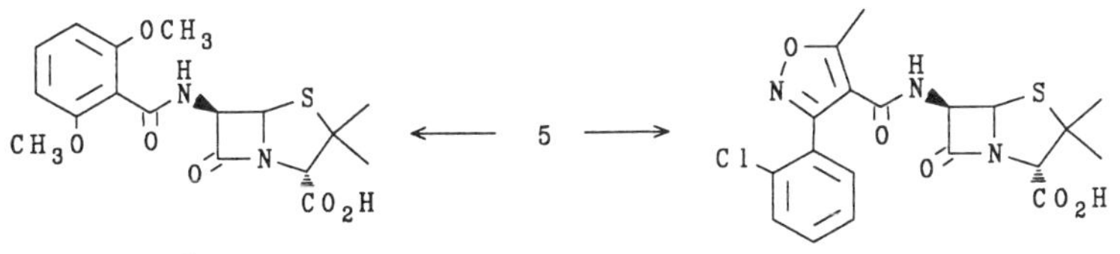

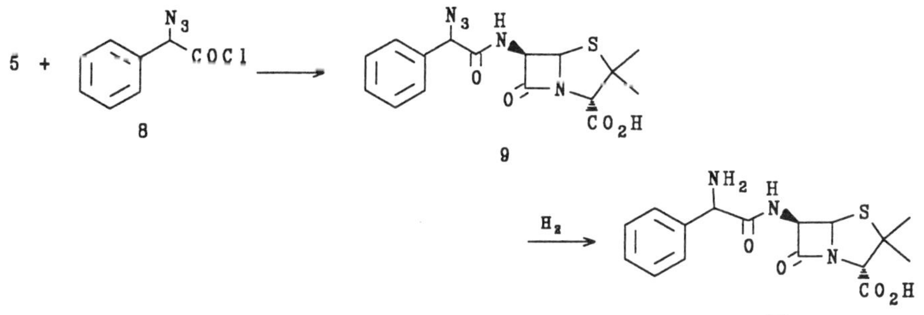

1

11

12

13

A

B

Highly polar water soluble organic compounds are often poorly absorbed from the GI tract. Reducing the polarity of such compounds results in increased solubility in lipid membranes and may thus increase oral absorption. The ubiquity of esterases in serum makes esters particularly well suited for converting acids to less polar derivatives; the enzymes should convert the circulating drug that has been absorbed to the free acid required for biological activity. Esters such as **A** from formaldehyde hydrate have been found particularly suitable for beta-lactams. Hydrolysis of the pivaloyl group leads to a derivative such as **B**, which spontaneously reverts to the free acid; hydrolysis of the other ester bond of course leads to the penicillin acid directly. The first step in one route to such compounds begins with the alkylation of the salt from benzyl penicillin (**1**) with chloromethyl pivalate to give the derivative **11**. This intermediate is then subjected to the side chain removal sequence outlined above (**2** → **5**) to afford the 6-APA derivative **12**. Acylation of **12** with the acid chloride from D-phenylglycine gives the orally active antibiotic **pivampicillin** (**13**).[5]

Penicillin as well as the first generation of semisynthetic analogues are mainly active against Gram-positive bacteria. The antibacterial spectrum is somewhat broadened, as noted above, by including functionality in the amide side chain. Coupling additional polar moieties onto the phenylglycine amine leads to further broadening of the antibacterial spectrum. Thus, acylation of ampicillin with the carbonyl chloride derivative **14** from imidazolone affords **azlocillin** (**15**).[6] In a similar fashion, condensation of the glycylamide **16** with the iminoether from 4-cyanopyridine leads to **pirbenicillin** (**17**).[7] Both of these compounds are active against *pseudomonas* as well as against the usual Gram-positive organisms.

A free carboxylic acid group also enhances the antibacterial spectrum in the penicillin series. Acylation of 6-APA (**5**) with the half-acid chloride **18** from benzyl phenylmalonate leads to the amide **19**. Removal of the benzyl protecting group by catalytic hydrogenation[8] or by enzymatic hydrolysis,[9] affords **carbencillin** (**20**). A

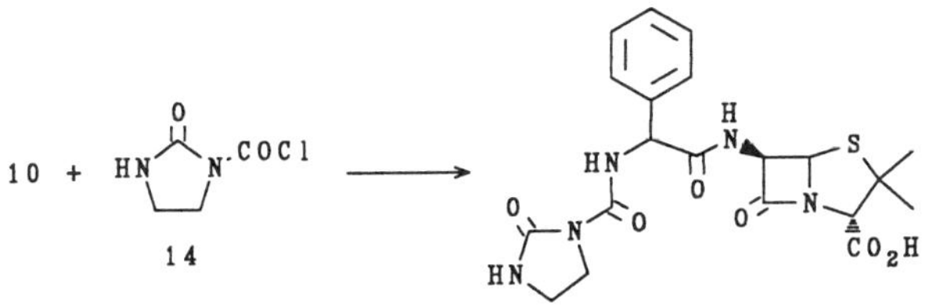

10 + 14 → 15

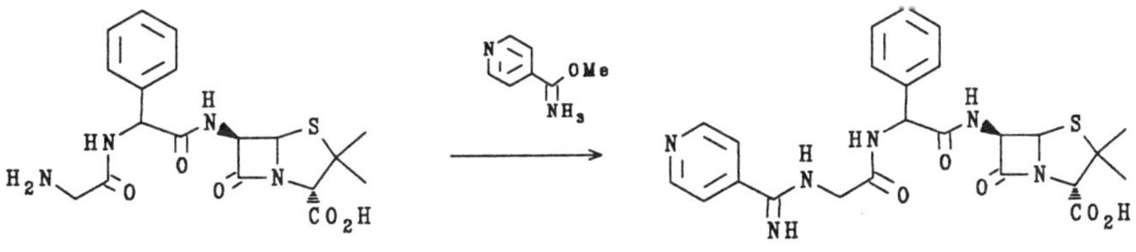

16 → 17

5 + $PhCH_2O_2C$–CH(Ph)–COCl (18) → (19) $\xrightarrow{H_2}$ (20)

(21)

NCH=O (22) $\xrightarrow{(COCl)_2}$ (23) $\xrightarrow{5}$ (24)

similar sequence starting with 3-thiophenylmalonic acid leads to the considerably more potent analogue **ticarcillin** (**21**).[10]

Antibacterial activity is retained when the relatively complex amide side chains are replaced by a simple heterocycle amidine. The required reagent **23** is prepared by reaction of azepine formamide (**22**) with oxalyl chloride. Condensation of **23** with 6-APA (**5**) leads to formation of the amidine and thus **amdinocillin** (**24**).[11]

The finding that the addition of a methoxyl group at the 6α-position significantly enhance resistance to bacterial beta-lactamase actually traces to the cephalosporin series with the discovery of a series of fermentation products, the cephamycins, which bear this substituent. The preparation of the semisynthetic cephalosporin derivative **cefoxitin** (**116**), which carries this substituent, will be found later in this chapter.

The preparation of an analogous penicillin derivative starts with protection of 6-APA (**5**) as its benzyl ester. Reaction of this product with formic acid in the presence of DCC gives the amide **25**. Treatment of **25** with phosgene in the presence of *N*-methylmorpholine converts the amide to the corresponding isocyanate with retention of stereochemistry. The combination of the adjacent carbonyl and the isocyanate group facilitate the formation of anions at the 6-position. Thus, reaction of **26** with methylmethoxycarbonyl disulfide in the presence of powdered potassium carbonate leads to the 6-thiomethyl ether **27**; the α stereochemistry of sulfur is a consequence of approach from the more open side of the molecule rather than retention of stereochemistry by the carbanion. Reaction of the isocyanate grouping with *p*-toluenesulfonic acid monohydrate achieves selective conversion of the isocyanate back to a primary amino group to give **28** without affecting the presumably more reactive azetidone. Treatment of **28** with mercuric chloride in methanol in the presence of base leads to fragmentation of the carbon–sulfur bond and formation of a transient carbocation. Compound **28** adds methanol, again from the open side, to give the key 6α-methoxy intermediate **29**.

The primary amino group at the 6-position is then acylated with the half-acid chloride **30** from benzyl thiophenemalonic acid to give the amide **31**. Reductive debenzylation of **31** affords the antibiotic **temocillin** (**32**).[12]

Some penicillin derivatives lacking nitrogen at the 6-position, in which the ring sulfur atom is oxidized to a sulfone, act as inhibitors of bacterial beta-lactamase. These agents would be used as adjuncts to beta-lactams since they have no antibacterial activity in their own right. A key reaction in the synthesis of each compound involves replacement of the amine at the 6-position and protection of this position as a mono- or dihalide. Thus, reaction of 6-APA (**5**) with nitrous acid gives the diazonium salt **33**; this compound is converted to the dibromide **34** on treatment with bromine. The ring sulfur is then oxidized with permanganate to the sulfone **35**. Hydrogenolysis of the product replaces the two bromine atoms by hydrogen to afford **sulbactam** (**36**).[13]

A subsequent analogue involves the substitution of a heterocyclic moiety onto one of the geminal methyl groups. Diazotization of 6-APA (**5**) in the presence of excess bromide ion gives the monobromo derivative **37**. Controlled oxidation of sulfur, for example, with periodate leads to the sulfoxide **38**. The carboxyl group is then protected as its *p*-nitrobenzyl ester, **39** a group often used in beta-lactam chemistry. The bromide at the 7-position is then removed by catalytic hydrogenation over palladium to give **40**. The thiazoline oxide ring-opening reaction that follows was first developed in work

5 → (1. BzCl; 2. HCO_2H) → 25 → ($COCl_2$) → 26 → ($MeSSCO_2Me$, K_2CO_3) → 27 → ($TSA \cdot H_2O$) → 28 → (MeOH, $HgCl_2$/Py) → 29

Bz-CH_2Ph

30 → (29) → 31 → (H_2) → 32

Bz-$CH_2C_6H_5$

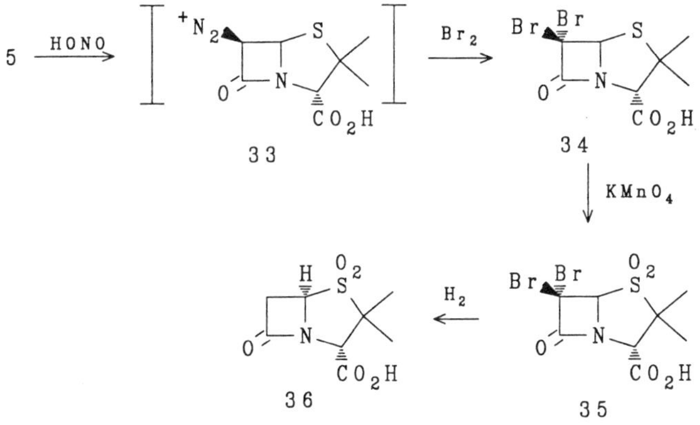

5 $\xrightarrow{HONO}$ [33] $\xrightarrow{Br_2}$ 34 $\xrightarrow{KMnO_4}$ 35 $\xrightarrow{H_2}$ 36

aimed at the conversion of penicillins to cephalosporins; a closely related ring opening is noted below in the discussion of this rearrangement. Thus, reaction of the sulfoxide **40** with 2-mercaptobenzothiazlole leads to the disulfide **41**. (The reaction may first involve formation of a ring opened vinyl sulfinic acid by initial abstraction of a proton on one of the geminal methyl groups.) Treatment of the disulfide with cuprous chloride proceeds via heterolytic cleavage of the disulfide bond. The thus generated sulfur radical then adds back to the vinyl group to reform the thiazoline ring. Capture of chloride present in the reaction medium completes the formation of the observed product **42**. The reasons for the preferential formation of the β-chloromethyl isomer are not immediately apparent.

Construction of the side chain heterocycle starts with the displacement of chlorine in **42** with sodium azide to afford the cumulene **43**. The ring oxygen is then oxidized to a sulfone, as shown above, by reaction with permanganate. The azide function in **44** undergoes 1,3-dipolar cycloaddition when treated with acetylene to form a 1,2,3-triazole ring. Saponification with dilute base then cleaves the nitrobenzyl protection group to afford the beta-lactamase inhibitor **tazobactam** (**45**).[14]

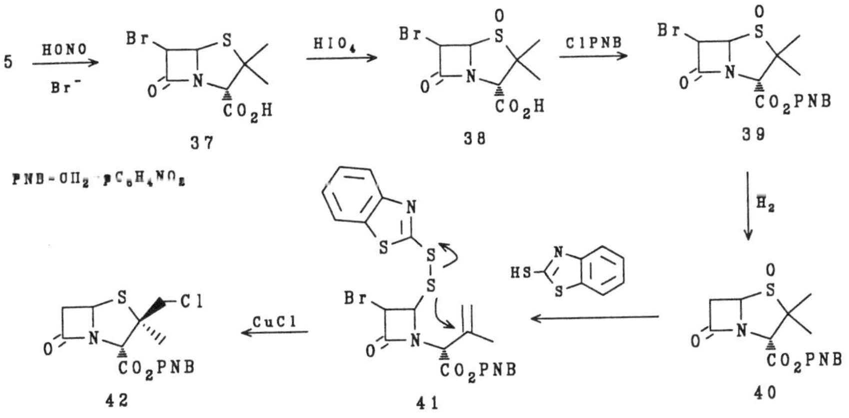

5 $\xrightarrow[Br^-]{HONO}$ 37 $\xrightarrow{HIO_4}$ 38 $\xrightarrow{ClPNB}$ 39 $\xrightarrow{H_2}$ 40 $\rightarrow$ 41 $\xrightarrow{CuCl}$ 42

$PNB = CH_2\text{-}pC_6H_4NO_2$

42 $\xrightarrow{NaN_3}$ 43 (CO_2PNB, $-N{=}N{=}N$) $\xrightarrow{KMnO_4}$ 44 (O_2, CO_2PNB, $-N{=}N{=}N$) $\xrightarrow[2.\ NaOH]{1.\ HC{\equiv}CH}$ 45 (O_2, CO_2H)

The search for new antibacterial agents in fermentation broths continued long after the identification of the cephalosporins. One result of this continuing research was the identification of a very potent, broad spectrum beta-lactam antibiotic that is unusually resistant to beta-lactamase. The structure of this compound, **thienamycin** (**63**), at first glance resembles that of a penicillin. It differs, however, in many respects; most markedly by the fact that the dimethylthiazolidine ring is replaced by pyrroline and the amine at the β C-6 position is replaced by an α-hydroxyethyl group. Stability problems with the pure compound, traceable to lability of the thioenol ether function, precluded its use as a drug. One synthesis of the far more stable amidine analogue, **imipenem** (**62**) illustrates the chemistry used for total synthesis of beta-lactam antibiotics.

Reaction of the ethyl ester (**46**) of acetonedicarboxylate with benzylamine in the presence of a molecular sieve gives the corresponding enamine **47**. Condensation of **47** with diketene gives the acylation product **48**, in which the olefin has shifted to form an imine. Treatment of **48** with sodium cyanoborohydride leads to stereospecific reduction of both the imine and keto functions, probably by way of the conjugated enol form, to give the aminoalcohol **49**. Reaction of **49** with acid leads to formation of the six-membered lactone; reaction with the alternate carbethoxy group would lead to an unstable butyrolactone. The benzyl protecting group on the amine is removed by catalytic hydrogenation to afford intermediate **50**. Reaction of **50** with benzyl alcohol in the presence of acid proceeds at the more reactive lactone carbonyl group to give the corresponding benzyl ester. Thus, amino acid **51** is obtained. The beta-lactam ring is then closed by internal amide formation catalyzed by dicyclohexylcarbodiimide, a reagent originally developed for just such reactions, to afford **52**. The lactam function is protected by conversion to its *tert*-butyldimethylsilyl (*t*-BDMS) derivative and the benzyl protecting group is removed reductively to give the beta-lactam **53**.

The sequence for adding the two additional carbon atoms for the fused ring starts by activation of the carboxyl group in **53** by conversion to an acylimidazole by reaction with CDI. Treatment of the activated malonate, Meldrum's acid (**54**), with this leads to the corresponding acylation product. Acidic workup of the product leads to generation of the malonic acid by hydrolysis of the actonide; the resulting acylmalonic acid decarboxylates to give the two-carbon addition product. The free carboxylic acid is then esterified to the *p*-nitrobenzyl derivative; hydrolysis of the silyl protecting group

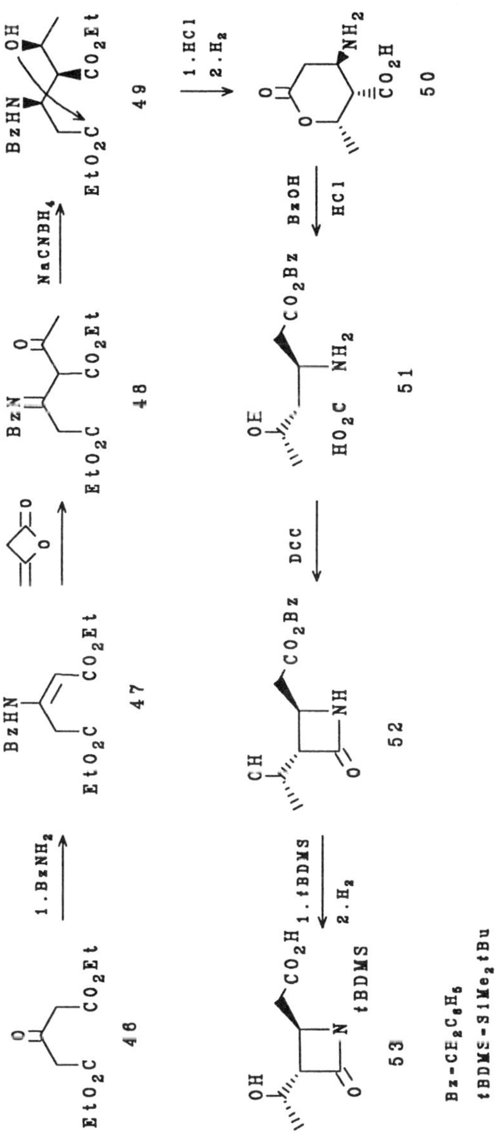

Bz = $CH_2C_6H_5$

tBDMS = $SiMe_2tBu$

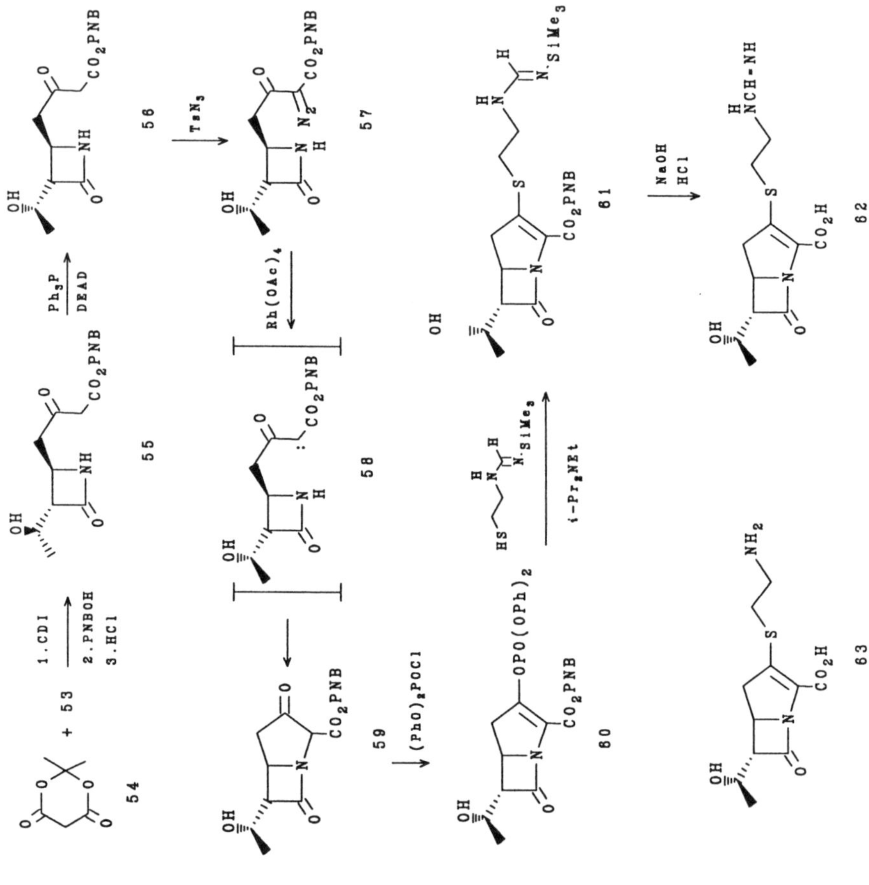
+ 53
1. CDI
2. PNBOH
3. HCl
54
55
56
57
58
59
60
61
62
63
Ph_3P
DEAD
TsN_3
$Rh(OAc)_4$
$(PhO)_2POCl$
i-Pr_2NEt
NaOH
HCl
CO_2PNB
CO_2H
$OPO(OPh)_2$
$N \cdot SiMe_3$
$NCH=NH$
NH_2

affords **55**. The stereochemistry of the hydroxyethyl group, which traces back to the cyanoborohydride reduction, is in fact the reverse of what is required for good antibacterial activity. This center can be conveniently inverted by a version of the Mitsonobu reaction in which hydroxide is used as the displacing nucleophile. Thus, the intermediate **56** is obtained. One of the key steps in building the fused ring involves reaction of the activated acetoacetate methylene in this compound with toluenesulfonyl azide to give the diazo intermediate **57**. Treatment of **57** with rhodium acetate leads to loss of nitrogen with consequent formation of carbene **58**; this reactive center inserts into the adjacent amide N–H bond to form a five-membered ring and thus the carbapenem (**59**).[15]

The first step in incorporation of the thioenol function consists in conversion of the ketone to the enol phosphate derivative **60** by reaction with diphenyl chlorophosphate. Cystamine amidine, which is required for the next step, cannot be used as such since it spontaneously cyclizes to a thiazoline. The silyl ether avoids that side reaction; thus, treatment of **60** with the silyl amidine leads to the thioenol ether **61** by addition–elimination. Removal of the protecting groups by sequential acid and base hydrolysis gives **62** as a racemate.[16] The chiral product **imipenem** can be obtained by the same scheme using a resolved intermediate.

2. CEPHALOSPORINS

Not too long after the widespread adoption of penicillin for the treatment of bacterial infections a related beta-lactam compound was isolated as a consequence of the continuing search for new drugs. This drug, cephalosporin C (**64**), differs in that a six- rather than a five-membered ring is fused onto the azetidone. Although the agent showed superior resistance to beta-lactamase and a broader spectrum of action than penicillin, its clinical use was precluded by its low potency and poor biopharmaceutic properties. The development of a method for efficiently removing the aminoadipate side chain provided the first method for preparing 7-aminocephalosporanic acid (7-ACA) (**67**), which serves the same role in this series as does 6-APA (**5**) in the penicillin series. (Attempts to remove the side chain directly yield less than 1% 7-ACA.)

The key reaction, based on a method for removing glutamate residues in peptides, involves conversion of the sole primary amine in the molecule to a diazo function. The most expeditious method consists of the reaction of **64** with nitrosyl chloride. The resulting diazo function in the product **65** can be displaced formally by oxygen from the enol form of the amide at the 7-position to form the iminolactone **66**; the reaction may involve spontaneous loss of nitrogen followed by capture of the resulting carbocation. Hydrolysis of the imine function in the product leads to one of the key intermediates in this series, 7-ACA (**67**).[17]

This intermediate, like 6-APA, incorporates a primary amine that can be coupled with a host of side chains. The presence of an additional reactive function, the allylic acetate, at the 3-position provides an additional center that can be modified. The observation that both types of modifications provided improved antibiotics has re-

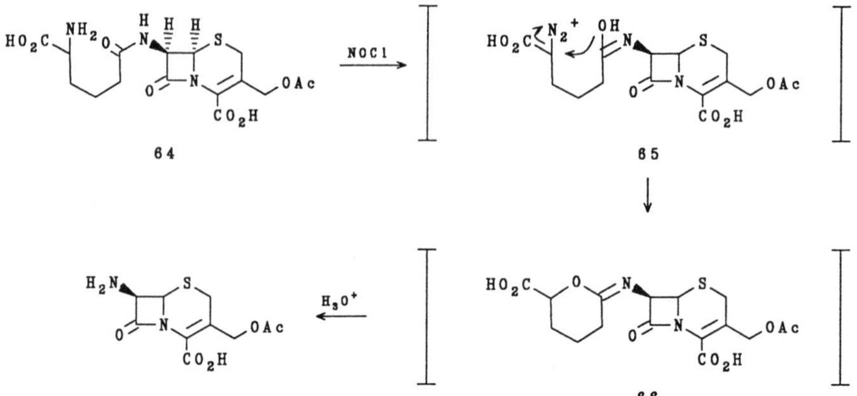

64 → (NOCl) → [65] → [66] → (H_3O^+) → 67

sulted in the synthesis of hundreds of analogues. The very few examples discussed below barely scratch the surface in this field. One of the earliest examples of a derivatized 7-ACA derivative, **cephalothin** (**68**), is still widely used as an antibiotic. Acylation of **67** with 2-thiophenylacetyl chloride gives the corresponding amide **68**. Heating **68** with pyridine leads to displacement of the allyl acetate by the basic nitrogen. The resulting product, **cephaloridine** (**69**), is isolated as the internal betaine.[18]

The order of the steps can be reversed. Thus, reaction of **67** with the tetrazole-thione **70** results in displacement of the allylic group by nucleophilic sulfur and formation of the intermediate **71**. This product is then acylated with acid chloride from the dichoroacetyl ester D-madellic acid. The protecting group on the side chain is then removed to give the antibiotic **cefamandole** (**72**).[19]

Incorporation of complex side chains at the 7-position based on alkyloximes of 2-aminothiazole-5-gyloxylamides have provided drugs with very wide antibacterial activity against hitherto resistant species such as *pseudomonas*. The preparation of one of the simpler side chains first involves formation of the methyl ether from the oxime obtained by nitrosation of methyl acetoacetate. Chlorination of the product, for example, with sulfuryl chloride give the intermediate **73**. The aminothiazole ring is then formed by reaction of this compound with thiourea to give **74**. The free acid **75**, is obtained by saponification of the product. The protected acid chloride **76** is obtained

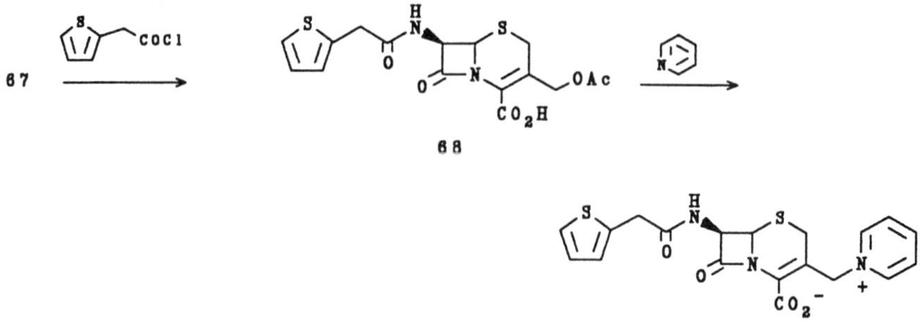

67 → (2-thienyl-COCl) → 68 → (pyridine) → 69

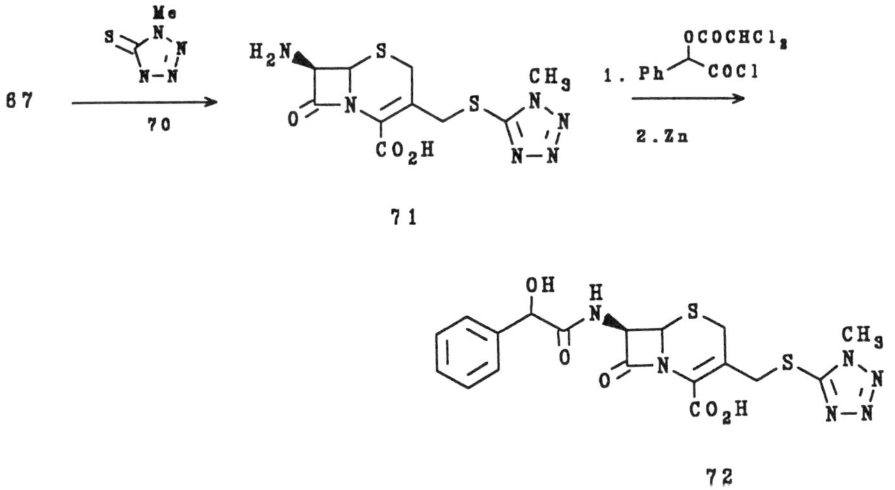

67 → (70) → 71 → (1. Ph(OCOCHCl$_2$)COCl; 2. Zn) → 72

by sequential acylation of the amino group with chloroacetyl chloride and then reaction with thionyl chloride.

In a typical example of the synthesis of one of these broad spectrum agents, the cefamandole intermediate **71** is first acylated with the protected thiazole acid chloride **76** to give the amide **77**. Removal of the protecting group by reaction of **77** with thiourea can be envisaged as involving initial displacement of chlorine on the side chain by sulfur to form an intermediate such as A; this compound then cyclizes to a second thiazole ring in the process cleaving the protecting group. Thus, **cefmenoxime** (**78**)[20] is obtained.

The preparation of a related compound first involves replacement of allyl oxygen in the *tert*-butylcarbonyloxy protected 7-ACA derivative (**79**) by nitrogen in azaindan to afford the betaine **80**. The protecting group is then removed by treatment with trifluoroacetic acid. Reaction with the free thiazole acid **75** in the presence of DCC then affords **cefpirone** (**81**).[21]

Acylation of the amino group at the 7-position markedly enhances oral absorption of beta-lactam antibiotics. The key step in the synthesis of these compounds is of course quite analogous to peptide coupling; the mixed-anhydride method has been

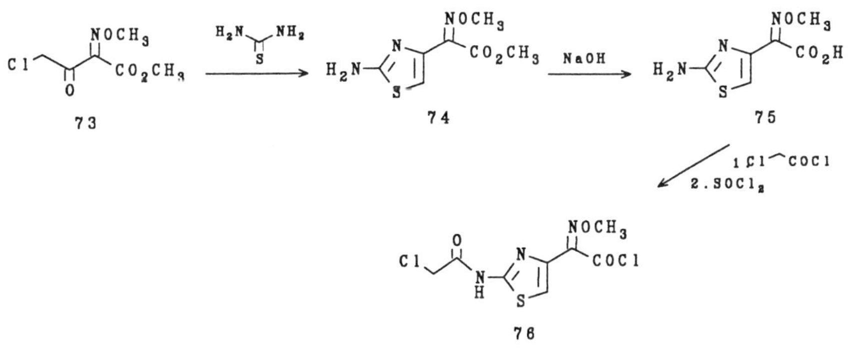

73 → ($H_2NC(S)NH_2$) → 74 → (NaOH) → 75 → (1. ClCH$_2$COCl; 2. $SOCl_2$) → 76

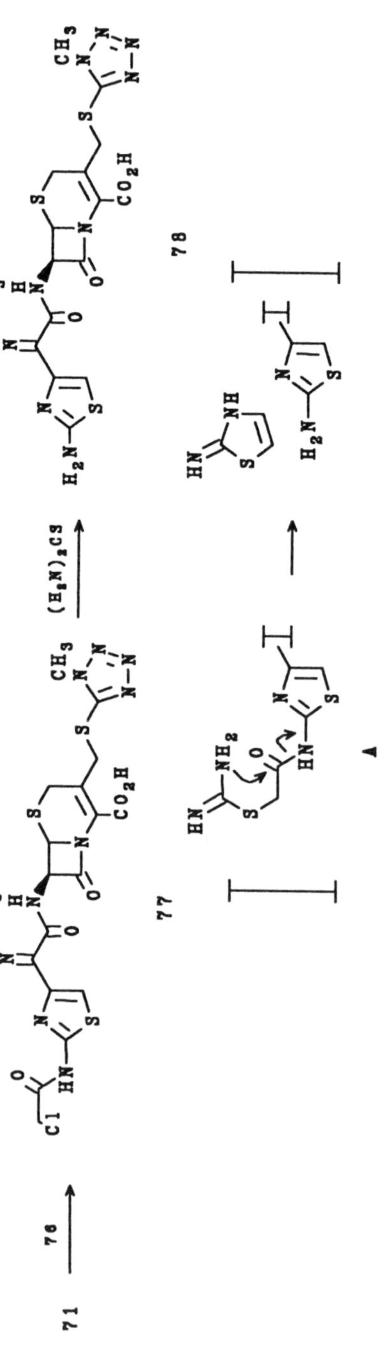

71 $\xrightarrow{76}$ 77 $\xrightarrow{(H_2N)_2CS}$ 78

A

79

BoC-$PhCH_2OCO$

80

1. TFA
2. 75/DCC

81

used extensively for preparing compounds in this series. Thus, reaction of 7-ACA (**67**) with the mixed anhydride from the *t* Boc derivative of D-phenylglycine and isobutyl chloroformate leads to the amide **82**. Treatment of **82** with trifluoroacetic acid leads to the free amine and thus **cephaloglycin** (**83**).[22] An increase in lipophilicity, by deleting the acetoxy group at the 3-position, further improves oral availability with no loss in antibiotic activity. Catalytic reduction of **83** over palladium on charcoal leads to hydrogenolysis of the allylic acetoxy group. Thus, the widely used antibiotic **cephalexin** (**84**)[23] is obtained.

The availability of benzyl penicillin (**1**) and penicillin V (**85**) in virtually tonnage quantity has been noted earlier. The observation that these compounds sometimes occur together with cephalosporins, combined with the fact that the latter may be viewed as a dehydro form of their five-membered counterparts, led to speculation that the antibiotics might be formed by a common pathway. The chemical work occasioned by this theory led to several efficient routes for the preparation of cephalosporin starting materials from penicillins.[24] In its simplest application, **85** is first oxidized to the corresponding sulfoxide **86**. Treatment of **86** with toluenesulfonic acid in refluxing xylene leads to ring opening to the sulfenic acid intermediate **87**. This intermediate cyclizes by addition to the conjugated double bond under the reaction conditions with loss of oxygen to afford the cephalosporin **88**.[25] The phenoxyacetamido side chain is then removed via its imino chloride by the same sequence of steps used for removing the phenylacetamide side chain in penicillin (**1** → **5**) to give 7-aminodeacetylcephalosporanic acid (7-ADCA) (**89**).

The deacetyl compound **89** is now used for direct production of **cephalexin** (**84**) as well as several other related agents that incorporate a similar amide side chain. For example, reaction of the protected 7-ADCA derivative **90**, with the *t*-Boc amide from D-*p*-hydroxyphenylglycine (**91**) by the mixed-anhydride method gives the amide **92**. Serial scission of the *t*-Boc group and the silyl ester affords the antibiotic **cefadroxyl** (**93**).[26] Exactly the same sequence starting with the Birch reduction product from D-phenylglycine leads to **cephradine** (**94**).[27]

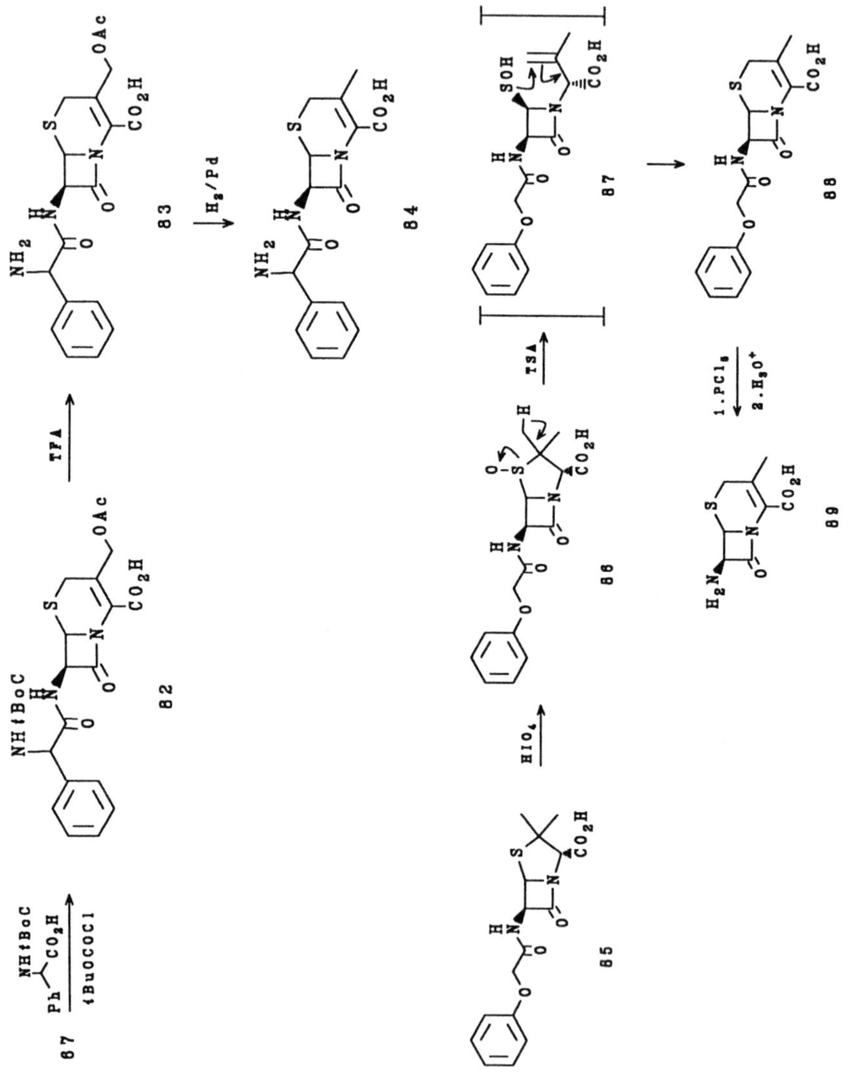

67 $\xrightarrow[\text{tBuOCOCl}]{\text{Ph-CH(NHtBoC)CO}_2\text{H}}$ 82 $\xrightarrow{\text{TFA}}$ 83 $\xrightarrow{\text{H}_2/\text{Pd}}$ 84

85 $\xrightarrow{\text{HIO}_4}$ 86 $\xrightarrow{\text{TSA}}$ [87] $\rightarrow$ 88 $\xrightarrow[\text{2. H}_3\text{O}^+]{\text{1. PCl}_5}$ 89

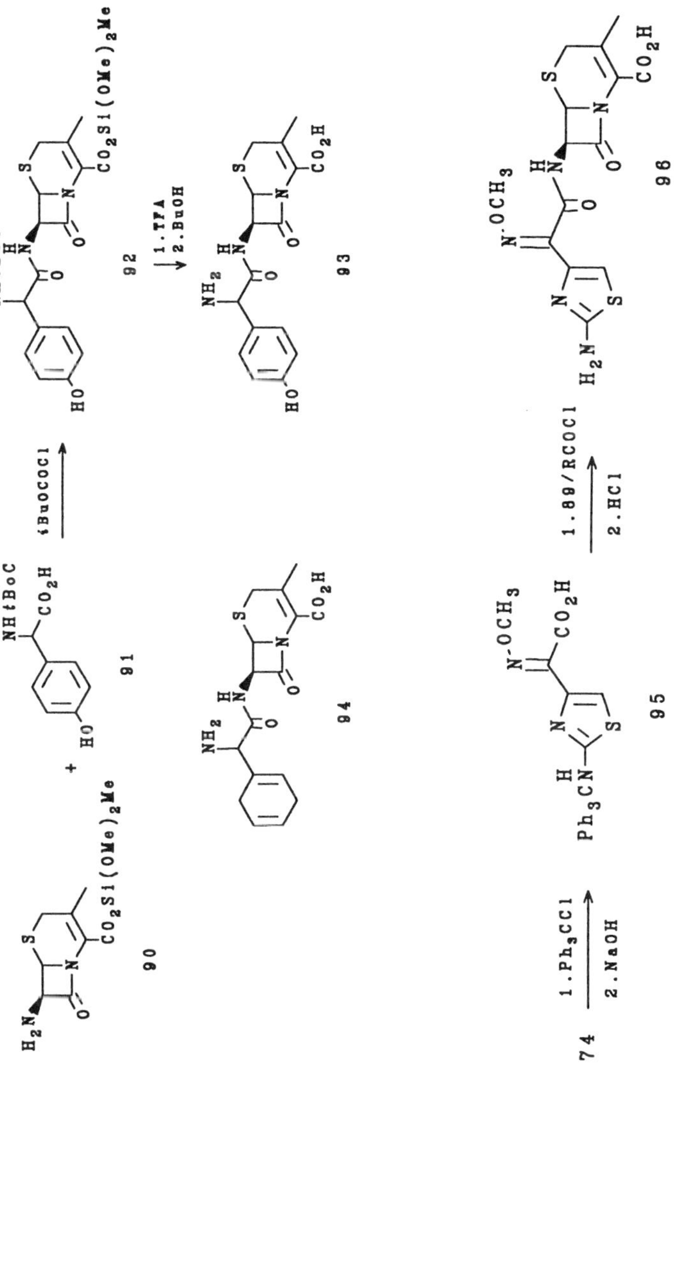

90 + **91** $\xrightarrow{t\text{BuOCOCl}}$ **92** $\xrightarrow{\text{1. TFA; 2. BuOH}}$ **93**

94

74 $\xrightarrow{\text{1. }Ph_3CCl\text{; 2. NaOH}}$ **95** $\xrightarrow{\text{1. 89/RCOCl; 2. HCl}}$ **96**

Acylation of the amine with the methoxime from aminothiazole-glyoxylate has a similar effect on broadening the antibacterial spectrum in the 3-methyl series. In this case, the amine group on the starting thiazole (**74**) is first protected by conversion to **95** by alkylation with triphenylmethyl chloride. Condensation of the acid by the mixed-anhydride method with 7-ADCA (**89**) leads to the corresponding amide. The trityl group is then removed to afford the antibiotic **cefetamet** (**96**).[28]

Very good antibiotic activity is, interestingly, retained when the substituent at the 3-position in fact consists of a heteroatom or even hydrogen. The preparation of these compounds relies on the chemistry involved in the penicillin to cephalosporin ring-enlargement reactions. One synthesis for the drug in which the 3-substituent consists of chlorine, **cefaclor** (**102**), starts with the sulfoxide **97** of penicillin V protected as its *p*-nitrobenzyl ester. Reaction of **97** with a chlorinating agent such as *N*-chlorosuccinimide in the presence of an acid scavenger probably first involves chlorination on sulfur to form a chlorosulfonium chloride species; this compound then ring opens with loss of hydrogen chloride to the unsaturated sulfenyl chloride **98**. Treatment of **98** with a Lewis acid then leads to Friedel–Crafts-like attack on the olefin to afford the cyclized product; the double bond concomitantly shifts to the exocyclic position to give **99**. Ozonization in the cold followed by cleavage of the ozonide affords the corresponding ring ketone; this compound exists virtually entirely as its conjugated enol; the sulfoxide is then reduced to afford the cephem (**100**). Reaction of **100** with phosphorus pentachloride converts the enol to the enol chloride and at the same time transforms the side chain amide to its imino chloride. The latter hydrolyzes on workup to afford the primary amine to give **101**. This product is then converted to the corresponding amide with D-phenylglycine by the standard scheme, using mixed anhydride coupling of *t*-Boc protected amino acid, to give **cefaclor** (**102**).[24]

The enol **100** can in principle serve as the intermediate for the antibiotic **cefroxadine**, **109**, in which the substituent at the 3 position consists of a methyl ether. The route differs from that above in that the key step involves ring opening of penicillin sulfoxide by mercaptobenzo-thiazole; application of that reaction was discussed

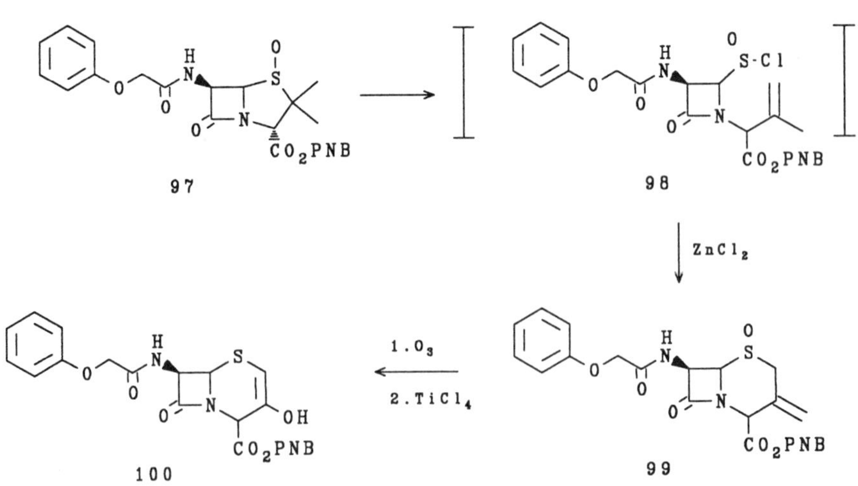

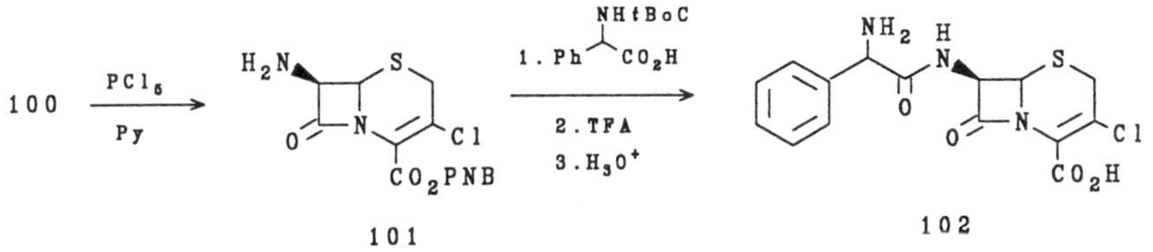

earlier in connection with the synthesis of the beta lactamase inhibitor, tazobactam **42**. In the present case, the action of that reagent on the benzhydryl ester **103** of penicillin V sulfoxide, affords the ring opened intermediate **104**. The superfluous carbon is next removed by ozonization to afford the corresponding β-ketoester **105** which again exists mainly as the enol; this is converted to its methyl ether, **106**, by treatment with diazomethane.

The reaction of this last intermediate with the non-nucleophilic base, 1,8-diazabicyclo[5.4.0]undec-7-ene (DBU) can be envisaged as involving first the formation of a carbanion on the terminal methyl group; this then attacks the disulfide with expulsion of the excellent leaving group, benzothiazolomercaptide. There is thus formed the 3-methoxy cephem **107** which still however bears the penicillin V side chain. That is then removed by the standard imino chloride route to give the key intermediate **108**. The free amine at the 7 position is then acylated in the standard fashion with the *t*-Boc derivative from (2,5-dihydrophenyl)glycine. Removal of the protecting groups leads to the antibiotic **cefroxadine (109)**.[29]

The enolic chlorine at the 3-position in **101** can be removed under reductive conditions such as zinc and acid to give a cephem (**110**) that bears hydrogen at this position. Acylation of **101** with the acid chloride from the aminotiazole intermediate **75** affords the antibiotic **ceftizoxime (111)**.[30]

The cephamycins comprise a set of beta-lactamase resistant fermentation products that consist of cephalosporins that include a methoxy group on the carbon bearing the

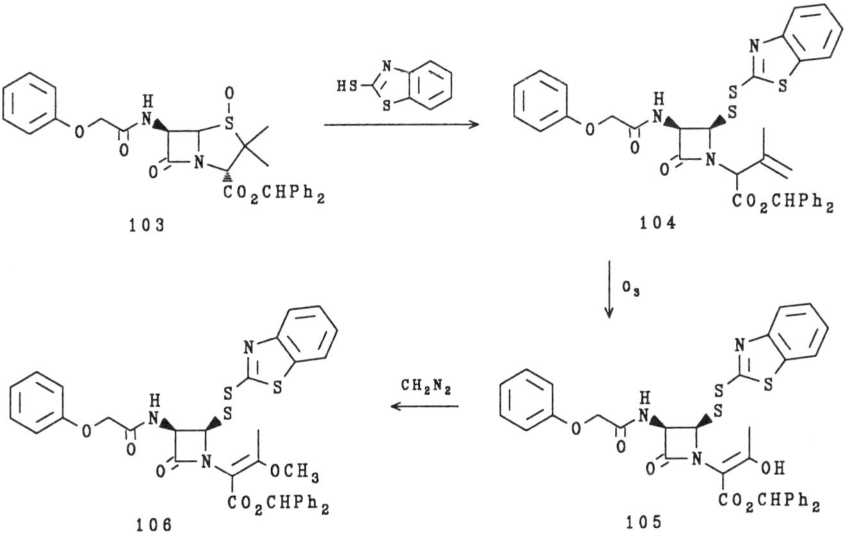

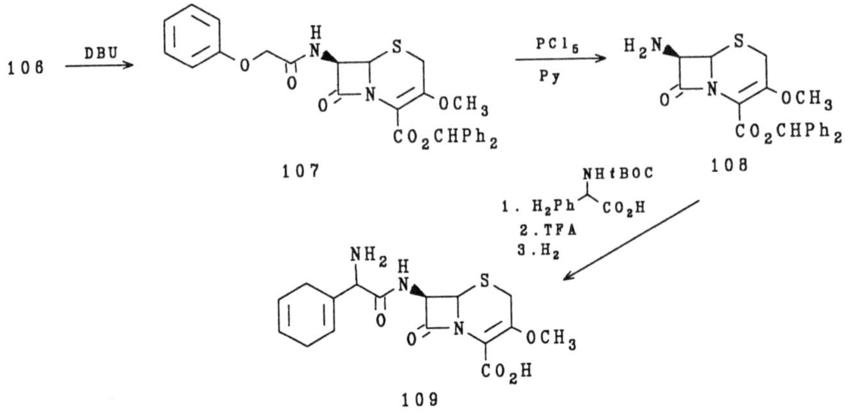

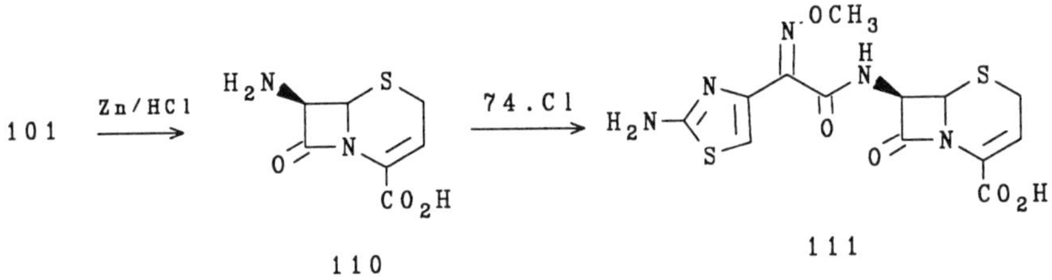

106 —DBU→ 107 —PCl_5, Py→ 108 —1. H_2Ph–CH(NHtBOC)–CO_2H; 2. TFA; 3. H_2→ 109

101 —Zn/HCl→ 110 —74.Cl→ 111

amine. These agents too were not suitable as drugs in their own right because of their poor biopharmaceutical properties. The reason for their poor properties was due to the circumstance that, like the original cephalosporins, the natural products occurred as acylation products with the highly polar aminoadipic acid. A number of methods were elaborated for direct introduction of the methoxyl group in the course of the search for an analogue suitable for use in the clinic; these methods are generally similar in concept to those discussed above in connection with the 6-methoxy penicillin **temocillin** (**32**). A method for direct exchange of side chains was developed when the amide with 2-(2-thienyl)acetic acid was identified as the clinical candidate. The preparation starts with acylation of the primary side chain amine in cephamycin C (**112**) with trichloroethoxycarbonyl chloride; reaction of this product with benzhydryl chloride and base converts both carboxylic acids to their benzhydryl esters to give **113**. This intermediate is then treated with 2-(2-thienyl)acetyl chloride in the presence of trimethylsilyl trifluoroacetamide to add the second amide on the amine at the 7-position to afford the reactive imide **114**. The trichloroethyl carbamate group is then very selectively cleaved by reaction with zinc metal. The newly liberated primary side chain amine in this transient intermediate **115** then reacts with the adipyl amide carbonyl and in effect displaces the side chain. Hydrogenolysis of the benzhydryl ester affords the antibiotic **cefoxitin** (**116**).[31]

The compounds obtained by replacement of ring sulfur by carbon, as in the case of penicillins, show somewhat improved antibiotic properties. A free radical based route has been described for the conversion of the fermentation derived cephalosporins to

112 $\xrightarrow{\text{1. RCOCl; 2. BHzCl}}$ 113 $\xrightarrow{\text{ClOCTh}}$ 114 $\xrightarrow{\text{Zn}}$ [115] $\xrightarrow{H_2}$ 116

A-CH_2OCONH_2

R-OCH_2CCl_3
BHz-Ph_2CH
Th-CH_2(2-thienyl)

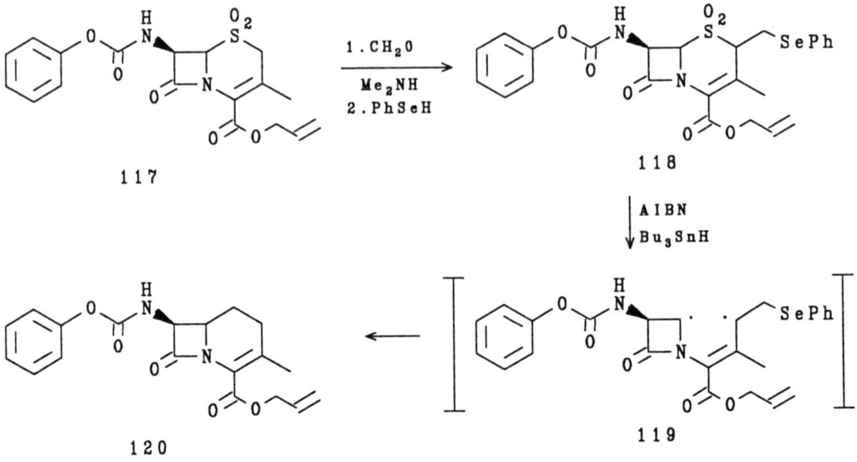

their carbocyclic derivatives. The first step in this sequence consists of condensation of the cephalosporin sulfone **117** with formaldehyde and dimethylamine; the initial product from the Mannich-like reaction consists of the exomethylene derivative at the position adjacent to the activating sulfone. The product is treated in situ with phenylselenol to give the Michael adduct **118**. When heated with the free radical initiator AIBN in the presence of tributyltin hydride this fragments with extrusion of sulfur dioxide; the reaction can be envisaged as leading to the formation of a diradical such as **119**. This species can then close to a six-membered ring; reductive loss of phenylselenol affords the observed carbacephem (**120**).[32]

The key sequence in a somewhat involved stereospecific synthesis of a carbacephem consists in construction of a chiral auxiliary. It is interesting to note that nitrogen is the only atom from the auxiliary in this molecule retained in the final product. Construction of this moiety starts with formation of the carbethoxy derivative **122** from L(+)-phenylglycine (**121**). Selective reduction of the free carboxyl group with borane·THF leads to the hydroxyester **123**. In a one-pot sequence, this compound is first cyclized to the corresponding oxazolidinone (**124**) by means of sodium hydride

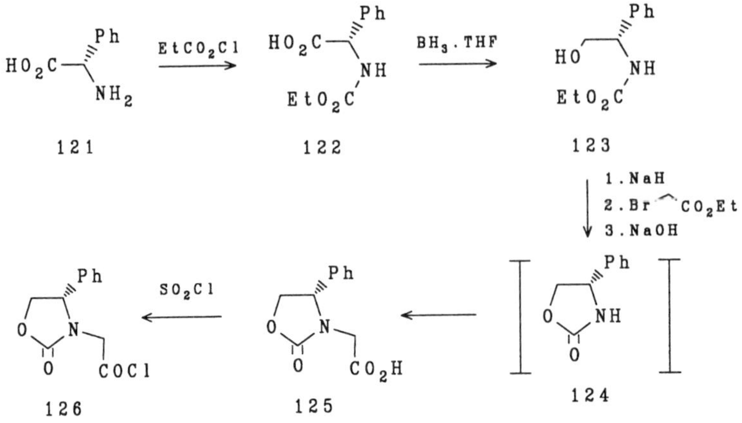

127 + 128 → 129 (H_2) → 130

Li/NH_3, ROH → 131

PhO‿COCl → 132

O_3 → 133

CO_2Mg / CO_2PNB, CDI → 134

DSO_2N_3 → 135

$Rh(RCO_2)_4$ → 136

$(PhO)_3PCl_2$ → 137 → 138

Bz-$CH_2C_6H_5$

PNB-$CH_2pC_6H_4NO_2$

and then alkylated with ethyl bromoacetate; saponification of the side chain affords the chiral acetic acid **125**. The carboxyl group is then activated by conversion to its acid chloride **126**.

The beta-lactam moiety is then formed by the 2 + 2 cycloaddition of a ketene with an imine. Thus, reaction of the acid chloride **126** with the benzylimine (**128**) from 3-furylacrolein in the presence of triethylamine goes directly to the azetidone **129**; in all probability via the ketene **127**. The reaction is remarkably stereoselective, giving the desired diastereomer in a ratio of 92:8. The double bond is reduced by catalytic hydrogenation to give **130**. Birch reduction using lithium in liquid ammonia with *tert*-butanol as a proton source cleaves both benzylic amine bonds; the benzyl protecting is thus lost directly while the oxazolidinone first opens to a carbamate that is cleaved under reaction conditions. This reaction affords the bare azetidone **131**. The primary amine is protected on an interim basis as its phenoxyacetamide (**132**) by reaction in situ with phenoxyacetyl chloride. Furyl groups are well-precedented latent carboxylic acids; ozonization of that moiety in **132** thus gives the corresponding acid **133** on oxidative workup of the ozonide.

The strategy for building the fused six-membered ring echoes the one used for the synthesis of **imipenem** (**62**), although the reagents differ significantly. Extension of the side chain that will form the new ring starts by activation of the terminal acid as its imidazolide by reaction with carbonyl diimidazole. This compound is then used to acylate the magnesium salt for the mono-*p*-nitrobenzyl ester of malonic acid. The resulting β-tricarbonyl compound decarboxylates on workup to afford the ketoester **134**. This compound is then diazotized with *p*-dodecylphenylsulfonyl azide (DSO_2N_3) to afford the intermediate **135**. Treatment of the diazo compound with rhodium tetraoctanoate leads to formation of the carbene by loss of nitrogen. This reactive species inserts into the amide N–H bond to form a fused piperidone ring; the product, **136**, exists largely as its enol tautomer, as in the case of the analogous cephem **100**. The phosphorus pentachloride commonly used for the next step is replaced in this case by dichlorotriphenyl phosphite; reaction with this reagent converts the enol to its chloride and serves to remove the amide at the 7-position via its imino chloride. Thus, the primary amine **137** is obtained. The D-phenylglycine side chain is then incorporated by the usual sequence, using in this case the amino acid as its ethyl acetoacetate enamine. The ester is then removed by treatment with zinc; exchange of the enamine with semicarbazide then removes that protecting group to afford **loracarbef** (**138**).[33]

3. MONOBACTAMS

The soil screening programs, prompted by the discovery of penicillin and streptomycin, led to the finding of many new classes of antibiotics beyond the beta-lactams discussed in this book. The great majority of these agents are elaborated as chemical defenses by molds and actinomycetes. The search for new antibiotics turned to other sources as the soil screening programs started yielding diminishing returns. This result led to the discovery of a promising antibacterial agent that was itself produced by the bacterium *Chromobacterium violaceum*. Structural determination revealed that this

139 → (1. $SO_2Cl/MeOH$; 2. NH_3) → 140 → ($BzOCOCl$) → 141 → ($MsCl$) → 142

142 → ($PySO_3$) → 143 → ($KHCO_3$; Bu_4NHSO_4) → 144 → (H_2) → 145

145 + 146 → (1. DCC; 2. TFA) → 147

148

agent surprisingly consisted of the unfused azetidone **148**. The natural product, as has often been the case in the beta-lactam series, was again not suitable as a drug.

One of the first compound to be introduced to the clinic, **aztreonam** (**147**), has been produced by total synthesis. Construction of the chiral azetidone starts with esterification of L-threonine via its acid chloride; treatment with ammonia leads to the corresponding amide **140**. The primary amino group in **140** is then protected as its carbobenzyloxy derivative **141**. Reaction of **141** with methanesulfonyl chloride affords the mesylate **142**. Treatment of **142** with the pyridine sulfur trioxide complex leads to formation of the *N*-sulfonated amide **143**. Potassium bicarbonate is sufficiently basic to ionize the very acidic proton on the amide; the resulting anion then displaces the adjacent mesylate to form the desired azetidone; the product is isolated as its tetrabutylammonium salt **144**. Catalytic hydrogenation over palladium removes the carbobenzyloxy protecting group to afford the free primary amine **145**.

The side chain in **145** mirrors those used in some of the more complex cephalosporins. Thus, reaction of the primary amino group in **145** with the half-ester **146** in the presence of DCC affords the corresponding amide. The protecting group on the side chain is then removed by treatment with trifluoroacetic acid to afford **aztreonam** (**147**).[34]

The configuration of the carbon bearing a carbamate in the monobactam **carumonam** (**157**) is interestingly the reverse of the one in the natural product or in **aztreonam**. The azetidone ring in this case is formed early in the synthesis by use of a 2 + 2 cycloaddition reaction. One component in this condensation consists of the imine **150** from a glyoxal ester with methyl valinate; this moiety serves as a chiral auxiliary. Reaction of **150** with the hypothetical ketene **149** from carbobenzyloxy glycine and *iso*-butyl chloroformate, leads to the azetidone **151** largely as a single chiral diastereomer. Treatment of this product with sodium carbonate then removes both ester groupings to afford **152**. The free hydroxyl is then converted to the carbamate **153**. The valine group on nitrogen is then converted to its iminium salt by anodic oxidation; hydrolysis with potassium carbonate cleaves that function to lead to the secondary amide **154**. This intermediate is then converted to its sulfonic acid derivative with sulfur trioxide and is hydrogenolyzed to give the primary amine **155**.

155 + 156 $\xrightarrow[\text{2. TFA}]{\text{1. DCC}}$ 157

The amide side chain for this compound also derives from cephalosporin chemistry. In much the same manner as shown above, amide formation between **156** and the primary amine **155**, followed by cleavage of the protecting group, affords **carumonam** (**157**).[35]

REFERENCES

1. Weissenburger, H.W.O.; Van der Hoven, M.G. *Recl. Trav. Chim. Pays-Bas* **1970**, *89*, 1081.
2. Doyle, F.P.; Hardy, K.; Nayler, J.H.C.; Soulal, M.J.; Stove, E.R.; Waddington, H.R.J. *J. Chem. Soc.* **1962**, 1453.
3. Doyle, F.P.; Hanson, J.C.; Long, A.A.W.; Nayler, J.H.C.; Stove, E.R. *J. Chem. Soc.* **1963**, 5838.
4. Ekstrom, B.A.; Sjoberg, B.O.H. Swed. Patent **1976**, 386900; *Chem. Abstr.* **1977**, *87*, 53266.
5. Godtfredsen, W.O. In *Chronicles of Drug Discovery*, Bindra, J.; Lednicer, D., Eds.; Wiley: New York; 1983, Vol. 2, p. 133.
6. Koenig, H.B.; Metzer, K.G.; Offe, H.A.; Schroek, W. *Eur. J. Med. Chem.* **1982**, *17*, 59.
7. Hamanaka, E.S.; Stam, J.G. So. African Patent **1973**, 7400509; *Chem. Abstr.* **1975**, *83*, 58808.
8. Cole, M.; Fullbrook, P. Br. Patent **1969**, 1160211; *Chem. Abstr.* **1970**, *72*, 41674.
9. Acred, P.; Brown, D.M.; Knudsen, E.T.; Robinson, G.M.; Sutherland, R. *Nature (London)* **1967**, *215*, 25.
10. Brain, E.G.; Nayler, J.H. U. S. Patent **1966**, 3282926 (see also *Chem. Abstr.* **1965**, *63*, 13269.)
11. Lund, F.: Tybring, L. *Nature (London)* **1972**, *236*, 135.
12. Bentley, P.H.; Clayton, J.P.; Boles, M.O.; Girven, R.J. *J. Chem. Soc. Perkin Trans. 1* **1979**, 2455.
13. Kapur, J.C.; Fasel, H.P. *Tetrahedron Lett.* **1985**, *26*, 3875.
14. Micetch, R.G.; Maita, S.N.; Spevak, P.; Hall, T.W.; Yamabe, S.; Ishida, N.; Tanaka, M.; Yamakazi, T.; Nakai, A.; Ogawa, K. *J. Med. Chem.* **1987**, *30*, 1469.
15. Mellilo, D.G.; Shinkai, I.; Liu, T.; Ryan, K.; Sletzinger, M. *Tetrahedron Lett.* **1980**, *21*, 2783.
16. Shinkai, I.; Reamer, R.A.; Hartner, F.W.; Liu, T.; Sletzinger, M. *Tetrahedron Lett.* **1982**, *23*, 4903.

17. Morin, R.B.; Jackson, B.G.; Flynn, E.E.; Roeske, R.W. *J. Am. Chem. Soc.* **1962**, *84*, 3400. Morin, R.B.; Jackson, B.G.; Flynn, E.E.; Roeske, R.W.; Andrews, S.L. *J. Am. Chem. Soc.* **1969**, *91*, 1396.
18. Chauvette, R.R.; Flynn, E.H.; Jackson, B.G.; Lavagino, E.R.; Morin, R.B.; Mueller, R.A.; Pioch, R.P.; Roeske, R.W.; Ryan, C.W.; Spencer, C.L.; VanHeyningen, E.M. *J. Am. Chem. Soc.* **1962**, *84*, 3401.
19. Guarini, J.R. U. S. Patent **1976**, 3903278.
20. Ochiai, M.; Morimoto, A.; Miyawaki, T.; Matsushita, Y.; Okada, T.; Natsugari, H.; Kida, M. *J. Antibiotics* **1981**, *39*, 171.
21. Hashimoto, M.; Shiozawa, S.; Aoki, M.; Watanabe, T. *Jpn. Kokai* **1990**, 2069483. *Chem. Abstr.* **1990**, *113*, 78017.
22. Spencer, J.L.; Flynn, E.H.; Roeske, R.W.; Siu, F.Y.; Chauvette, R.R. *J. Med. Chem.* **1966**, *9*, 746.
23. Ryan, C.W.; Simon, R.L.; VanHeyningen, E.M. *J. Med. Chem.* **1969**, *12*, 310.
24. For a succinct discussion see Gorman, M.; Chauvette, R.R.; Kukolja, S. In *Chronicles of Drug Discovery*, Bindra, J.; Lednicer, D., Eds.; Wiley: New York, 1983, Vol. 1, p.49.
25. Morin, R.B.; Jackson, B.G.; Mueller, R.A.; Lavagnino, E.R.; Scanlon, W.B.; Andrews, S.L. *J. Am. Chem. Soc.* **1969**, *91*, 1401.
26. Ishimaru, T.; Kodama, Y. German Offen. **1973**, 2263861. *Chem. Abstr.* **1973**, *79*, 78826.
27. Dolfini, J.E.; Applegate, H.E.; Bach, G.; Basch, J.; Bernstein, J.; Schwartz, J.; Wiesenborn, F.L. *J. Med. Chem.* **1971**, *14*, 117.
28. Boucourt, R.; Heymes, R.; Lutz, A.; Penasse, L.; Perronet, J. *Tetrahedron* **1978**, *34*, 2233.
29. Woodward, R.B.; Bickel, H. U. S. Patent **1979**, 4147864. *Chem. Abstr.* **1979**, *91*, 74633.
30. Takaya, T.; Takasugi, H.; Tsuji, K.; Chiba, T. German Offen. **1978**, 2810922. *Chem. Abstr.* **1979**, *90*, 204116.
31. Christensen, B.G. In *Chronicles of Drug Discovery*, Bindra, J.; Lednicer, D., Eds.; Wiley: New York; 1983, Vol. 1, p. 223.
32. Chris, L. Eur. Patent Appl. **1990**, 359540. *Chem. Abstr.* **1990**, *113*, 131868.
33. Bodurow, C.C,; Boyer, B.D; Brennan, J.; Burnell, C.A.; Burks, J.E.; Carr, M.A.; Doecke, C.W.; Ekrich, T.M.; Fischer, T.M.; Gardner, J.P.; Graves, B.J; Hines, P.; Hoying, R.C.; Jackson, B.G.; Kinnick, M.D.; Kochert, C.D.; Lewis, J.S.; Luke, W.D.; Moore, L.L.; Morin, J.M.; Nist, R.L.; People, W.T.; Prather, D.E.; Sparks, D.L.; Vladuchick, W.C. *Tetrahedron Lett.* **1989**, *30*, 2321.
34. Cimarusti, C.M. In *Chronicles of Drug Discovery*, Lednicer, D., Ed.; ACS Books: Washington, DC, 1993, Vol. 3; p.239.
35. Hashiguchi, S.; Maeda, Y.; Kishimoto, S.; Ochiai, M. *Heterocycles* **1986**, *24*, 2273.

CHAPTER 15

HETEROCYCLES FUSED TO OTHER HETEROCYCLIC RINGS

A sizeable number of endogenous compounds that play a key role in regulation of various life processes consists of fused heterocycles. Two familiar examples include the purines that not only form part of DNA and RNA but also provide the backbone for nicotinamide adenine dinucleotide phosphate (NADP), which is involved in metabolic electron transport, and the pteridines involved in the folic acid cycles. This circumstance, combined with the enormous structural diversity available among fused heterocycles, has led to their being intensively examined as a source for drugs. A sizcable number of these agents has shown sufficient biological activity to lead to their investigation in clinical trials.

1. TWO FUSED FIVE-MEMBERED RINGS

A pyrrolopyrroline carboxylic acid NSAID once again illustrates the structural tolerance in this therapeutic area. The synthesis of this agent starts with Friedel–Crafts acylation of 2-thiomethylpyrrole (**1**) with anisoyl chloride to afford the ketone **2**. The sulfide is then converted to a better leaving group by oxidation to a sulfone (**3**) by treatment with peracid. The remaining carbon atoms required for the fused ring are added by means of a spiro-cyclopropyl substituted Meldrum's acid. Reaction of the anion from the pyrrole leads to ring opening of the cyclopropane ring to give **4** and, in effect, addition of a four carbon chain on nitrogen.

Methanolysis of the product leads to exchange of the acetonide with the alcohol and thus formation of the methyl ester **5**. The anion obtained by treatment of the malonate with base then undergoes internal displacement of the methylsulfone with consequent formation of the fused pyrroline ring. The ester groups in the product **6** are then saponified; the resulting malonic acid decarboxylates on warming to afford **anilorac** (**7**).[1]

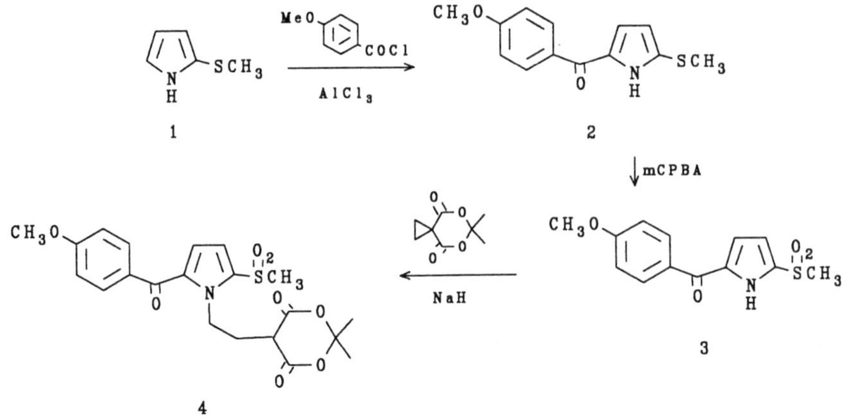

A perhydrofuranopyrrolidine is described as an analgesic that acts by some undefined non-opioid mechanism. Catalytic hydrogenation of the furan ring in the diol **8** leads to the tetrahydro derivative **9**; the method of reduction as well as the subsequent formation of a fused bicyclic product suggest that the substituents are in the cis configuration. The hydroxyl groups are then activated toward displacement by conversion to their *p*-toluenesulfonate esters to give **10**. Reaction of **10** with benzylamine leads to the cyclic bis-alkylation product and thus formation of the pyrrolidine ring. The benzyl protecting group in product **11** is then removed by hydrogenolysis over palladium to give the free secondary amine **12**. Acylation of this group with benzoyl chloride then affords **octazamide** (**13**).[2]

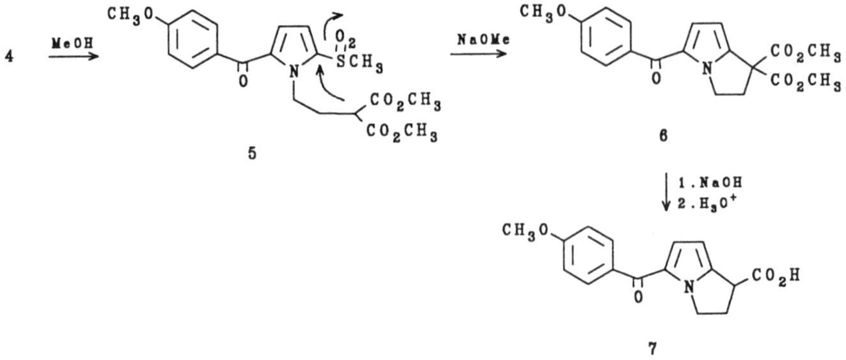

The imidazothiazole **tetramisole** (**18**) was originally developed as an antihelminstic agent. The levorotatory isomer **levamisole** (**21**) was found to be twice as potent as the racemate although both have been used as veterinary antinematodal drugs. Interestingly, reaction of 2-aminothiazoline (**14**) with phenacyl bromide, occurs on the ring nitrogen to afford **15**. The exocyclic imine is then acylated by means of acetic anhydride to afford **16**. Reduction of the ketone with sodium borohydride leads to the benzylic alcohol **17**. Treatment of **17** with thionyl chloride leads to ring closure to **18** with simultaneous loss of the acetyl group. Thus, tetramisole (**18**)[3] is obtained.

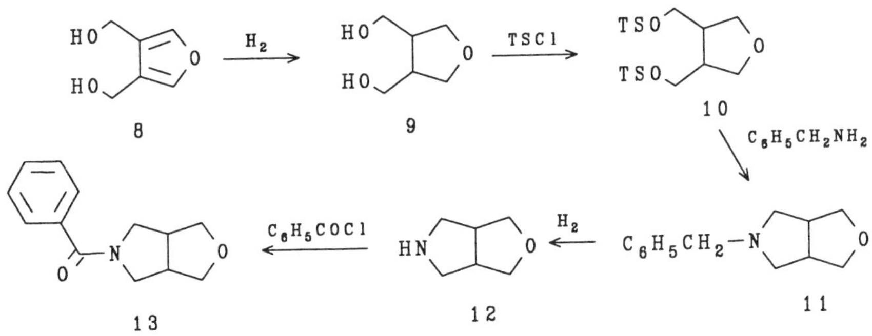

Subsequent investigation showed that these compounds also have pronounced activity as modulators of the immune system in humans; consequently, levamisole (**21**) now forms part of a multidrug protocol for treating colon cancer. This enantiomer can be prepared stereospecifically by starting with the resolved diamine **19** of known absolute configuration. Reaction with carbon disulfide leads to the mercaptoimidazoline **20**. The remaining ring is then closed by treatment of **20** with 1,2-dibromoethane. The absolute configuration of levamisole (**21**) follows from that of the starting diamine and the fact that none of the transformation involve the chiral center.[4]

A related scheme starts with the conversion of the exocyclic amino group in 2-methylaminoimidazoline to a good leaving group by conversion to its nitramine derivative **22**. Reaction of **22** with phenylethanolamine leads to displacement of the nitramine by the primary amine on the reagent and formation of the substitution product **23**. This compound is then cyclized with concentrated sulfuric acid to give an imidazoimidazole probably via the benzylic carbocation. Thus, **imafen** (**24**),[5] a compound described as an antidepressant, is obtained.

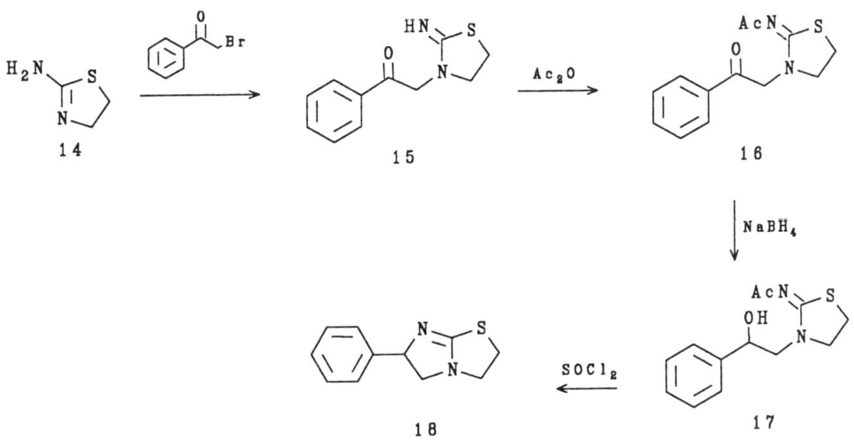

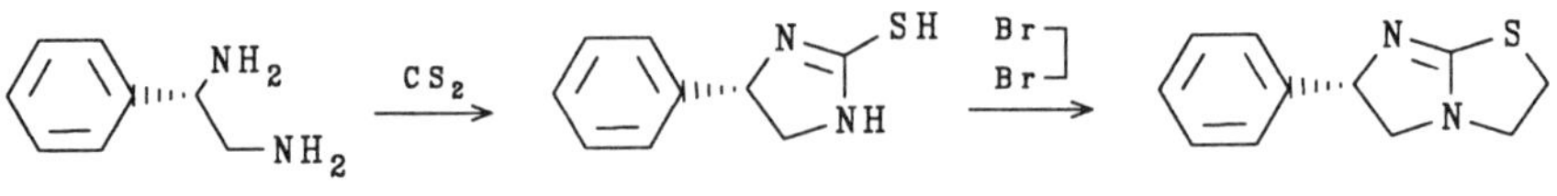

19 → (CS_2) → 20 → ($Br(CH_2)_2Br$) → 21

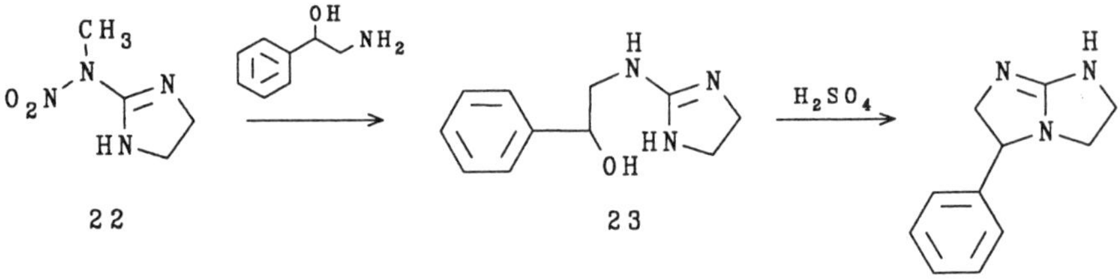

22 → 23 → (H_2SO_4) → 24

2. FIVE-MEMBERED HETEROCYCLES FUSED TO SIX-MEMBERED RINGS

A. Compounds with Two Heteroatoms

The majority of available antihypertensive drugs fall into well-recognized structural and thus pharmacological classes; the furanopyridine **cicletanine** (**28**) constitutes a new structural class that reduces blood pressure by decreasing peripheral resistance by a mechanism that is still under investigation. The synthesis of **28** starts with the addition of *p*-chlorophenylmagnesium bromide to the complex pyridine carboxaldehyde **25**. Treatment of **26** with strong acid proceeds initially to hydrolysis of the acetonide protecting group to afford the triol **27**. The benzhydryl carbinol then probably goes to the corresponding carbocation. Capture of the adjacent hydroxyl forms the dihydrofuran ring and thus cicletanine (**28**).[6]

The modification of the Polonovski reaction, which involves reaction of a pyridine *N*-oxide with phosphorus oxychloride, has been used extensively for the introduction of chlorine at the 2-position. A modification of this reaction allows the introduction of a nitrile group. Thus reaction of the *N*-oxide **30** from oxidation of the 2-arylpyridine **29** with potassium cyanide in the presence of dimethyl sulfate leads to the cyanopyridine **31**. The classic Polonovski reaction involves the intermediacy of an *O*-acetate, a role served by an *O*-phosphoryl species in the chlorination; the case at hand can be rationalized by invoking an *O*-sulfated intermediate. The nitrile is then reduced to the

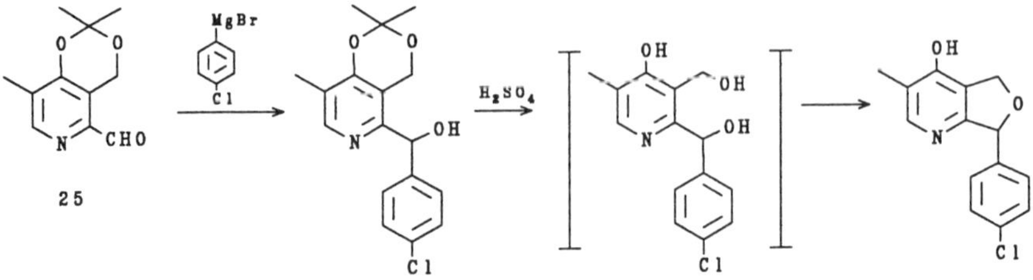

25 → 26 → (H_2SO_4) → [27] → 28

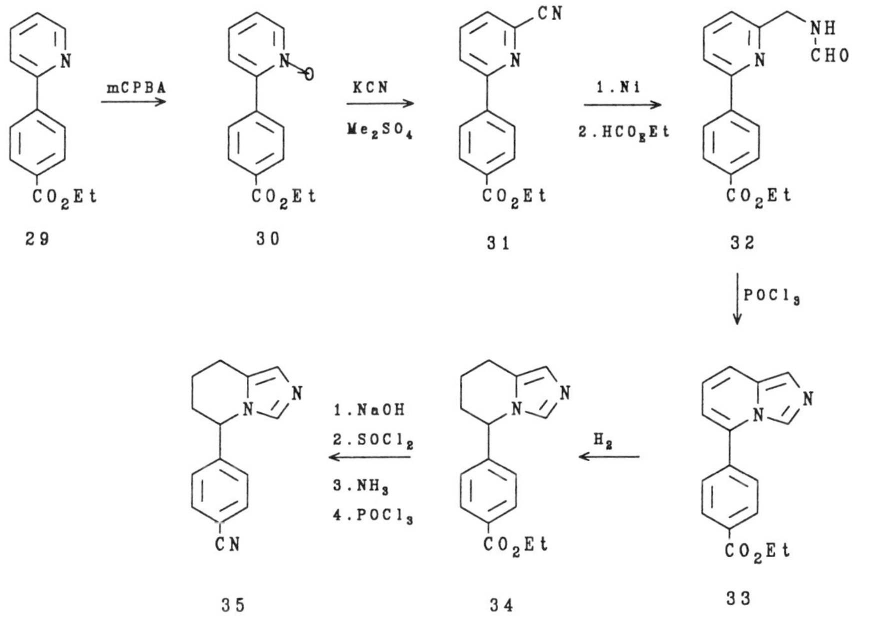

primary amine and is converted to the formamide **32** by exchange with methyl formate. Reaction of **32** with phosphorus oxychloride probably initially proceeds to the imino chloride; this compound cyclizes onto the pyridine ring to give the imidazopyridine **33**. The pyridine ring is then reduced by catalytic hydrogenation to give the intermediate **34**. The carbethoxy group on the pendant benzene ring is converted to its amide by sequential saponification, conversion to the acid chloride, and reaction with ammonia. Treatment of this compound with phosphorus oxychloride leads to dehydration to a nitrile.[7] This product, **fadrazole** (**35**), is a rather unusual steroid aromatase inhibitor (see glutethimide, Chapter 9).

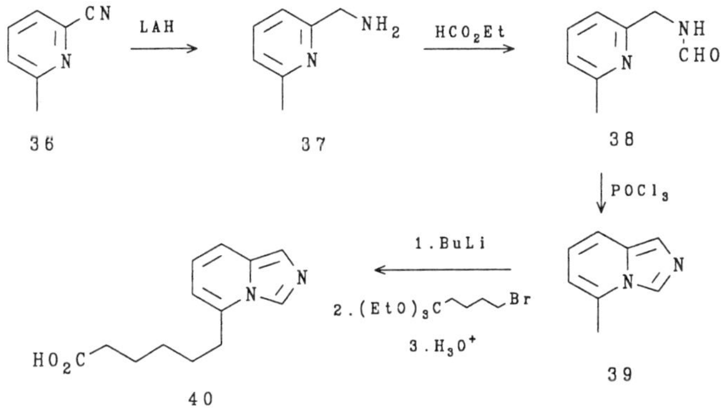

A fully unsaturated imidazopyridine acts as a platelet aggregation inhibitor; the presence of the fatty acid side chain suggests a possible interaction with arachidonic acid cascade products. Construction of the fused heterocyclic system parallels the one described above. The cyanopicoline **36** is thus reduced to the amine **37**, and this compound is converted to formamide **38**; cyclization with the ubiquitous phosphorus oxychloride gives the key intermediate **39**. Reaction of **39** with butyllithium abstracts a proton on the methyl group to afford the corresponding carbanion; treatment with the diethyl orthoester from 5-bromopentanoic acid gives the corresponding alkylated product. Hydrolysis of the orthoester group gives the carboxylic acid and thus **pirmagrel** (**40**).[8]

Condensation of a 2-aminopyridine with an α-haloketone provides an alternative method for building the imidazopyiridine. For example, reaction of 2-aminopicoline **41** with *p*-methylphenacyl bromide leads directly to the imidazopyridine **42**. The overall transformation can be rationalized by assuming initial alkylation on ring nitrogen; imine formation followed by bond reorganization then forms the imidazole. Treatment of the product **43** with formalin and dimethylamine leads to the Mannich base **44**. The dimethylamino group is then activated toward displacement by conversion to the quaternary salt by alkylation with methyl iodide. Reaction with potassium cyanide leads to the acetonitrile **45**. The nitrile is then hydrolyzed to the corresponding carboxylic acid and this is then taken on to the dimethylamide.[9] This product, **zolpidem** (**46**), is an anxiolytic agent that interestingly interacts with the same receptor as the structurally quite distinct benzodiazepines.

A very recent method for building the fused heterocyclic system for **ticlopidine** (**53**) starts by formation of the mesylate **48** from cyanopyridone (**47**) by treatment with methanesulfonyl chloride. The reaction of this compound with butylthioglycolate can be envisaged as first involving displacement of the mesylate by the thiol group to the ether **49**; internal aldol-type addition to the nitrile will lead to the observed product **50** after bond reorganization. The amino group is then converted to its diazonium salt with nitrous acid and this compound is reduced by reaction with hypophosphinic acid. Treatment of **51** with aqueous base then hydrolyzes both the ester and the acetyl group of nitrogen. The resulting acid is then decarboxylated by heating in the presence of copper powder to give the piperidinothiophene **52**.[10] Acylation of **52** with *o*-benzoyl chloride followed by reduction of the amide will afford ticlopidine (**53**).

The first published scheme for the preparation of this antithrombotic agent describes the alkylation of the fully unsaturated version **54** of the same heterocycle with

41 → [42] → 43; 43 (CH_2-O, Me_2NH) → 44; 44 (1. MeI, 2. KCN) → 45; 45 (1. H_2SO_4, 2. $SOCl_2$, 3. Me_2NH) → 46

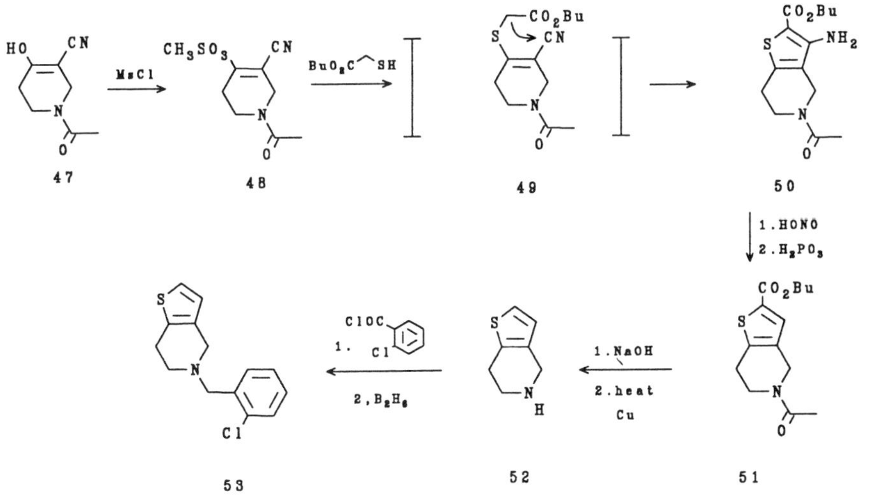

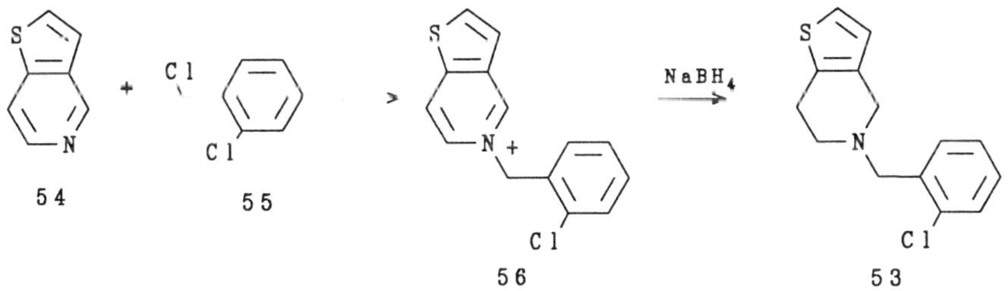

2α-dichlorotoluene (**55**) to give the quaternary salt **56**. Reaction of **56** with sodium borohydride selectively reduces the charged ring to give ticlopidine (**53**).[11]

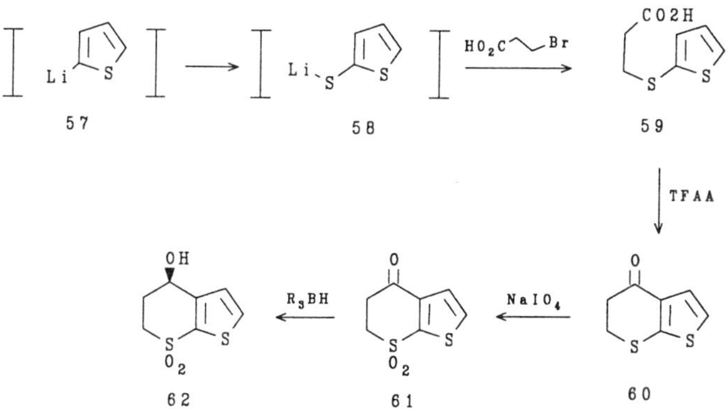

The sulfonamido group constitutes an important moiety in the classical carbonic anhydrase inhibitor diuretic agents such as **acetazolamide** (see Chapter 8). The observed reduction in ocular pressure by such agents in glaucoma patients led to the development of more specific agents based on the thiophenothiopyran nucleus. The enantioselective synthesis of this agent starts with reaction of the product **57** from lithiation of thiophene with sulfur to give intermediate **58**. This compound is reacted in situ with 3-bromopropionic acid to afford the alkylation product **59**. Treatment with trifluoroacetic anhydride (TFAA) leads to Friedel–Crafts type cyclization to the bicyclic compound **60**. Sodium periodate acid selectively oxidizes sulfur in the dihydrothiopyranone ring to give the sulfone **61**. Reduction of the carbonyl group with a chiral oxazaborolidine hydride affords the corresponding alcohol **62** in high enantiomeric excess though in the opposite configuration from that desired in the final product.

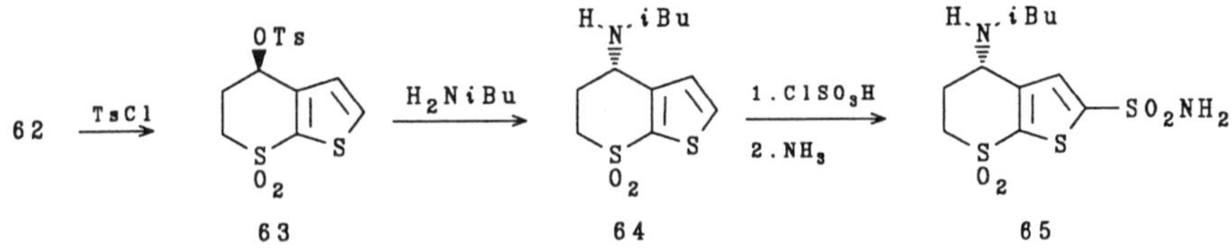

The alcohol is then converted to the tosylate **63**; displacement with isobutylamine gives the amine derivative **64** of the opposite configuration. The all important sulfonamide group is then introduced by first reacting the compound with chlorosulfonic acid. Treatment of the thus obtained sulfonyl chloride with ammonia gives the ocular carbonic anhydrase inhibitor **sezolamide** (**65**).[12]

B. Compounds with Three Heteroatoms

The reaction used for the construction of the starting material for a pyrrolopyridine involves the standard pyrimidine chemistry discussed in Chapter 9. Thus condensation of the substituted cyanoacetate **66** with acetamidine leads to the aminopyrimidinol **67**. Treatment of **67** with acid leads to hydrolysis of the acetal and formation of the transient free aldehyde **68**. This compound then undergoes internal imine formation to afford the fused heterocycle **69**. The enol is then converted to the chloride (**70**) in the usual way by reaction with phosphorus oxychloride. Displacement of chlorine with benzylamine leads to the muscle relaxant **rolodine** (**71**).[13]

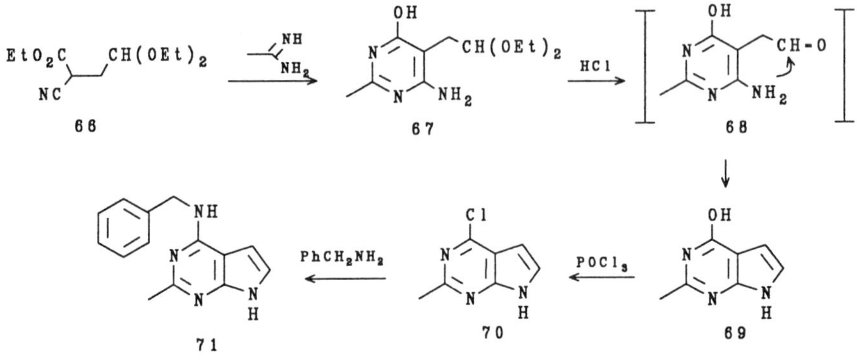

The utility of modified pyrimidines for the treatment of viral diseases and cancer, (see Chapter 9) applies as well to purine heterocycles. **Dezaguanine** (**79**), in which one of the pyrimidine nitrogen atoms is replaced by carbon has, for example, been tested as a cancer chemotherapeutic agent. The aminoketone **72** comes from the acetone dicarboxylic ester by successive nitrosation and reduction of the nitroso derivative. Reaction of **72** with potassium isothiocyanate can be envisioned as leading initially to the thiourea **73**. This compound cyclizes spontaneously to the imidazothione **74** under reaction conditions. Treatment of **74** with liquid ammonia leads to preferential exchange of the unconjugated ester group to give the amide **75**; the greater reactivity can be attributed to the higher electron density at this carbonyl group. This compound is then desulfurized by reaction with Raney nickel to give the imidazole **76**.

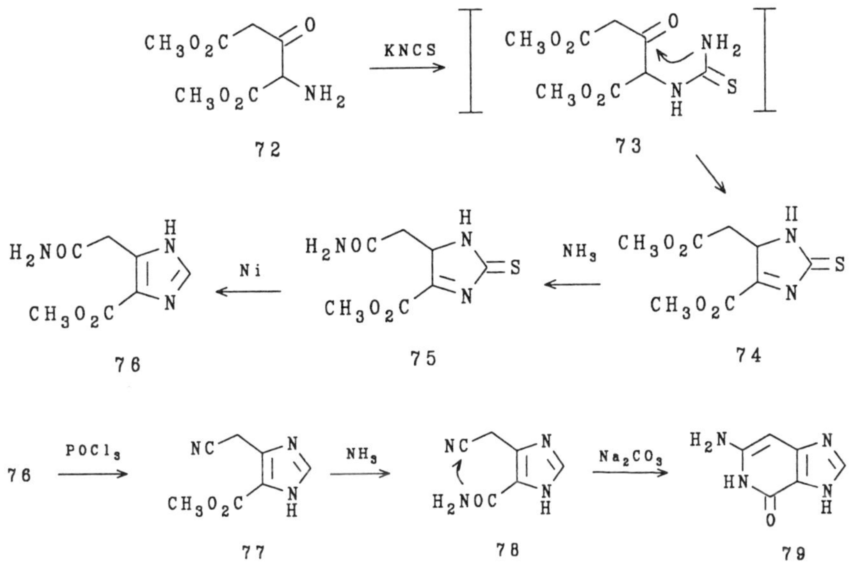

Treatment of **76** with phosphorus oxychloride then serves to dehydrate the amide group to a nitrile to give **77**. The remaining ester group is then converted to the amide **78** by reaction with ammonia under more strenuous conditions. This function cyclizes onto the nitrile when treated with sodium carbonate to afford dezaguanine (**79**).[14]

The ring system that provides the core for the cardiotonic agent **isomazole** (**86**) may be viewed as a pyridine analogue of a benzimidazole; the chemistry used to form this ring system is in fact quite analogous to the one discussed in Chapter 10. The preparation of the requisite benzoic acid first involves conversion of a phenol to a thiophenol using the thiocarbamate interchange reaction. Thus, the phenol **80** is acylated with dimethylaminothioformyl chloride to give the carbamate **81**. This product undergoes O to S aryl migration on heating to give the isomeric carbamate **82**. Treatment of **82** with aqueous base then removes the carbamate and at the same time serves to saponify the ester. Reaction of **82** with methyl iodide in the presence of base alkylates both the thiophenol and the carboxyl group; a second saponification restores the free acid to give **83**. This product is then condensed with the diaminopyridine analogue **84** of *o*-phenylenediamine to yield the substituted imidazopyridine **85**.

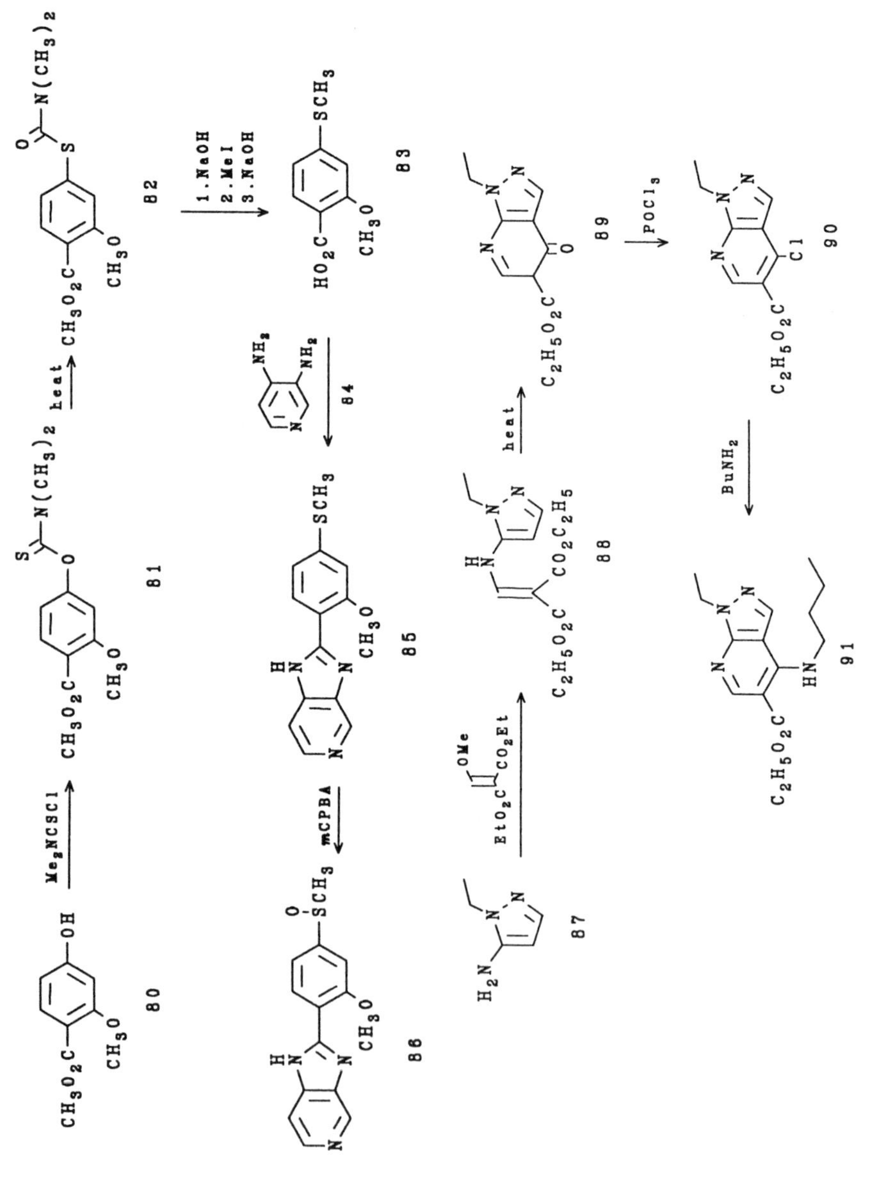

80 $\xrightarrow{Me_2NCSCl}$ 81 $\xrightarrow{heat}$ 82 $\xrightarrow{1. NaOH;\ 2. MeI;\ 3. NaOH}$ 83

83 $\xrightarrow{84}$ 85 $\xrightarrow{mCPBA}$ 86

87 $\xrightarrow{EtO_2C(CO_2Et)C{=}CHOMe}$ 88 $\xrightarrow{heat}$ 89 $\xrightarrow{POCl_3}$ 90 $\xrightarrow{BuNH_2}$ 91

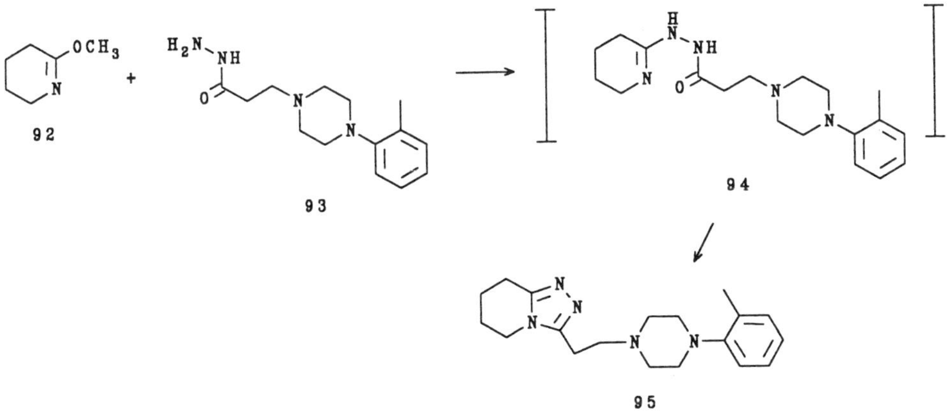

92 93 94 95

Oxidation of the thioether in this compound with peracid under controlled conditions gives the sulfoxide and thus isomazole (**86**).[15]

Condensation of aminopyrrazole **87** with ethyl methoxymethylene malonate proceeds to give the corresponding imine **88**, probably by an addition–elimination sequence. One of the carbethoxy groups cyclizes onto the heterocyclic ring when this intermediate is heated in diphenyl ether to give the fused pyridone **89**, which is shown as its keto tautomer; this reaction is quite analogous to the quinolone synthesis discussed in Chapter 11. Reaction of this intermediate with phosphorus oxychloride leads to the enol chloride **90**. This reactive group is then displaced by butylamine to give **cartazolate** (**91**),[16] a compound described as an antidepressant.

A seeming complex triazolopiperidine is prepared by a surprisingly simple two-component condensation. Reaction of the hydrazide **93** with the iminoether from 2-piperidone (**92**) can be rationalized by assuming initial formation of an intermediate such as **94** by an addition–elimination sequence. The second step involves attack on the carbonyl group by the now quite basic ring amidine nitrogen. The observed product **dapiprazole** (**95**),[17] displays α-adrenergic blocking as well as antipsychotic activity.

Benzene and thiophene rings can of course often be interchanged in biologically active agents. The very broad structural latitude consistent with NSAID activity is by

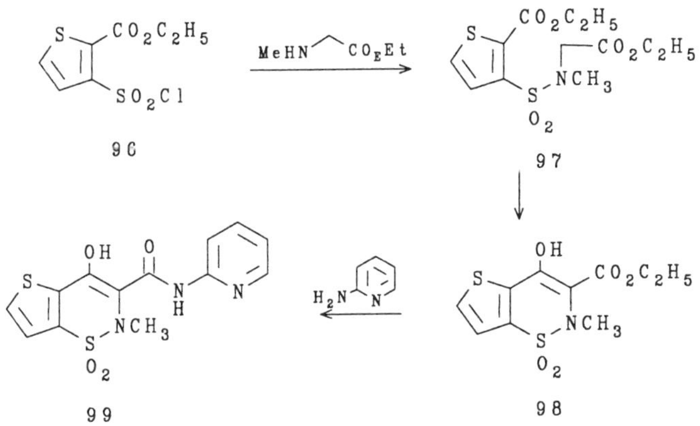

96 97 98 99

now also a familiar theme. Preparation of the fused thiophene counterpart of the NSAID piroxicam starts with the reaction of the thiophene **96**, itself the product of a multistep sequence with ethyl *N*-methylglycinate to give the sulfonamide **97**. Treatment of **97** with base leads to intramolecular Claisen condensation and thus formation of the β-ketoester **98**. Amide–ester interchange with 2-aminopyridine completes the synthesis of **tenoxicam** (**99**).[18]

C. Compounds with Four Heteroatoms

1. Purines

The recognition of the biological activity of the simple purines, which are exemplified by caffeine, predates the formal study of pharmacology by several centuries. Infusions such as coffee and tea that contain these bases, which are also called methylxanthines, have of course long been used for their CNS stimulant properties. There is also some anecdotal evidence for the bronchodilating activity of caffeine. The use of **theophylline**, **105**, which lacks the imidazole *N*-methyl group of caffeine, in the treatment of asthma is of more recent origin and was derived initially from empirical observations. It is now recognized that **105** is an inhibitor of the hydrolysis of cyclic esters of nucleotides, mainly cyclic adenosine monophosphate (cAMP), and thus prolongs the relaxant action of this mediator. The first sequence in one of the syntheses of this compound consists of typical pyrimidine chemistry. Thus amide–ester interchange between *N,N′*-dimethyl urea (**100**) and ethyl carboxamidoacetate leads to the acylurea **101**. This compound cyclizes to the aminopyrimidone **102** on treatment with base. The second nitrogen substituent on the ring is introduced by treating **102** with nitrous acid to afford the nitroso derivative **103**. This compound is readily reduced to the diamine **104** under any of several conditions; for example, reaction with zinc in mineral acid. The fused imidazole ring is then closed by condensing the diamine with formic acid, or some formic acid equivalent such as ethyl orthoformate, to give theophylline (**105**).[19]

Although very effective, **theophylline** (**105**) has a number of significant drawbacks as a drug, not the least of which is its very narrow therapeutic window: minimally

HN-CH3
O=
HN-CH3
100

EtO2C
H2NOC

CH3-N
O
NH O NH2
CH3
101

NaOH

CH3-N
O
O
N NH2
CH3
102

HONO

CH3-N
O
O
N NH2
CH3
NO
103

Zn
HCl

CH3-N
O
O
N NH2
CH3
NH2
104

HCO2H

CH3-N
O
O
N
CH3
H
N
N
105

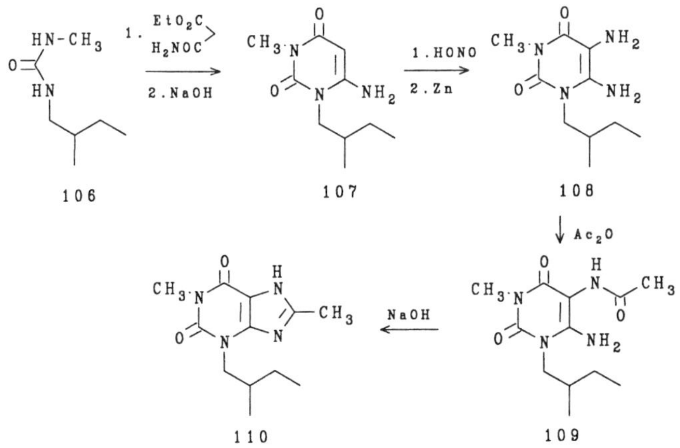

effective blood levels are in the range of 10 μg/mL, while toxic signs begin at twice this concentration. One approach to improving the biological properties of drugs involves increasing lipophilicity and thus presumably bioavailability. The synthesis of a more lipophilic analogue of theophylline substitutes the isopentyl substituted urea **106** in the previous scheme to give the pyrimidinone **107**; initial ester interchange at the *N*-methyl nitrogen is probably guided by the less hindered milieu about this center. Sequential nitrosation and reduction of the resulting nitroso compound gives the diamine **108**. The remaining carbon atom is introduced in stepwise fashion in this case. Reaction with acetic anhydride leads to **109**. This compound cyclizes to **verofylline** (**110**) on reaction with base.[20]

Acylation of the theophylline diamine intermediate **104** with phenylacetyl chloride followed by cyclization of the thus obtained amide leads to the product **111**, which now includes a quite lipophilic benzyl group on the imidazole ring. The molecule is then provided with a basic side chain, probably to improve water solubility. The anion from **111** is first alkylated with bromochloroethane to afford the chloroethyl product

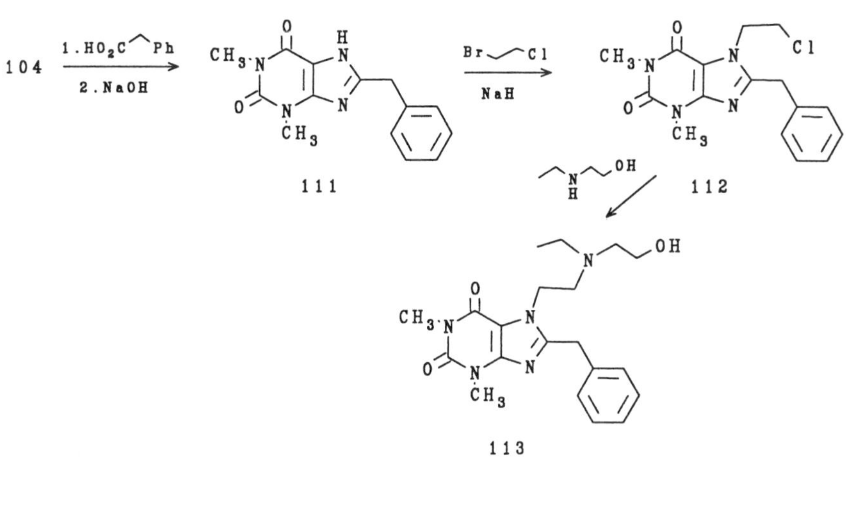

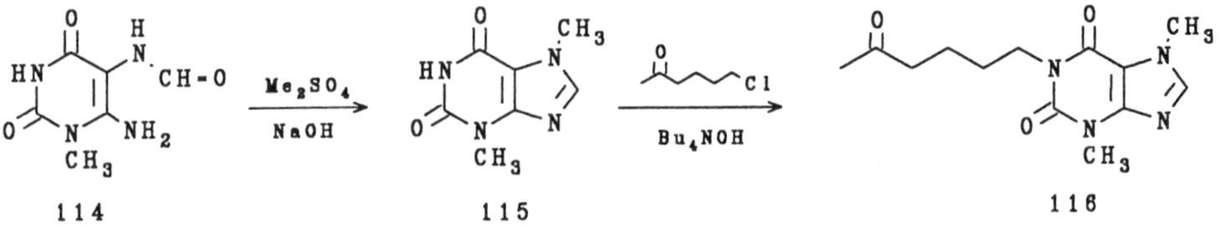

114 → (Me_2SO_4, NaOH) → 115 → (Bu$_4$NOH) → 116

112. Displacement of chlorine with N-ethylethanolamine affords the bronchodilator **bamifylline (113)**.[21]

Substituting a fatty side chain on the imide nitrogen leads to a compound used mainly as a vasodilator. One preparation for the readily available starting material, theobromine (**115**), involves reaction of the formamido aminopyridone **114** with dimethyl sulfate in the presence of base. The first step is thought to consist of methylation on the formamide nitrogen; this compound cyclizes to theobromine (**115**) under reaction conditions. The anion from the imide nitrogen is then alkylated with 6-chlorohexan-2-one to give **pentoxifylline (116)**.[22]

Because of their close structural relation to endogenous compounds involved in metabolism, purines have provided the nucleus for a sizeable number of antimetabolites used in the treatment of cancer. **Mercaptopurine (120)** was one of the first of this class of drugs to be used in the clinic. One synthesis for the starting purine, hypoxanthine (**119**), starts with the formylation of aminoimidazole (**117**) with a mixture of formic acid and acetic anhydride to give **118**. This compound cyclizes to hypoxanthine with the surprisingly mild base sodium bicarbonate. Reaction of **119** with phosphorus pentasulfide converts the carbonyl to its sulfur equivalent.[23] The product mercaptopurine (**120**) is shown as its enol tautomer. Treatment of hypoxanthine (**119**) with phosphorus oxychloride converts it to the corresponding chloro derivative **121**. The reactive chlorine atom is then displaced by a highly substituted mercaptoimidazole to give **azathioprine (122)**.[24] This somewhat more widely used drug is also employed as an immunosuppressant for organ implant procedures; compound **122** is in fact converted metabolically to mercaptopurine.

The route to two adenines substituted on the imidazole ring both rely on incorporating this substituent prior to closing the fused five-membered ring. The first several

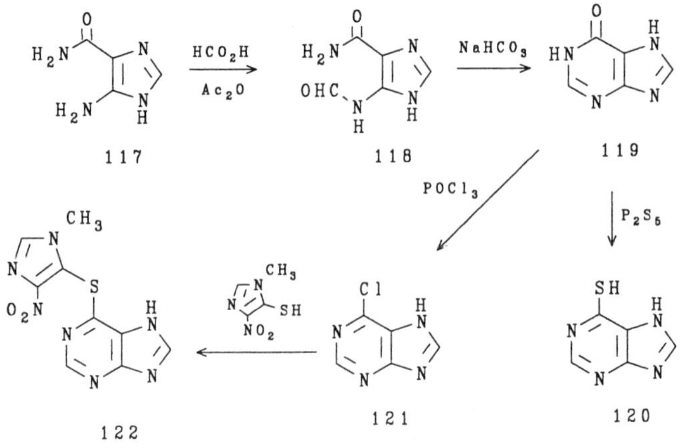

117 → (HCO_2H, Ac_2O) → 118 → ($NaHCO_3$) → 119 → (P_2S_5) → 120

119 → ($POCl_3$) → 121 → 122

steps in the synthesis of the natural product from mushrooms **eritadenine** (**128**) consists of constructing the requisite chiral side chain starting from the butyrolactone **123**, which is obtained by degradation of a sugar. Reaction with the sodium phthalimide leads to ring opening of the reactive lactone with displacement of the ester oxygen by nitrogen; the phthalimide group is then removed by treatment of the intermediate with hydrazine to afford **124**. The amine group is then allowed to displace chlorine in the 2-amino-4-chloro-3-nitropyrimidine to afford **125**. Catalytic hydrogenation of **125** leads to reduction of the nitro group and formation of **126**. Acylation of this triamine with formic acid proceeds at the more basic of these groups. The acetonide protecting group hydrolyzes under the reaction conditions to give the observed product, the amide **127**. This compound cyclizes to the desired imidazole on reaction with base to give a purine. Thus, eritadenine (**128**), a compound that showed hypolipidemic activity in various test systems,[25] is obtained.

Treatment of the symmetrical triaminopyrimidine **129** with sulfuryl chloride ties up the two adjacent amines in a thiadiazole ring, protecting these groups from attack in subsequent reactions. Reaction of the product **130** with ortho difluorinated benzylamine (**131**) results in replacement of the pyrimidine amino group by amine in the reagent, most likely by an addition–elimination sequence to afford **132**. This amino group is then converted to the formamide **133** with formic acid. Exposure of the product to Raney nickel leads to loss of sulfur and formation of the transient intermediate **134**. This compound cyclizes to a purine under reaction conditions to yield **arprinocid** (**135**),[26] a compound used as a poultry coccidiostat.

While the cancer chemotherapeutic agent **fludarabine** (**142**) includes a sugar moiety characteristic of endogenous nucleotides, the hydroxyl group at the 3-position in the furan ring has the unnatural arabinose configuration; the presence of fluorine on

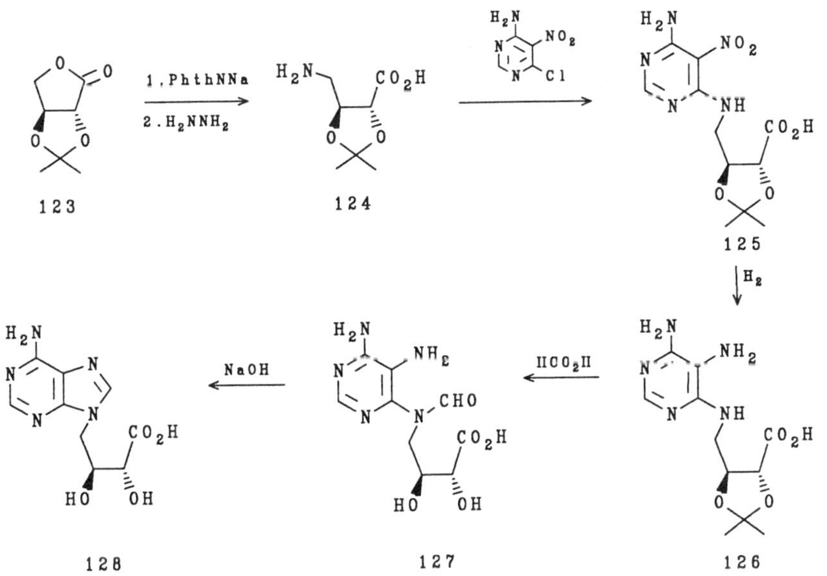

1. PhthNNa
2. H_2NNH_2
123
124
125
H_2
126
HCO_2H
127
NaOH
128

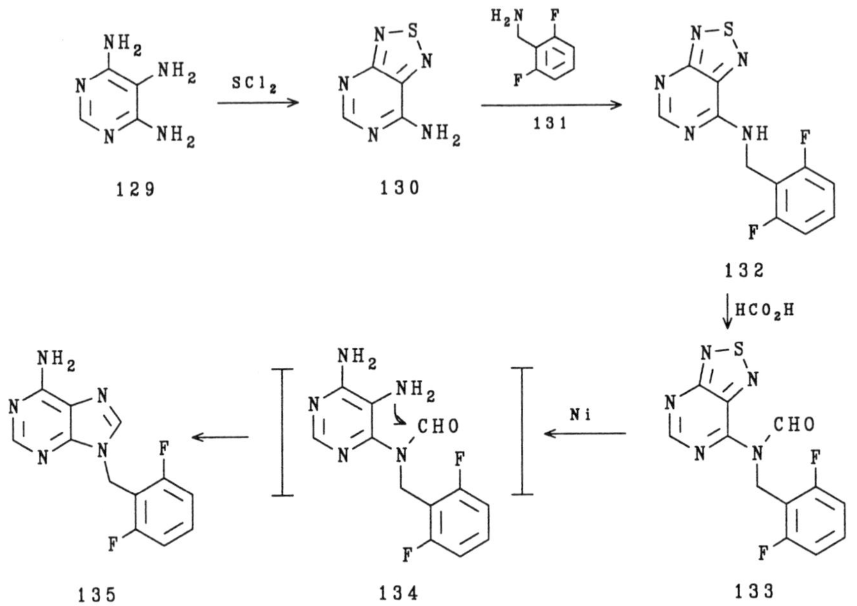

the purine nucleus marks a further change from the natural purines. The synthesis begins by reacting polyaminopyrimidine (**136**) with formamide to form diaminopurine (**137**), which happens to be itself a metabolic inhibitor. The symmetry of the starting

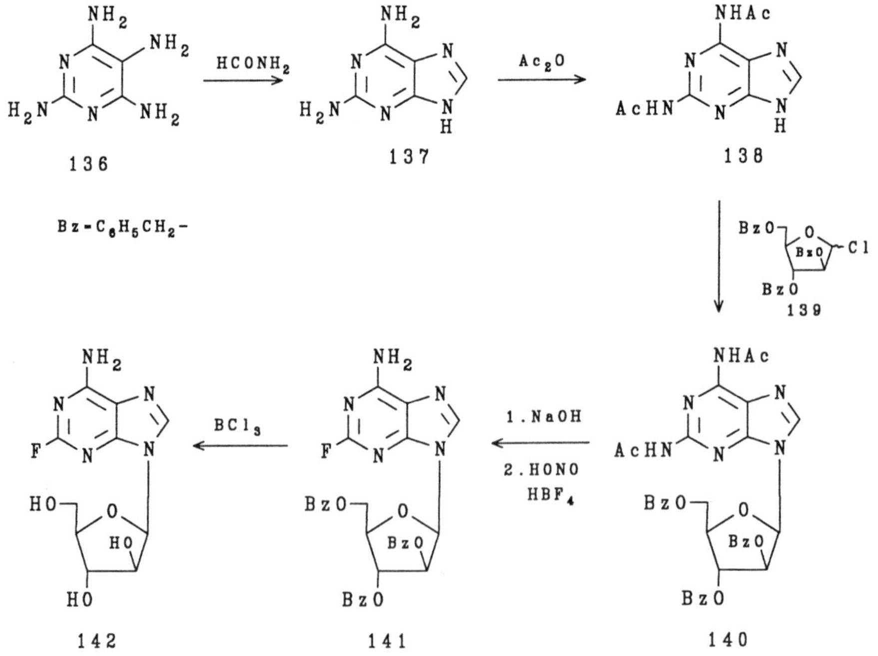

$Bz = C_6H_5CH_2-$

material permits only a single product. Treatment with acetic anhydride gives the corresponding diacetylated derivative **138**. Displacement of halogen in the fully benzylated arabinoside chlorosugar **139** results in glycosylation of the purine and formation of **140**. The amide groups are then removed by treatment of this product with base. Reaction of the product with nitrous acid in the presence of fluoroboric acid leads initially to formation of the diazonium salt at the 2-position; this compound is displaced by fluorine to give the fluorinated derivative **141**. The benzyl protecting groups are then cleaved by means of boron trichloride to afford fludarabine (**142**).[27]

The rationale for the use of modified nucleosides for the treatment of cancer and viral disease relies on the hope that cancer cells or virally infected cells will mistake the modified compounds for the natural substrates and incorporate them into a metabolic pathway. The altered structure of the false substrate, it is hoped, will then bring this process to a halt and result in cell death. This strategy has met particular success with antiviral agents that consist of guanines glycosylated with open-chain sugar surrogates. The antiherpes drug **acyclovir** (**148**),[28] was the first of its type to gain approval. In a current synthesis, the side chain synthon **146** is prepared by acylation of dioxolane (**145**) with acetyl chloride to give the ring opened product. Reaction of **145** with the diacetate **144** from guanine (**143**) in the presence of *p*-toluenesulfonic acid leads to the glycoside-like derivative **147** by exchange of the acetal with purine nitrogen. Treatment of **147** with methanolic ammonia at ambient temperature affords acyclovir (**148**).[29]

The addition of a hydroxymethyl group, which presumably increases the resemblance to a sugar, gives a compound that is mainly effective against cytomegalo viruses (CMV). The protected triol **151** for this compound is prepared by a stepwise reaction of epichlorohydrin (**149**) with two equivalents of the anion from benzyl alcohol; in a β-blocker-like sequence, the first equivalent leads to the glycidic ether **150**; this compound opens to **151** with a second equivalent of alkoxide. Reaction of **151** with formaldehyde and hydrogen chloride gives the chloromethyl ether **152**. Treatment of

143 —Ac_2O→ 144

145 —CH_3COCl→ AcO⌒O⌒OAc 146

144 + 146 —TSA→ 147 —NH_3/MeOH→ 148

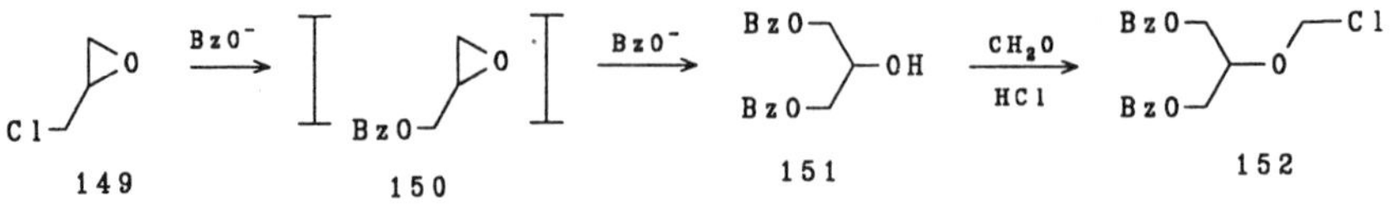

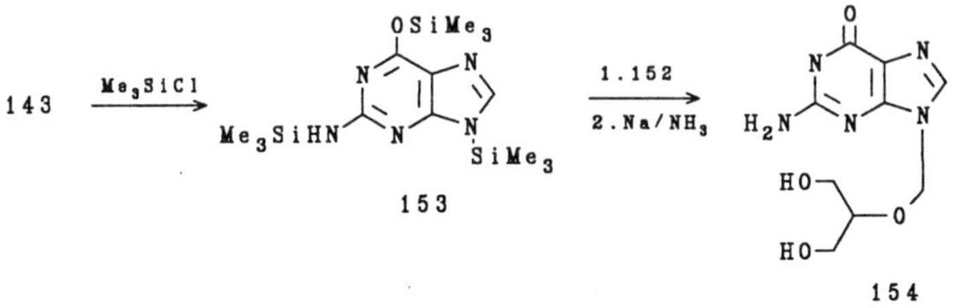

the tris-trimethylsilyl derivative **153** from guanine with the reactive intermediate **152** leads to the glycoside-like intermediate. The benzyl groups are then removed reductively by means of sodium in liquid ammonia to afford **ganciclovir** (**154**).[30]

2. Compounds with an N,N-Linkage

The pyrrazolopyrimidine **allopurinol** (**158**) was originally developed as a false substrate for the enzyme hypoxanthine oxidase, which was responsible for the fast metabolism of mercaptopurine. Use of this drug in the clinic revealed that this compound inhibits metabolism of other xanthines in addition to hypoxanthine (**119**). This generalized xanthine oxidase inhibiting activity made **158** a useful drug for treating gout, as the symptoms of this disease are caused by the accumulation of uric acid, the end product of xanthine metabolism. One synthesis of this purine-like compound begins with addition–elimination of hydrazine to ethoxymethylene malononitrile (**155**); this initially formed adduct cyclizes under reaction conditions by addition of the terminal hydrazine nitrogen to the nitrile, to form the pyrrazole **156**. The remaining cyano group is then hydrolyzed to the corresponding amide **157** with concentrated sulfuric acid. Reaction of **157** with formamide supplies the last carbon atom and leads to formation of allopurinol (**158**).[31]

Heterocycles related to 2-aminopyrimidine have been investigated as diuretic agents because these compounds tend to be less potassium depleting than the classical sulfonamides. The six-membered ring in one of these examples, **bemitradine** (**167**), is formed in a typical pyrimidine synthesis by condensation of the ketoester **159** with guanidine. The amino group in the product **160**, shown as its keto tautomer, is then protected as its formamide **161**. Reaction with phosphorus oxychloride then converts

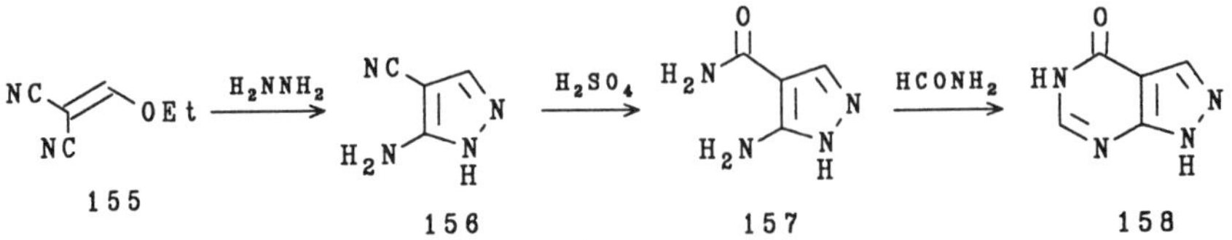

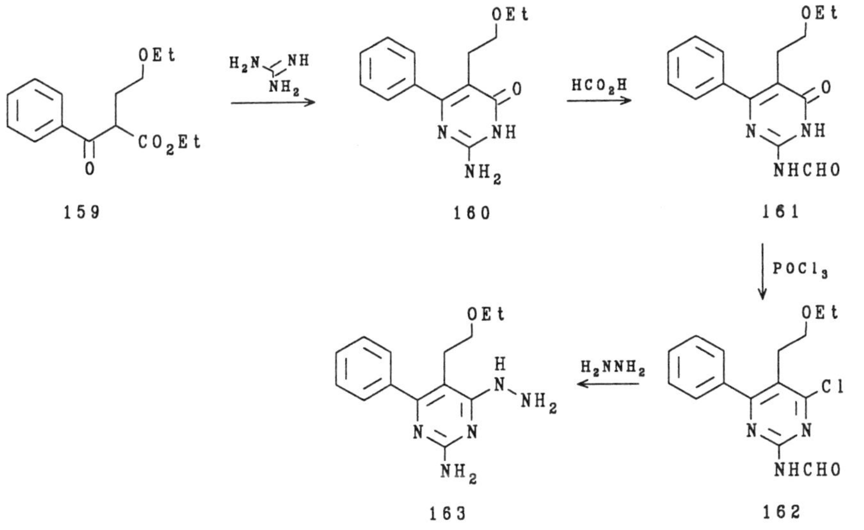

159 → ($H_2NC(=NH)NH_2$) → 160 → (HCO_2H) → 161 → ($POCl_3$) → 162 → (H_2NNH_2) → 163

the enol to the chloro derivative **162**. Replacement of halogen by hydrazine then affords the key intermediate **163**.

The fused five-membered ring is then formed by condensation with ethyl orthoformate to give the pyrrazolopyrimidine **164**. This compound rearranges to the isomeric compound **167** on heating. The sequence can be rationalized by assuming initial fragmentation of the pyrimidine bond to give intermediate **165**; simple rotation of the triazole moiety leads to **166**. Ring closure will then afford the observed product **bemitradine** (**167**).[32]

The starting material **168** for a pyrrazolopyridazine can be obtained by treating the corresponding enol, which is simply the condensation product of methylmaleic anhydride and hydrazine, with phosphorus oxychloride. Reaction of **168** with piperazine leads to displacement of the sterically more accessible chlorine to afford

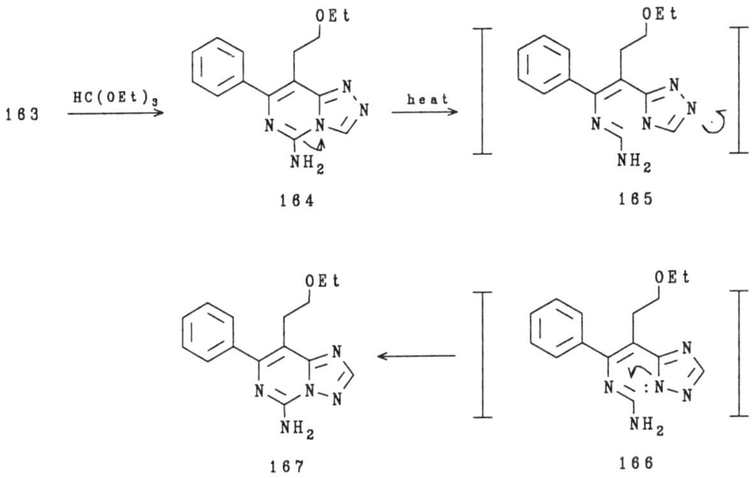

163 → ($HC(OEt)_3$) → 164 → (heat) → [165] → [166] → 167

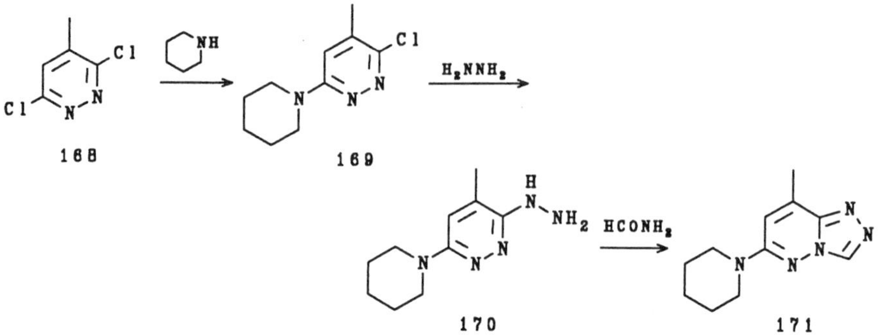

the alkylation product **169**. Treatment of **169** with hydrazine leads to replacement of the remaining halogen and formation of **170**. The missing carbon is in this case supplied by formamide to afford **zindotrine (171)**,[33] a compound that shows activity as a bronchodilator.

3. TWO FUSED SIX-MEMBERED RINGS

A. Compounds Related to Methotrexate

The pivotal role of folates in purine synthesis and consequently cell growth has been addressed earlier, as has the use of folate antagonists as cytotoxic agents. **Methotrexate** (**176**), one of the first modified pteridines investigated as a folate antagonist, is still used quite extensively in the chemotherapy of cancer and to a minor extent in other indications calling for cytotoxic agents.

One synthesis for **176** involves the three component condensation of polyaminopyrimidine (**136**), *N*-methylglutamyl-*p*-aminobenzamide (**172**), and 2,3-dibromopropionaldehyde (**173**) in the presence of potassium iodide. The initial step can be envisaged as condensation of the aldehyde with the aminopyrimidine to give an intermediate such as **174**, assuming that the reaction starts with formation of the pteridine. Alkylation of the amine on the PABA moiety by the bromide **174** will then lead to the observed product **175**. The reactions may, however, start with alkylation of the amine on PABA by the bromoaldehyde. Air oxidation of the dihydro ring in the initial product completes the synthesis of **methotrexate** (**176**).[34]

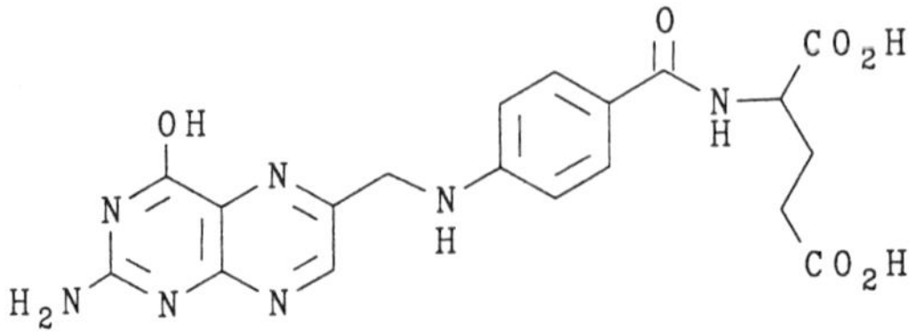

Folic Acid

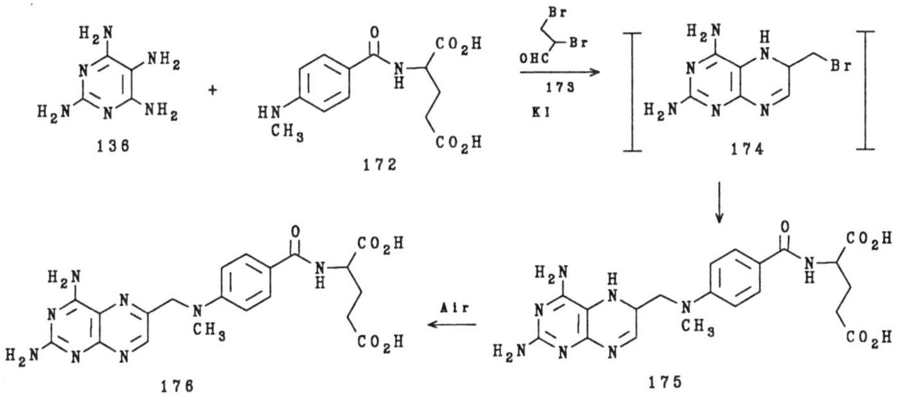

Replacement of the nitrogen para to the carboxyl on PABA by carbon leads to a compound that has one less site for potential metabolic cleavage. Reaction of the bromomethyl pteridine **174**, obtainable by suitable modification of the scheme above, with triphenylphosphine, leads to the phosphonium salt **177**. Condensation of the ylide from treatment of this salt with butyllithium with *p*-carbethoxypropiophenone (**178**) gives the coupling product **179**. The superfluous double bond is then reduced by catalytic hydrogenation; the pyrazine ring is reduced as well in an undesired side reaction to give **180**. Oxidation with hydrogen peroxide restores this ring to its fully saturated form (**181**).

Saponification of product **181** with base then gives the corresponding acid (**182**). Dicyclohexylcarbodiimide (DCC) mediate coupling with diethyl glutamate gives the desired amide. A second round of saponification affords the free carboxylic acid and thus **edatrexate** (**183**).[35]

Synthesis of the analogue in which one of the pteridine nitrogen atoms is omitted starts with reaction of aminopyrimidone **184** with 2-bromomalonaldehyde. This condensation may be envisoned as involving initial condensation of the aldehyde with the electron rich enamide carbon; imine formation then completes the formation of the fused pyridine ring to afford the product **185.** The amine at position 2 is then protected as its pivaloyl amide, **186**, by reaction with pivaloyl chloride. Palladium catalyzed coupling of halogen in the intermediate with mono-trimethylsilyl acetylene gives the product **187** which now contains the side chain carbon atoms. The silyl protecting group is then removed by exposure of the compound to fluorine anions.

The surrogate PABA ring is then added in a second palladium catalyzed coupling step. Thus, reaction of the acetylene **188** with iodobenzene **189** leads to the intermediate **190** which now includes the complete carbon skeleton. The acetylene is then reduced by catalytic hydrogenation; the pyridine ring is reduced to its tetrahydro derivative in the same step. Saponification removes the ethyl esters as well as the pivaloyl amide to give the folate antagonist **lometrexol, 191**.[36]

A related compound lacking the glutamide residue still shows considerable antifolate activity. The synthesis starts with Knoevnagel-like condensation of benzaldehyde **192** with ethyl acetoacetate. Catalytic reduction of the product gives the substituted acetoacetate **193**. Reaction of the product with aminopyrimidine **194** follows a similar

174 $\xrightarrow{Ph_3P}$ 177 $\xrightarrow[BuLi]{178}$ 179 $\xrightarrow{H_2}$ 180 $\xrightarrow{H_2O_2}$ 181 $\xrightarrow{NaOH}$ 182 $\xrightarrow[2.\ NaOH]{1.\ H_2N\text{–}CH(CO_2Et)CH_2CH_2CO_2Et,\ DCC}$ 183

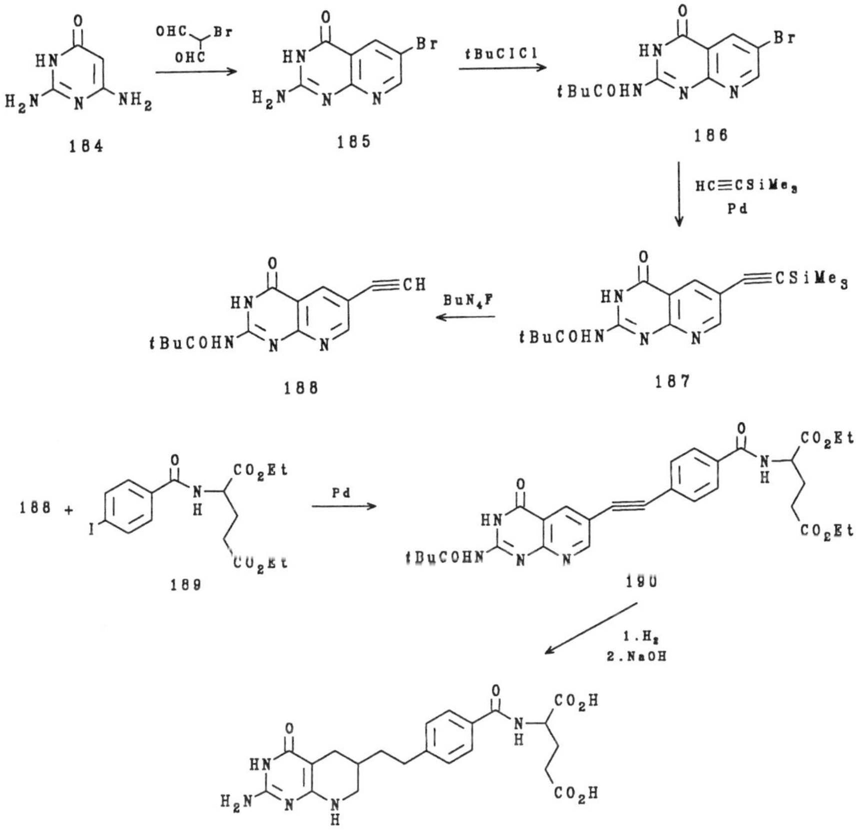

course to the one discussed above; the product **195**, in this case, however, is an amide due to the higher oxidation state of one of the carbonyl groups. The sequence for reducing this function starts by reaction with phosphorous oxyxchloride to give the chloride **196**. Hydrogenolysis of the product over palladium removes the halogen to afford **piritrexim** (**197**).[37]

B. Other Fused Heterocyclic Compounds

The compounds that follow, in contrast to those in Section 3.A, do not have any unifying theme, be it biological or chemical. They are consequently considered simply in order of the increasing number of heteroatoms present in the bicyclic nucleus.

The well-established antiarrhythmic agent **disopyramide** (**198**) constitutes the starting material for a cyclized version of this drug, which is somewhat more potent than the parent. Catalytic hydrogenation of **198** in the presence of acid results in selective reduction of the pyridine ring to give the corresponding piperidine. Treatment of this intermediate with acetic anhydride affords the acetamide **199**. Reaction of **199** with base leads to attack of the nitrogen on the secondary side chain amide on the

192 → 1. NaOEt, 2. H_2 (CO_2Et) → 193 → 194 → 195 → $POCl_3$ → 196 → H_2 → 197

198 → 1. H_2, 2. Ac_2O → 199 → NaOEt → 200

newly introduced amide carbonyl with formation of the cyclic amidine. Thus, the antiarrhythmic agent **actisomide** (**200**)[38] is obtained.

Pyridones form an integral part of the pharmacophore in the cardiotonic agents amrinone and milrenone (see Chapter 9). The biologic activity is retained when the ring is fused onto pyridine. Condensation of the ammonia enamine **201** from acetylacetone with the ethyl ester of propargilic acid can be visualized as proceeding through initial conjugate addition of the enamine to the triple bond to give a transient intermediate such as **202**. Ester–amide interchange will then lead to cyclization to the observed pyridone **203**. The methyl group on this ring is activated both by the adjacent enamide and vinyl ketone. Reaction with Bredereck's reagent then gives the aminoformylated derivative **204**. Treatment with ammonium acetate probably results in initial displacement of the dimethylamino group by ammonia to give the primary enamine; this closes to a pyridine ring in a Hantsch-like reaction. Thus, **medorinone** (**205**) is obtained.[39]

The general method used to form the 4-pyridone ring, which characterizes the quinolone antibiotics, can also be used to form such a ring fused onto other heterocyles. Addition–elimination of aminopicoline (**206**) with ethoxymethylenecyanoacetate gives the adduct **207** (no stereochemistry should be ascribed to the depiction). This compound cyclizes to the fused heterocycle **208** on heating in diphenyl ether. The presence of a methyl group on the other ortho position renders moot the question of C versus N cyclization. Reaction of this product with sodium azide converts the nitrile to a tetrazole.[40] Thus, **pemirolast** (**209**), one of a sizeable group of tetrazoles that show mediator release inhibiting and thus antiallergic activity in experimental animals, is obtained.

The initial quinolone antibiotic nalidixic acid (Chapter 11) consisted of a pyridone fused onto a pyridine ring. Retention of activity in this series in the face of yet another extra nitrogen atom in the fused ring is thus not completely unexpected. Construction of the starting material for an aza analogue begins by pyrrolidine displacing the more labile chlorine in the dichloropyrimidine **210** to give **211**. Displacement of the

NH_2 CO_2Et EtO_2C H_2N O O HN O O

201 202 203

Me_2N^+-$CHNMe_2$

O O HN NH_4OAc HN N $(CH_3)_2N$ O

205 204

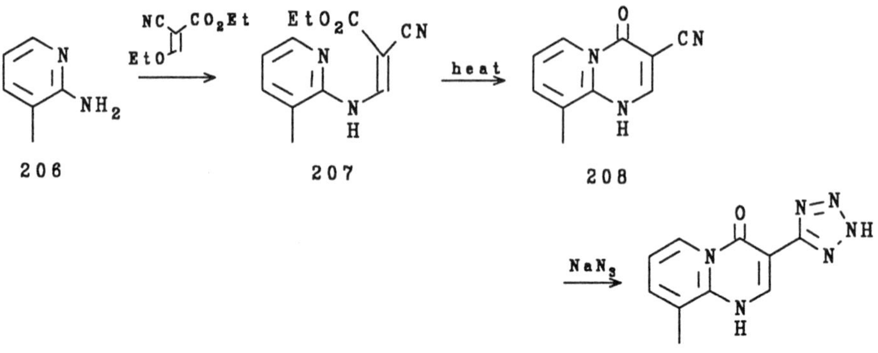

206 207 208

209

remaining halogen by ammonia gives the primary amine **212**. The quinolone ring is then fused on in the usual way by addition–elimination with ethoxymethylenemalonate followed by thermal ring closure to give **213**. The secondary amine on the pyrimidone ring is then alkylated with ethyl iodide. Saponification of the ester completes the synthesis of the antibacterial agent **piromidic acid**, (**214**).[41]

Displacement of the very labile chlorine on chloronitropyridine (**215**) with the monoethyl urethane from hydrazine gives the substitution product. The protecting group is then removed by successive saponification and decarboxylation of the thus formed carbamic acid to give the bidendate derivative **216**. Condensation of **216** with phenylacetic acid probably leads first to the reaction with the more basic nitrogen to form a hydrazide; this compound then cyclizes to the amidine **217**. Dehydrogenation with manganese dioxide completes the synthesis of the antifungal agent **triafungin** (**218**).[42]

One of the first so-called potassium sparing, non-thiazide diuretic agents contains a pterdine nucleus. This observation is reflected in the use of the pterdine starting material tetraaminopyrimidine (**136**) in the synthesis. Thus, reaction of benzaldehyde

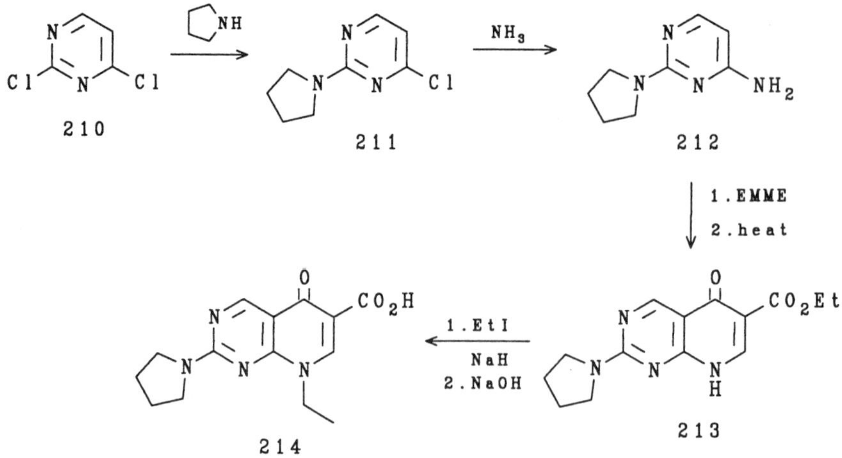

210 211 212

214 213

with this polyamine and potassium cyanide leads to the formation of the cyanohydrin-like α-aminonitrile **219** from reaction of the most basic amine. Treatment of this intermediate with base leads to addition of the amine to the nitrile to give the dihydropteridine **220**. Simple exposure to air leads to dehydrogenation and formation of **triamterene** (**221**).[43]

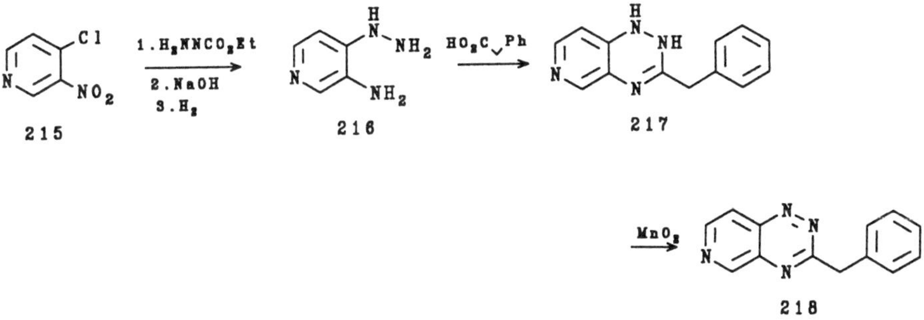

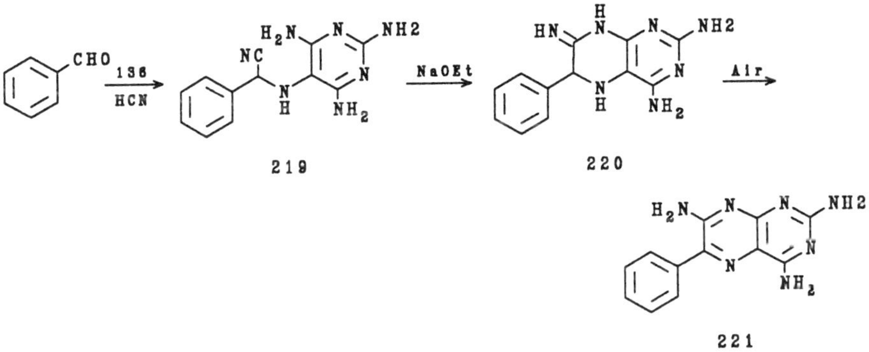

4. HETERODIAZEPINES

The interchangeability of unsaturated rings is illustrated particularly well in compounds derived from the anxiolytic agent chlordiazepoxide (Chapter 12); analogues in which the fused benzene ring in the prototype benzodiazepines is replaced by heterocycles such as pyrrazole or thiophene show activity that is equivalent or superior. The chemistry used to prepare these analogues quite closely parallels those used to prepare the benzene fused compounds. In one of the simpler examples, Friedel–Crafts acylation of chloropyrrazole **222** with *m*-chlorobenzoyl chloride affords the pyrrazolophenone **223**. The initial step in the reaction of this compound with ethylenediamine probably involves displacement of halogen in the heterocycle by one of the amino groups in the reagent. Imine formation with the second amino group closes the seven-membered ring to afford the anxiolytic agent **zometapine** (**224**).[44]

Reaction of the pyrrazole **225** with nitric and sulfuric acid results in formation of the nitro derivative **226**. The carboxylic acid is then converted to its acid chloride and this compound is used to acylate benzene in a Friedel–Crafts reaction. The nitro group in the resulting pyrrazolophenone **227** is then reduced to give the amine **228**. Following classical benzodiazepine chemistry, this aminobenzophenone counterpart is then treated with ethyl glycinate in pyridine. The hypothetical initially formed imine **229** cyclizes to the diazepinone ring under reaction conditions. Thus, the minor tranquilizer **ripazepam** (**230**)[45] is obtained.

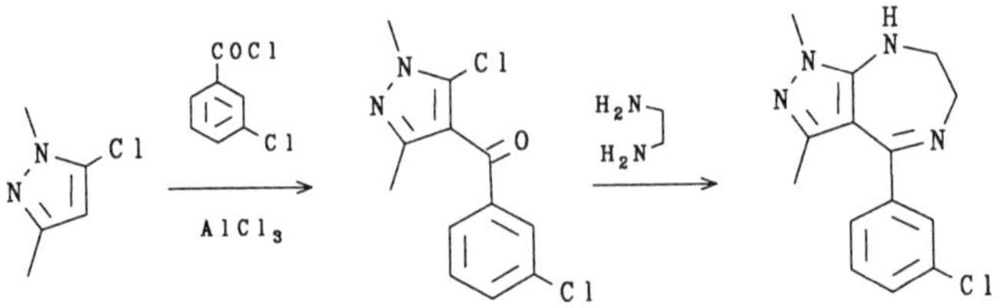

Another common approach to the seven-membered ring involves introducing nitrogen at a late stage. The requisite phenone **232** is obtained by Friedel–Crafts acylation of aminopyrrazole **231** with *o*-fluorobenzoyl chloride. The acetamide protecting group is then removed and the thus obtained secondary amine is acylated with chloroacetyl chloride to give the chloroacetamide **233**. Nitrogen is then introduced by displacement of the reactive chlorine with sodium azide. Catalytic hydrogenation reduces the azide to a primary amine; the resulting product **234** spontaneously cyclizes to form an azepinone to give **zolazepam** (**235**).[46]

The large increase in potency, which is obtained by fusing an additional heterocyclic ring onto the seven-membered ring, is observed in the heterodiazepines as well. The required diazepinone **237** is prepared from the thiophenophenone **236** by exactly the same sequence as the one used above. Construction of the fused triazole ring follows

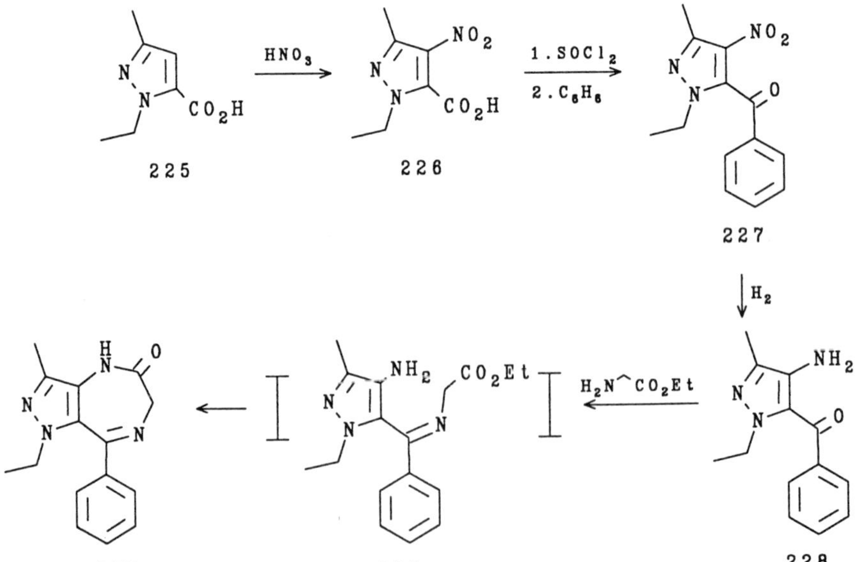

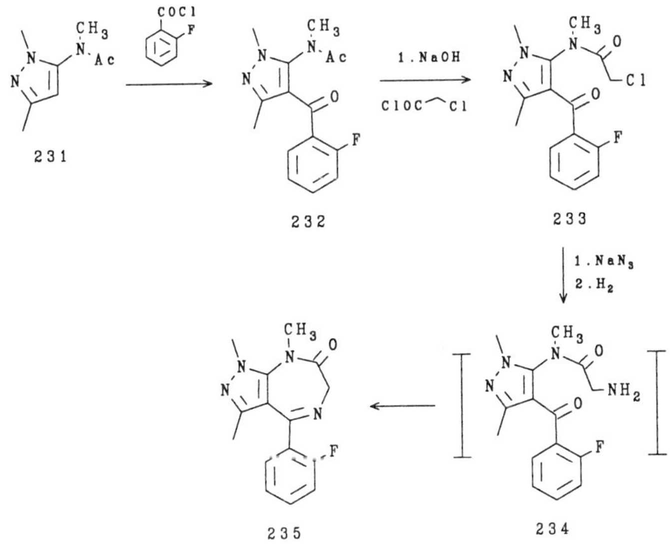

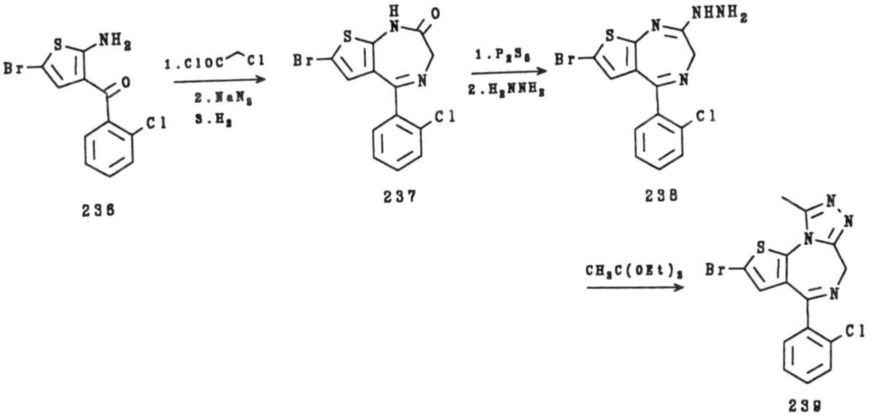

231 232 233 234 235

236 237 238 239

the method used for triazolam (Chapter 12). The amide in **237** is then converted to the corresponding thioamide by treatment with phosphorus pentasulfide; addition–elimination of hydrazine then leads to the *N*-aminoamidine **238**. Condensation of **238** with ethyl orthoacetate closes the last ring and affords the very potent anxiolytic agent **brotizolam** (**239**).[47]

5. HETEROCYCLIC COMPOUNDS WITH THREE OR MORE RINGS

A small group of polycyclic compounds defy ready classification on a chemical or, for that matter, pharmacological basis. It must be assumed that they showed enough

240 + 241 → (K_2CO_3) 242 → (1. H_2 2. TsCl) 243 → (H_2NBz) 244 → (H_2) 245

$Bz = C_6H_5CH_2-$

promise in various experimental test systems to be at least considered for clinical trials. None, however, seems to have been commercialized as a drug.

The clinical and commercial success of the antidepressant compound **fluoxetine** (Chapter 2; Prozac) engendered considerable work in other laboratories. A benzodioxan based compound that shows similar activity shares only a few structural features with the prototype. The benzodioxan nucleus **242** is formed by an alkylation reaction between the fluorocatechol (**240**) and the derivative **241** from meso, and hence achiral, butanetetrol. The benzyl protecting groups are then removed by hydrogenation over palladium, and the thus obtained diol converted to **243** by reaction with toluenesulfonyl chloride. Treatment of **243** with benzylamine leads to bis-alkylation on the same nitrogen to form a pyrrolidine ring and thus the tricyclic compound **244**. A second hydrogenolysis step then leads to **fluparoxan** (**245**).[48]

The starting material for the tricyclic NSAID **meseclazone** (**249**) consists, appropriately, of chlorosalicylic acid (**246**), which has NSAID activity in its own right. Reaction of the acid with hydroxylamine gives the hydroxamic acid **247**. Treatment of **247** with the diethyl acetal from 3-chlorobutyraldehyde gives the derivative **248**, which is in effect a carbinolamine derivative of the aldehyde. Exposure to mild base results in formation of the final ring by displacement of the terminal side chain chlorine by the hydroxylamine oxygen.[49] It is very probable that the product meseclazone (**249**) is actually converted to the salicylate **246** in vivo.

Compounds derived from indole have been extensively investigated as potential psychoactive drugs as noted earlier. The construction of a tricyclic indole derivative starts with benzyltrimethylammonium hydroxide catalyzed Michael addition of

246 → (NH_2OH) 247 → ($(EtO)_2HC$–CH₂CH(Cl)CH₃) 248 → (NaOH) 249

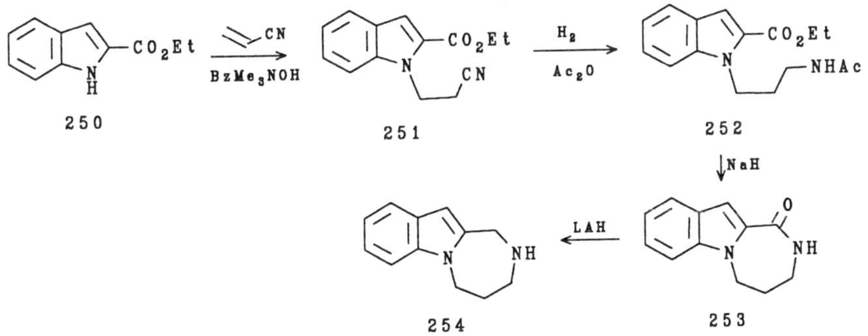

2-carbethoxyindole (**250**) to acrylonitrile to give the adduct **251**. In one of several alternatives for the next step, the nitrile is then hydrogenated in the presence of acetic anhydride to afford the acetamide **252**. Reaction of **252** with sodium hydride leads to formation of an anion on amide nitrogen; this anion then attacks the carbethoxy group to form the diazepinone ring. The imide presumably hydrolyzes on workup to give **253** as the observed product. The amide carbonyl group is then reduced with lithium aluminum hydride to give the antidepressant **azepindole** (**254**).[50]

Reaction of methylaminoindole (**255**) with ethylene oxide leads to ring opening of the oxirane with consequent formation of the hydroxyalkylated product **256**. This compound undergoes Friedel–Crafts-like ring closure on treatment with strong acid to give the tricyclic derivative **257**. Alkylation of the secondary amino group with 3-(3-chloropropyl)pyridine gives **gevotroline** (**258**),[51] a compound that shows antipsychotic activity.

A rather more complex tetracyclic indole based compound lowers blood pressure by selective blockade of α_1-adrenergic receptors. Reaction of the anion from indole **259** with butyrolactone leads to scission of the carbon–oxygen bond in the reagent and formation of the alkylated product **260**. The acid is then cyclized onto the adjacent 2-position to give the ketone **261** by treatment with a Lewis acid such as polyphosphoric acid. Reaction with bromine then leads to the brominated ketone **262**. This compound is then subjected to reductive alkylation with ethylenediamine and sodium borohydride. The reaction may involve either initial imine formation or displacement of halogen. Intermediate **263**, which would form if the latter prevails, will then undergo

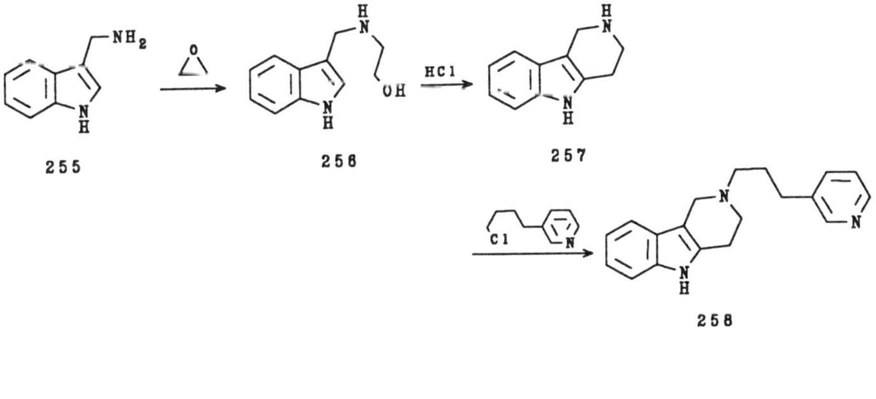

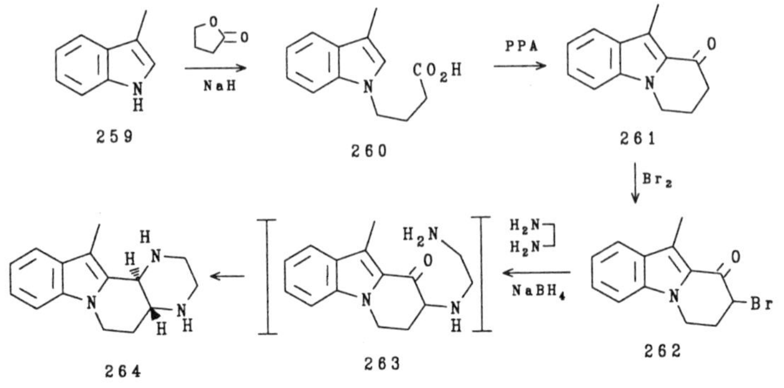

259 260 261 262 263 264

reductive alkylation at the carbonyl group; thermodynamic control of the final reduction step would account for the formation of the observed trans fused product **264**.

The very different steric milieu of the 2 nitrogen atoms is illustrated by the fact that reaction of the piperazine **264** with isopropyl bromide proceeds selectively at the more open amine to give the monoalkylated product **265**. The remaining secondary amine is then converted to the more reactive anion; treatment of **265** with ethyl bromide affords **atiprosin** (**266**).[52]

The very versatile synthon, isatin (**267**), provides starting material for a fused heterocyclic compound that has been investigated as an antiallergic agent on the strength of its activity as mediator release inhibitor. Oxidation of the hydrazone from isatin with mercuric oxide gives the diazo derivative **268**. Treatment of this reactive intermediate with the dipolarophile propynal leads to 1,3-dipolar cycloaddition and thus formation of the *spiro*-pyrrazole **269**. This intermediate undergoes spontaneous rearrangement to a planar conjugated isomer. Migration of the indolone bond leads to ring enlargement to the quinolone **270**. Reduction of the aldehyde with sodium borohydride gives the corresponding carbinol and thus **pirquinozol** (**271**).[53]

Although the synthesis of the tricyclic CNS agent **piquindone** (**277**) does not involve an indole, a partly reduced form of this moiety is imbedded in the final structure. The route to this compound starts with alkylation of the acetal **272** from 3-acetylpyridine with methyl iodide to give the corresponding quaternary salt. This compound is reduced to the tetrahydropyridine on treatment with sodium borohydride. Hydrolysis of the acetal protecting group affords the conjugated ketone **273**. Reaction of **273** with the anion from ethyl malonate initially gives the transient Michael adduct **274**; the reversible nature of this reaction assures the formation of the thermodynamically favored trans (diequatorial) product. The anion from the acetyl methyl group then

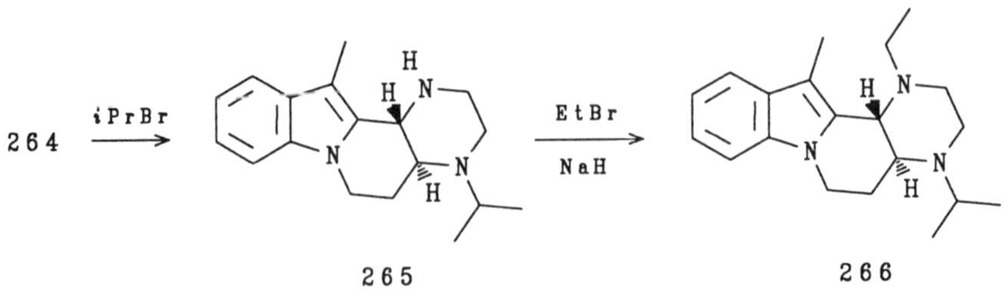

264 → (iPrBr) 265 → (EtBr, NaH) 266

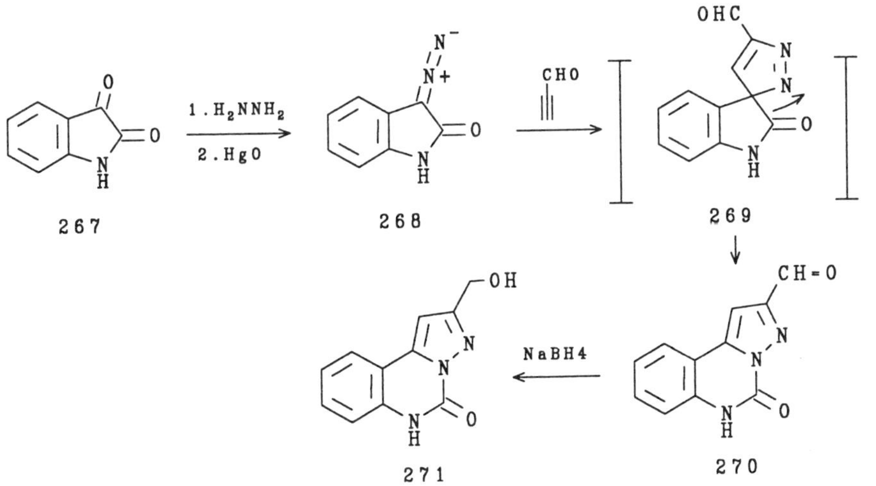

attacks one of the carbethoxy groups to form a cyclohexanone to give **275** as the isolated product. The free acid obtained on hydrolysis of the ester then decarboxylates to give the β-diketone **276**. In a classic application of the Knorr pyrrole synthesis, the diketone is then allowed to react with 2-aminopentan-3-one. Since the latter is unstable, it is generated in situ by reduction of the nitrosation product from diethylketone. Thus, piquindone (**277**),[54] a compound that displays antipsychotic activity, is obtained.

The pyrimidinoquinoline ring system provides the nucleus for yet another antiallergic mediator release inhibitor. Knoevnagel condensation of the nitrobenzaldehyde **278** with cyanoacetamide gives the cinnamide **279** (no stereochemistry implied by the depiction). The nitro group is then reduced by treatment with iron in acetic acid. Exposure of the product **280** to base leads to addition of the aniline nitrogen to the nitrile with consequent formation of the quinoline **281**. The last ring is then formed by reaction of the amino amide with diethyl oxalate. This compound then affords the pyrimidone ring and thus **pirolate** (**282**).[55]

The stereoselective nature of drug action should not be unexpected in view of the fact that all structures with which these compounds interact, receptors and enzymes,

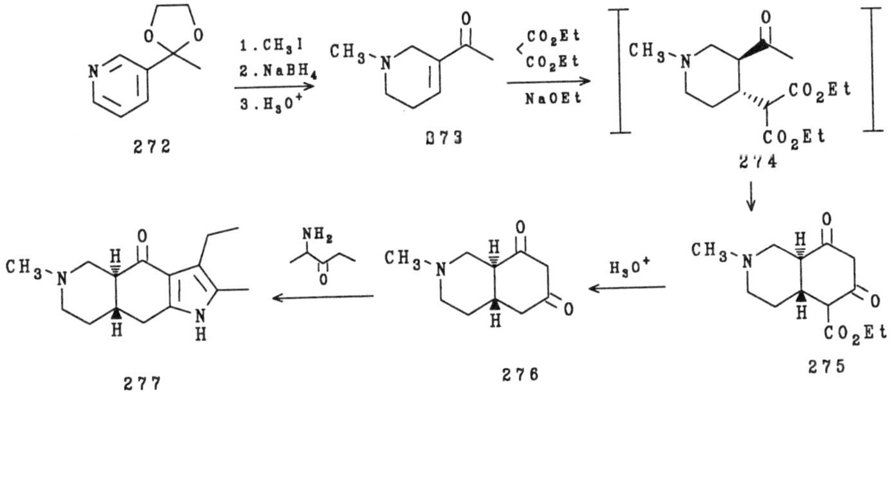

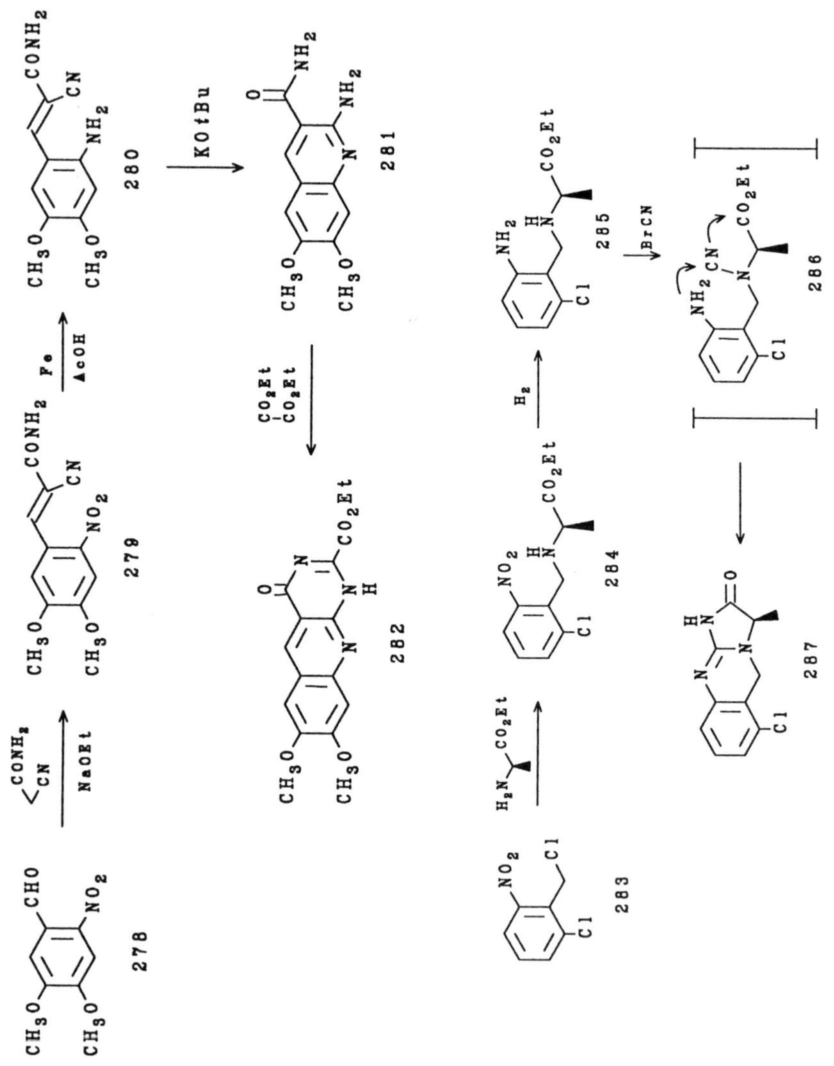

278
NaOEt
279
Fe
AcOH
280
KOtBu
281
282
283
H_2
284
285
BrCN
286
287

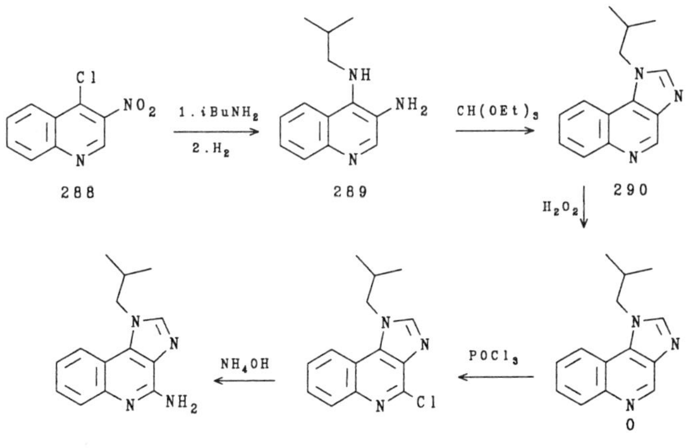

are composed of chiral amino acids. The synthesis of the platelet aggregation inhibitor **quazinone** (**286**) thus incorporates the chiral moiety leading to the active isomer in an early step. Thus, displacement of halogen in the benzyl chloride **283** with D-alanine affords the alkylated derivative **284** in chiral form. The nitro group is then reduced to the corresponding aniline **285**. Reaction of this product with cyanogen bromide leads to reaction at the more basic, aliphatic amine to give the transient cyanamide **286**. On heating **286** reacts further; the aniline nitrogen adds to the newly introduced cyano group; the thus formed imine then displaces ethoxide from the adjacent ester to form the imidazolone ring. The end product of this cascade is the imidazo quinazoline **quazinone** (**287**).[56]

Involvement of the immune system in a host of diseases has led to a search for compounds that modulate this system and thus hopefully change the course of the disease. An angular imidazoquinoline has shown some activity as an immune modulator. The synthesis of this compound starts with nucleophilic aromatic displacement of chlorine in the quinoline **288** by isobutylamine. Reduction of the nitro group then affords the 3,4-diamine **289**. The imidazole ring is then formed in the usual way by reaction with ethyl orthoformate to afford the tricyclic intermediate **290**. Introduction of the requisite amino group at the 2-position starts by oxidation of quinoline nitrogen with hydrogen peroxide to the *N*-oxide **291**. Treatment with phosphorus oxychloride leads to the 2-chloro derivative **292** by the familiar variant on the Polonovski reaction. Ammonolysis of **292** in aqueous ammonia affords **imiquimod** (**293**).[57]

REFERENCES

1. Franco, F.; Greenhouse, R.; Muchowski, J.M. *J. Org. Chem.* **1982**, *47*, 1682.
2. Miller, A.D. U. S. Patent **1976**, 3975532. *Chem. Abstr.* **1976**, *85*, 177393.
3. Raeymaekers, A.H.M.; Allewijn, F.T.N.; Vanderberk, J.; Demoen, P.J.A.; Van Offenwert, T.T.; Janssen, P.A.J. *J. Med. Chem.* **1966**, *9*, 545.

4. Raeymaekers, A.H.M.; Roevens, L.F.C; Janssen, P.A.J. *Tetrahedron Lett.* **1967**, 1467.
5. Van Gelder, J.H.L.; Raeymaekers, A.H.M.; Roevens, L.F.C; Van Laerhoven, W.J. U. S. Patent **1976**, 3925383. *Chem. Abstr.* **1976**, *84*, 180264.
6. Esanu, A. Belgium Patent **1982**, 891797. *Chem. Abstr.* **1982**, *97*, 127623.
7. Browne, L.J. Eur. Patent Appl. **1985**, 165904. *Chem. Abstr.* **1986**, *105*, 6507.
8. Ford, N.F.; Browne, L.J.; Campbell, T.; Gemendem, C.; Goldstein, R.; Gude, C.; Wasley, J.W.F. *J. Med. Chem.* **1985**, *28*, 164.
9. Dimsdale, M.; Friedmann, J.C.; Prenez, A.; Sauvanet, J.P.; Thenot, J.P.; Zirkovic, B. *Drugs Future* **1987**, *12*, 777.
10. Yamakawa, K.; Sato, K.; Sugame, T. *Jpn. Kokai* **1994**, 06271582. *Chem. Abstr.* **1994**, *122*, 292840.
11. Maffrand, J.P.; Elloy, F. *Eur. J. Med. Chem.* **1974**, *9*, 483.
12. Jones, T.K.; Mohan, J.J.; Xavier, L.C.; Blacklock, T.J.; Mathre, D.J.; Sohar, P.; Jones, E.T.T.; Reamer, R.A.; Roberts, F.E.; Grabowski, E.J.J. *J. Org. Chem.* **1991**, *56*, 763.
13. West, R.A.; Beauchamp, L. *J. Org. Chem.* **1961**, *26*, 3809.
14. Cook, P.D.; Rousseau, R.J.; Mian, A.M.; Dea, P.; Meyer, R.B.; Robins, R.K. *J. Am. Chem. Soc.* **1976**, *98*, 1492.
15. Robertson, D.W.; Beedle, E.E.; Krushinski, J.H.; Pollock, G.D.; Wilson, H.; Wyss, V.L.; Hayes, J.S. *J. Med. Chem.* **1985**, *28*, 717.
16. Hoehn, H.; Deuzel, T. German Offen. **1971**, 2123318. *Chem. Abstr.* **1971**, *76*, 59619.
17. Silvestrini, B.; Baiocchi, L. German Offen. **1979**, 2915318. *Chem. Abstr.* **1980**, *92*, 111060.
18. Hromatka, O.; Binder, D.; Pfister, R.; Zeller, P. German Offen. **1976**, 2537070. *Chem. Abstr.* **1976**, *85*, 63077.
19. Traube, W. *Berichte* **1900**, *33*, 3035.
20. Diamond, J. German Offen. **1977**, 2713389. *Chem. Abstr.* **1978**, *88*, 22984.
21. De Ridder, R.R. Fr. Demande **1981**, 2483922. *Chem. Abstr.* **1982**, *96*, 199432.
22. Mohler, W.; Soder, A. *Arzneim.-Forsch.* **1971**, *21*, 1159.
23. Elion, G.B.; Hitchings, G.H. *J. Am. Chem. Soc.* **1952**, *74*, 411.
24. Hitchings, G.H.; Elion, G.B. U. S. Patent **1962**, 3056785. *Chem. Abstr.* **1963**, *58*, 5701.
25. Kamiya, T.; Saito, Y.; Hashimoto, M.; Seki, H. *Tetrahedron Lett.* **1969**, 4729.
26. Tull, R.J.; Hartman, G.T.; Wienstock, L.M. U. S. Patent **1978**, 4098787. *Chem. Abstr.* **1978**, *89*, 215435.
27. Montgomery, J.A. U. S. Patent Appl. **1979**, 962107. *Chem. Abstr.* **1980**, *92*, 22778.
28. Schaeffer, H.J. German Offen. **1976**, 2539963.
29. Muzuno, Y.; Takeuchi, K.; Kaneko, C.; Matsumoto, H.; Yamada, K. *Jpn. Kokai* **1988**, 63107982. *Chem. Abstr.* **1988**, *109*, 149564.
30. Verheyden, J.P.H. In *Chronicles of Drug Discovery*, Lednicer, D., Ed.; ACS Books: Washington DC, 1993, Vol. 3, p. 299.
31. Robins, R.K. *J. Am. Chem. Soc.* **1956**, *78*, 784.
32. Heilman, R.D.; Rorig, K.J. *Drugs Future*, **1985**, *10*, 299.
33. Lewis, J.; Shea, P.J. U. S. Patent **1979**, 4136182. *Chem. Abstr.* **1979**, *90*, 145958.

34. Seeger, D.R.; Cosulich, D.B.; Smith, J.M.; Hultquist, M.E. *J. Am. Chem. Soc.* **1949**, *71*, 1753.
35. Piper, J. R.; Johnson, C.A.; Otter, G.M.; Sirotnak, F.M. *J. Med. Chem.* **1992**, *35*, 3002.
36. Taylor, E.C.; Kuhnt, D.; Shih, C.; Rinzel, S.M.; Grindley, G.B.; Barredo, J.; Jannatipour, M.; Moran, R.G. *J. Med. Chem.* **1992**, *35*, 4450.
37. Grivsky, E.M.; Sigel, C.W.; Duch, D.S.; Nichol, C.A. Eur. Patent Appl. **1981**, 21292. *Chem. Abstr.* **1981**, *94*, 208899.
38. Chorvat, R.J.; Prodan, K.A.; Adelstein, G.W.; Rydzewski, R.M.; McLaughlin, K.T.; Stamm, M.H.; Frederick, L.G.; Schniep, H.C.; Stickney, J.L. *J. Med. Chem.* **1985**, *28*, 1285.
39. Singh, B.; Lesher, G.Y. *J. Heterocylc. Chem.* **1990**, *27*, 2085.
40. Juby, P.F. U. S. Patent **1978**, 4122274. *Chem. Abstr.* **1979**, *90*, 103998.
41. Minami, S.; Shono, T.; Matsumoto, J. *Chem. Pharm. Bull.* **1971**, *19*, 1426.
42. Wright, G.C.; Bayless, A.V.; Gray, J.E. German Offen. **1975**, 2427382. *Chem. Abstr.* **1975**, *82*, 171087.
43. Pachter, I.J. *J.Org. Chem.* **1963**, *28*, 1191.
44. DeWald, H.A.; Lobbenstael, S.J. So. African Patent **1975**, 7307696. *Chem. Abstr.* **1976**, *84*, 59594.
45. Nordin, I.C. U. S. Patent **1971**, 3553207. *Chem. Abstr.* **1971**, *75*, 5972.
46. DeWald, H.A.; Lobbenstael, S.J.; Butler, D.E. *J. Med. Chem.* **1977**, *20*, 1562.
47. Weber, K.H.; Bauer, A.; Danneberg, P.; Kuhn, F.J. German Offen. **1976**, 2410030. *Chem. Abstr.* **1976**, *84*, 31148.
48. Kitchin, J.; Cherry, P.C.; Pipe, A.J.; Crame, A.J.; Borthwick, A.D. Br. Patent Appl. **1985**, 2157691. *Chem. Abstr.* **1986**, *105*, 208896.
49. Reisner, D.B.; Ludwig, B.J.; Bates, H.M.; Berger, F.M. German Offen. 2010418. *Chem. Abstr.* **1970**, 120644.
50. Reynolds, B.; Carson, J. German Offen. 1928726. *Chem. Abstr.* **1970**, *72*, 55528.
51. Abou-Gharbia, M.; Patel, U.R.; Webb, M.B.; Moyer, J.A.; Andree, T.H.; Muth, E.A. *J. Med. Chem.* **1987**, *30*, 1818.
52. Jirkovsky, I.; Santroch, G.; Baudy, R.; Oshiro, G. *J. Med. Chem.* **1987**, *30*, 388.
53. Vogt, B.R. German Offen. 2726389. *Chem. Abstr.* **1978**, *88*, 121240.
54. Coffen, D.L; Hengartner, H.; Katonak, D.A.; Mulligan, M.E.; Burdick, D.C.; Olsen, G.L.; Todaro, L.J. *J. Org. Chem.* **1984**, *49*, 5109.
55. Althuis, T.H.; Czuba, L.J.; Hess, H.J.; Kadin, S.B. German Offen. 2418498. *Chem. Abstr.* **1975**, *82*, 73015.
56. Chodnekar, M.S.; Kaiser, A. German Offen. **1979**, 2832138.
57. Gertser, J.F. Eur. Patent Appl. **1985**, 145340 *Chem. Abstr.* **103**, *103*, 196090.

CROSS INDEX OF COMPOUNDS

REACTION INDEX

Note: Page numbers denote location of flow schemes on which reactions appear.

SUBJECT INDEX

Note: **Boldface** entries represent syntheses.

UNIVERSITY OF GLAMORGAN
PRIFYSGOL MORGANNWG